Studies in Mechanobiology, Tissue Engineering and Biomaterials

Volume 24

More information about this series at https://link.springer.com/bookseries/8415

Gerhard Sommer · Kewei Li ·
Daniel Ch. Haspinger · Raymond W. Ogden
Editors

Solid (Bio)mechanics: Challenges of the Next Decade

A Book Dedicated to Professor Gerhard A. Holzapfel

Springer

Editors
Gerhard Sommer
Institute of Biomechanics
Graz University of Technology
Graz, Austria

Daniel Ch. Haspinger
Institute of Biomechanics
Graz University of Technology
Graz, Austria

Kewei Li
Columbia University Irving Medical Center
New York, NY, USA

Raymond W. Ogden
School of Mathematics and Statistics
University of Glasgow
Glasgow, UK

ISSN 1868-2006 ISSN 1868-2014 (electronic)
Studies in Mechanobiology, Tissue Engineering and Biomaterials
ISBN 978-3-030-92338-9 ISBN 978-3-030-92339-6 (eBook)
https://doi.org/10.1007/978-3-030-92339-6

This Springer imprint is published by the registered company Springer Nature Switzerland AG
The registered company address is: Gewerbestrasse 11, 6330 Cham, Switzerland

To Gerhard A. Holzapfel on the occasion of his 60th birthday

Photo by Lunghammer—TU Graz

Preface

Biomechanics is a challenging subject that studies living systems through the development, extension, and application of mechanics to better understand biological and medical phenomena. Nonlinear continuum mechanics, multiscale modeling, advanced 3D imaging techniques along with unprecedented predictive computational power enable us to solve cutting-edge biomechanical problems nowadays with rather complex patient-specific anatomies. Interdisciplinary approaches that address phenomena at the nano, micro-, or macro-levels are often required to solve such problems. Mechanics, natural sciences such as biophysics, biochemistry, and mathematics, as well as biomedical, tissue, healthcare, computer, and biomolecular engineering sciences must be combined with medicine to tackle real-life problems.

Professor Gerhard A. Holzapfel has been an international leader in the field of biomechanics for at least the past two decades through scientific contributions, editorships, conference organization, mentoring, and doctoral training. He is widely known for his outstanding contributions in the field of nonlinear solid (bio)mechanics and the constitutive and computational modeling of fiber-reinforced materials and soft biological tissues. Professor Holzapfel has dealt intensively with the cardiovascular system in health and disease such as aneurysm and aortic dissection, and with therapeutic interventions such as balloon angioplasty and stenting, the development of continuum theories, computational methods, and simulations and experiments in biomechanics and mechanobiology of soft biological materials. This book on *Solid (Bio)mechanics: Challenges of the Next Decade* is dedicated to Professor Gerhard A. Holzapfel on the occasion of his 60th birthday and his eminent career achievements.

Professor Gerhard A. Holzapfel received his master's degree in civil engineering in 1985 and obtained his PhD in mechanical engineering from Graz University of Technology (TU Graz) in 1990. In 1991, he traveled to Shenyang in the northeast of the People's Republic of China to work as a visiting scholar at an institution that is currently part of Shenyang University. He then received a Schrödinger Scholarship from the Austrian Science Fund to work as a post-doctoral fellow at the Division of Applied Mechanics, Department of Mechanical Engineering, Stanford University, CA, USA, with the late Professor Juan C. Simo from 1993 to 1995. He completed his habilitation in mechanics at Vienna University of Technology, Austria, in 1996.

From May 1987 to November 2004, he was an assistant at the Institute of Strength of Materials and Associate Professor at the Institute of Structural Analysis, TU Graz, Austria. He was Professor of Biomechanics at KTH Royal Institute of Technology in Stockholm, Sweden for 9 years (7 years as Adjunct Professor) until 2013. Since 2007, he has been a Full Professor of biomechanics and the head of the institute at TU Graz. He is also the International Chair of Biomechanics (adjunct professorship) at the Norwegian University of Science and Technology (NTNU) in Trondheim and Visiting Professor at the School of Mathematics and Statistics, University of Glasgow, Scotland.

Professor Holzapfel has authored a well-known graduate textbook entitled *Nonlinear Solid Mechanics. A Continuum Approach for Engineering'* (John Wiley & Sons), and he co-edited seven books and five special issues in journals. He contributed chapters to 25 other books, published 230+ peer-reviewed journal articles, 65+ conference proceedings (full papers), and contributed 490+ oral or poster presentations to several conferences all over the world. He is the co-founder and co-editor-in-chief of the international scientific journal *Biomechanics and Modeling in Mechanobiology* (Springer Nature) since the first issue published in June 2002. He supervised 2 habilitation theses, 18 post-doctoral fellows, and 30 doctoral students.

He received many awards and honors, such as the Schrödinger Scholarship for post-doctoral training at Stanford University (1993–1994), the Austrian Start-Prize 1997 from the Austrian Science Fund, the highest Austrian award for young scientists, the Josef-Krainer Würdigungspreis 2003 for exceptional achievements in the field of biomechanics, and the Erwin Schrödinger Prize 2011 from the Austrian Academy of Sciences for lifetime achievements by Austrians in the fields of mathematics and natural sciences. He has been listed as a Highly Cited Researcher in Engineering selected by ISI Web of Science, Thomson Reuters and listed as *The World's Most Influential Scientific Minds: 2014*. He is a Founding Fellow of EAMBES (2012), the European Alliance for Medical and Biological Engineering & Science, and he was elected as a corresponding member of the Austrian Academy of Sciences (2012), a member of the Academia Europaea (2014), a fellow of the European Mechanics Society (2015), a member of the World Council on Biomechanics (2018), and an ordinary member of the European Academy of Sciences and Arts (2019). It is noteworthy that, in 2021, he received the William Prager Medal of the Society of Engineering Science (USA) and the Warner T. Koiter Medal of the American Society of Mechanical Engineers (USA).

This book provides a comprehensive and timely overview of the latest developments in the field of biomechanics and extensive knowledge of tissue structure, function, and modeling. It contains 19 chapters, each of which begins with a personal dedication to Professor Holzapfel. We hope that the book is useful not only for those working as a graduate student, researcher, and (bio)engineer in the challenging field of biomechanics, but also for those in other fields such as biomedical engineering, biophysics, mechanical engineering, materials science, applied mathematics, and medicine.

We have invited some of the most prominent scientists within Professor Holzapfel's research fields to contribute to this book with their latest research results.

This book includes some state-of-the-art advances such as constitutive modeling and computational simulation of biological tissues under physiological and pathological conditions, and mechanical characterization of a range of biological tissues for purposes such as computational simulation and surgical planning. It covers innovative and cutting-edge studies of cardiovascular tissues such as arteries, heart, valvular tissue, but also on other tissues and organs such as thrombus, brain tumor, muscle, liver, kidney, and stomach, to name a few. It also highlights some of the biggest challenges biomechanics will face over the next decade and how we can better prepare for them.

Without the considerable effort and support of all authors, this book would not exist. We would like to express our sincere gratitude to everyone who contributed to this book. Finally, we would like to thank Dr. Leontina Di Cecco, Senior Editor of Applied Sciences and Engineering, and her publishing team at Springer for their support and encouragement in publishing this book.

Graz, Austria — Gerhard Sommer
New York, USA — Kewei Li
Graz, Austria — Daniel Ch. Haspinger
Glasgow, Scotland — Raymond W. Ogden
July 2021

Contents

About the Editors

Asst.Prof. Gerhard Sommer finished his Master's Thesis at Graz University of Technology (TUG) in the field of soft tissue biomechanics under the supervision of Christian A.J. Schulze-Bauer and Gerhard A. Holzapfel in 2003. He then began his Ph.D. Thesis in the field of biomechanics supervised by Gerhard A. Holzapfel, where he continued to explore the mechanical properties of human soft tissues, especially of the cardiovascular system. In 2007, together with Gerhard A. Holzapfel, he founded the Institute of Biomechanics at TUG, where he was responsible for setting up the biomechanics laboratory. In October 2008 he completed his Ph.D. From 2013 to 2018 he was Senior Postdoctoral Researcher and became Assistant Professor in 2018. He has published more than 40 articles in international journals and supervised several Master's and Ph.D. students.

Dr. Kewei Li is a Postdoctoral Research Scientist at the Columbia University Irving Medical Center, New York, NY, United States. He earned his Ph.D. in Mechanical Engineering under the supervision of Professor Wei Sun at the University of Connecticut in 2013. After a brief internship at Dassault Systèmes Simulia Corporation, he moved to Austria in 2014 and worked as a (postdoctor) University Assistant at Graz University of Technology with Professors Gerhard A. Holzapfel and Ray W. Ogden. He taught undergraduate and graduate courses in the fields of statics, strength of materials, and computational biomechanics. His research interests include cardiovascular biomechanics, nonlinear elasticity, and finite element analysis. He has published 18 research papers in peer-reviewed journals and has been an Editorial Board member for *Scientific Reports* since 2019.

Dr.techn. Daniel Ch. Haspinger completed his master's and doctoral studies with a strong background in biomechanics at Graz University of Technology, Austria, under the supervision of Prof. Gerhard A. Holzapfel. Working as a teaching and research assistant at the Institute of Biomechanics offered him the rare opportunity to work and learn from some of the world's leading scientists in the field of cardiovascular biomechanics. He has published several peer-reviewed journal articles, presented his research findings at various international scientific conferences, and received the Ing.

F. Schmiedl Award from the City of Graz. His research aims to provide better insights into possible causes and consequences of physiological and pathological adaptation mechanisms in the cardiovascular system from an experimental and computational perspective.

Prof. Raymond W. Ogden A mathematics graduate from Cambridge University. He is George Sinclair Professor of Mathematics at the University of Glasgow. His research interests include nonlinear elasticity theory and its applications to the mechanics of rubberlike materials, soft biological tissues and fibre-reinforced materials. He has published more than 250 papers in international journals, and several books. He is a Fellow of the Royal Society of London and has received numerous awards for his research, including the IUTAM/Elsevier Rodney Hill Prize in Solid Mechanics and the Timoshenko Medal of the American Society of Mechanical Engineers.

Arterial Biomechanics in Health and Disease

Multiscale Experimental Characterization and Computational Modeling of the Human Aorta

Misael Dalbosco, Daniel Ch. Haspinger, Kewei Li, Sae-Il Murtada, Anna Pukaluk, Malte Rolf-Pissarczyk, Selda Sherifova, and Gerhard Sommer

We came to know Professor Gerhard A. Holzapfel at different times from different parts of the world. But we all came to Graz with the same purpose: to learn from the master of biomechanics and to study and work in this thrilling field. Along the way, we have met new colleagues, friends, and also encountered new problems in a completely different cultural environment. Professor Holzapfel himself is aware of such cultural differences and provided us with extraordinary help and guidance beyond the professional level, especially for some of us from afar. We express our profound gratitude to Professor Holzapfel for the wonderful opportunity to be part of his team and his tremendous help both at the personal and professional levels during our times in Graz and beyond. His generous encouragement, helpful guidance, and, most importantly, persistent faith in us, were and are of invaluable help for our scientific careers now and in the future. He will be remembered not only as a master of biomechanics but also as a fine individual, always delightful to be around.

Abstract Advanced imaging techniques, novel experimental approaches and sophisticated computational modeling frameworks to characterize and simulate the mechan-

---

M. Dalbosco · D. Ch. Haspinger · A. Pukaluk · M. Rolf-Pissarczyk · S. Sherifova · G. Sommer (✉)
Institute of Biomechanics, Graz University of Technology, Stremayrgasse 16-2, 8010 Graz, Austria
e-mail: sommer@tugraz.at

K. Li
Department of Surgery, Vagelos College of Physicians and Surgeons, Columbia University Irving Medical Center, 630 West 168th Street, New York, NY 10032, USA

S.-I. Murtada
Department of Biomedical Engineering, School of Engineering and Applied Science, Yale University, 55 Prospect Street, New Haven, CT 06511, USA

G. Sommer et al. (eds.), *Solid (Bio)mechanics: Challenges of the Next Decade*, Studies in Mechanobiology, Tissue Engineering and Biomaterials 24,
https://doi.org/10.1007/978-3-030-92339-6_1

ical behavior of soft biological tissues have dramatically improved in the past decades. Particularly, the advancing of multiphoton microscopy and other imaging techniques has enabled a detailed three-dimensional visualization of the underlying microscopic structure of various biological tissues including arterial walls. In addition, mechanical testing combined with sophisticated microscopy techniques allowed us to quantify the tissue microstructural reorganization and the mechanical response under large deformation simultaneously. Multiscale constitutive models incorporating detailed microstructural information such as the 3D dispersion of collagen fibers in the extracellular matrix and experimentally-derived tissue material properties have been developed and employed in the computational simulations of human aortic tissues under various (patho)physiological conditions. Thus, in this chapter, we review some of the most critical advances and developments in experimental approaches and computational modeling strategies to characterize the mechanical behavior of human aortic tissue. In addition, we discuss future challenges to improve our understanding of the aortic tissue and its related pathologies.

1 Introduction

Human aortas, similar to other soft biological tissues, are characterized by their distinct hierarchical structure reaching through multiple scales where nanoscale molecular phenomena are expressed in macroscale mechanical responses (Cranford and Buehler 2010). Thus, multiscale approaches have been employed to deepen our understanding of the relations between the structure, mechanics and (patho)physiological processes of the organ. Advances in microscopy techniques allowed a deeper understanding of the tissue organization (see, e.g., Dingemans et al. 2000; O'Connell et al. 2008), and provided invaluable insights for constitutive modeling.

The healthy aortic wall consists of three distinct layers, i.e., the intima, media, and adventitia. Each layer is characterized by a unique content and organization of its constituents, serves different functions and exhibits specific mechanical responses. In a young healthy adult, the intima consists primarily of a single layer of endothelial cells, supported by a thin basal lamina reinforced by network-like collagen. The media is a complex network of smooth muscle cells (SMCs) and extracellular matrix (ECM) components such as collagen, elastin, and proteoglycans; whereas the adventitia hosts fibroblasts and fibrocytes embedded in an ECM mainly composed of collagen fibers.

Among the ECM components, collagen is a ubiquitous protein found in all aortic layers and is responsible for tissue strength (Holzapfel 2008), although its organization differs in each layer: intimal collagen builds a carpet-like network, circumferentially oriented undulating collagen fibers are found within the media, and the adventitia is primarily composed of thick and wavy bundles of axially oriented collagen fibers (Niestrawska et al. 2016). Elastin forms fenestrated, concentric sheets dividing the media into lamellae, as well as interlamellar elastic fibers connecting

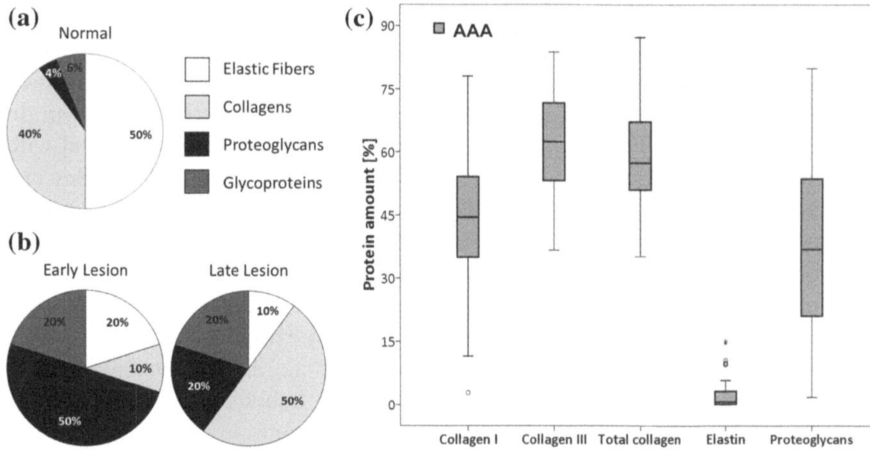

Fig. 1 Composition of the arterial ECM in (a) health and (b), (c) disease. The progression of atherosclerosis (b) is associated with an initial percentage decrease in collagen and elastic fibers contents, as well as an increase in proteoglycans and glycoproteins at early stages, followed by a relative increase in collagen accompanied by elastic fibers loss and reduced proteoglycan content at later stages. Relative elastin loss and increased proteoglycan content is also reported in aneurysms (c). Panels (a) and (b) are reproduced from Wight (2018), (c) from Tanios et al. (2015) with permission

them (Dingemans et al. 2000; O'Connell et al. 2008). Such fibers are also observed in the human aortic adventitia, albeit with a different organization (Koch et al. 2014). Proteoglycans consist of a core protein with one or more glycosaminoglycan chains covalently attached to it (Couchman and Pataki 2012; Schaefer and Schaefer 2010). In human aortas, the major proteoglycan is versican, localized in the media and intima, whereas small leucine-rich proteoglycans such as decorin or biglycan are commonly found within collagen fibers in all aortic layers (Kellerher et al. 2004; Wagenseil and Mecham 2009). Collagen and elastin contents and their arrangements were reported to change due to various factors such as aging (Maurel et al. 1987), aneurysm formation (Carmo et al. 2002; Menashi et al. 1987; Rizzo et al. 1989), or atherosclerosis (Katsuda et al. 1992; Murata et al. 1986). Furthermore, the amount of proteoglycans was increased in the atherosclerotic (Wight 2018) and aneurysmatic (Tanios et al. 2015) aortic wall compared with the healthy wall (Fig. 1).

Vascular SMCs are reported to have a spindle-like shape with an approximate length of 60 μm and a width of 5 μm (Luo et al. 2016), surrounded by elastic lamellae, a dense network of interlamellar elastic fibers and struts, as well as collagen fibers (O'Connell et al. 2008). Current microscopic examinations have shown that SMCs exhibit similar helical orientation patterns in the axial-circumferential plane as the medial collagen fibers with a small additional radial tilt (Holzapfel et al. 2002a; Horný et al. 2010; Luo et al. 2016; O'Connell et al. 2008). The relative SMC content varies along the arterial tree and, in general, increases towards the periphery (Humphrey

2002). Inside SMCs, the mechanical contraction takes place via so-called contractile units, i.e., actomyosin complexes, which consist of thin rope-like actin filaments and thick myosin filaments that slide relative to each other in a ratchet-like manner. SMCs play a significant role in short- and long-term changes of the arterial wall. In the homeostatic state, they are partially contracted providing the blood vessel with a certain tautness. In response to an altered blood pressure or blood flow, they can contract or relax in order to modulate the stiffness of large arteries or the luminal diameter of small arteries to maintain homeostasis (Milnor 1990; Humphrey 2002). However, sustained changes to the loading conditions or pathological alterations, e.g., atherosclerosis or aneurysms, may trigger an increased expression of growth factors leading to phenotypic changes of SMCs, i.e., they exhibit reduced contractility and start synthesizing collagen in an attempt to maintain the mechanical integrity of the arterial wall (Li et al. 1998; Li and Xu 2000; Lehoux et al. 2006; Gomez and Owens 2012). Furthermore, dysfunctional smooth muscle contractility, which usually occurs prior to changes in the ECM properties, is associated with several pathological conditions in the aortic wall. There is, therefore, a pressing need to quantify and assess their contractile functions, where experimental techniques as well as mathematical models play critical roles.

The changes in the ECM components and SMCs in healthy and diseased aortas suggest substantial changes on the organ's mechanical properties. Experimental and computational investigations of the structural changes and the related mechanical behavior are thus indispensable for a better understanding of aortas in health and disease. To gain a deeper insight into the mechanical behavior of the human aortic wall, earlier mechanical experiments aimed at only capturing the response of the tissue under different loading modes (e.g., Mohan and Melvin 1982, 1983). Later, reduced sample preparation time and faster imaging methods enabled correlating the tissue architecture with its mechanical properties (e.g., Niestrawska et al. 2016, 2019; Schriefl et al. 2015; Weisbecker et al. 2012). For example, imaging data from multiphoton microscopy (MPM) were utilized to mathematically describe the organization of collagen fibers in healthy and diseased aortas, and to obtain structural parameters (Niestrawska et al. 2016, 2019; Schriefl et al. 2013) for the constitutive models (Gasser et al. 2006; Holzapfel et al. 2015). Nowadays, the tissue's mechanical response and microstructural changes under loading can also be investigated simultaneously (Chow et al. 2014; Cavinato et al. 2017). Nevertheless, the alterations in the nanostructure in response to the macroscopical load remain largely elusive.

A similar development was seen in constitutive modeling: early models for soft biological tissues were phenomenological in nature, e.g. the classical Fung hyperelastic model (Fung 1993), which has been successfully employed for computational simulations of soft tissues (Li and Sun 2010; Sun et al. 2010). Phenomenological models can fit specific loading modes of measurable tissue behavior such as the stress versus strain relationship. However, the model parameters have no physical meaning. As a result, the set of parameters that could fit to a specific experimental data is often not unique, making it difficult to compare the differences in tissue behavior using these parameters. In contrast, microstructurally-motivated constitutive models can incorporate the mechanical properties of individual constituents and assume that the overall mechanical behavior of the tissue derives from a combination of their responses. Such constitutive models have gained a tremendous momentum in the past two decades (Holzapfel et al. 2000a; Gasser et al. 2006; Stylianopoulos and Barocas 2007a; Weisbecker et al. 2015; Li et al. 2016, 2018) and have been employed extensively in the computational modeling and simulation of human aortic tissues under various (patho)physiological conditions (Holzapfel et al. 2002b; Joldes et al. 2016; Liang et al. 2018). However, the precise description of the 3D distribution of collagen and elastic fibers in a constitutive model together with the consideration of the smooth muscle contractility pose formidable challenges albeit considerable simplifications and idealizations. Moreover, since some pathological conditions (e.g., aneurysms and aortic dissections) are characterized by damage and failure of micro-constituents, incorporation of such dissipative effects at different length-scales also has to be addressed in constitutive models.

In this chapter we provide an overview of multiscale experimental findings and computational modeling approaches for the passive and active behavior of the aorta as well as dissipative phenomena. In Sect. 2, we focus on the passive mechanical behavior of human aortas, and start with a review of experimental findings connecting the macro-, micro- and nanoscale investigations. Then, we present the development of (hyperelastic) multiscale constitutive models from the basic ones with parallel collagen fibers to the more advanced ones with dispersed collagen fibers. In Sect. 3, we discuss the importance of vascular SMCs with respect to the mechanical response of arteries, present the recent advances in experimental quantification of the multiaxial contractile SMC properties, and their incorporation in multi-physical constitutive models. In Sect. 4, we focus on the dissipative phenomena, namely, viscoelasticity, damage and failure. In particular, we present the related experimental findings considering the multiscale microstructural organization of both collagen and elastic fibers, followed by a summary of various computational approaches to the multiscale modeling of such phenomena. Finally, in Sect. 5 we discuss some of the key challenges that have to be addressed in order to improve our knowledge of the mechanical behavior of aortic tissue.

2 Passive Mechanical Behavior

In general, the passive stress-stretch response of the human aorta exhibits nonlinear stiffening, pseudoelasticity and anisotropy with the formation of remarkably small hysteresis upon cyclic loading-unloading (Holzapfel 2006; Schmid et al. 2005; Sherifova et al. 2019; Vande Geest et al. 2004). Out of numerous ECM components, type I and III fibrillar collagen is mainly responsible for the nonlinear stiffening and anisotropy of the tissue, while elastin accounts for its elasticity. Moreover, aortic tissues are capable of withstanding high stresses and strains thanks to these proteins (Holzapfel and Ogden 2018). In short, the mechanics of a tissue is governed by its microstructure—at the very least.

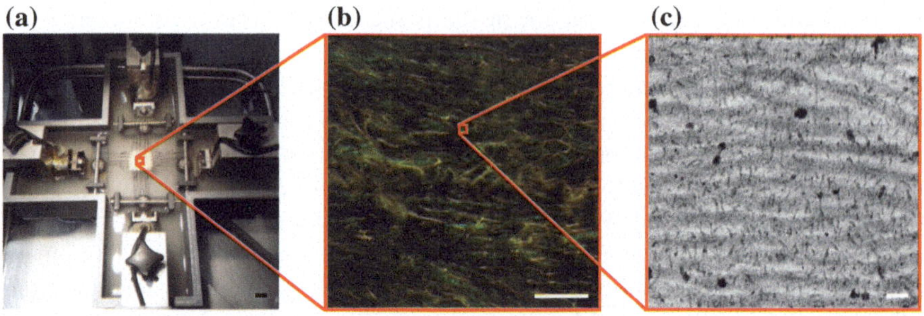

Fig. 2 Example of a multiscale experimental approach: (a) the biaxial extension test of an aortic wall is accompanied by (b) MPM and (c) followed by electron microscopy. Multiphoton microscopy (b) reveals the second-harmonic generation signal of collagen fibers (green) and autofluorescence of elastin fibers (yellow). Transmission electron microscopy (c) shows light gray collagen fibrils connected with spindle-like, dark gray proteoglycans (Wittgenstein 2018 with permission). Scale bars denote (a) 1 cm, (b) 100 μm, (c) 100 nm. Pukaluk et al., unpublished

As introduced in Sect. 1, (patho)physiological processes can lead to changes in the content, structure and orientation of various ECM components at different scales. In turn, these structural changes can have various mechanical manifestations such as stiffening or loss of anisotropy, and multiscale approaches are essential to their understanding. Figure 2 illustrates an example of such an experimental approach, where the macroscale mechanical properties of aortas at different atherosclerotic stages are investigated in connection with the macro-, micro- and nanoscale structural information. Current generations of multiscale constitutive models, which can incorporate these details, are of great importance for realistic simulations of the aortic tissues behavior under distinct (patho)physiological conditions. In this section, therefore, some important developments in experimental methods and computational modeling strategies for characterizing the passive mechanical behavior of human aortic tissues are reviewed.

2.1 Experimental Findings

Since the proteins of the ECM drive the macroscopic behavior of the aorta and define the above-mentioned mechanical properties (Wagenseil and Mecham 2009), one of the experimental approaches is to compare the stress-strain responses of intact and protein deficient tissues (Sherifova and Holzapfel 2020). Alternatively, promising methods for the investigation of the mechanical role of certain proteins include the enzymatic digestion of specific proteins and mechanical testing of the tissues before and after enzymatic treatment. For example, circumferential strips from the human thoracic media subjected to uniaxial extension with gradual collagenase treatment revealed a progressive change of their mechanical behavior from highly nonlinear to nearly linear without a distinct softening, whereby the collagen digested aortic tissue

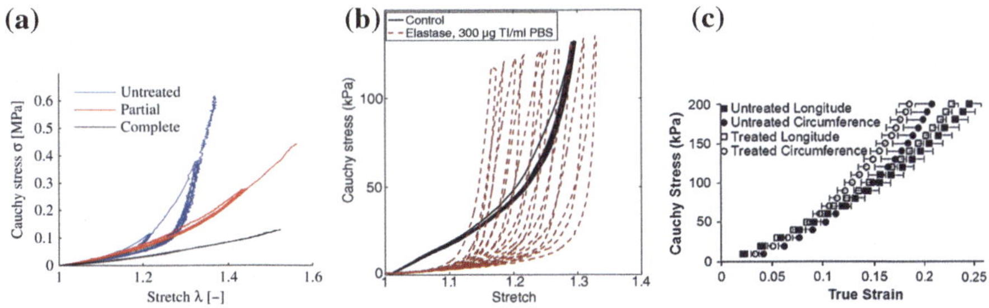

Fig. 3 (a) Effect of collagen, (b) elastin, and (c) glycosaminoglycan digestion on the stress-stretch response of the aortic tissue. Figure **(a)** reprinted from Weisbecker et al. (2013), **(b)** from Schriefl et al. (2015), and **(c)** from Mattson et al. (2017) with permission

lost its characteristic stiffening behavior (Fig. 3a) (Weisbecker et al. 2013). Similar findings were reported for enzymatically treated medial and adventitial strips from human abdominal aortas (Schriefl et al. 2015). In contrast, uniaxial extension experiments on elastin-degraded human thoracic (Weisbecker et al. 2013) and abdominal (Schriefl et al. 2015) aortas suggested that elastin plays a substantial role in maintaining the integrity of the tissue. Specifically, elastase treated samples showed continuous softening with remanent deformations during the subsequent loading and unloading cycles (Fig. 3b), and the stiffness at low deformations/loads was lower in elastin-digested tissues compared to control samples. Although collagen and elastin dominate the tissue content and mechanics, proteoglycans deserve further attention (Sherifova and Holzapfel 2020). The explicit mechanical roles of proteoglycans in arterial mechanics remain elusive, however, Mattson et al. (2017) showed that removal of glycosaminoglycans, the main constituent of proteoglycans, caused earlier stiffening indicated by transition points located at significantly lower strains on the nonlinear stress-strain curves obtained during equibiaxial extension tests (Fig. 3c).

A further promising experimental approach for the investigation of the aortic properties at macro- and microscales involves simultaneous mechanical testing and microscopy imaging. This approach allows the quantification of organizational changes in the tissue components under different loadings. For example, MPM combined with biaxial extension tests on porcine aortas revealed that medial collagen was engaged throughout loading, whereas adventitial collagen only engaged at higher loads (Chow et al. 2014). However, in layer-specific tests, collagen in the porcine adventitia gradually straightened under load with no noted delay (Li et al. 2019). In contrast, prominent elastin engagement was mainly observed at lower strains (Chow et al. 2014). Furthermore, equibiaxial loading caused no significant realignment of collagen or elastic fibers in contrast to non-equibiaxial loading, which caused a realignment of collagen fibers in both layers but not of elastic fibers (Chow et al. 2014). In general, elastic fibers were observed to be less undulated than collagen fibers in the media of rabbit aortas at physiological load (Sugita and Matsumoto 2017). In another study, bulge inflation tests were coupled with MPM to investi-

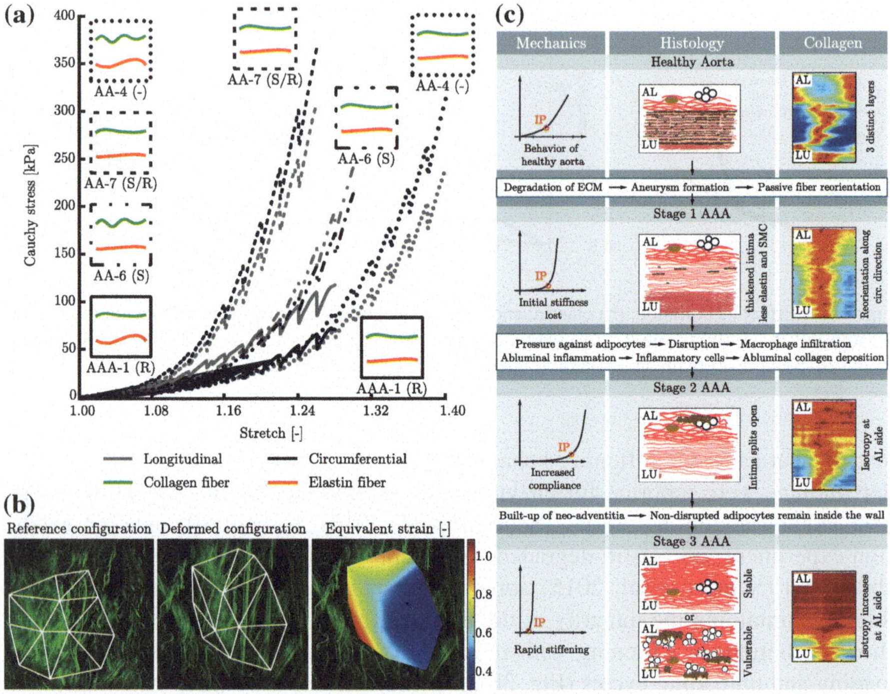

Fig. 4 Global mechanical properties and microstructure of the human aortas: (a) representative mechanical responses of equibiaxial tests (successful (-), interrupted due to the stiffness (S), rupture (R) or stiffness and rupture (S/R)) on the human abdominal media with the schematic representation of collagen and elastic fibers illustrating their waviness at the reference and maximum achieved stretch configurations (Pukaluk et al., unpublished); (b) local equivalent strain field on the human abdominal adventitia subjected to 1.1 global equibiaxial stretch (Pukaluk et al., unpublished); (c) changes in the mechanical behavior of AAA tissue in the course of pathogenesis accompanied by changes in content and collagen fiber orientation (reproduced from Niestrawska et al. 2019 with permission)

gate adventitial collagen in human thoracic aortas (Cavinato et al. 2017). This study showed that dense and undulated collagen bundles in the unloaded state tended to progressively straighten at higher pressure, and initially thick bundles of collagen transitioned into a network of thinner bundles. Ongoing investigations in this regard utilize planar biaxial extension tests coupled with MPM to investigate structural alterations of collagen and elastin in human abdominal aortas (Pukaluk et al., unpublished). Specifically, notable differences in the structural reorganization between the loaded media and adventitia suggest distinct stretching mechanisms reaching beyond different fiber orientations in these layers in the unloaded state, which the waviness of medial collagen and elastic fibers might explain, see Fig. 4a. Moreover, the application of fluorescent beads could help to understand the local strain generated as a response to the global stretch (Fig. 4b).

As previously explained in Sect. 1, content and structure of ECM components are affected by age and diseases. Not surprisingly, the passive mechanical behavior also changes due to aging and pathological conditions (Holzapfel et al. 2004; Vande Geest et al. 2006; Åstrand et al. 2011). For example, Haskett et al. (2010) reported that with advanced age the degree of fiber alignment increased and tissues showed a stiffer response under planar biaxial tests, leading to the assumption that age-related aortic stiffening is also associated with fiber alignment. Deeper insights into the microstructure and the related mechanics of healthy and diseased, i.e. aneurysmatic, human abdominal aortas were reported by Niestrawska et al. (2016, 2019). They found loss of the characteristic three-layered wall structure, significantly higher out-of-plane dispersion of collagen fibers as well as stiffer behavior of the aneurysmatic samples compared with control tissues. The authors also noted a large variability in the mechanical and structural data among the aneurysmatic tissues. Subsequently, Niestrawska et al. (2019) explained this variability with a three-stage disease progression hypothesis, where they related the changes in the in-plane dispersion of collagen fibers with the altered mechanics of the aortic wall (Fig. 4c). More specifically, fiber reorientation towards the circumferential direction in stage I correlated with decreased initial stiffness. Subsequently, in stage II, almost isotropic fiber dispersion at the abluminal side led to increased compliance, and finally, in stage III increased isotropic fiber dispersion throughout the aortic wall thickness correlated with rapid stiffening (Fig. 4c). Niestrawska et al. (2016) also reported a decreased waviness of collagen fibers in aneurysm walls, however, fiber waviness quantification and its correlation to the mechanics of the human aorta has not been reported to date. Further microstructural and mechanical investigations and related quantification in this regard may show rewarding considering that higher stiffness of the aortic tissue is associated with decreased undulation in collagen fibers for porcine aortas (Zeinali-Davarani et al. 2015).

Regarding ultrastructural investigations, several studies focused on the mechanical properties of ECM components at the nanoscale level. Examples include but are not limited to stretching of a collagen monomer by Sun et al. (2002), bending of a collagen fibril by Dutov et al. (2016), bending and indentation of an elastic fiber by Koenders et al. (2009) and stretching of a glycosaminoglycan molecule by Haverkamp et al. (2005). Furthermore, measurements on healthy and atherosclerotic human abdominal aortas provided the local tissue stiffness of elastic lamellae, interlamellar zones as well as different locations of the atherosclerotic plaque such as the fibrous cap, calcification zone, and lipid pool (Rezvani-Sharif et al. 2019). Nevertheless, these studies did not relate the mechanical behavior of single molecules or fibrils to micro- or macromechanics of the tissue. Attempting to narrow this gap between scales, combined planar biaxial tests and electron tomography investigations performed by our group enabled a deeper insight into arrangement of collagen fibrils and proteoglycans inside the aortic layers at different stretches. Preliminary results show that proteoglycans reorient and increase their cross section in response to higher stretch (Pukaluk et al., unpublished). Other scale-bridging experiments on the human aorta were conducted by Lindeman et al. (2010), where the adventitia of healthy and aneurysmatic abdominal aortas was probed by atomic force microscopy cantilevers at the levels of collagen fibrils (nanoscale) and fibers (microscale). The healthy aorta behaved as a coherent network and mechanical forces were distributed over the tissue, in contrast to the aneurysmatic tissue. These findings emphasize that multiscale mechanical experiments that bridge the nanoscale to higher scales are critical for a better understanding of the (patho)physiological tissue mechanics.

2.2 Constitutive Modeling

Inspired by the pioneering work of Lanir (1983, 2018), soft fibrous tissues have been treated as a fiber-reinforced composite material. The total strain-energy function (SEF) of the material is obtained by a summation of tissue constituents' strain energy under finite deformation. Within this framework, Holzapfel's group proposed the *so-called* HGO model for arterial tissues with collagen fibers predominately distributed around the mean direction within each family (Holzapfel et al. 2000a). In this model, the arterial tissues were treated as a multilayered composite material, and each layer consisted of a ground substance and two families of collagen fibers for the mechanical characterization of their passive behavior. Briefly, following the multiplicative decomposition (Flory 1961; Ogden 1978) of the deformation gradient $\mathbf{F}$ into a volumetric (dilatational) part $J^{1/3}\mathbf{I}$ and an isochoric (distortional) part $\overline{\mathbf{F}} = J^{-1/3}\mathbf{F}$, with $J = \det \mathbf{F} > 0$, the SEF Ψ of the arterial tissue per unit volume in the reference configuration is usually written in a decoupled form for an efficient computational implementation, i.e.

$$\Psi = \Psi_{\text{vol}} + \Psi_{\text{iso}}, \tag{1}$$

where Ψ_{vol} describes the volumetric deformation of the aortic tissue and Ψ_{iso} the isochoric deformation. Since arterial tissues are usually treated as incompressible materials, the volumetric part of the SEF is used as a penalty function, and it is convenient to adopt a form available in the finite element program. For example, the following form could be employed if FEAP (Taylor 2013) is used as the finite element solver,

$$\Psi_{\text{vol}} = \frac{K}{4}(J^2 - 1 - 2\ln J), \tag{2}$$

where K is a penalty parameter.

The isochoric contribution can be further decomposed into two parts,

$$\Psi_{\text{iso}} = \Psi_{\text{g}} + \Psi_{\text{f}}, \tag{3}$$

where Ψ_{g} denotes the isochoric strain energy of the ground substance, which is assumed to be isotropic and to depend only on the modified first invariant $\bar{I}_1 = \text{tr}\overline{\mathbf{C}}$ of the modified right Cauchy–Green tensor $\overline{\mathbf{C}} = \overline{\mathbf{F}}^{\text{T}}\overline{\mathbf{F}}$, and Ψ_{f} represents the total isochoric strain energy of collagen fibers. Since it has been shown that the non-collagenous ground substance of arterial tissue (Weisbecker et al. 2013) exhibits a nearly linear stress–strain response, the neo-Hookean hyperelastic model is often used for Ψ_{g}, i.e.,

$$\Psi_{\mathrm{g}}(\bar{I}_1) = \frac{\mu}{2}(\bar{I}_1 - 3), \tag{4}$$

where the constant μ (> 0) is the shear modulus.

To account for the mechanical behavior of the collagen fibers in the arterial wall, the HGO model proposed by Holzapfel et al. (2000a) defines the following exponential SEF for two families of fibers, i.e.

$$\Psi_{\mathrm{f}} = \frac{k_1}{2k_2} \sum_{i=4,6} \left(e^{k_2(\bar{I}_i - 1)^2} - 1 \right), \tag{5}$$

where k_1 and k_2 are the material parameters associated with the collagen fibers, with a dimension of stress and without dimension, respectively. $\bar{I}_4$ and $\bar{I}_6$ are modified counterparts of the squared stretches $I_4 = \mathbf{C} : \mathbf{M}_1 \otimes \mathbf{M}_1$ and $I_6 = \mathbf{C} : \mathbf{M}_2 \otimes \mathbf{M}_2$ along the mean fiber directions, $\mathbf{M}_1$ and $\mathbf{M}_2$, of two symmetrically arranged collagen fiber families. The two families are assumed to be the same type within each layer of arterial tissue. If either fiber stretch in the mean direction is less than unity ($I_4 < 1$ or $I_6 < 1$), then that fiber family is excluded from Ψ_{f}. Elastin in the arterial wall could be treated as part of the ground substance, or it could also be incorporated in the constitutive model in the form of elastic fibers in a similar manner (Rolf-Pissarczyk et al. 2021).

This 'simple' and yet robust constitutive model (5) captures the mechanical behavior of the essential constituents—ground substance and collagen fibers—of arterial tissue and has been extensively employed in the computational modeling not only of arterial tissues, but also other soft fibrous tissues (Mao et al. 2016). The HGO model initially included two families of collagen fibers, and it was later extended to include four families of fibers by Humphrey and colleagues for a better fit to experimental data of some arterial tissues (Baek et al. 2007).

The HGO model was very efficient in modeling soft biological tissues with collagen fibers predominately distributed around a mean direction within a family or soft tissues with families of parallel collagen fibers such as mitral valve chordae tendineae (Zuo et al. 2016). However, as shown in Sect. 2.1, the distribution of collagen fibers in each layer of the arterial wall is often dispersed within each family in both healthy and aneurysmal conditions (see Fig. 4c). To account for such dispersion in a constitutive equation, two approaches, namely the 'angular integration' (AI) approach and the 'generalized structure tensor' (GST) approach, have been widely used.

The AI approach was originally proposed by Lanir (1983, 2018) in 1983, where the strain energy of a single fiber is only a function of the fiber stretch. Then, an integration of the single fiber SEF over all the fiber directions weighted by a continuous probability density function (PDF) yields the total fiber strain energy Ψ_{f} (Lanir 1983, 2018). The pioneering work of Lanir has gained growing interest in the past decades, and there have been numerous constitutive models based on this approach since it was proposed in 1983 (Holzapfel and Ogden 2015).

The fiber contribution to the total SEF can be evaluated by using a numerical integration of the weighted single fiber SEF over the unit sphere, known as the micro-sphere based model, see Alastrué et al. (2009) and references therein. However, if the fiber dispersion is accounted for as a summation of a finite number of discrete fiber contributions, then it is referred to as the 'discrete fiber dispersion' (DFD) model, see Li et al. (2018). While a numerical integration over the sphere or hemisphere is required in the AI approach, this is not needed in the DFD model since the fibers are treated in a discrete manner.

At first, we describe the AI approach for modeling soft fibrous tissues. Briefly, we introduce unit Cartesian basis vectors $\mathbf{E}_1$, $\mathbf{E}_2$, and $\mathbf{E}_3$, and then any fiber direction $\mathbf{N}$ within a 3D dispersed fiber family in terms of spherical polar angles Θ and Φ relative to $\mathbf{E}_1$, $\mathbf{E}_2$, and $\mathbf{E}_3$ reads,

$$\mathbf{N} = \sin\Theta\cos\Phi\mathbf{E}_1 + \sin\Theta\sin\Phi\mathbf{E}_2 + \cos\Theta\mathbf{E}_3. \tag{6}$$

With that, the fiber contribution Ψ_f to the total SEF in the AI model is obtained by integrating the weighted single fiber strain energy $\Psi_n(I_4)$ of each fiber direction $\mathbf{N}$ over the unit sphere $\mathbb{S}^2 = \{(\Theta, \Phi) \mid \Theta \in [0, \pi], \Phi \in [0, 2\pi]\}$, i.e.,

$$\Psi_\mathrm{f} = \frac{1}{4\pi}\int_{\mathbb{S}^2} \rho(\Theta, \Phi)\Psi_n(I_4)\sin\Theta\mathrm{d}\Theta\mathrm{d}\Phi, \tag{7}$$

where the PDF $\rho(\Theta, \Phi)$ is the probability density of the fibers in the direction $\mathbf{N}(\Theta, \Phi)$ in the reference configuration. The PDF $\rho(\Theta, \Phi)$ in Eq. (7) must satisfy the normalization condition

$$\frac{1}{4\pi}\int_{\mathbb{S}^2} \rho(\Theta, \Phi)\sin\Theta\mathrm{d}\Theta\mathrm{d}\Phi = 1. \tag{8}$$

The computational implementation of the SEF Eq. (7) requires two-dimensional integration over the unit sphere at each Gauss point during a finite element simulation. Because of the complicated nature of the integrand in Eq. (7), a numerical integration over the sphere is often employed.

Also referred to as the GOH model in the literature, the GST approach was proposed by Holzapfel's group (Gasser et al. 2006) in 2006 where the contribution of the entire fiber family (Ψ_f) to the total SEF is 'wrapped' in a generalized structure tensor. The extension of the HGO model (5) to the GOH model is obtained simply by replacing $\bar{I}_i$ in Eq. (5) by $\bar{I}_i^\star$ (Holzapfel and Ogden 2015), i.e.,

$$\bar{I}_i^\star = \kappa\bar{I}_1 + (1 - 3\kappa)\bar{I}_i, \quad i = 4, 6, \tag{9}$$

where κ is the parameter quantifying the rotationally symmetric dispersion of fibers around the mean direction. A widely used PDF for such a fiber dispersion is the von

Mises distribution over the unit sphere,

$$\rho(\Theta, \Phi) = 4\sqrt{\frac{b}{2\pi}} \frac{\exp[2b(\mathbf{N} \cdot \mathbf{M})^2]}{\mathrm{erfi}(\sqrt{2b})}, \tag{10}$$

where b is the concentration parameter describing how closely the fibers are distributed around the mean fiber direction $\mathbf{M}$ of the fiber family, $\mathrm{erfi}(x) = -\mathrm{i}\,\mathrm{erf}(\mathrm{i}x)$ denotes the imaginary error function, and $\mathrm{erf}(x)$ is the standard error function. Since the microstructural data showed a non-symmetric dispersion of collagen fibers in the arterial wall (Schriefl et al. 2012), the GOH model was then extended to include this information in Holzapfel et al. (2015).

Because of the slenderness and waviness of collagen fibers in the reference configuration, it is usually assumed that they do not have any compressive strength and will buckle if compressed. For strictly incompressible materials with 3D fiber dispersion such as arterial tissues, there will always be some fibers under compression and others under tension under general finite deformation. Thus, the fibers under compression within a subset of the unit sphere should be excluded when computing the total strain energy of the tissue. However, there is some confusion in the literature and in finite element implementation of this tension–compression 'switch' of the fibers as highlighted in Holzapfel and Ogden (2015). In addition, it is not strictly correct to exclude the entire fiber family from the SEF of the tissue when the mean fiber direction is under compression as it is proposed in the GOH model, because some fibers are still under tension even if the mean direction is under compression. Eliminating the contribution of those fibers under tension could underestimate the total strain energy of the tissue unless the contribution of those fibers is negligible under some deformation states.

To circumvent this problem, a more accurate method was proposed by Holzapfel and Ogden (2015) for exclusion of compressed fibers within a dispersion, which was later extended and implemented in FEAP and illustrated with some representative examples in Li et al. (2016). Unfortunately, however, the computational cost for the numerical integration of the weighted single fiber SEF over a subset of the unit hemisphere $\mathbb{S} = \{(\Theta, \Phi) \mid \Theta \in [0, \pi], \Phi \in [0, \pi]\}$ in the model was very high. To overcome this costly numerical difficulty and reduce the computational time, the new DFD model (Li et al. 2018) was proposed by the Holzapfel group recently.

In the DFD model (Li et al. 2018), the dispersion of the collagen fibers in each fiber family was accounted for in a discrete manner. More specifically, the authors discretized the unit sphere into a finite number $2m$ of elementary areas $\Delta\mathbb{S}_n$, $n = 1, \ldots, 2m$, see Fig. 5b for an example of such discretization of the unit sphere with $2m = 180$ spherical triangles where m represents the number of elementary areas over a half of the unit sphere. Note that only the elementary areas over one half of the unit sphere were used in the constitutive model. Without loss of generality, a rotationally symmetric fiber dispersion (10) was used for an illustration of the method. The representative fiber direction (Θ_n, Φ_n) at the centroid of each elementary area and an elementary fiber density ρ_n was defined as

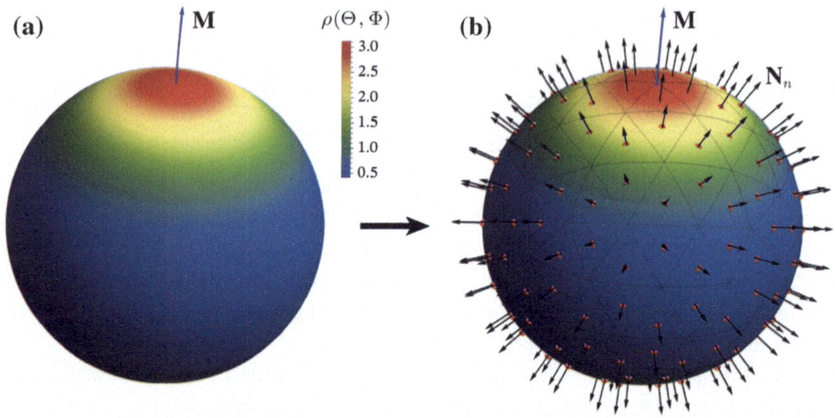

Fig. 5 (a) Contour plot of the PDF $\rho(\Theta, \Phi)$ defined over the unit sphere by the von Mises distribution with mean fiber direction $\mathbf{M}$ and a concentration parameter $b = 1.0$, see Eq. (10); (b) example of a triangular discretization of the unit sphere with $2m = 180$ representative fiber directions $\mathbf{N}_n$ (black arrows) defined at the centroids (red dots) of the spherical triangles. Reprinted from Li et al. (2018) with permission

$$\rho_n = \frac{1}{2\pi} \int_{\Delta\mathbb{S}_n} \rho(\Theta, \Phi) \sin\Theta \mathrm{d}\Theta \mathrm{d}\Phi, \qquad n = 1, \ldots, m. \tag{11}$$

It was required for the PDF $\rho(\Theta, \Phi)$ to be an integrable function over each $\Delta\mathbb{S}_n$. If this is not satisfied, then, the elementary area could be further discretized so that the PDF becomes integrable over each elementary area. The elementary fiber density ρ_n must satisfy the normalization condition, i.e.,

$$\sum_{n=1}^{m} \rho_n = 1, \tag{12}$$

which is the discrete counterpart of Eq. (8) over the unit hemisphere. With a discretized unit hemisphere, the SEF for the total contribution of each fiber family was given as

$$\Psi_\mathrm{f} = \sum_{n=1}^{m} \rho_n \Psi_n(I_{4n}), \tag{13}$$

where $I_{4n} = \mathbf{C} : \mathbf{N}_n \otimes \mathbf{N}_n$ and $\mathbf{N}_n$ was defined at the centroid of each spherical triangle via Eq. (6) with $\Theta = \Theta_n$ and $\Phi = \Phi_n$; see the black arrows in Fig. 5b. In order to exclude the fibers under compression within a dispersion, Li et al. (2018) set Ψ_n to zero if $I_{4n} < 1$. This model has been implemented in FEAP and verified with three numerical examples, results of which were consistent with the previously published data in Li et al. (2016). However, Li et al. (2018) achieved these results with a significantly reduced computational time—224 times faster.

Another approach for modeling the collagen fiber dispersion in the arterial wall is the fiber network model (Hadi et al. 2012; Stylianopoulos and Barocas 2007a). Therein, the 3D fiber dispersion is accounted for by generating a network of collagen fibers within a representative volume element (RVE) either randomly (Lake et al. 2012) or based on the imaging data of the tissue. Then, the macroscopic Cauchy stress tensor of the collagen fibers is obtained in terms of the microscopic stress tensor by using the volume-averaging theory (Stylianopoulos and Barocas 2007b). Since the collagen fibers are explicitly represented in those models, fiber damage and fiber–fiber interaction can also be readily incorporated into the model, see, e.g., Hadi et al. (2012). Future measurement of detailed 3D collagen fiber architecture in arterial tissue will certainly improve the accuracy of this approach.

Recently, Holzapfel and Ogden (2020a) proposed a new fiber dispersion model considering the orientation of collagen fiber cross-links relative to the fiber direction and cross-link density on the mechanical response of the tissue. The model was able to predict the stiffening of the artery as the cross-linking density increased. However, since the mechanical properties of cross-links and their interactions with the collagen fibers are largely unknown, the model will need to be refined and extended as new mechanical and structural data become available in the next decade.

3 Active Mechanical Behavior

The mechanical properties of the vascular wall have been associated with various cardiovascular diseases and have mainly been attributed to the ECM. However, more recently the intramural cells, particularly SMCs, have been found to play critical roles in the physiological state of the vascular wall (Li et al. 2020). Two leading risk factors of cardiovascular diseases are hypertension and arterial stiffness, both which are influenced by the SMCs (Lacolley et al. 2017; Spronck et al. 2020). Increased blood pressure can result from an increase in cardiac output or an increase in vascular resistance. The former is the amount of blood pumped by the heart per minute and is the product of the heart rate and the stroke volume, which is the amount of blood pumped out of the heart. The vasculature plays an important role in regulating blood flow through vital properties such as the elastic compliance of arteries as well as the blood capacitance of the veins, both which depend on the SMC active tone. Vascular resistance is mainly regulated by the peripheral blood vessels and is inversely proportional to the fourth power of the inner radius, which makes it highly dependent on the SMC active tone.

In addition, arterial stiffness has been demonstrated to be influenced by the vascular tone and shown to have a significant and complex relationship to the pathophysiological state of the vessel (Lacolley et al. 2017). The arterial stiffness can be clinically measured through the pulse wave velocity PWV and can be calculated through the Moens–Korteweg equation

$$\text{PWV} = \sqrt{\frac{E(P, C)h(P, C)}{2r(P, C)\rho}}, \tag{14}$$

where E is the intrinsic (linear) material stiffness, h the wall thickness, r the inner radius of the vessel and ρ is the density of the tissue. All metrics, except perhaps the wall density, are strongly influenced by the average pressure P and the vasoactive SMC tone C (Humphrey 2021a; Murtada and Humphrey 2018). Thus, there is a significant importance to experimentally quantify and model the active mechanical behavior of arteries to better understand the pathophysiological processes of cardiovascular diseases.

3.1 Experimental Methods

Two well-known ex vivo experimental techniques to quantify the active tissue-level vascular tone in arteries are the uniaxial test of intact vascular ring segments and the biaxial pressure-inflation test of intact vessels. Both methods can be used to assess the vasoactive response, albeit under different loading conditions, and are discussed next.

3.1.1 Active Tension and Vascular Tone

The active force generated by vascular SMCs depends on the dynamic interactions between intracellular myosin motors and actin filaments, which result in filament sliding and SMC contractility. The activation of myosin motors occurs through phosphorylation of the regulatory myosin light-chains, which is regulated by the calcium-dependent phosphorylating myosin light-chain kinase (MLCK). There are several ways to increase intracellular calcium through, e.g., pharmacomechanical and electromechanical pathways (Somlyo and Somlyo 1994). In pharmacomechanical pathway activation, agonists bind to specific guanine nucleotide-binding protein (G-protein) receptors located on the SMC membrane that work as molecular switches activating SMC intracellular pathways. These pathways trigger a range of functions in the SMCs, including the contractile response of the actin-myosin filaments (actomyosin). Herein, protocols where the SMC active tone and vascular response, through ex vivo pharmacomechanical activation are at focus.

In a uniaxial ring test, the active force is generated by the mounted vessel ring segments between two pins recorded through a force transducer attached to one pin, while the distance between the pins are controlled by a micrometer attached to the other pin. The vessel ring segment can be contracted under two different uniaxial boundary conditions in the circumferential direction: isometric contraction where the length is held constant while the force is recorded, and isotonic contraction where the force is held constant while change in length is recorded. The active tension (in N/m) can be calculated and compared by normalizing the active force to the axial width of the ring segment. Further, the active tension can be normalized to the wall thickness or to the thickness of SMC-rich media to calculate the active stress generated by the SMCs. The normalization is useful when comparing active tone in arteries from different regions in the vascular tree (elastic, muscular, resistance). One of the main advantages of this test is that several ring samples can be studied from a single vessel segment, which is useful when assessing, e.g., pharmacological dose-response relationships.

During an isometric contraction test, the magnitude and the rate of active tension can be studied for different types of agonists and concentration. The active tension magnitude is a useful metric to compare the maximal active tone produced by the vessel, and can be related to the number of active and load-bearing cross-bridges in the SMCs (Murtada et al. 2010a; Stålhand et al. 2008). The rate can be correlated to the recruitment and the kinetics of the load-bearing cross-bridges (Hai and Murphy 1988).

An alternative approach to measure the velocity of the attached cross-bridges is through isotonic quick-release experiments. In this testing method, an arterial ring segment is first contracted isometrically at a fixed length. After the active force has reached a steady-state value, the arterial ring is released quickly against a fixed lower force while the change in length is recorded (Murtada et al. 2010a). The evolution of the length change can be studied through Hill's three-element muscle model, and described by two phases, where an initial elastic recoil is followed by an isotonic and almost linear change in length (Chun and Seow 2013). The experiment can be repeated for different isotonic loads and the velocity of the length-shortening during the second phase can be calculated for each load, upon which a nonlinear relationship between the isotonic load and the shortening velocity can be observed. This relationship was described in 1938 (Hill 1938) with a hyperbolic equation using muscle heat energetic measurements as

$$(F + a)(v + b) = b(F_{\max} + a), \tag{15}$$

where $F_{\max}$ is the maximal isometric force, F is the constant isotonic force, v is the velocity, and a and b are fitting parameters. Equation (15) accurately predicts the relationship between muscle force and velocity with exceptions at the lower (near zero) and higher ($F = F_{\max}$) force ranges. In addition, it can be demonstrated that a similar hyperbolic equation is useful for the velocity at the cross-bridge level by utilizing multistate kinetics models of actomyosin interactions with some necessary assumptions (Chun and Seow 2013). Such an approach makes isotonic ring tests a unique ex vivo technique to study cross-bridge kinetics and ATP hydrolysis at the tissue level (Edman 1975, 1979; Pate and Cooke 1989).

A uniaxial ring test does not reflect the physiological in vivo boundary conditions of a vessel since only the circumference is regulated while the radial and axial lengths are free. A different method to study the active (and passive) response of the vessel under more physiological-like loading conditions is to mount an intact vessel in a pressure-myograph. The vessel can be fixed with thin sutures and cannulated onto two glass pipettes or needles that are connected to computer-controlled linear actuators or micrometers to regulate the axial vessel length. The luminal pressure is controlled via pressure transducers and a pressure-controlled reservoir, and the pressure-diameter response of the vessel is measured during inflation and deflation at a set fixed axial length with the pressure-myograph. The measurements can be done with and without activation of SMCs, and the difference in response can be used to study the contractile magnitude at different pressures and axial stretches (Caulk et al. 2019; Zulliger et al. 2004). While the passive pressure-diameter response does not display a significant difference between inflation and deflation (hysteresis) after proper mechanical preconditioning (Ferruzzi et al. 2013), the active response may display a difference (Greenwald et al. 1982).

Comparable to the force development during uniaxial isometric contraction in a ring segment, the active change in the outer diameter of a mounted vessel at isobaric and axially isometric boundary conditions can be studied in response to different pharmacological stimuli. The agonist is commonly added to the adventitial bath and diffuses to the SMCs in the media, which is different to uniaxial ring test where the SMCs in the medial layer are in direct contact with the bath solution and no diffusion is needed. This results in a faster development of active tension during a uniaxial ring test than the change in diameter during a biaxial isobaric test (Murtada et al. 2016a, b). In addition to the change in the outer diameter, the change in the inner radius a and the circumferential wall stress σ_θ in response to different stimuli can be calculated under the assumption of incompressibility. Using Laplace's equation for cylinders gives

$$\sigma_\theta = \frac{Pa(P, C)}{h(P, C)}, \tag{16}$$

where P is the distending pressure, h is the deformed thickness and C is the degree of smooth muscle contractility (Humphrey 2021a; Murtada and Humphrey 2018).

The active change in the diameter during isobaric contraction is a combined effect of the SMC active response and the nonlinear passive ECM response, and where a mathematical model is necessary to separate the different responses. In addition to the vessel contraction induced by the actomyosin interactions, it is important to note that the contractile SMC response may influence the passive mechanical response of the vessel. An alternative to isobaric and axial isometric contraction is biaxial isometric vessel contraction, where the pressure is increased to maintain the diameter during contraction (Caulk et al. 2019). Thereby, the pressure increase can be used to quantify the active SMC tone and to compare the isometric contraction response under biaxial and uniaxial loading.

3.1.2 Multiaxial Stretch-Sensitivity

Vascular SMCs are spindle-shaped cells with intracellular actomyosin contractile units mainly aligned with its long axis (Devine and Somlyo 1971) and are oriented along the circumferential direction (O'Connell et al. 2008). When the vessel wall is deformed in the circumferential direction, the arrangement and the overlap of the actin and myosin filaments change depending on the chronic pressure and the ECM properties of the vessel (Gordon et al. 1966; Murtada et al. 2012). It is well-known that there exists an optimal circumferential stretch for maximal active force production for all muscle types, which has been associated with the overlap and arrangement of actin and myosin filaments (Huxley and Niedergerke 1954; Mulvany and Warshaw 1979).

The uniaxial ring test provides a straight-forward experimental approach to assess the circumferential stretch-sensitivity on the active tension development in arterial ring. The sarcomeres in the skeletal muscle exhibit a well-defined striated acto-myosin organization with repeated bi-polar myosin thick filaments overlapped by hexagonally arranged actin filaments. The vascular SMCs lack such an organization, and contractile units consist of both side- and bi-polar myosin filaments surrounded by a network as well as bundles of actin filaments (Cooke et al. 1989; Craig and Megerman 1977). Interestingly, the optimal circumferential stretch in the SMCs is significantly higher than in the skeletal muscle, which can be related to the 'smooth' intracellular actomyosin arrangement in SMCs in terms of both filament overlap and filament orientation dispersion (Haspinger et al. 2018). Using a phenomenological description of the filament overlap through either a Gaussian or a parabolic function, researchers successfully captured the circumferential stretch-sensitivity in vascular SMCs (Murtada et al. 2012; Rachev and Hayashi 1999; Stålhand et al. 2008; Zulliger et al. 2004).

The different loading conditions may result in significantly different axial and radial stretches at similar values of circumferential stretch for the uniaxial and biaxial testing techniques. For example, if an arterial ring is stretched to $\lambda_\theta = 1.5$ in the circumferential direction, stretches in the radial and axial directions would result in $\lambda_r = \lambda_z = 1/\sqrt{1.5} = 0.816$ assuming incompressibility. This leads to an unphysiological value of the axial stretch taking into account $\lambda_z > 1.0$ in healthy arteries. However, if an artery is biaxially loaded to $\lambda_\theta = \lambda_z = 1.5$ through the applied pressure and the axial stretch, the radial stretch is nearly half the value compared to the uniaxial case ($\lambda_r = 0.444$ vs. $\lambda_r = 0.816$). The influence of the axial stretch and, more importantly, of the radial compressive stretch on the active tension production in vascular SMCs has not been well-studied.

In addition, the filament lattice spacing, i.e., the distance between the actin and myosin filaments, has been demonstrated to have a strong influence on the active force generation as well as cross-bridge kinetics of the actomyosin complex (Arheden et al. 1987; Edman 1999; Williams et al. 2010). Recently, the combined effect of the actomyosin filament overlap and the filament lattice spacing was investigated by using a multiaxial and multiscale mathematical model by Murtada et al. (2017). It was reported that when the SMC model only consider the filament overlap, the model correctly predicts the circumferential stretch-sensitivity observed in uniaxial ring tests, but significantly underestimates the active change in the diameter observed in vessels under isobaric—axially isometric conditions. When the model was extended to include the filament lattice spacing, both uniaxial and biaxial contraction tests were accurately predicted. The multiaxial model suggested that a dynamic myosin filament length description dependent on the filament lattice spacing was necessary, and that the actomyosin contractile units acted mainly in the radial-circumferential plane during isobaric—axially isometric contraction. This emphasized the importance of the physiological loading conditions and the biaxial stretch-sensitive SMC contractility in blood vessels.

3.2 Mathematical Modeling and Related Computational Aspects

Next, we provide a brief overview of different classes of computational models that have been developed to numerically investigate the effects of smooth muscle contraction onto the mechanical response of the arterial wall as they have progressed historically with a focus on more recent constitutive models incorporating microstructural, mechanical, chemical and electrical features.

In 1938, the pioneering work by Hill (1938) introduced the idea that the experimentally obtained active mechanical behavior of a striated muscle can be mathematically described by a two-component system, which consists of a contractile element (damper), governed by a characteristic evolution law such as Eq. (15), in series with a purely elastic element (spring). To capture the entirely passive mechanical response of the relaxed muscle tissue, an additional elastic element (spring) needs to be added in parallel to the series elements, which is known as Hill's three-element model (Fung 1993). In fact, this rheological model hallmarks the fundamental base for various constitutive models developed and used over the past 80 years to analyze and predict the mechanical behavior of skeletal muscles, the heart muscle, and soft tissue rich of SMCs (Hatze 1977; Gestrelius and Borgström 1986; Wexler et al. 1997; Ettema and Meijer 2000; Lloyd and Besier 2003; Lichtwark and Wilson 2005).

While in the early stages, a coherent microstructural interpretation of the individual components in Hill's three-element model was missing, many improvements have been made. Supported by recent structural multiscale investigations (Ye et al. 2014), nowadays, the contractile element is associated with the active force generation due to the relative sliding between the actin and myosin filaments, i.e., the sliding filament theory. In the context of vascular SMC contraction, the series element is assumed to represent the intrinsic elasticity of the actomyosin cross-bridges, the dense bodies that connect the individual actomyosin complexes, and the focal adhesion complexes that provide a physical link between the actomyosin complexes and the ECM (Ye et al. 2014; Murtada et al. 2016b). The parallel spring captures the purely passive elastic response of the ECM (collagen and elastic fibers, and proteoglycans) and the intracellular structural components (cortical actin network, microtubules, and intermediate filaments).

By applying the principle of virtual power (Germain 1973), it can be shown that the total stress $\boldsymbol{\sigma}$ in Hill's three-element model is additively decomposed into a passive part $\boldsymbol{\sigma}_{\text{passive}}$ resulting from the parallel spring, and an active part $\boldsymbol{\sigma}_{\text{active}}$ generated by the contractile element (being equal to the stress in the serial spring). Assuming finite hyperelasticity, we start from the premise that there exists a Helmholtz free-energy potential Ψ, which we can also be additively decomposed into an active part Ψ_{active} accounting for the energy stored in the elastic actomyosin cross-bridges (the serial spring), and a passive contribution Ψ_{passive} representing the structural and mechanical properties of the passive ECM components in the arterial wall (as described in Sect. 2.2), i.e. $\Psi = \Psi_{\text{active}} + \Psi_{\text{passive}}$ from which we can derive the expressions for stresses $\boldsymbol{\sigma}_{\text{active}}$ and $\boldsymbol{\sigma}_{\text{passive}}$. In the following, we provide a brief overview of frequently used phenomenological material models and microstructurally-motivated mechanochemical models that capture essential characteristics of the active mechanical contribution of SMCs in the arterial wall.

Numerous experimental studies revealed that vascular smooth muscle-rich tissue exhibits a parabolic or bell-shaped active length-tension behavior and that the muscle tone strongly depends on vasoactive chemical agents, such as the intracellular free calcium concentration (Dobrin 1973a, b; Cox 1975; Horowitz et al. 1996; Zulliger et al. 2004; Murtada et al. 2012). Based on these findings, together with the assumption that SMCs are mainly oriented along the vessel's circumferential direction, various similar phenomenological models considering the smooth muscle tone have been developed and utilized. Thereby, the active part of the SEF and the corresponding stress tensor, respectively, were expressed in terms of a normalized parabolic, a Gaussian, or a logarithmic shape function $f(\lambda_\theta)$, which depends on the circumferential smooth muscle stretch with $\max(f(\lambda_\theta)) = 1$. Furthermore, the active stress depends on a stress-like activation constant $\sigma_{\text{active}}^{\max}([\text{Ca}^{2+}]_{\text{i}})$ that corresponds to the maximum possible active stress for a specific intracellular calcium concentration $[\text{Ca}^{2+}]_{\text{i}}$. Thus,

$$\boldsymbol{\sigma}_{\text{active}} = \sigma_{\text{active}}^{\max}([\text{Ca}^{2+}]_{\text{i}})\, f(\lambda_\theta)\, \mathbf{e}_\theta \otimes \mathbf{e}_\theta, \tag{17}$$

where it is assumed that the mean SMC orientation coincides with the circumferential direction $\mathbf{e}_\theta$ causing only circumferential stresses (Rachev and Hayashi 1999; Zulliger et al. 2004; Carlson and Secomb 2005; Gleason et al. 2008; Yosibash and Priel 2012; Sepahi et al. 2017; Gilbert et al. 2019). Among others, this relatively simple class of activation models allowed (Rachev and Hayashi 1999) to show that under physiological conditions, the smooth muscle contraction together with the consideration of residual stresses leads to a reduced transmural stress gradient in the arterial wall, which is supported by experimental findings of Matsumoto et al. (1996).

More recently, constitutive models also considered the experimentally observed multiaxial smooth muscle response by defining the active stress as a function of both, the axial stretch λ_z and the circumferential stretch λ_θ of the tissue, e.g., by adding the term $\sigma_{\text{active}}^{\max} f(\lambda_z)\, \mathbf{e}_z \otimes \mathbf{e}_z$ to Eq. (17), resulting in active biaxial stresses, see Wagner and Humphrey (2011), Agianniotis et al. (2012), Huo et al. (2012), and Chen and Kassab (2017). Although these models allow to capture and analyze the macroscopic active tissue response based on a calcium-related activation constant, they are purely phenomenological and do not consider the underlying kinematics and kinetics at the filament level, i.e., the relative sliding between actin and myosin responsible for muscle contraction. Thus, these models cannot describe the smooth muscle contraction dynamics, as experimentally observed in isometric contraction tests or quick-release tests.

3.2.1 Microstructurally-Motivated Mechanochemical Models

Another class of frequently used models, usually referred to as mechanochemical models, is based on a kinetic model for cross-bridge interactions proposed by Hai and Murphy (1988). Thereby, the myosin heads (or cross-bridges) inside the contractile units are assumed to act independently and to be in one of the four different functional states: (A) detached and unphosphorylated, (B) detached and phosphorylated, (C) attached and phosphorylated, and (D) attached and unphosphorylated, which is also known as the latch-state, whereby the cross-bridge cycling is described by the transition between states (B) and (C). Hai and Murphy (1988) modeled the possible transitions between these states through the following set of ordinary differential equations, i.e.

$$\begin{bmatrix} \dot{n}_\mathrm{A} \\ \dot{n}_\mathrm{B} \\ \dot{n}_\mathrm{C} \\ \dot{n}_\mathrm{D} \end{bmatrix} = \begin{bmatrix} -k_1 & k_2 & 0 & k_7 \\ k_1 & -k_2-k_3 & k_4 & 0 \\ 0 & k_3 & -k_4-k_5 & k_6 \\ 0 & 0 & k_5 & -k_6-k_7 \end{bmatrix} \begin{bmatrix} n_\mathrm{A} \\ n_\mathrm{B} \\ n_\mathrm{C} \\ n_\mathrm{D} \end{bmatrix}, \tag{18}$$

where n_i are the chemical fractions of the myosin heads being in state i={A, B, C, D}, $\dot{n}_i$ are the corresponding time derivatives, and $k_1, \ldots, k_7$ are constants describing the rate of myosin phosphorylation ($k_1 = k_6$), dephosphorylation ($k_2 = k_5$), attachment (k_3), and detachment (k_4, k_7). Thereby, the calcium-dependent myosin phosphorylation $k_1([\mathrm{Ca}^{2+}]_\mathrm{i})$ is the only regulatory mechanism in this model and is usually assumed to be an S-shaped function of the intracellular calcium concentration based on the Michaelis–Menten kinetics (Yang et al. 2003a). Note that only attached myosin heads can transmit forces. Hence, the generated active stress by the contractile units and the energy stored in the elastic cross-bridges are proportional to the fraction of myosin heads in the states (C) and (D), respectively, to the average elastic stiffness and elongation of the attached cross-bridges, as well as to the relative amount of actin and myosin filament overlap that is responsible for the bell-shaped active length-tension behavior.

Based on this concept, Yang et al. (2003a, b) developed an integrated model for smooth muscle activation consisting of an electrochemical and a mechanochemical subsystem validated by voltage-clamp experiments on isolated cells. Therein, the electrochemical subsystem—a combination of a Hodgkin–Huxley type membrane model and a fluid compartment model—numerically describes how an extracellular electrical stimulus causes membrane depolarization that ultimately leads to an increased intracellular free calcium concentration by the release of calcium ions from the sarcoplasmic reticulum that ultimately enters the parameter k_1 in Eq. (18). The mechanochemical subsystem—a combination of the Hai and Murphy model and a modified version of Hill's three-element model—reflects the coupling between the active force generation by the contractile elements and the partly viscoelastic properties of the actomyosin complexes and their attachments with the cell wall.

While the mechanochemical coupling in Yang et al. (2003a, b) is rather intuitively stated than based on energetic functionals, more recent approaches use a nonlinear continuum mechanical hyperelastic formulation that allows to derive the governing equations directly based on the balance laws and thermodynamic considerations, e.g., see Stålhand et al. (2008) for a one-dimensional and Stålhand et al. (2011, 2016) for a three-dimensional mechanochemical continuum model considering smooth muscle contraction under finite strains. Nevertheless, a frequent assumption is that strains at the cellular level are small, which enables expressing the stretch of a contractile unit as $\lambda_{\mathrm{cu}} = 1 + \varepsilon_{\mathrm{cb}} + \varepsilon_{\mathrm{fs}}$, where $\varepsilon_{\mathrm{fs}}$ is a strain-like variable associated with the relative filament sliding between actin and myosin (Murtada et al. 2010a; Stålhand and Holzapfel 2016). Hence, by interpreting the connections between the individual myosin heads and the actin binding sites as elastic springs, the free energy stored in all the elastic cross-bridges in the vascular SMCs takes on the form

$$\Psi_{\mathrm{active}} = \frac{\mu_{\mathrm{cb}}}{2}(n_{\mathrm{C}} + n_{\mathrm{D}}) f(\varepsilon_{\mathrm{fs}})(\lambda_{\mathrm{cu}} - 1 - \varepsilon_{\mathrm{fs}})^2, \tag{19}$$

where $f(\varepsilon_{\mathrm{fs}})$ may be a Gaussian or a parabolic function of the relative filament sliding $\varepsilon_{\mathrm{fs}}$ that modulates the shape of the active length-tension behavior, the parameter μ_{cb} corresponds to the stiffness of the elastic cross-bridges, and n_{C} and n_{D} denote the percentage of attached and load-bearing cross-bridges, respectively (Murtada et al. 2010a; Stålhand and Holzapfel 2016). The filament sliding is usually obtained by an evolution law based on Eq. (15), where the filament sliding velocity is proportional to the resulting stress acting upon it. However, based on the homogeneity assumption enforcing that the stretch of the contractile unit λ_{cu} equals the tissue stretch λ together with the fact that the additive decomposition of the contractile unit stretch λ_{cu} is only valid for small strains, the multiplicative decomposition of the stretch $\lambda_{\mathrm{cu}} = \lambda_{\mathrm{cb}}\lambda_{\mathrm{fs}}$ or the deformation gradient $\mathbf{F}_{\mathrm{cu}} = \mathbf{F}_{\mathrm{cb}}\mathbf{F}_{\mathrm{fs}}$ would be the appropriate general kinematic framework for soft biological tissues, such as arteries (Stålhand et al. 2011, 2016). Motivated by Yang et al. (2003a), Sharifimajd and Stålhand (2014) established a thermodynamically and physiologically consistent framework for modeling smooth muscle contraction by combining cell membrane excitability and kinetics of myosin phosphorylation with the mechanics of muscle contraction.

The advantage of this type of models is that they are straightforward to implement in a finite element environment, while being able to reproduce the experimentally obtained mechanical response of, e.g., an artery undergoing an isometric contraction. Figure 6 depicts the results of a finite element simulation of such an experiment performed on a medial strip cut from the medial layer of the human abdominal aorta in the circumferential direction, using an implementation of a modified GOH model version (Holzapfel et al. 2015) along with Eqs. (18) and (19). Figure 6a shows the solutions of the Hai and Murphy model, as introduced in Eq. (18) for a realistic intracellular calcium transient, while Fig. 6b–d illustrate the passive and

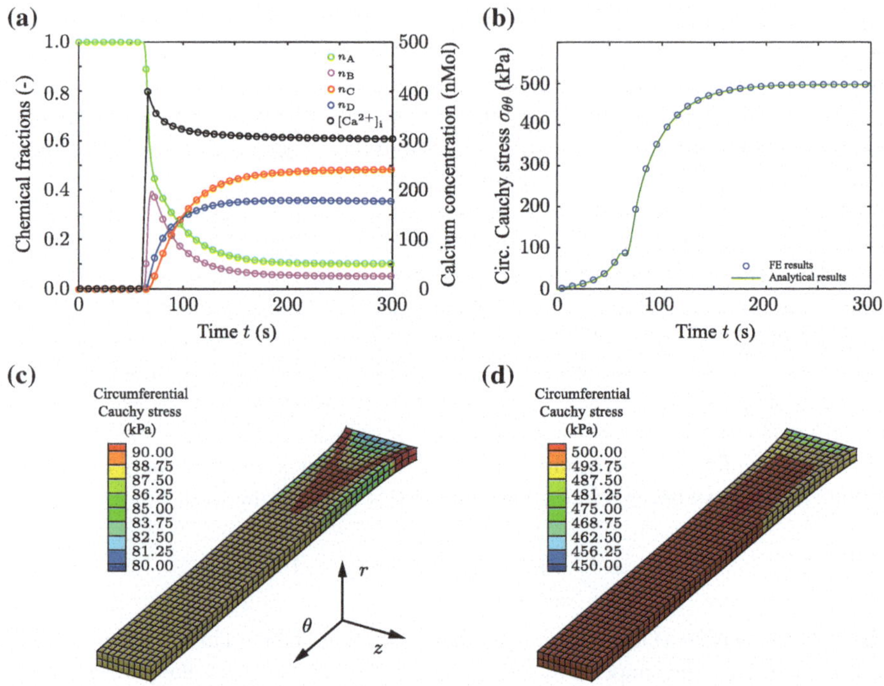

Fig. 6 Numerical simulation of an isometric contraction experiment performed on a human abdominal aortic medial strip: (a) time dependent intracellular calcium concentration $[\mathrm{Ca}^{2+}]_\mathrm{i}$ triggering the evolution of the myosin head fractions n_i being in the states $i = \{\mathrm{A, B, C, D}\}$; (b) circumferential Cauchy stress (active and passive response)—circles represent finite element (FE) results, solid curves denote the analytical solution; (c), (d) corresponding three-dimensional distribution of the circumferential Cauchy stresses visualized for one half of the medial strip (c) after applying a circumferential pre-stretch of 1.28 (at $t = 60\,\mathrm{s}$) and (d) after reaching the maximum isometric stress (at $t = 300\,\mathrm{s}$). Reproduced from Haspinger et al. (2018) with permission

active circumferential Cauchy stress response that is experienced and generated by the arterial tissue during the isometric contraction as a function of time, as well as the circumferential stress distribution of the purely passive medial strip after applying a uniaxial stretch of 1.28 and at the fully contracted state. The model parameters therein were fitted to the scaled experimental data and SMCs were assumed to be aligned in the circumferential direction.

Active constitutive models of the type as introduced in Eq. (19) in combination with a passive constitutive model, as introduced in Sect. 2.2, have been frequently used in computational models of arteries and were further developed. For example, Murtada et al. (2010b) considered the orientation dispersion of actomyosin complexes around a certain mean direction, Murtada et al. (2012) demonstrated a microstructurally-motivated filament overlap function $f(\varepsilon_\mathrm{fs})$ and a filament sliding evolution law based on Eq. (15), which finally allows to describe the experimentally observed length-tension and isotonic force-velocity behavior. Recently, modifications of the model presented in Murtada et al. (2012) were used by Murtada and Holzapfel (2014) to numerically analyze the impact of temporal alterations of wall thickness and calcium concentration on the stress distribution in pig carotid arteries. Furthermore, in Murtada et al. (2016b) the model was used to investigate the adaptation of the active tone in the mouse descending thoracic aorta when it is subjected to acute changes in loading. Haspinger et al. (2018) used a modified version to analyze how alterations of the smooth muscle orientation and the actomyosin filament overlap affect the mechanical behavior of the human abdominal aorta and how they are interrelated using finite element analyses.

In general, the three-dimensional orientation of SMCs, denoted by a vector $\mathbf{M}$ in the reference configuration, can easily be incorporated in Eq. (19) by setting $\lambda_{\mathrm{cu}} = \sqrt{I_{\mathrm{cu}}}$ with $I_{\mathrm{cu}} = \mathbf{C} : \mathbf{M}_{\mathrm{cu}} \otimes \mathbf{M}_{\mathrm{cu}}$, where $\mathbf{C} = \mathbf{F}^{\mathrm{T}}\mathbf{F}$ is the right Cauchy–Green tensor (Haspinger et al. 2018), as introduced in Sect. 2.2. Schmitz and Böl (2011) also developed and implemented a continuum mechanical steady-state model of smooth muscle activation into a finite element framework allowing to consider the three-dimensional SMC orientation inside the arterial tissue. However, they have used a rather phenomenological explanation of the active length-tension behavior, similar to Rachev and Hayashi (1999). Böl et al. (2012) further developed their approach to consider the temporal evolution and spatial distribution of both the calcium concentration and the resulting active stresses in a carotid artery employing Fick's second law of diffusion.

Novel experimental approaches on many different length scales and improved imaging techniques provide valuable data that drive the development of more detailed microstructurally-motivated multiphysical computational models to gain better insights into many physiological and pathological vascular tissue adaptation mechanisms, whereby the SMCs are thought to play a significant role. Adaptation mechanisms that ultimately affect an artery's mechanical behavior may occur at different length scales and may cause or result in structural, functional, mechanical, and chemical alterations (Humphrey 2002; Wang et al. 2009). Hence, to capture these essential interrelations, computational models of the arterial wall must consider and couple the multiphysical phenomena at the nano, micro, and macro levels, i.e. they must be multiscale (Hayenga et al. 2011; Murtada et al. 2017; Coccarelli et al. 2018). In this context, Murtada et al. (2017) presented a new multiscale mathematical model of the arterial contractility accounting for structural and functional constituents at molecular, cellular, and tissue levels. Therein, the artery is described by an anisotropic model of the ECM in combination with a novel model of the smooth muscle contractility being sensitive to microscale filament overlap and filament lattice spacing, which enables to predict both uniaxial and biaxial experimental contraction data accurately using one set of material parameters.

In conclusion, computational models of the active arterial response have evolved over the past 80 years from purely phenomenological models to thermodynamically consistent models that couple the electrical, chemical, and mechanical characteristics involved in smooth muscle contraction directly linked to the individual cellular constituents. However, several open issues require further considerations, e.g., the chemical and mechanical interactions between endothelial cells, ECM components, and SMCs in a multiscale context aiming to better understand the underlying mechanisms and actions of short- and long-term in vivo tissue adaptations from a biomechanical point of view. The implementation of these multiphysical multiscale models into finite element programs, combined with patient-specific geometries, material parameter sets, and boundary conditions, is thought to advance computer-assisted clinical diagnosis, risk assessment, and intervention in the future (Famaey et al. 2013; Laubrie et al. 2020).

4 Damage, Viscoelasticity and Failure

There is compelling experimental evidence that the macroscopic passive mechanical behavior of healthy large arteries such as the aorta is characterized by very little dissipation under cyclic loading, at least in a macroscopic sense (Amabili et al. 2019a; Holzapfel et al. 2000a). Therefore, they are often categorized as 'elastic' arteries that contain large amounts of elastin as opposed to 'muscular' arteries that are small and highly viscoelastic vessels located in the periphery of the circulatory system. Nevertheless, as we discuss in Sect. 4.1, dissipative effects such as viscoelasticity, damage and failure may become relevant especially under pathological conditions, e.g., aneurysms and aortic dissections where the elastic fibers are degraded. In addition, considering the rather large variations in tissue strength reported by the biomechanical studies and recent experimental findings on failure and damage, it has been proposed that an effective and appropriate material failure criterion for aortic tissues should take the organization of different microstructural constituents and their influence at different length-scales into account (Sherifova and Holzapfel 2019). Therefore, multiscale characterization and modeling of dissipative phenomena is crucial for a better understanding of disease progression as well as related catastrophic events.

We then emphasize the importance of multiscale approaches in characterizing these phenomena in soft biological tissues, with a particular focus on the aorta. First, we summarize the experimental findings on the multiscale contributions of elastic and collagen fibers to these events in Sect. 4.1. Next, in Sect. 4.2, we continue with different multiscale approaches to their modeling including classical continuum models, which incorporate features of the aortic microstructure as well as models based on RVEs and homogenization. We note that there might not be a single multiscale approach suitable for all applications, since different approaches may be more effective in modeling different pathological findings related to the aorta.

4.1 Experimental Findings

Albeit the low energy loss in response to cyclic loading (Amabili et al. 2019a), elastic arteries still exhibit stress relaxation and creep. Zou and Zhang (2011) found intact porcine aortic wall to exhibit the largest amount of stress relaxation, followed by decellularized ECM and purified elastin under biaxial loading, whereas all groups showed negligible creep response. Others reported a significant creep under uniaxial loading of elastin degraded porcine (Kratzberg et al. 2009) and human aortas (Schriefl et al. 2015). With decreasing water content, the amount of stress relaxation was reported to decrease for bovine aortas under uniaxial loading (Shahmirzadi et al. 2013), but increase for elastin purified porcine aortas under biaxial loading (Wang et al. 2018a). The amount of stress relaxation also increased with increased glycation (Wang et al. 2018b). It is important to note that not only elastin but properly crosslinked elastic fibers are required for a low energy loss in response to cyclic loading (Kim et al. 2017). Elastic fiber degradation can occur with aging (Schlatmann and Becker 1977) and accompanies various pathological conditions (Sherifova and Holzapfel 2020). Therefore, a viscoelastic characterization of large arteries in health and disease requires further investigations.

Weisbecker et al. (2013) degraded elastin in the aortic tissue strips with elastase treatment, and subjected them to consecutive uniaxial loading-unloading cycles. The substantial softening of the mechanical response, see Fig. 7a, was attributed to the fact that the collagen fibers could slide against one another in an inadequately connected network. Chow et al. (2013) reported that the changes in the mechanical response during elastin degradation were observed earlier in the axial direction compared with the circumferential direction, due to the collagen fibers, which are mainly oriented in the circumferential direction, which may slow down the enzymatic degradation of the elastic fibers that are oriented parallel to them. The rate of elastin degradation increased when elastase was added under biaxial loading, although the final elastin content was similar when the tissue was elastase treated with and without loading. It has been reported, however, that elastase can also degrade denaturated collagen as well as fibril-bridging proteoglycans (Kronick and Sacks 1994; van de Lest et al. 1995). Schriefl et al. (2015) developed a treatment protocol with trypsin inhibitor to overcome these limitations, and reported a similar softening behavior, as shown in Fig. 7a. In addition, the softening was observed only at higher loads and it did not influence the loading-unloading curves at low loads upon partial elastin degradation, see Fig. 7b.

Taken together, these results indicate that properly cross-linked elastic fibers not only ensure low energy loss, but can also have a significant influence on the (patho)physiological remodeling of arteries under in vivo loading conditions. Elastic fiber degradation may lead to earlier or extensive damage accumulation in collagen fibers under in vivo cyclic loading. Subsequently, this is likely to trigger and/or accelerate a cascade of events leading to pathological formations, especially when combined with disruptions to the biological repair mechanisms where intramural cells play a vital role. For example, decreased undulation of collagen fibers (Niestrawska et al. 2016) and fibrils (Jones et al. 2020) as well as disrupted d-periodicity of fibrils (Jones et al. 2020) were reported for abdominal aortic aneurysms. Hence, more efforts are needed to uncover the explicit multiscale roles of elastic and collagen fibers on the organ metabolism and complex interactions between cellular and extracellular constituents for a better understanding of (patho)physiological damage

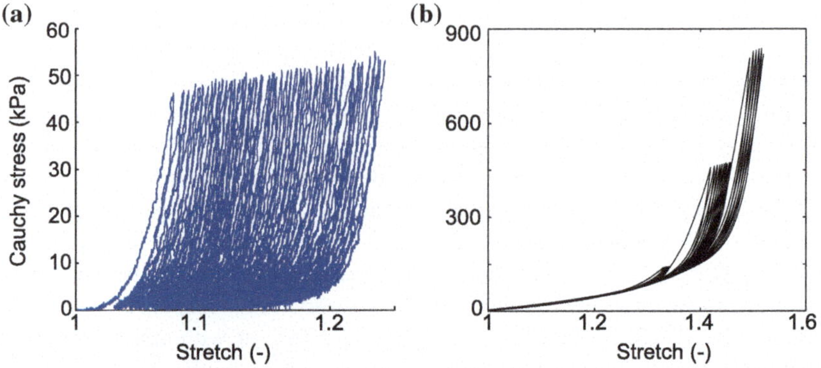

Fig. 7 Elastase-induced strain softening in response to cyclic quasi-static uniaxial loading of a strip from (a) a human thoracic aortic media after complete elastin degradation (reproduced from Weisbecker et al. 2013 with permission), and (b) a human abdominal aortic media after partial elastin degradation (reproduced from Schriefl et al. 2015 with permission). Note that the softening behavior is limited to the large strain regime and there is no permanent deformation after unloading, see (b), which is in contrast to (a)

mechanisms. Studies of collagen fiber damage in tendons and arteries at the molecular level provide invaluable insight into some of these mechanisms, some of which are summarized below.

Zitnay et al. (2017) utilized collagen hybridizing peptides (CHP) to quantify damage due to monotonous and cyclic uniaxial loading in rat tail tendons (RTTs). Significantly more damage accumulation was observed under cyclic loading compared with monotonous loading at both quasi-static and physiological strain rates to the same strain level with the respective strain rate. Furthermore, when RTTs were subjected to cyclic loading until failure at different strain rates, the percent of collagen damage increased with the number of cycles to failure regardless of the strain rate. At all strain rates, first loading cycles caused considerable damage, and damage accumulation was strongly correlated with the creep strain (Zitnay et al. 2020). These findings indicate that molecular damage is a strain-dependent process regardless of the strain rate or strain history. In addition, an increased number of cycles to failure observed at higher loading rates despite the faster damage accumulation (Zitnay et al. 2017, 2020) indicate greater resistance to both fatigue and monotonous loading at these rates (Zitnay et al. 2020). Although the possible influence of cyclic loading or the strain rate on collagen damage at the molecular level in arteries has not yet been fully elucidated, rate dependent differences in strength were also reported by Mohan and Melvin (1982, 1983) for the aortas, which deserve further investigation.

Under monotonous uniaxial loading, damage to collagen at the molecular level in RTTs accumulated before apparent macroscale damage and the onset of a significant damage accumulation corresponded with to this yield point of the stress-strain curves, steadily increasing afterwards (Zitnay et al. 2017). The authors identified intermolecular sliding of collagen α-chains as the main triple helix unfolding mech-

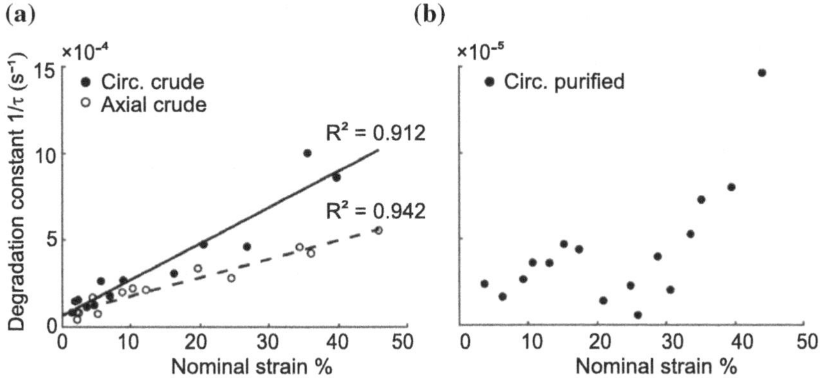

Fig. 8 Degradation constant $1/\tau$ versus nominal strain of uniaxially stretched porcine common carotid arteries immersed in (a) crude and (b) purified bacterial collagenase. While the response in crude collagenase (a) can be represented linearly, the response in purified collagenase (b) is better approximated by a two-phase V-shaped curve. Filled and open circles denote circumferential and axial specimens, respectively. Note the scale difference in the y-axis. Reproduced from Gaul et al. (2018b) with permission

anism using molecular dynamics simulations. Similar damage mechanisms were reported in sheep cerebral arteries under quasi-static loading (Converse et al. 2018; Marino et al. 2019) where significant damage accumulation was also closely related with the yield point. Such a failure mechanism, in which the shear deformation results in intermolecular sliding during tensile loading, explains at least in part the macroscopic and microscopic observations from the rupture zones that have been reported for porcine (Brunet et al. 2019) and human aortas (Sherifova et al. 2019). In addition, shear deformation was identified as the main contributor to crack propagation in inflation tests for human aortic rings (Rabin et al. 2020) as well as for porcine common carotid artery segments (Brunet et al. 2021).

It is worth noting that the amount of damage accumulation from the yield point until failure is influenced by the failure behavior. Recently, Lin et al. (2020) reported a significantly lower damage accumulation for the significantly weaker tendons, which are known for their 'brittle-like' failure as well as higher valency and density of their collagen crosslinks that may prevent intermolecular sliding. While the stronger tendons were subjected to a stronger plastic deformation (Herod et al. 2016; Lin et al. 2020), which was visible from the long plateau region in the stress-strain curve and allowed an accumulation of damage, the weaker tendons failed abruptly. Regardless of these differences, the molecular damage mechanisms in different tendons due to mechanical loading were essentially the same, and the significant onset of collagen damage accumulation in different tendons still corresponded to the yield point at the applied loading rates, despite the known biochemomechanical differences (Lin et al. 2020).

Strain-mediated damage to collagen in arterial tissue was also studied with other methods. Gaul et al. (2018a) subjected circumferential strips from the media of porcine common carotid arteries to three different strain levels in crude bacterial collagenase solution. Mechanical experiments were combined with histology, second-harmonic generation (SHG) imaging, small angle light scattering (SALS) and small angle X-ray scattering (SAXS) to identify changes in the collagen fiber alignment as well as strain-induced collagen degradation. Results showed highly aligned fibers at the intermediate strain level, whereas the fibers were more dispersed at lower and higher strain levels. Furthermore, both the unloaded and highly strained fibers were degraded preferentially, with the intermediate level of strain acting as a protective mechanism to degradation. It is important to note that Gaul et al. (2018b) reported different degradation responses when the experiments were performed with crude and purified collagenase. While the former, see Fig. 8a, could be represented linearly, the latter, see Fig. 8b, showed a multiphase response, where the initial increase in the degradation rate was followed by a low degradation and a subsequent steep increase. The authors attributed this difference in the degradation response at the low strain levels to a combination of the gradient of collagen crimp across the vessel wall and the alteration of the mechanical contribution of the matrix material by partial degradation by crude collagenase. In a follow up study, Gaul et al. (2020) subjected tubular segments to pressure inflation tests at a range of physiological pressures in the presence of purified collagenase. Despite the complete collagen degradation, vessels with a pressure of 60 mmHg showed no failure as the remaining matrix could bear the load. However, when the vessels were exposed to either 60 or 100 mmHg of collagenase for 45 min to allow degradation and then 180 mmHg, then vessels degraded under lower pressure showed a significantly shorter time to failure according to their previous observations. These findings also suggest that despite the increased degradation at low pressure levels, the tissue is unlikely to fail without the subsequent increase in pressure.

It has been shown that the failure properties of thoracic and abdominal aneurysms as well as healthy aortas are direction and location dependent (Angouras et al. 2019; Liu et al. 2019; Myneni et al. 2020; Peña et al. 2019; Polzer et al. 2020; Sherifova and Holzapfel 2019, and references therein). Although changes in the ratio of the ECM constituents may provide a partial explanation for the location dependence of strength considering the changes along the aortic tree (Wolinsky and Glagov 1969), evidence suggests that the preferred orientation and dispersion of the collagen fibers are primarily responsible for the anisotropic failure properties in the circumferential-axial plane (Liu et al. 2019; Sherifova et al. 2019). Molecular damage accumulation in collagen preceding failure was observed primarily in the collagen fibers along the loading direction (Converse et al. 2018; Marino et al. 2019), and damage accumulation was significantly less in axial strips compared with circumferential strips (Gaul et al. 2018b). In addition, the lamellar structure of the aorta explains the reduced strength in response to radial tensile loading and interlamellar shear (Korenczuk et al. 2019; Sommer et al. 2016). Therefore, peeling, direct tension, and interlamellar shear tests are invaluable for understanding crack initiation and propagation in aortic dissections, and tests in combination with various imaging methodologies require a multimodal and multiscale approach to modeling dissections.

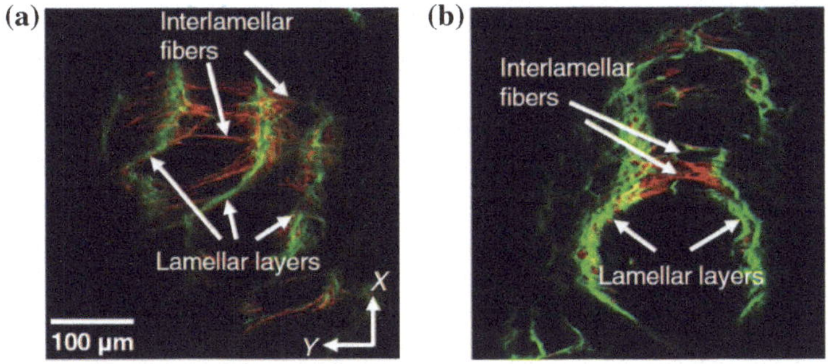

Fig. 9 Multiphoton microscopy images at the dissection front during peeling in (a) the axial and (b) the circumferential direction. Elastin and collagen are represented in green and red, respectively. Reproduced from Yu et al. (2020) (https://creativecommons.org/licenses/by-nc/4.0/)

Using a combined optical coherence tomography and a digital image correlation in connection with radial tensile tests, Giuseppe et al. (2021) identified an early strain localization at the location of the crack onset in direct tension tests. Preliminary tests showed a 'sharper' localization in aneurysmatic human ascending aortas than in healthy porcine descending thoracic aortas, however, whether this observation could be connected to microstructural changes in aneurysms awaits confirmation. Such strain localizations combined with microstructural discontinuities and in vivo loading conditions can lead to catastrophic events. In a combined experimental-computational study in angiotensin-II-infused mice, Aslanidou et al. (2020) reported correlations between excessive mechanical strain near the branches of the abdominal aorta and the locations of microstructural defects, detected by histopathology and contrast agent microleaks into the aortic wall, which may subsequently propagate into macroscopic dissections.

Peeling tests on strips of purified elastin and porcine aortic media in the circumferential and axial directions showed that the peeling force and the energy release rate were higher in both directions for the intact aortic media compared with the purified elastin, and the axial direction was associated with a higher energy release rate in both materials (Wang et al. 2021b; Yu et al. 2020). The MPM images in Fig. 9 show clearly how both interlamellar collagen and elastin of several lamellar units are employed in the axial peeling (Fig. 9a), whereas the peeling in the circumferential direction (Fig. 9b) is more confined. Interestingly, Yu et al. (2020) found that the depletion of the connections between medial collagen fibrils, GAGs, increased dissection energy and force compared to non-treated samples. These findings confirm that although collagen plays the dominant role, the contribution of elastin to the interlamellar bonding strength should not be neglected. Wang et al. (2021a) studied the influence of glycation using peeling and direct tension tests. The increase in the rate of energy release was significant after the 7 day exposure to glucose compared to that non-treatment group followed by an insignificant gradual increase on days 14 and 21 compared to the 7-day group. The failure stress in direct tension tests also increased essentially after 7 days, but a significant increase compared to the untreated samples was not observed until after the 14 days of treatment. Additionally, the overall sample thickness, hence the interlamellar space, decreased with glycation, whereas the failure strain increased. The reduced tendency to delaminate with elastin glycation provides a partial explanation for the inverse relationship between diabetes and the risk of aortic dissection, which has been observed both clinically (Avdic et al. 2018; Patel et al. 2018) and experimentally (Tong et al. 2020).

4.2 Mathematical Modeling and Related Computational Aspects

The modeling of dissipative phenomena in biological tissues, as opposed to common engineering materials, poses particular challenges arising both from the intricate microstructure and the living characteristic of the material. From a purely physiological point of view, tissues may be regarded as 'healthy' or 'injured/diseased'. However, in the field of biomechanics, there seems to be no clear consensus on the meaning of the term 'damage' of a biological tissue from the microstructural point of view (Holzapfel and Fereidoonnezhad 2017). Another challenge comes from the fact that all living tissues have the ability to heal or remodel to some extent. Therefore, 'damage' and dissipation in biological tissues may not be regarded simply as a one-way (entropic) path, as normally done for engineering materials. Nevertheless, several multiscale models incorporating microscopic features of the tissue have been developed to account for dissipative phenomena in arteries, and they improved our understanding of the overall mechanical behavior of the tissue.

Since 'elastic' arteries exhibit low dissipation under cyclic loading (Amabili et al. 2019a), only a few viscoelastic multiscale models were applied specifically to the aorta. With respect to arteries, in general, Holzapfel et al. (2002a) applied the viscoelastic version of the HGO model (Holzapfel and Gasser 2001) in conjunction with the FEM to model the dissipation in a segment of a residually-stressed left ascending coronary artery subjected to static and dynamic loading conditions. The predicted dissipation due to cyclic blood pressure within the cardiac cycle was low and quite insensitive to frequency. Importantly, the study also presented an early example of the identification of structurally-motivated model parameters, e.g., the mean fiber angle from histological data. A recent example regarding viscoelasticity of the aorta is the computational study of Amabili et al. (2019b), which combines the GOH model (Gasser et al. 2006) with the integer order and fractional versions of the generalized Maxwell viscoelastic model. Upon fitting the experimental data published by Amabili et al. (2019a) to both versions, the authors reported similar accuracy, and concluded that the integer order should be preferred as it has fewer parameters. Other examples of multiscale viscoelastic models applied to arteries were published, e.g., by Nguyen et al. (2007) and Peña et al. (2010, 2011).

A number of multiscale models incorporating damage phenomena within the arterial wall have been developed focusing primarily on collagen and elastic fibers—constituents that dominate the passive mechanical behavior of the tissue, see Sect. 2. Starting from the classical continuum damage theory, Balzani et al. (2004) modified the anisotropic term of the HGO model (Holzapfel et al. 2000a) to account for damage of the collagen fibers, e.g., rupture of cross-links or fibrils. The proposed incompressible SEF had the form

$$\Psi(\mathbf{C}, \mathbf{M}_i, D_i) = \Psi_\mathrm{g}(\mathbf{C}) + \sum_{i=1}^{2} \left[(1 - D_i)\Psi_{\mathrm{f}_i}(\mathbf{C}, \mathbf{M}_i)\right], \tag{20}$$

where Ψ_g and Ψ_{f_i} are the SEFs of the ground substance and of the undamaged ith fiber family, respectively. The latter is a function of both the right Cauchy–Green tensor $\mathbf{C}$ and the mean fiber directions $\mathbf{M}_i$, $i = 1, 2$. The scalar damage variable D_i indicates the level of damage of the ith fiber family, ranging from 0 (undamaged) to 1 (completely damaged).

The model proposed in Eq. (20) was applied to simulate overstretching of an atherosclerotic artery by Schröder et al. (2005), and Balzani et al. (2006) included the effect of residual stresses. In both cases, the authors pointed out that the resulting damage fields should be taken qualitatively due to a lack of suitable experimental data. This issue was later addressed by Balzani et al. (2012), in which two models with different choices of the anisotropic strain energy in Eq. (20) were fitted to experimental data from cyclic uniaxial tests in the supraphysiological domain, providing good results. Therein, the effect of rotationally symmetric fiber dispersion was also taken into account by employing the generalized structural tensor approach (Gasser et al. 2006), as well as the possibility of permanent deformations after damage. Schriefl et al. (2015) used one of the constitutive models proposed by Balzani et al. (2012) to fit uniaxial data from cyclic uniaxial loading of intact and layer-specific elastase-treated samples of human abdominal aortas, with good correlations. In the same study, the authors argue that the lack of softening in collagenase-treated samples indicates that including damage of the elastin ground substance might be unnecessary, even though it is also a common modeling approach, see below.

Aforementioned works used an exponential function to describe the evolution of the damage variable D_i, whereas Schmidt et al. (2014) proposed a microstructurally-motivated function based on the sliding filament theory. The authors also incorporated microscopic features such as collagen fibril overlap and dispersion of proteoglycan orientation. The model was fitted to the experimental data from Balzani et al. (2012), but in this case, the use of a structurally-motivated damage function provided some insights on microscopic phenomena. More specifically, the results indicated that the statistical distribution of proteoglycan orientation may have the most significant influence on the damage behavior. The model thus offers a good framework for the investigation of microscopic damage in combination with the novel techniques discussed in Sect. 4.1. Rodríguez

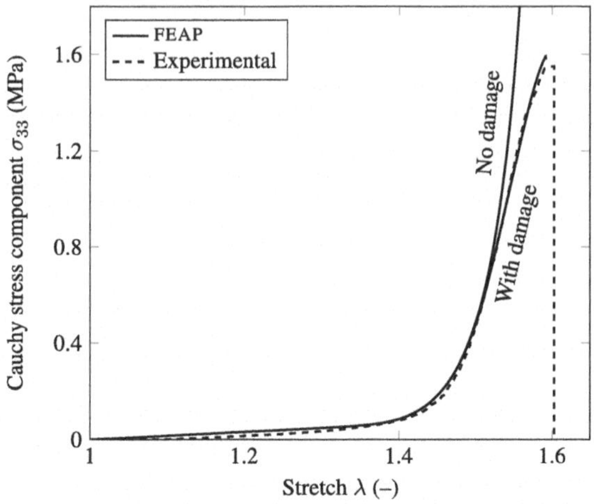

Fig. 10 Influence of damage on the uniaxial response of a medial strip from a human thoracic aorta. Comparison between experimental data (dashed curve) and FEM solutions (solid curves) indicates that accounting for damage is crucial to adequately capture the response of the tissue under large strains. Reprinted from Li and Holzapfel (2019), with permission from Elsevier

et al. (2006) accounted for damage of the elastin-dominated ground substance following the classical approach by multiplying Ψ_g in Eq. (20) with $(1 - D_g)$, where D_g is the scalar damage variable of the ground substance, and applied the model in a series of numerical examples. With regard to arteries, Li and Holzapfel (2019) incorporated the ground substance and the collagen fiber damage into the SEF in a way similar to Eq. (20), where the fiber dispersion was accounted for by using the DFD model, as introduced by Li et al. (2018), instead of the generalized structure tensor approach. A particular application of the model is shown in Fig. 10, where the predicted model response is clearly in good agreement with experimental data from a uniaxial test on a medial strip of a human thoracic aorta. Rolf-Pissarczyk et al. (2021) further expanded the DFD model of Li et al. (2018) by adding a so-called degradation parameter to the specific strain-energy term of the dispersed elastic fibers. The model was then employed to investigate the influence of radially-oriented damaged elastic fibers on the stress distribution in aortic dissection. The results showed that the stiffness contribution of the elastic fibers may have only a limited influence on the stress distribution at higher pressures. Furthermore, as a possible mechanism triggering the dissection propagation, the authors suggested that the media is delaminated by the stretch applied to the dissection flap. More specifically, the flap is stretched due to the cyclic dilatation of the false and the true lumen and, consequently, pulled apart from the intact wall.

To account for damage in the arterial tissue, another family of multiscale models employs the pseudo-elastic approach introduced by Ogden and Roxburgh (1999). The incompressible SEF for a pseudo-elastic material has the form

$$\Psi(\mathbf{C}, \eta) = \eta \tilde{\Psi}(\mathbf{C}) + \phi(\eta), \tag{21}$$

where $\tilde{\Psi}$ is the SEF of the undamaged material and $\phi(\eta)$ is the damage function that depends on the scalar damage variable η, which ranges from 1 (undamaged) to 0 (completely damaged). Building on this approach, Peña and Doblaré (2009) substituted $\tilde{\Psi}$ in Eq. (21) with the HGO SEF (Holzapfel et al. 2000a) and obtained good fits to experimental data from tests of human vaginal tissues and ovine vena cava samples. With regard to arteries, Weisbecker et al. (2012) combined the GOH model (Gasser et al. 2006) with Eq. (21) to account for the rotationally symmetric fiber dispersion, and fitted a large data set from intact and layer-specific healthy abdominal and thoracic aortas subjected to cyclic uniaxial loading, with a good correlation. Fereidoonnezhad et al. (2016) have further improved the model by including the possibility of inelastic strains and fitted the same data set of Weisbecker et al. (2012) in which some samples showed permanent deformations after unloading. More recently, Holzapfel and Ogden (2020b) modified an HGO-based model (Holzapfel and Ogden 2020a) to account for the pseudo-elastic behavior of the collagen fibers only, and the model provided a good fit to uniaxial experimental data from an RTT fiber.

Residual deformations in arteries are typically related to supraphysiological loading, e.g., after balloon angioplasty (Holzapfel et al. 2000b), and seldom associated with larger arteries such as the aorta. However, they can also occur as a result of microstructural defects, as mentioned in Sect. 4.1. In this context, several models have been developed to describe permanent deformations of smaller arteries, some of which have been mentioned in the previous paragraph. For completeness, we briefly introduce some models that employ the classic elastic-plastic decomposition of the deformation gradient to account for the inelastic behavior of the arterial tissue. For instance, Gasser and Holzapfel (2002) extended the HGO model following this approach, and the model was used to simulate a balloon angioplasty in a left anterior descending coronary artery (Gasser and Holzapfel 2007). The results showed non-negligible shear stresses in the interface between the media and the adventitia, which speaks against the use of plane-stress elements in angioplasty simulations. Nevertheless, normal stress components dominated both during and after balloon inflation. Other models using an elasto-plastic formulation similar to that of arterial tissue have been proposed, e.g., by Zhu et al. (2014) and Marino et al. (2019).

A different multiscale approach to model microscopic phenomena in arteries is based on RVEs and homogenization. By including a geometrical model—the RVE—of the microstructure and its simulation using the FEM, not only the macroscopic (homogenized) response is obtained via the volumetric averaging (see Sect. 2.2), but also the microscopic phenomena in the RVE are examined. In this context, Shah et al. (2014) improved the model introduced by Stylianopoulos and Barocas (2007a) to include fiber damage by imposing a critical fiber stretch threshold (λ_{crit}) and excluding it from the volumetric average when that threshold is reached. Uniaxial and biaxial tests of arterial tissues were simulated with the model, and homogenized curves were compared with experimental data. At the microscopic level, results show that uniaxial loading leads to a considerable reorientation of the collagen fibers. Additional microstructural features, e.g., elastin fibers and interlamellar connections were incorporated to the RVE with their corresponding failure stretches by Witzenburg et al. (2017). Rocha et al. (2018) proposed a unified framework for multiscale models built from RVEs consisting of a network of trusses, and performed several numerical experiments. Importantly, the authors addressed the influence of the chosen multiscale boundary conditions, an important issue that was overlooked in previous works (Stylianopoulos and Barocas 2007a; Shah et al. 2014; Witzenburg et al. 2017).

Pursuing a similar strategy, Thunes et al. (2016) constructed an RVE of the media composed of two elastin sheets and interlamellar ground substance reinforced with dispersed collagen fibers modeled as embedded truss elements. The model was improved by Thunes et al. (2018), where the critical fiber stretch, similar to Shah et al. (2014) and Witzenburg et al. (2017), was incorporated. Macroscopic results of the uniaxial simulations were compared with experimental data from uniaxial tests on healthy and aneurysmatic aortic tissues. At the microstructural level, results indicate that collagen fiber failure is the dominant cause for the macroscopic softening of the tissue under large strains. A similar micro-macro analysis—albeit with different multiscale boundary conditions—was performed by Maiti et al. (2020) for biaxial loading considering different biaxiality ratios. Based on their results, the authors argued that the tissue strength under biaxial loading can be assessed from uniaxial data in the longitudinal and circumferential directions. However, the works by Thunes et al. (2016, 2018) and Maiti et al. (2020) lack a detailed discussion of the chosen multiscale boundary conditions, an issue that was recently addressed in Dalbosco et al. (2021).

Roccabianca et al. (2014a, b) presented a finite element-based continuum approach to study the influence of pooled GAGs in the media on the intramural stress distribution by introducing a swelling-related fixed charge density. Motivated by experimental findings on the microscale, they modeled a rectangular slab of the media with a central GAG inclusion. They found significant stress concentrations around the GAG pools and intralamellar Donnan swelling pressures caused by them. This approach was later extended by Ahmadzadeh et al. (2018, 2019), where the authors introduced a particle-based computational model to study the role of pooled GAGs in the initiation and propagation of dissection. To model the growth and coalescence of locally pooled GAGs, the classical concept of a scalar damage variable was applied as in Eq. (20), and an evolution equation was formulated. As already shown by Rausch et al. (2017), this approach enabled the investigation of the aortic wall rupture, since material discontinuities were also modeled in a straightforward way.

Damage accumulation can lead to aortic wall delamination, where the classical continuum damage theory encounters challenges as numerical techniques are required to deal with the related displacement discontinuities. Such techniques are of particular interest when modeling the initiation and propagation of aortic dissection or the rupture of the aortic wall. Gasser and Holzapfel (2006) addressed this issue by combining the cohesive crack concept with the partition of unity FEM to model a peeling test of the aortic wall where its underlying microstructure was incorporated. Later, Ferrara and Pandolfi (2008, 2010), Wang et al. (2015, 2017, 2018) and Yu et al. (2020) used a similar approach. In particular, Yu et al. (2020) created two medial strips that are connected via discrete interlamellar collagen fibers used to model the failure of the aorta during a peeling test. The orientation and statistical distribution of collagen fibers between the strips was based on histological data, and if the failure threshold of a discrete collagen fiber has been

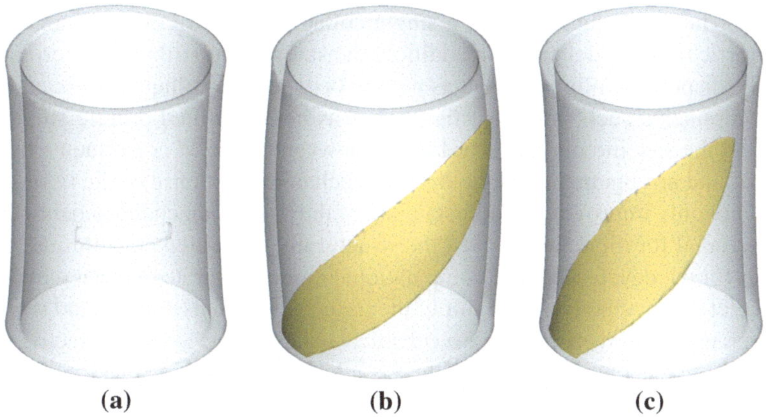

Fig. 11 Propagation of aortic dissection starting from an intimal tear, at instants (a) (initial), (b) and (c). Reprinted from Gültekin et al. (2019) (https://creativecommons.org/licenses/by/4.0/)

reached, the respective fiber has been removed from the simulation. They reported an avalanche-like failure during peeling caused by the local accumulation of the strain energy followed by a cascade failure of the interlamellar collagen fibers.

Another computational approach to modeling fracture is the phase-field approach that was first applied to arterial walls with a focus on the aorta by Gültekin et al. (2016). Within this context, Gültekin et al. (2018) subsequently investigated the numerical aspects of anisotropic failure and found that an energy-based criterion is favorable. Later, Gültekin et al. (2019) modeled the aorta as an idealized cylinder with multiple layers and a prescribed tear to study the dissection propagation with the proposed phase-field approach. As illustrated in Fig. 11, the computational results demonstrate clearly the damage zone around the prescribed tear. The damage zone follows a helical pattern around the aorta, which is aligned with the fiber orientation, similar to the clinical observations, as shown by Tse at al. (2011).

5 Discussion

After introducing the ECM organization in the aorta and the role of SMCs in the organ function, we reviewed the findings on the passive mechanical behavior of the human aortic tissue in Sect. 2 from experimental and computational aspects. Despite the significant progress in the imaging techniques and experimental methods, some critical information about the tissue microstructure that can play substantial roles in its accurate modeling remain elusive. For example, imaging techniques such as SHG allow visualization of the collagen fiber distribution only in a small region of the specimen, and not the 3D collagen fiber architecture throughout the arterial wall. The distribution of collagen fiber waviness in the aortic wall is also largely unknown. The mechanical data of a single collagen or elastic fiber obtained from human aortic tissue is deficient. In addition, the statistical distribution of cross-links within the dispersed collagen fibers as well as between the collagen fibers and other tissue components such as the elastic fibers was not quantified. Future studies can help reveal the specific multiscale relations between the ECM constituents in terms of content and arrangement and the overall mechanical response of the tissue. Such data will not only improve our understanding of the tissue's mechanical behavior, but are also crucial for the refinement of the structurally motivated constitutive models as well as for the development of computational modeling strategies that may enable us to predict the mechanical behavior of the tissue based on its content and structure.

Further, we focused on the active mechanical behavior due to the contractile ability of SMCs in Sect. 3. Quantification of the vascular tone in the human aorta is difficult to conduct due to its nature. Animal models are helpful to assess the active biomechanical properties of the aortic wall, and here we briefly reviewed two common approaches—uniaxial ring test and biaxial vessel test—that can be used to quantify SMC contractile function in the aorta. Both tests have their advantages and disadvantages and should be chosen with careful consideration of the multiaxial properties of the SMC contractile function. In particular, this point is crucial for functions where mechanical stretch-sensitivity may play an important role, e.g., in active hysteresis, active wall stiffness and vasodilation. It is challenging to assess the basal tone in the aorta, and the reviewed testing approaches quantify the maximum range of the contractile potential in the aorta. However, utilizing computational models including both the active and passive properties have been found to be valuable to better understand the effect and biomechanical roles of the SMC active tone in the aorta.

Finally, in Sect. 4, we summarized the experimental findings on and computational approaches to the multiscale modeling of the aorta regarding viscoelasticity, damage and failure. Although we focused on the passive aspects of these processes, it must be noted that damage or degradation of aortic constituents in vivo is not purely mechanical. On the contrary, intramural cells play a substantial role in these processes via growth and remodeling triggered by biochemomechanical cues. Experimental identification of the many related mechanisms, as well as the development of multiscale models incorporating these biological effects, pose additional challenges for the future. For example, constrained mixture models developed by Humphrey and Rajagopal (2002) include the kinetics of production and removal of individual constituents to model disease progression. These models can describe the macroscopic deformation by taking into account the microstructure of aortic constituents and the influence of biochemical processes on various scales. Such models, recently reviewed by Humphrey (2021b), can provide great insight with in silico studies and can be modified to describe the experiments as data becomes available.

Acknowledgements We gratefully acknowledge the collaboration with Heimo Wolinski from the Institute of Molecular Biosciences at the University of Graz and the IMB-Graz Microscopy Core Facility for the multiphoton images in Sect. 2; and Dagmar Kolb and Gerd Leitinger from the Gottfried Schatz Research Center and the Center for Medical Research, Medical University of Graz, for the transmission electron microscopy images, which are presented in Sect. 2. This chapter was partially financially supported by the Lead Project on 'Mechanics, Modeling and Simulation of

Aortic Dissection' granted by Graz University of Technology, Austria; the Project on 'Multiscale Biomechanical Investigation of Human Aortas' financed by Austrian Science Funds (FWF) with grant no. P30260; and the Project 'Does time heal all wounds? Damage in blood vessels' supported by the Austrian Science Funds (FWF) with grant no. I4545.

References

Agianniotis, A., Rachev, A., Stergiopulos, N.: Active axial stress in mouse aorta. J. Biomech. **45**, 1924–1927 (2012)

Ahmadzadeh, H., Rausch, M.K., Humphrey, J.D.: Particle-based computational modelling of arterial disease. J. R. Soc. Interface **15**, 20180616 (2018)

Ahmadzadeh, H., Rausch, M., Humphrey, J.D.: Modeling lamellar disruption within the aortic wall using a particle-based approach. Sci. Rep. **9**, 15320 (2019)

Alastrué, V., Martinez, M.A., Doblaré, M., Menzel, A.: Anisotropic micro-sphere-based finite elasticity applied to blood vessel modelling. J. Mech. Phys. Solids **57**, 178–203 (2009)

Amabili, M., Balasubramanian, P., Bozzo, I., Breslavsky, I.D., Ferrari, G.: Layer-specific hyperelastic and viscoelastic characterization of human descending thoracic aortas. J. Mech. Behav. Biomed. Mater. **99**, 27–46 (2019a)

Amabili, M., Balasubramanian, P., Breslavsky, I.: Anisotropic fractional viscoelastic constitutive models for human descending thoracic aortas. J. Mech. Behav. Biomed. Mater. **99**, 186–197 (2019b)

Angouras, D.C., Kritharis, E.P., Sokolis, D.P.: Regional distribution of delamination strength in ascending thoracic aortic aneurysms. J. Mech. Behav. Biomed. Mater. **98**, 58–70 (2019)

Arheden, H., Arner, A., Hellstrand, P.: Force-velocity relation and rate of ATP hydrolysis in osmotically compressed skinned smooth muscle of the guinea pig. J. Muscle Res. Cell Motil. **8**, 151–160 (1987)

Aslanidou, L., Ferraro, M., Lovric, G., Bersi, M.R., Humphrey, J.D., Segers, P., Trachet, B., Stergiopulos, N.: Co-localization of microstructural damage and excessive mechanical strain at aortic branches in angiotensin-II-infused mice. Biomech. Model. Mechanobiol. **19**, 81–97 (2020)

Åstrand, H., Stålhand, J., Karlsson, J., Karlsson, M., Sonesson, B., Länne, T.: In vivo estimation of the contribution of elastin and collagen to the mechanical properties in the human abdominal aorta: effect of age and sex. J. Appl. Physiol. **110**, 176–187 (2011)

Avdic, T., Franzén, S., Zarrouk, M., Acosta, S., Nilsson, P., Gottsäter, A., Svensson, A., Gudbjörnsdottir, S., Eliasson, B.: Reduced long-term risk of aortic aneurysm and aortic dissection among individuals with type 2 diabetes mellitus: a nationwide observational study. J. Am. Heart Assoc. **7**, e007618 (2018)

Baek, S., Gleason, R.L., Rajagopal, K.R., Humphrey, J.D.: Theory of small on large: potential utility in computations of fluid-solid interactions in arteries. Comput. Meth. Appl. Mech. Eng. **196**, 3070–3078 (2007)

Balzani, D., Schröder, J., Gross, D.: A simple model for anisotropic damage with applications to soft tissues. Proc. Appl. Math. Mech. **4**, 236–237 (2004)

Balzani, D., Schröder, J., Gross, D.: Simulation of discontinuous damage incorporating residual stresses in circumferentially overstretched atherosclerotic arteries. Acta Biomater. **2**, 609–618 (2006)

Balzani, D., Brinkhues, S., Holzapfel, G.A.: Constitutive framework for the modeling of damage in collagenous soft tissues with application to arterial walls. Comput. Meth. Appl. Mech. Eng. **213–216**, 139–151 (2012)

Böl, M., Schmitz, A., Nowak, G., Siebert, T.: A three-dimensional chemo-mechanical continuum model for smooth muscle contraction. J. Mech. Behav. Biomed. Mater. **13**, 215–229 (2012)

Brunet, J., Pierrat, B., Maire, E., Adrien, J., Badel, P.: A combined experimental-numerical lamellar-scale approach of tensile rupture in arterial medial tissue using X-ray tomography. J. Mech. Behav. Biomed. Mater. **95**, 116–123 (2019)

Brunet, J., Pierrat, B., Adrien, J., Maire, E., Curt, N., Badel, P.: A novel method for in vitro 3D imaging of dissecting pressurized arterial segments using X-ray microtomography. Exp. Mech. **61**, 147–157 (2021)

Carlson, B.E., Secomb, T.W.: A theoretical model for the myogenic response based on the length-tension characteristics of vascular smooth muscle. Microcirculation **12**, 327–338 (2005)

Carmo, M., Colombo, L., Bruno, A., Corsi, F.R., Roncoroni, L., Cuttin, M.S., Radice, F., Mussini, E., Settembrini, P.G.: Alteration of elastin, collagen and their cross-links in abdominal aortic aneurysms. Eur. J. Vasc. Endovasc. Surg. **23**, 543–549 (2002)

Caulk, A.W., Humphrey, J.D., Murtada, S.I.: Fundamental roles of axial stretch in isometric and isobaric evaluations of vascular contractility. J. Biomech. Eng. **141**, 0310081–03100810 (2019)

Cavinato, C., Helfenstein-Didier, C., Olivier, T., du Roscoat, S.R., Laroche, N., Badel, P.: Biaxial loading of arterial tissues with 3D in situ observations of adventitia fibrous microstructure: a method coupling multi-photon confocal microscopy and bulge inflation test. J. Mech. Behav. Biomed. Mater. **74**, 488–498 (2017)

Chen, H., Kassab, G.S.: Microstructure-based constitutive model of coronary artery with active smooth muscle contraction. Sci. Rep. **7**, 1–15 (2017)

Chow, M.J., Choi, M., Yun, S.H., Zhang, Y.: The effect of static stretch on elastin degradation in arteries. PLoS ONE **8**, e81951 (2013)

Chow, M.J., Turcotte, R., Lin, C.P., Zhang, Y.: Arterial extracellular matrix: a mechanobiological study of the contributions and interactions of elastin and collagen. Biophys. J. **106**, 2684–2692 (2014)

Chun, Y., Seow, J.: Hill's equation of muscle performance and its hidden insight on molecular mechanisms. J. Gen. Physiol. **142**, 561–573 (2013)

Coccarelli, A., Edwards, D.H., Aggarwal, A., Nithiarasu, P., Parthimos, D.: A multiscale active structural model of the arterial wall accounting for smooth muscle dynamics. J. R. Soc. Interface **15**, 20170732 (2018)

Converse, M.I., Walther, R.G., Ingram, J.T., Li, Y., Yu, S.M., Monson, K.L.: Detection and characterization of molecular-level collagen damage in overstretched cerebral arteries. Acta Biomater. **67**, 307–318 (2018)

Cooke, P.H., Fay, F.S., Craig, R.: Myosin filaments isolated from skinned amphibian smooth muscle cells are side-polar. J. Muscle Res. Cell Motil. **10**, 206–220 (1989)

Couchman, J.R., Pataki, C.A.: An introduction to proteoglycans and their localization. J. Histochem. Cytochem. **60**, 885–897 (2012)

Cox, R.H.: Arterial wall mechanics and composition and the effects of smooth muscle activation. Am. J. Physiol. **229**, 807–812 (1975)

Craig, R., Megerman, J.: Assembly of smooth muscle myosin into side-polar filaments. J. Cell Biol. **75**, 990–996 (1977)

Cranford, S., Buehler, M.J.: Materiomics: biological protein materials, from nano to macro. Nanotechnol. Sci. Appl. **3**, 127–148 (2010)

Dalbosco, M., Carniel, T.A., Fancello, E.A., Holzapfel, G.A.: Multiscale numerical analyses of arterial tissue with embedded elements in the finite strain regime. Comput. Methods Appl. Mech. Eng. **381**, 113844 (2021)

Devine, C.E., Somlyo, A.P.: Thick filaments in vascular smooth muscle. J. Cell Biol. **49**, 636–649 (1971)

Dingemans, K.P., Teeling, P., Lagendijk, J.H., Becker, A.E.: Extracellular matrix of the human aortic media: an ultrastructural histochemical and immunohistochemical study of the adult aortic media. Anat. Rec. **258**, 1–14 (2000)

Dobrin, P.B.: Influence of initial length on length-tension relationship of vascular smooth muscle. Am. J. Physiol. **225**, 664–670 (1973a)

Dobrin, P.B.: Isometric and isobaric contraction of carotid arterial smooth muscle. Am. J. Physiol. **225**, 659–663 (1973b)

Dutov, P., Antipova, O., Varma, S., Orgel, J.P.R.O., Schieber, J.D.: Measurement of elastic modulus of collagen type I single fiber. PLoS ONE **11**, e0145711 (2016)

Edman, K.A.: Mechanical deactivation induced by active shortening in isolated muscle fibres of the frog. J. Physiol. **246**, 255–275 (1975)

Edman, K.A.: The velocity of unloaded shortening and its relation to sarcomere length and isometric force in vertebrate muscle fibres. J. Physiol. **291**, 143–159 (1979)

Edman, K.A.: The force bearing capacity of frog muscle fibres during stretch: its relation to sarcomere length and fibre width. J. Physiol. **519**, 515–526 (1999)

Ettema, G.J., Meijer, K.: Muscle contraction history: modified Hill versus an exponential decay model. Biol. Cell. **83**, 491–500 (2000)

Famaey, N., Sloten, J.V., Kuhl, E.: A three-constituent damage model for arterial clamping in computer-assisted surgery. Biomech. Model. Mechanobiol. **12**, 123–136 (2013)

Fereidoonnezhad, B., Naghdabadi, R., Holzapfel, G.A.: Stress softening and permanent deformation in human aortas: continuum and computational modeling with application to arterial clamping. J. Mech. Behav. Biomed. Mater. **61**, 600–616 (2016)

Ferrara, A., Pandolfi, A.: Numerical modelling of fracture in human arteries. Comput. Methods Biomech. Biomed. Eng. **11**, 553–567 (2008)

Ferrara, A., Pandolfi, A.: A numerical study of arterial media dissection processes. Int. J. Fract. **166**, 21–33 (2010)

Ferruzzi, J., Bersi, M.R., Humphrey, J.D.: Biomechanical phenotyping of central arteries in health and disease: advantages of and methods for murine models. Ann. Biomed. Eng. **41**, 1311–1330 (2013)

Flory, P.J.: Thermodynamic relations for highly elastic materials. Trans. Faraday Soc. **57**, 829–838 (1961)

Fung, Y.C.: Biomechanics. Mechanical Properties of Living Tissues, 2nd edn. Springer, New York (1993)

Gasser, T.C., Holzapfel, G.A.: A rate-independent elastoplastic constitutive model for (biological) fiber-reinforced composites at finite strains: continuum basis, algorithmic formulation and finite element implementation. Comput. Mech. **29**, 340–360 (2002)

Gasser, T.C., Holzapfel, G.A.: Modeling the propagation of arterial dissection. Eur. J. Mech. A-Solid **25**, 617–633 (2006)

Gasser, T.C., Holzapfel, G.A.: Finite element modeling of balloon angioplasty by considering overstretch of remnant non-diseased tissues in lesions. Comput. Mech. **40**, 47–60 (2007)

Gasser, T.C., Ogden, R.W., Holzapfel, G.A.: Hyperelastic modelling of arterial layers with distributed collagen fibre orientations. J. R. Soc. Interface **3**, 15–35 (2006)

Gaul, R., Nolan, D., Lally, C.: The use of small angle light scattering in assessing strain induced collagen degradation in arterial tissue ex vivo. J. Biomech. **81**, 155–160 (2018a)

Gaul, R., Nolan, D., Ristori, T., Bouten, C., Loerakker, S., Lally, C.: Strain mediated enzymatic degradation of arterial tissue: insights into the role of the non-collagenous tissue matrix and collagen crimp. Acta Biomater. **77**, 301–310 (2018b)

Gaul, R.T., Nolan, D.R., Ristori, T., Bouten, C.V.C., Loerakker, S., Lally, C.: Pressure-induced collagen degradation in arterial tissue as a potential mechanism for degenerative arterial disease progression. J. Mech. Behav. Biomed. Mater. **109**, 103771 (2020)

Germain, P.: The method of virtual power in continuum mechanics. Part 2: microstructure. SIAM J. Appl. Math. **25**, 556–574 (1973)

Gestrelius, S., Borgström, P.: A dynamic model of smooth muscle contraction. Biophys. J. **50**, 157–169 (1986)

Gilbert, R.R., Grafenhorst, M., Hartmann, S., Yosibash, Z.: Simulating the temporal change of the active response of arteries by finite elements with high-order time-integrators. Comput. Mech. **64**, 1669–1684 (2019)

Giuseppe, M.D., Zingales, M., Pasta, S., Avril, S.: In vitro measurement of strain localization preceding dissection of the aortic wall subjected to radial tension. Exp. Mech. **61**, 119–130 (2021)

Gleason, R.L., Dye, W.W., Wilson, E., Humphrey, J.D.: Quantification of the mechanical behavior of carotid arteries from wild-type, dystrophin-deficient, and sarcoglycan-δ knockout mice. J. Biomech. **41**, 3213–3218 (2008)

Gomez, D., Owens, G.K.: Smooth muscle cell phenotypic switching in atherosclerosis. Cardiovasc. Res. **95**, 156–164 (2012)

Gordon, A.M., Huxley, A.F., Julian, F.J.: Tension development in highly stretched vertebrate muscle fibres. J. Physiol. **184**, 143–169 (1966)

Greenwald, S.E., Newman, D.L., Denyer, H.T.: Effect of smooth muscle activity on the static and dynamic elastic properties of the rabbit carotid artery. Cardiovasc. Res. **16**, 86–94 (1982)

Gültekin, O., Dal, H., Holzapfel, G.A.: A phase-field approach to model fracture of arterial walls: theory and finite element analysis. Comput. Methods Appl. Mech. Eng. **312**, 542–566 (2016)

Gültekin, O., Dal, H., Holzapfel, G.A.: Numerical aspects of anisotropic failure in soft biological tissues favor energy-based criteria: a rate-dependent anisotropic crack phase-field model. Comput. Methods. Appl. Mech. Eng. **331**, 23–52 (2018)

Gültekin, O., Hager, S.P., Dal, H., Holzapfel, G.A.: Computational modeling of progressive damage and rupture in fibrous biological tissues: application to aortic dissection. Biomech. Model. Mechanobiol. **18**, 1607–1628 (2019)

Hadi, M.F., Sander, E.A., Barocas, V.H.: Multiscale model predicts tissue-level failure from collagen fiber-level damage. J. Biomech. Eng. **134**, 091005 (2012)

Hai, C.M., Murphy, R.A.: Cross-bridge phosphorylation and regulation of latch state in smooth muscle. J. Appl. Physiol. **254**, C99–C106 (1988)

Haskett, D., Johnson, G., Zhou, A., Utzinger, U., Vande Geest, J.: Microstructural and biomechanical alterations of the human aorta as a function of age and location. Biomech. Model. Mechanobiol. **9**, 725–736 (2010)

Haspinger, D.C., Murtada, S.I., Niestrawska, J.A., Holzapfel, G.A.: Numerical analyses of the interrelation between extracellular smooth muscle orientation and intracellular filament overlap in the human abdominal aorta. Z. Angew. Math. Mech. **98**, 2198–2221 (2018)

Hatze, H.: Myocybernetic control model of skeletal muscle. Biol. Cell. **20**, 103–119 (1977)

Haverkamp, R.G., Williams, M.A., Scott, J.E.: Stretching single molecules of connective tissue glycans to characterize their shape-maintaining elasticity. Biomacromolecules **6**, 1816–1818 (2005)

Hayenga, H.N., Thorne, B.C., Peirce, S.M., Humphrey, J.D.: Ensuring congruency in multiscale modeling: towards linking agent based and continuum biomechanical models of arterial adaptation. Ann. Biomed. Eng. **39**, 2669 (2011)

Herod, T.W., Chambers, N.C., Veres, S.P.: Collagen fibrils in functionally distinct tendons have differing structural responses to tendon rupture and fatigue loading. Acta Biomater. **42**, 296–307 (2016)

Hill, A.V.: The heat of shortening and the dynamic constants of muscle. Proc. R. Soc. Lond. B **126**, 136–195 (1938)

Holzapfel, G.A.: Determination of material models for arterial walls from uniaxial extension tests and histological structure. J. Theor. Biol. **238**, 290–302 (2006)

Holzapfel, G.A.: Collagen in arterial walls: biomechanical aspects. In: Fratzl, P. (ed.) Collagen. Structure and Mechanics, pp. 285–324. Springer, Heidelberg (2008)

Holzapfel, G.A., Fereidoonnezhad, B.: Modeling of damage in soft biological tissues. In: Payan, Y., Ohayon, J. (eds.) Biomechanics of Living Organs. Hyperelastic Constitutive Laws for Finite Element Modeling, pp. 101–123. Academic, New York (2017)

Holzapfel, G.A., Gasser, T.C.: A viscoelastic model for fiber-reinforced composites at finite strains: continuum basis, computational aspects and applications. Comput. Methods Appl. Mech. Eng. **190**, 4379–4403 (2001)

Holzapfel, G.A., Ogden, R.W.: On the tension-compression switch in soft fibrous solids. Eur. J. Mech. A/Solids **49**, 561–569 (2015)

Holzapfel, G.A., Ogden, R.W.: Biomechanical relevance of the microstructure in artery walls with a focus on passive and active components. Am. J. Physiol. Heart Circ. Physiol. **315**, H540–H549 (2018)

Holzapfel, G.A., Ogden, R.W.: An arterial constitutive model accounting for collagen content and cross-linking. J. Mech. Phys. Solids **136**, 103682 (2020a)

Holzapfel, G.A., Ogden, R.W.: A damage model for collagen fibres with an application to collagenous soft tissues. Proc. R. Soc. Lond. A **476**, 20190821 (2020b)

Holzapfel, G.A., Gasser, T.C., Ogden, R.W.: A new constitutive framework for arterial wall mechanics and a comparative study of material models. J. Elast. **61**, 1–48 (2000a)

Holzapfel, G.A., Schulze-Bauer, C.A.J., Stadler, M.: Mechanics of angioplasty: wall, balloon and stent. In: Casey, J., Bao, G. (eds.) Mechanics in Biology, pp. 141–156. The American Society of Mechanical Engineers (ASME), New York (2000b). AMD-Vol. 242/BED-Vol. 46

Holzapfel, G.A., Gasser, T.C., Stadler, M.: A structural model for the viscoelastic behavior of arterial walls: continuum formulation and finite element analysis. Eur. J. Mech. A/Solids **21**, 441–463 (2002a)

Holzapfel, G.A., Stadler, M., Schulze-Bauer, C.A.J.: A layer-specific three-dimensional model for the simulation of balloon angioplasty using magnetic resonance imaging and mechanical testing. Ann. Biomed. Eng. **30**, 753–767 (2002b)

Holzapfel, G.A., Sommer, G., Regitnig, P.: Anisotropic mechanical properties of tissue components in human atherosclerotic plaques. J. Biomech. Eng. **126**, 657–665 (2004)

Holzapfel, G.A., Niestrawska, J.A., Ogden, R.W., Reinisch, A.J., Schriefl, A.J.: Modelling non-symmetric collagen fibre dispersion in arterial walls. J. R. Soc. Interface **12**, 20150188 (2015)

Horný, L., Kronek, J., Chlup, H., Žitný, R., Veselý, J., Hulan, M.: Orientations of collagen fibers in aortic histological section. Bull. Appl. Mech. **6**, 25–29 (2010)

Horowitz, A., Menice, C.B., Laporte, R., Morgan, K.G.: Mechanisms of smooth muscle contraction. Physiol. Rev. **76**, 967–1003 (1996)

Humphrey, J.D.: Cardiovascular Solid Mechanics. Cells, Tissues, and Organs. Springer, New York (2002)

Humphrey, J.D.: Mechanisms of vascular remodeling in hypertension. Am. J. Hypertens. **34**, 432–441 (2021a)

Humphrey, J.D.: Constrained mixture models of soft tissue growth and remodeling - Twenty years after. J. Elast. (2021b). https://doi.org/10.1007/s10659-020-09809-1

Humphrey, J.D., Rajagopal, K.R.: A constrained mixture model for growth and remodeling of soft tissues. Math. Model. Methods Appl. Sci. **12**, 407–430 (2002)

Huo, Y., Cheng, Y., Zhao, X., Lu, X., Kassab, G.S.: Biaxial vasoactivity of porcine coronary artery. Am. J. Physiol. Heart Circ. Physiol. **302**, H2058–H2063 (2012)

Huxley, A.F., Niedergerke, R.: Structural changes in muscle during contraction; interference microscopy of living muscle fibers. Nature **173**, 971–973 (1954)

Joldes, G.R., Miller, K., Wittek, A., Doyle, B.: A simple, effective and clinically applicable method to compute abdominal aortic aneurysm wall stress. J. Mech. Behav. Biomed. Mater. **58**, 139–148 (2016)

Jones, B., Tonniges, J.R., Debski, A., Albert, B., Yeung, D.A., Gadde, N., Mahajan, A., Sharma, N., Calomeni, E.P., Go, M.R., Hans, C.P., Agarwal, G.: Collagen fibril abnormalities in human and mice abdominal aortic aneurysm. Acta Biomater. **110**, 129–140 (2020)

Katsuda, S., Okada, Y., Minamoto, T., Oda, Y., Matsui, Y., Nakanishi, I.: Collagens in human atherosclerosis. Immunohistochemical analysis using collagen type-specific antibodies. Arterioscler. Thromb. Vasc. Biol. **12**, 494–502 (1992)

Kelleher, C.M., McLean, S.E., Mecham, R.P.: Vascular extracellular matrix and aortic development. Curr. Top. Dev. Biol. **62**, 153–188 (2004)

Kim, J., Staiculescu, M.C., Cocciolone, A.J., Yanagisawa, H., Mecham, R.P., Wagenseil, J.E.: Crosslinked elastic fibers are necessary for low energy loss in the ascending aorta. J. Biomech. **61**, 199–207 (2017)

Koch, R.G., Tsamis, A., D'Amore, A., Wagner, W.R., Watkins, S.C., Gleason, T.G., Vorp, D.A.: A custom image-based analysis tool for quantifying elastin and collagen micro-architecture in the wall of the human aorta from multi-photon microscopy. J. Biomech. **47**, 935–943 (2014)

Koenders, M.M.J.F., Yang, L., Wismans, R.G., van der Werf, K.O., Reinhardt, D.P., Daamen, W., Bennink, M.L., Dijkstra, P.J., van Kuppevelt, T.H., Feijen, J.: Microscale mechanical properties of single elastic fibers: the role of fibrillin-microfibrils. Biomaterials **30**, 2425–2432 (2009)

Korenczuk, C.E., Dhume, R.Y., Liao, K.K., Barocas, V.H.: Ex vivo mechanical tests and multiscale computational modeling highlight the importance of intramural shear stress in ascending thoracic aortic aneurysms. J. Biomech. Eng. **141**, 121010 (2019)

Kratzberg, J.A., Walker, P.J., Rikkers, E., Raghavan, M.L.: The effect of proteolytic treatment on plastic deformation of porcine aortic tissue. J. Mech. Behav. Biomed. Mater. **2**, 65–72 (2009)

Kronick, P.L., Sacks, M.S.: Matrix macromolecules that affect the viscoelasticity of calfskin. J. Biomech. Eng. **116**, 140–145 (1994)

Lacolley, P., Regnault, V., Segers, P., Laurent, S.: Vascular smooth muscle cells and arterial stiffening: relevance in development, aging, and disease. Phys. Rev. **97**, 1555–1617 (2017)

Lake, S.P., Hadi, M.F., Lai, V.K., Barocas, V.H.: Mechanics of a fiber network within a non-fibrillar matrix: model and comparison with collagen-agarose co-gels. Ann. Biomed. Eng. **40**, 2111–2121 (2012)

Lanir, Y.: Constitutive equations for fibrous connective tissues. J. Biomech. **16**, 1–12 (1983)

Lanir, Y.: Multi-scale structural modeling of soft tissues: mechanics and mechanobiology. J. Elast. **129**, 7–48 (2018)

Laubrie, J.D., Mousavi, J.S., Avril, S.: A new finite-element shell model for arterial growth and remodeling after stent implantation. Int. J. Numer. Method Biomed. Eng. **36**, e3282 (2020)

Lehoux, S., Castier, Y., Tedgui, A.: Molecular mechanisms of the vascular responses to haemodynamic forces. J. Intern. Med. **259**, 381–392 (2006)

van de Lest, C.H., Versteeg, E.M., Veerkamp, J.H., van Kuppevelt, T.H.: Digestion of proteoglycans in porcine pancreatic elastase-induced emphysema in rats. Eur. Respir. J. **8**, 238–245 (1995)

Li, C., Xu, Q.: Mechanical stress-initiated signal transductions in vascular smooth muscle cells. Cell. Signal. **12**, 435–445 (2000)

Li, G., Wang, M., Caulk, A.W., Cilfone, N.A., Gujja, S., Qin, L., Chen, P.Y., Chen, Z., Yousef, S., Jiao, Y., He, C., Jiang, B., Korneva, A., Bersi, M.R., Wang, G., Liu, X., Mehta, S., Geirsson, A., Gulcher, J.R., Chittenden, T.W., Simons, M., Humphrey, J.D., Tellides, G.: Chronic mTOR activation induces a degradative smooth muscle cell phenotype. Clin. Investig. Med. **130**, 1233–1251 (2020)

Li, H., Mattson, J.M., Zhang, Y.: Integrating structural heterogeneity, fiber orientation, and recruitment in multiscale ECM mechanics. J. Mech. Behav. Biomed. Mater. **92**, 1–10 (2019)

Li, K., Holzapfel, G.A.: Multiscale modeling of fiber recruitment and damage with a discrete fiber dispersion method. J. Mech. Phys. Solids **126**, 226–244 (2019)

Li, K., Sun, W.: Simulated thin pericardial bioprosthetic valve leaflet deformation under static pressure-only loading conditions: implications for percutaneous valves. Ann. Biomed. Eng. **38**, 2690–2701 (2010)

Li, K., Ogden, R.W., Holzapfel, G.A.: Computational method for excluding fibers under compression in modeling soft fibrous solids. Eur. J. Mech. A/Solids **57**, 178–193 (2016)

Li, K., Ogden, R.W., Holzapfel, G.A.: A discrete fibre dispersion method for excluding fibres under compression in the modelling of fibrous tissues. J. R. Soc. Interface **15**, 20170766 (2018)

Li, Q., Muragaki, Y., Hatamura, I., Ueno, H., Ooshima, A.: Stretch-induced collagen synthesis in cultured smooth muscle cells from rabbit aortic media and a possible involvement of angiotensin II and transforming growth factor-β. J. Vasc. Res. **35**, 93–103 (1998)

Liang, L., Liu, M., Martin, C., Sun, W.: A deep learning approach to estimate stress distribution: a fast and accurate surrogate of finite-element analysis. J. R. Soc. Interface **15**, 20170844 (2018)

Lichtwark, G.A., Wilson, A.M.: A modified Hill muscle model that predicts muscle power output and efficiency during sinusoidal length changes. J. Exp. Biol. **208**, 2831–2843 (2005)

Lin, A.H., Allan, A.N., Zitnay, J.L., Kessler, J.L., Yu, S.M., Weiss, J.A.: Collagen denaturation is initiated upon tissue yield in both positional and energy-storing tendons. Acta Biomater. **118**, 153–160 (2020)

Lindeman, J.H., Ashcroft, B.A., Beenakker, J.W., van Es, M., Koekkoek, N.B., Prins, F.A., Tielemans, J.F., Abdul-Hussien, H., Bank, R.A., Oosterkamp, T.H.: Distinct defects in collagen microarchitecture underlie vessel-wall failure in advanced abdominal aneurysms and aneurysms in Marfan syndrome. Proc. Natl. Acad. Sci. USA **107**, 862–865 (2010)

Liu, M., Dong, H., Lou, X., Iannucci, G., Chen, E.P., Leshnower, B.G., Sun, W.: A novel anisotropic failure criterion with dispersed fiber orientations for aortic tissues. J. Biomech. Eng. **98**, 111002 (2019)

Lloyd, D.G., Besier, T.F.: An EMG-driven musculoskeletal model to estimate muscle forces and knee joint moments in vivo. J. Biomech. **36**, 765–776 (2003)

Luo, T., Chen, H., Kassab, G.S.: 3D reconstruction of coronary artery vascular smooth muscle cells. PLoS ONE **11**, e0147272 (2016)

Maiti, S., Thunes, J.R., Fortunato, R.N., Gleason, T.G., Vorp, D.A.: Computational modeling of the strength of the ascending thoracic aortic media tissue under physiologic biaxial loading conditions. J. Biomech. **108**, 109884 (2020)

Mao, W., Li, K., Sun, W.: Fluid-structure interaction study of transcatheter aortic valve dynamics using smoothed particle hydrodynamics. Cardiovasc. Eng. Technol. **7**, 374–388 (2016)

Marino, M., Converse, M.I., Monson, K.L., Wriggers, P.: Molecular-level collagen damage explains softening and failure of arterial tissues: a quantitative interpretation of CHP data with a novel elasto-damage model. J. Mech. Behav. Biomed. Mater. **97**, 254–271 (2019)

Matsumoto, T., Tsuchida, M., Sato, M.: Change in intramural strain distribution in rat aorta due to smooth muscle contraction and relaxation. Am. J. Physiol. Heart Circ. Physiol. **271**, H1711–H1716 (1996)

Mattson, J.M., Turcotte, R., Zhang, Y.: Glycosaminoglycans contribute to extracellular matrix fiber recruitment and arterial wall mechanics. Biomech. Model. Mechanobiol. **16**, 213–225 (2017)

Maurel, E., Shuttleworth, C.A., Bouissou, H.: Interstitial collagens and ageing in human aorta. Virchows Arch. A **410**, 383–390 (1987)

Menashi, S., Campa, J.S., Greenhalgh, R.M., Powell, J.T.: Collagen in abdominal aortic aneurysm: typing, content, and degradation. J. Vasc. Surg. **6**, 578–582 (1987)

Milnor, W.R.: Cardiovascular Physiology. Oxford University Press, Oxford (1990)

Mohan, D., Melvin, J.W.: Failure properties of passive human aortic tissue. I – uniaxial tension tests. J. Biomech. **15**, 887–902 (1982)

Mohan, D., Melvin, J.W.: Failure properties of passive human aortic tissue. II – biaxial tension tests. J. Biomech. **16**, 31–44 (1983)

Mulvany, M.J., Warshaw, D.M.: The active tension-length curve of vascular smooth muscle related to its cellular components. J. Gen. Physiol. **74**, 85–104 (1979)

Murata, K., Motayama, T., Kotake, C.: Collagen types in various layers of the human aorta and their changes with the atherosclerotic process. Atherosclerosis **60**, 251–262 (1986)

Murtada, S., Holzapfel, G.A.: Investigating the role of smooth muscle cells in large elastic arteries: a finite element analysis. J. Theor. Biol. **358**, 1–10 (2014)

Murtada, S., Kroon, M., Holzapfel, G.A.: A calcium-driven mechanochemical model for prediction of force generation in smooth muscle. Biomech. Model. Mechanobiol. **9**, 749–762 (2010a)

Murtada, S., Kroon, M., Holzapfel, G.A.: Modeling the dispersion effects of contractile fibers in smooth muscles. J. Mech. Phys. Solids **58**, 2065–2082 (2010b)

Murtada, S.C., Arner, A., Holzapfel, G.A.: Experiments and mechanochemical modeling of smooth muscle contraction: significance of filament overlap. J. Theor. Biol. **297**, 176–186 (2012)

Murtada, S.I., Humphrey, J.D.: Regional heterogeneity in the regulation of vasoconstriction in arteries and its role in vascular mechanics. Adv. Exp. Med. Biol. **1097**, 105–128 (2018)

Murtada, S.I., Ferruzzi, J., Yanagisawa, H., Humphrey, J.D.: Reduced biaxial contractility in the descending thoracic aorta of fibulin-5 deficient mice. J. Biomech. Eng. **138**, 051008 (2016a)

Murtada, S.I., Lewin, S., Arner, A., Humphrey, J.D.: Adaptation of active tone in the mouse descending thoracic aorta under acute changes in loading. Biomech. Model. Mechanobiol. **15**, 579–592 (2016b)

Murtada, S.I., Humphrey, J.D., Holzapfel, G.A.: Multiscale and multiaxial mechanics of vascular smooth muscle. Biophys. J. **113**, 714–727 (2017)

Myneni, M., Rao, A., Jiang, M., Moreno, M.R., Rajagopal, K.R., Benjamin, C.: Segmental variations in the peel characteristics of the porcine thoracic aorta. Ann. Biomed. Eng. **48**, 1751–1767 (2020)

Nguyen, T.D., Jones, R.E., Boyce, B.L.: Modeling the anisotropic finite-deformation viscoelastic behavior of soft fiber-reinforced composites. Int. J. Solids Struct. **44**, 8366–8389 (2007)

Niestrawska, J.A., Viertler, C., Regitnig, P., Cohnert, T.U., Sommer, G., Holzapfel, G.A.: Microstructure and mechanics of healthy and aneurysmatic abdominal aortas: experimental analysis and modeling. J. R. Soc. Interface **13**, 20160620 (2016)

Niestrawska, J.A., Regitnig, P., Viertler, C., Cohnert, T.U., Babu, A.R., Holzapfel, G.A.: The role of tissue remodeling in mechanics and pathogenesis of abdominal aortic aneurysms. Acta Biomater. **88**, 149–161 (2019)

Ogden, R.W.: Nearly isochoric elastic deformations: application to rubberlike solids. J. Mech. Phys. Solids **26**, 37–57 (1978)

Ogden, R.W., Roxburgh, D.G.: A pseudo-elastic model for the Mullins effect in filled rubber. Proc. R. Soc. Lond. A **455**, 2861–2877 (1999)

O'Connell, M.K., Murthy, S., Phan, S., Xu, C., Buchanan, J., Spilker, R., Dalman, R.L., Zarins, C.K., Denk, W., Taylor, C.A.: The three-dimensional micro- and nanostructure of the aortic medial lamellar unit measured using 3D confocal and electron microscopy imaging. Matrix Biol. **27**, 171–181 (2008)

Pate, E., Cooke, R.: A model of crossbridge action: the effects of ATP, ADP and Pi. J. Muscle Res. Cell Motil. **10**, 181–196 (1989)

Patel, K., Zafar, M.A., Ziganshin, B.A., Elefteriades, J.A.: Diabetes mellitus: is it protective against aneurysm? A narrative review. Cardiology **141**, 107–122 (2018)

Peña, E., Doblaré, M.: An anisotropic pseudo-elastic approach for modelling Mullins effect in fibrous biological materials. Mech. Res. Commun. **36**, 784–790 (2009)

Peña, E., Alastrue, V., Laborda, A., Martíne, M.A., Doblaré, M.: A constitutive formulation of vascular tissue mechanics including viscoelasticity and softening behaviour. J. Biomech. **43**, 984–989 (2010)

Peña, J.A., Martínez, M.A., Peña, E.: A formulation to model the nonlinear viscoelastic properties of the vascular tissue. Acta Mech. **217**, 63–74 (2011)

Peña, J.A., Marínez, M.A., Peña, E.: Failure damage mechanical properties of thoracic and abdominal porcine aorta layers and related constitutive modeling: phenomenological and microstructural approach. Biomech. Model. Mechanobiol. **18**, 1709–1730 (2019)

Polzer, S., Man, V., Vlachovský, R., Kubícek, L., Kracík, J., Staffa, R., Novotnỳ, T., Burša, J., Raghavan, M.: Failure properties of abdominal aortic aneurysm tissue are orientation dependent. J. Mech. Behav. Biomed. Mater. **114**, 104181 (2020)

Rabin, J., Siddiqui, A., Gipple, J., Taylor, B., Scalea, T.M., Haslach, H.W.: Minor aortic injury may be at risk of progression from uncontrolled shear stress: an in-vitro model demonstrates aortic lesion expansion. Trauma (2020). https://doi.org/10.1177/1460408620957426

Rachev, A., Hayashi, K.: Theoretical study of the effects of vascular smooth muscle contraction on strain and stress distributions in arteries. Ann. Biomed. Eng. **27**, 459–468 (1999)

Rausch, M., Karniadakis, G., Humphrey, J.D.: Modeling soft tissue damage and failure using a combined particle/continuum approach. Biomech. Model. Mechanobiol. **16**, 1–13 (2017)

Rezvani-Sharif, A., Tafazzoli-Shadpour, M., Avolio, A.: Mechanical characterization of the lamellar structure of human abdominal aorta in the development of atherosclerosis: an atomic force microscopy study. Cardiovasc. Eng. Technol. **10**, 181–192 (2019)

Rizzo, R.J., McCarthy, W.J., Dixit, S.N., Lilly, M.P., Shively, V.P., Flinn, W.R., Yao, J.S.: Collagen types and matrix protein content in human abdominal aortic aneurysms. J. Vasc. Surg. **10**, 365–373 (1989)

Roccabianca, S., Ateshian, G.A., Humphrey, J.D.: Biomechanical roles of medial pooling of glycosaminoglycans in thoracic aortic dissection. Biomech. Model. Mechanobiol. **13**, 13–25 (2014a)
Roccabianca, S., Bellini, C., Humphrey, J.D.: Computational modelling suggests good, bad and ugly roles of glycosaminoglycans in arterial wall mechanics and mechanobiology. J. R. Soc. Interface **11**, 20140397 (2014b)
Rocha, F.F., Blanco, P.J., Sánchez, P.J., Feijóo, R.A.: Multi-scale modelling of arterial tissue: linking networks of fibres to continua. Comput. Methods Appl. Mech. Eng. **341**, 740–787 (2018)
Rodríguez, J.F., Cacho, F., Bea, J.A., Doblaré, M.: A stochastic-structurally based three dimensional finite-strain damage model for fibrous soft tissue. J. Mech. Phys. Solids **54**, 864–886 (2006)
Rolf-Pissarczyk, M., Li, K., Fleischmann, D., Holzapfel, G.A.: A discrete approach for modeling degraded elastic fibers in aortic dissection. Comput. Methods Appl. Mech. Eng. **373**, 113511 (2021)
Schaefer, L., Schaefer, R.M.: Proteoglycans: from structural compounds to signaling molecules. Connect. Tissue Res. **339**, 237–246 (2010)
Schlatmann, T.J.M., Becker, A.E.: Histologic changes in the normal aging aorta: implications for dissecting aortic aneurysm. Am. J. Cardiol. **39**, 13–20 (1977)
Schmid, F., Sommer, G., Rappolt, M., Schulze-Bauer, C.A.J., Regitnig, P., Holzapfel, G.A., Laggner, P., Amenitsch, H.: In situ tensile testing of human aortas by time-resolved small angle X-ray scattering. Synchrotron Radiat. **12**, 727–733 (2005)
Schmidt, T., Balzani, D., Holzapfel, G.A.: Statistical approach for a continuum description of damage evolution in soft collagenous tissues. Comput. Methods Appl. Mech. Eng. **278**, 41–61 (2014)
Schmitz, A., Böl, M.: On a phenomenological model for active smooth muscle contraction. J. Biomech. **44**, 2090–2095 (2011)
Schriefl, A.J., Zeindlinger, G., Pierce, D.M., Regitnig, P., Holzapfel, G.A.: Determination of the layer-specific distributed collagen fiber orientations in human thoracic and abdominal aortas and common iliac arteries. J. R. Soc. Interface **9**, 1275–1286 (2012)
Schriefl, A.J., Wolinski, H., Regitnig, P., Kohlwein, S.D., Holzapfel, G.A.: An automated approach for three-dimensional quantification of fibrillar structures in optically cleared soft biological tissues. J. R. Soc. Interface **10**, 20120760 (2013)
Schriefl, A.J., Schmidt, T., Balzani, D., Sommer, G., Holzapfel, G.A.: Selective enzymatic removal of elastin and collagen from human abdominal aortas: uniaxial mechanical response and constitutive modeling. Acta Biomater. **17**, 125–136 (2015)
Schröder, J., Balzani, D., Gross, D.: Aspects of modeling and computer simulation of soft tissues: applications to arterial walls. Mat.-wiss. u. Werkstofftechn. **36**, 795–801 (2005)
Sepahi, O., Radtke, L., Debus, S.E., Düster, A.: Anisotropic hierarchic solid finite elements for the simulation of passive-active arterial wall models. Comput. Math. Appl. **74**, 3058–3079 (2017)
Shah, S.B., Witzenburg, C., Hadi, M.F., Wagner, H.P., Goodrich, J.M., Alford, P.W., Barocas, V.H.: Prefailure and failure mechanics of the porcine ascending thoracic aorta: experiments and a multiscale model. J. Biomech. Eng. **136**, 021028 (2014)
Shahmirzadi, D., Bruck, H.A., Hsieh, A.H.: Measurement of mechanical properties of soft tissues in vitro under controlled tissue hydration. Exp. Mech. **53**, 405–414 (2013)
Sharifimajd, B., Stålhand, J.: A continuum model for excitation-contraction of smooth muscle under finite deformations. J. Theor. Biol. **355**, 1–9 (2014)
Sherifova, S., Holzapfel, G.A.: Biomechanics of aortic wall failure with a focus on dissection and aneurysm: a review. Acta Biomater. **99**, 1–17 (2019)
Sherifova, S., Holzapfel, G.A.: Biochemomechanics of the thoracic aorta in health and disease. Prog. Biomed. Eng. **2**, 032002 (2020)
Sherifova, S., Sommer, G., Viertler, C., Regitnig, P., Caranasos, T., Smith, M.A., Griffith, B.E., Ogden, R.W., Holzapfel, G.A.: Failure properties and microstructure of healthy and aneurysmatic human thoracic aortas subjected to uniaxial extension with a focus on the media. Acta Biomater. **99**, 443–456 (2019)

Somlyo, A.P., Somlyo, A.V.: Signal transduction and regulation in smooth muscle. Nature **372**, 231–236 (1994)

Sommer, G., Sherifova, S., Oberwalder, P.J., Dapunt, O.E., Ursomanno, P.A., DeAnda, A., Griffith, B.E., Holzapfel, G.A.: Mechanical strength of aneurysmatic and dissected human thoracic aortas at different shear loading modes. J. Biomech. **49**, 2374–2382 (2016)

Spronck, B., Ferruzzi, J., Bellini, C., Caulk, A.W., Murtada, S.I., Humphrey, J.D.: Aortic remodeling is modest and sex-independent in mice when hypertension is superimposed on aging. Am. J. Hypertens. **38**, 1312–1321 (2020)

Stålhand, J., Holzapfel, G.A.: Length adaptation of smooth muscle contractile filaments in response to sustained activation. J. Theor. Biol. **397**, 13–21 (2016)

Stålhand, J., Klarbring, A., Holzapfel, G.A.: Smooth muscle contraction: mechanochemical formulation for homogeneous finite strains. Prog. Biophys. Mol. Biol. **96**, 465–481 (2008)

Stålhand, J., Klarbring, A., Holzapfel, G.A.: A mechanochemical 3D continuum model for smooth muscle contraction under finite strains. J. Theor. Biol. **268**, 120–130 (2011)

Stålhand, J., McMeeking, R.M., Holzapfel, G.A.: On the thermodynamics of smooth muscle contraction. J. Mech. Phys. Solids **94**, 490–503 (2016)

Stylianopoulos, T., Barocas, V.H.: Multiscale, structure-based modeling for the elastic mechanical behavior of arterial walls. J. Biomech. Eng. **129**, 611–618 (2007a)

Stylianopoulos, T., Barocas, V.H.: Volume-averaging theory for the study of the mechanics of collagen networks. Comput. Methods Appl. Mech. Eng. **196**, 2981–2990 (2007b)

Sugita, S., Matsumoto, T.: Multiphoton microscopy observations of 3D elastin and collagen fiber microstructure changes during pressurization in aortic media. Biomech. Model. Mechanobiol. **16**, 763–773 (2017)

Sun, W., Li, K., Sirois, E.: Simulated elliptical bioprosthetic valve deformation: implications for asymmetric transcatheter valve deployment. J. Biomech. **43**, 3085–3090 (2010)

Sun, Y.L., Luo, Z.P., Fertala, A., An, K.A.: Direct quantification of the flexibility of type I collagen monomer. Biochem. Biophys. Res. Commun. **295**, 382–386 (2002)

Tanios, F., Gee, M.W., Pelisek, J., Kehl, S., Biehler, J., Grabher-Meier, V., Wall, W.A., Eckstein, H.H., Reeps, C.: Interaction of biomechanics with extracellular matrix components in abdominal aortic aneurysm wall. Eur. J. Vasc. Endovasc. Surg. **50**, 167–174 (2015)

Taylor, R.L.: FEAP – A Finite Element Analysis Program, Version 8.4 User Manual. University of California at Berkeley, Berkeley, California (2013)

Thunes, J.R., Pal, S., Fortunato, R.N., Phillippi, J.A., Gleason, T.G., Vorp, D.A., Maiti, S.: A structural finite element model for lamellar unit of aortic media indicates heterogeneous stress field after collagen recruitment. J. Biomech. **49**, 1562–1569 (2016)

Thunes, J.R., Phillippi, J.A., Gleason, T.G., Vorp, D., Maiti, S.: Structural modeling reveals microstructure-strength relationship for human ascending thoracic aorta. J. Biomech. **71**, 84–93 (2018)

Tong, J., Xin, Y.F., Xu, X., Yang, F., Zhang, Z.: Effect of diabetes mellitus on the dissection properties of the rabbit descending thoracic aortas. J. Biomech. **100**, 109592 (2020)

Tse, K.M., Chiu, P., Lee, H., Ho, P.: Investigation of hemodynamics in the development of dissecting aneurysm within patient-specific dissecting aneurismal aortas using computational fluid dynamics (CFD) simulations. J. Biomech. **44**, 827–836 (2011)

Vande Geest, J.P., Sacks, M.S., Vorp, D.A.: Age dependency of the biaxial biomechanical behavior of human abdominal aorta. J. Biomech. Eng. **126**, 815–822 (2004)

Vande Geest, J.P., Sacks, M.S., Vorp, D.A.: The effects of aneurysm on the biaxial mechanical behavior of human abdominal aorta. J. Biomech. **39**, 1324–1334 (2006)

Wagenseil, J.E., Mecham, R.P.: Vascular extracellular matrix and arterial mechanics. Physiol. Rev. **89**, 957–989 (2009)

Wagner, H.P., Humphrey, J.D.: Differential passive and active biaxial mechanical behaviors of muscular and elastic arteries: basilar versus common carotid. J. Biomech. Eng. **133**, 051009 (2011)

Wang, L., Roper, S.M., Luo, X.Y., Hill, N.A.: Modelling of tear propagation and arrest in fibre-reinforced soft tissue subject to internal pressure. J. Eng. Math. **95**, 249–265 (2015)
Wang, L., Roper, S.M., Hill, N.A., Luo, X.: Propagation of dissection in a residually-stressed artery model. Biomech. Model. Mechanobiol. **16**, 139–149 (2017)
Wang, L., Hill, N.A., Roper, S.M., Luo, X.: Modelling peeling- and pressure-driven propagation of arterial dissection. J. Eng. Math. **109**, 227–238 (2018)
Wang, N., Tytell, J.D., Ingber, D.E.: Mechanotransduction at a distance: mechanically coupling the extracellular matrix with the nucleus. Nat. Rev. Mol. Cell Biol. **10**, 75–82 (2009)
Wang, R., Yu, X., Gkousioudi, A., Zhang, Y.: Effect of glycation on interlamellar bonding of arterial elastin. Exp. Mech. **61**, 81–94 (2021a)
Wang, R., Yu, X., Zhang, Y.: Mechanical and structural contributions of elastin and collagen fibers to interlamellar bonding in the arterial wall. Biomech. Model. Mechanobiol. **20**, 93–106 (2021b)
Wang, Y., Hahn, J., Zhang, Y.: Mechanical properties of arterial elastin with water loss. J. Biomech. Eng. **140**, 041012 (2018a)
Wang, Y., Li, H., Zhang, Y.: Understanding the viscoelastic behavior of arterial elastin in glucose via relaxation time distribution spectrum. J. Mech. Behav. Biomed. Mater. **77**, 634–641 (2018b)
Weisbecker, H., Pierce, D.M., Regitnig, P., Holzapfel, G.A.: Layer-specific damage experiments and modeling of human thoracic and abdominal aortas with non-atherosclerotic intimal thickening. J. Mech. Behav. Biomed. Mater. **12**, 93–106 (2012)
Weisbecker, H., Viertler, C., Pierce, D.M., Holzapfel, G.A.: The role of elastin and collagen in the softening behavior of the human thoracic aortic media. J. Biomech. **46**, 1859–1865 (2013)
Weisbecker, H., Unterberger, M.J., Holzapfel, G.A.: Constitutive modelling of arteries considering fibre recruitment and three-dimensional fibre distribution. J. R. Soc. Interface **12**, 20150111 (2015)
Wexler, A.S., Ding, J., Binder-Macleod, S.A.: A mathematical model that predicts skeletal muscle force. IEEE Trans. Biomed. Eng. **44**, 337–348 (1997)
Wight, T.N.: A role for proteoglycans in vascular disease. Matrix Biol. **71–72**, 396–420 (2018)
Williams, C.D., Regnier, M., Daniel, T.L.: Axial and radial forces of cross-bridges depend on lattice spacing. PLoS Comput. Biol. **6**, e1001018 (2010)
Wittgenstein, A.: Ultrastructural investigation of biaxially loaded human aortic tissue. Master's thesis, Graz University of Technology, Biomedical Engineering (2018)
Witzenburg, C.M., Dhume, R.Y., Shah, S.B., Korenczuk, C.E., Wagner, H.P., Alford, P.W., Barocas, V.H.: Failure of the porcine ascending aorta: multidirectional experiments and a unifying microstructural model. J. Biomech. Eng. **139**, 031005 (2017)
Wolinsky, H., Glagov, S.: Comparison of abdominal and thoracic aortic medial structure in mammals. Deviation of man from the usual pattern. Circ. Res. **25**, 677–686 (1969)
Yang, J., Clark, J.W., Jr., Bryan, R.M., Robertson, C.: The myogenic response in isolated rat cerebrovascular arteries: smooth muscle cell model. Med. Eng. Phys. **25**, 691–709 (2003a)
Yang, J., Clark, J.W., Jr., Bryan, R.M., Robertson, C.S.: The myogenic response in isolated rat cerebrovascular arteries: vessel model. Med. Eng. Phys. **25**, 711–717 (2003b)
Ye, G.J.C., Nesmith, A.P., Parker, K.K.: The role of mechanotransduction on vascular smooth muscle myocytes cytoskeleton and contractile function. Anat. Rec. **297**, 1758–1769 (2014)
Yosibash, Z., Priel, E.: Artery active mechanical response: high order finite element implementation and investigation. Comput. Methods Appl. Mech. Eng. **237**, 51–66 (2012)
Yu, X., Suki, B., Zhang, Y.: Avalanches and power law behavior in aortic dissection propagation. Sci. Adv. **6**, eaaz1173 (2020)
Zeinali-Davarani, S., Wang, Y., Chow, M.J., Turcotte, R., Zhang, Y.: Contribution of collagen fiber undulation to regional biomechanical properties along porcine thoracic aorta. J. Biomech. Eng. **137**, 051001 (2015)
Zhu, Y., Kang, G., Kan, Q., Yu, C.: A finite viscoelastic-plastic model for describing the uniaxial ratchetting of soft biological tissues. J. Biomech. **47**, 996–1003 (2014)

Zitnay, J.L., Li, Y., Qin, Z., San, B.H., Depalle, B., Reese, S.P., Buehler, M.J., Yu, S.M., Weiss, J.A.: Molecular level detection and localization of mechanical damage in collagen enabled by collagen hybridizing peptides. Nat. Commun. **8**, 14913 (2017)

Zitnay, J.L., Jung, G.S., Lin, A., Qin, Z., Li, Y., Yu, S.M., Buehler, M.J., Weiss, J.A.: Accumulation of collagen molecular unfolding is the mechanism of cyclic fatigue damage and failure in collagenous tissues. Sci. Adv. **6**, eaba2795 (2020)

Zou, Y., Zhang, Y.: The orthotropic viscoelastic behavior of aortic elastin. Biomech. Model. Mechanobiol. **10**, 613–625 (2011)

Zulliger, M.A., Rachev, A., Stergiopulos, N.: A constitutive formulation of arterial mechanics including vascular smooth muscle tone. Am. J. Physiol. Heart Circ. Physiol. **287**, H1335–H1343 (2004)

Zuo, K., Pham, T., Li, K., Martin, C., He, Z., Sun, W.: Characterization of biomechanical properties of aged human and ovine mitral valve chordae tendineae. J. Mech. Behav. Biomed. Mater. **62**, 607–618 (2016)

Computational Modeling of Flow and Thrombus Formation in Type B Aortic Dissection: The Influence of False Lumen Perfused Side Branches

Chlöe H. Armour, Claudia Menichini, Lydia Hanna, Richard G. J. Gibbs, and Xiao Y. Xu

I first met Gerhard at the World Congress of Biomechanics in Sapporo in 1998; I remembered chatting to him on the beautiful Hokkaido University campus, and we have met on numerous occasions since. Although my research has been mainly focused on blood flow and mass transport in the cardiovascular system, I followed Gerhard's work closely and have the greatest admiration for his seminal work on the modeling of arterial biomechanics. I was absolutely delighted when he invited me to participate in the LEAD project workshop on 'Mechanics, Modeling and Simulation of Aortic Dissection', in Riegersburg, October 2019, where I learnt about this ambitious research program at TU Graz, and had the honor to join the International Advisory Board of the LEAD program. For this reason, I chose to contribute a chapter on aortic dissection. I wish Gerhard every success in this exciting program and I look forward to a fruitful collaboration in the future.

Xiao Yun

Abstract Type B aortic dissection (TBAD) is a complex disease in which a secondary channel of blood flow, known as the false lumen (FL), forms due to a tear in the aortic wall. Common treatment methods for TBAD are medical management or thoracic endovascular aortic repair (TEVAR), but their long-term outcomes are difficult to predict. Anatomical studies have assessed the influence of morphological features on disease progression and treatment outcome, and the number of side branches perfused by the FL has been identified as an important factor. By simulating

C. H. Armour · C. Menichini · X. Y. Xu (✉)
Department of Chemical Engineering, Imperial College London, South Kensington Campus, London SW7 2AZ, UK
e-mail: yun.xu@imperial.ac.uk

L. Hanna · R. G. J. Gibbs
Regional Vascular Unit, St Mary's Hospital, Imperial College Healthcare NHS Trust, London, UK

G. Sommer et al. (eds.), *Solid (Bio)mechanics: Challenges of the Next Decade*, Studies in Mechanobiology, Tissue Engineering and Biomaterials 24,
https://doi.org/10.1007/978-3-030-92339-6_2

hemodynamics and thrombus formation in a post-TEVAR model, this study showed that the presence of an FL fed inferior mesenteric artery increased wall shear stress by 25.1% in the near branch area, and reduced abdominal thrombus formation by 15.3%. Additionally, hemodynamic simulations of a pre-TEVAR model showed that accounting for 11 intercostal arteries perfused by the FL led to an increase of 13.4% in peak systolic velocity through the primary entry tear. Our results demonstrate the importance of including FL perfused side branches in quantitative analysis of hemodynamics in TBAD, and provide a mechanistic understanding of the role of these branches in FL thrombosis.

1 Introduction

Aortic dissection is the most common acute disease of the aorta (McMahon and Squirrell 2010), affecting approximately 3–5 per 100 000 individuals, and up to 35 per 100 000 of higher-risk individuals (Nienaber et al. 2016). A dissection occurs when an initial tear is formed in the intimal layer of the aortic wall, causing blood to enter the aortic wall. The pressure and flow of the blood into the wall forces the tear to propagate further into the medial layer and ultimately results in separation of the wall layers, leading to a secondary channel of blood flow, the false lumen (FL). Additional re-entry tears may form in the distal aorta providing a path for blood to return to the true lumen (TL). According to the Stanford classification system, a type B aortic dissection (TBAD) is one in which the primary tear is in the descending aorta, as opposed to a type A dissection in which the primary tear is in the ascending aorta or aortic arch (Salameh and Ratchford 2016).

A TBAD is treated either medically or endovascularly. Medical management involves the administration of antihypertensive drugs which aims to control the aortic blood pressure and reduce the stress on the dissected wall. Thoracic endovascular aortic repair (TEVAR) involves the insertion of a stent graft into the TL, to cover the primary entry tear thereby cutting off blood flow to the FL, and to expand and provide needed support to the often-compressed TL. Ultimately, any treatment of TBAD aims to achieve complete false lumen thrombosis (FLT), meaning the blood within the FL has clotted and there is no longer a diversion of blood away from the TL. Studies have shown complete FLT to be associated with the greatest positive prognosis (Trimarchi et al. 2013). Interestingly, a stable yet fully patent FL has shown similar prognosis trends to complete thrombosis (Tsai et al. 2007). However, partial thrombosis of the false lumen has been shown to lead to a greater risk of late onset complications and higher mortality (Qin et al. 2012; Girish et al. 2016).

Currently, treatment choice is driven by the state of the dissection upon diagnosis. A dissection is classified as complicated if there are any additional problems at the time of diagnosis, including end-organ ischemia, rapid expansion of the aorta, consistent pain, and refractory hypertension. Such complications can ultimately lead to major organ failure due to a lack of TL blood flow or aortic rupture due to a significantly weakened vessel wall. The primary treatment method for uncomplicated

TBAD is currently medical management, and with this 90% of patients survive to hospital discharge (Clough and Nienaber 2015). However, this decreases to only 50% for complicated cases (Riambau et al. 2017). Therefore, TEVAR has been promoted as the preferred primary treatment for complicated cases requiring emergency intervention, over open-heart surgery due to improved mortality rates with TEVAR (Fattori et al. 2008; Zeeshan et al. 2010; Steuer et al. 2011).

Recent studies including the INSTEAD (Nienaber et al. 2014) and the ADSORB trials (Brunkwall et al. 2014) have also shown the efficacy of TEVAR in promoting FLT and positive aortic remodeling in uncomplicated TBAD cases. Despite this, there remains uncertainties regarding which patients should receive TEVAR and within what time scale. Furthermore, although it is a minimally invasive procedure, performing TEVAR carries additional risks—3–10% of TEVAR patients experience a stroke (Buth et al. 2007); stent coverage of the left subclavian artery or intercostal arteries may result in spinal cord ischemia, paraparesis and paraplegia (Rizvi et al. 2009; Zipfel et al. 2013); the development of stent-induced new entry tears or retrograde type A dissections may require further re-intervention. Therefore, it may not be beneficial to consider TEVAR as the primary treatment for all uncomplicated cases.

For these reasons it is desirable to gain a deeper understand into the progression of TBAD, and to devise a method for quantitative assessment of individual patient's risk for developing FLT or experiencing FL expansion. This would allow for treatments to be tailored to specific patients, giving the opportunity for early intervention with TEVAR to prevent late onset complications, while avoiding increased risk for patients that would not necessarily need TEVAR. Several anatomical studies have aimed to do just this, and have identified morphological features for both medical and TEVAR patients such as tear number, size and location (Khoynezhad et al. 2010; Evangelista et al. 2012; Qin et al. 2012; Tolenaar et al. 2013a, b; Kitamura et al. 2015; Kotelis et al. 2016; Chen et al. 2017), and aorta diameter and area (Marui et al. 2007; Qin et al. 2012; Kim et al. 2014; Ray et al. 2017; Sailer et al. 2017) to correlate with aortic remodeling (either FLT, or FL reduction or expansion). A parameter identified in multiple studies is the number of false lumen perfused side branches, which has been linked to aortic expansion and complications in medical patients (Kamman et al. 2017; Sailer et al. 2017) and reduced FLT in TEVAR patients (Qin et al. 2012; Tolenaar et al. 2014; Ge et al. 2017; Liu 2018).

The use of computational fluid dynamics (CFD) simulations allows for the hemodynamic condition of TBAD to be assessed and can help to understand why morphological factors, such as FL perfused branches, impact disease progression. As thrombus formation is noted to be influenced by morphological factors the computational modeling and prediction of thrombosis is also desirable and has been a focus of several studies.

Thrombus formation, encompassing the recruitment, activation and adhesion of platelets to form blood clots, is a complex multi-stage biological process, which involves many species and biochemical reactions. Thus, to model such a process presents significant challenges, and various, mainly kinetic based, studies have addressed specific stages of the process with a range of assumptions. Simple single species models have focused on certain stages of thrombosis—Wootton et al. (2001) modeled the reaction and adhesion of platelets in an idealized stenosed vessel, while Anand et al. (2005) presented a comprehensive model of the extrinsic pathway of the coagulant cascade, in which 29 species were modeled. Others have selected a limited number of species to model, included the combination of platelets, platelet agonists, prothrombin, thrombin, and anti-thrombin, modeled by Sorensen et al. (1999) in a 2D microdomain, and then by Goodman et al. (2005) in a series of idealized contracting and expanding vessels. Goodman et al. (2005) also incorporated the influence of thrombus formation on flow, by adjusting velocity patterns at each time step. Leiderman and Fogelson (2011) then presented a kinetic model of platelets in multiple states, while also incorporating the influence of thrombus formation on the flow field through modification of the momentum equation, in an idealized 2D geometry. While these studies provide valuable insight into thrombus formation, their focus on the micro/cellular level makes it challenging to apply to an anatomically realistic model where macro-scale hemodynamic behavior is also important.

To simulate thrombus formation on a realistic scale in combination with the flow field, the hemodynamics-based model originally developed by Menichini and Xu (2016) and extended to anatomically realistic geometries (Menichini et al. 2016, 2018; Armour et al. 2020) can be utilized. The most recent version of the model is presented in this chapter and is employed to study the influence of FL perfused side branches on hemodynamics and thrombus formation in patient-specific settings.

2 Methodology

2.1 Mathematical Model for Thrombus Formation and Growth

Thrombosis is modeled through a hemodynamics-based mechanism. Due to computational cost of modeling hemodynamics and biological processes at a patient-specific level, several assumptions are made to simplify the thrombosis process. Primarily, only a limited number of species are modeled: resting and activated platelets, coagulant (representative of all species involved in the coagulation process) and bound platelets (representative of thrombus). Both resting platelets (RP) and activated platelets (AP) are modeled through a standard convection-diffusion-reaction transport equation, i.e.

$$\frac{\partial c_i}{\partial t} + v \cdot \nabla c_i = D_{\mathrm{p}} \nabla^2 c_i + S_i, \qquad i = \mathrm{AP}, \mathrm{RP}, \tag{1}$$

where c_i is the concentration of species i, D_{p} is the diffusivity of the platelets ($1.6 \times 10^{-13}\,\mathrm{m^2 s^{-1}}$), and S_i is a source term. The concentrations [AP] and [RP] are normalized against their initial values, and thus represent a probability of encountering that relevant species, rather than an exact value of concentration. The source S_i is the sum of two reactions, which control the conversion of platelets from the resting state to the activated state. The first reaction r_1 represents the activation of RPs due to their exposure to other APs. The second reaction r_2 represents the activation of RPs due to exposure to thrombin (a substance produced by damaged endothelium cells to initiate the activation of platelets). Thus,

$$r_1 = k_1[\mathrm{AP}][\mathrm{RP}], \qquad r_2 = k_2[\mathrm{RP}]\mathrm{RRT}, \tag{2}$$

where $k_1 = 0.15\,\mathrm{s}^{-1}$ and $k_2 = 0.5\,\mathrm{s}^{-1}$. As thrombin is not modeled, r_2 is scaled by the relative residence time (RRT), as it is assumed that thrombin concentrations would be high in areas of flow recirculation and therefore high residence time. In the model, residence time (RT) is modeled as an unreactive tracer. Thus,

$$\frac{\partial \mathrm{RT}}{\partial t} + v \cdot \nabla \mathrm{RT} = D_{\mathrm{RT}} \nabla^2 \mathrm{RT} + 1, \tag{3}$$

where D_{RT} is the self-diffusivity of blood ($1.14 \times 10^{-11}\,\mathrm{m^2 s^{-1}}$). Due to the unit increase of RT with each unit increase of time, RT will continuously rise throughout the simulation. Therefore, RRT is calculated by normalizing RT with respect to the cardiac cycle.

While these two reactions represent common paths of platelet activation, it is also known that high shear rates induce platelet activation—high shear is often found near and within the tears. To account for any activation that would have occurred when the dissection was diagnosed and during the time before the model begins, a background activation level of 5% of the inlet RP concentration is assumed.

The coagulant (C) is modeled through a shear-dependent diffusive mechanism, to dually incorporate the convective dominant nature of the transport of coagulant in the bulk and the diffusive dominant nature of the transport of coagulant in the near wall and low velocity areas at the site of thrombus formation, i.e.

$$\frac{\partial C}{\partial t} = D_{C_{\mathrm{eff}}} \nabla^2 C + k_C \varphi_{C\,\mathrm{control}}[\mathrm{AP}] - k_{C2}(1 - \varphi_{\dot{\gamma}})\varphi_C[\mathrm{AP}], \tag{4}$$

$$D_{C_{\mathrm{eff}}} = \varphi_{\dot{\gamma}} D_c, \tag{5}$$

$$\varphi_{C\,\mathrm{control}} = \varphi_{\mathrm{BP}}, \tag{6}$$

$$\varphi_{\dot{\gamma}} = \frac{\overline{\dot{\gamma}_t}^2}{\overline{\dot{\gamma}}^2 + \overline{\dot{\gamma}_t}^2}, \tag{7}$$

where $D_c = 10^{-8}\,\mathrm{m^2 s^{-1}}$. The use of $\varphi_{\dot{\gamma}}$ ensures that when shear rates are high the coagulant is consumed $-\varphi_{\dot{\gamma}}$ tends to 0 and the coagulant is carried into the blood stream, unable to deposit on a surface. In this study, a shear rate threshold $\dot{\gamma}_t$ of $50\,\mathrm{s}^{-1}$ was set. When shear rates are low $\varphi_{\dot{\gamma}}$ tends to 1, leaving the transport of coagulant to be diffusion dominated. $\varphi_{\dot{\gamma}}$ is also used to control the production of bound platelets (BP), modeled as a growth only static variable, to ensure that thrombus cannot form unrealistically in high shear regions:

$$\frac{\partial \mathrm{BP}}{\partial t} = k_{\mathrm{BP}} \varphi_{\mathrm{BP\,control}} \varphi_{\dot{\gamma}} [\mathrm{AP}], \qquad \varphi_{\mathrm{BP\,control}} = \varphi_C \varphi_{\mathrm{RRT}} \varphi_{\mathrm{AP}}. \tag{8}$$

Based on previous sensitivity tests (Menichini and Xu 2016) which showed the extent of thrombosis to be independent of kinetic constants k_{BP} (12 nmol L^{-1} s^{-1}), k_C (16 nmol L^{-1} s^{-1}) and k_{C2} (6 nmol L^{-1} s^{-1}), the kinetic constants have been set to artificially accelerate the process of thrombosis, so that simulations can be completed in a reasonable time frame. The production of coagulant initially is controlled through a flux boundary condition that is driven by the local time-average wall shear stress (TAWSS) and BP concentration. Thus,

$$D_{C_{\mathrm{eff}}} \left(\frac{\partial C}{\partial n} \right)_{\mathrm{wall}} = \begin{cases} k_{C_{\mathrm{wall}}} & \text{if TAWSS} < 0.15\,\mathrm{Pa} \text{ and BP} \leq 200\,\mathrm{nM}, \\ 0 & \text{otherwise,} \end{cases} \tag{9}$$

where $k_{C_{\mathrm{wall}}} = 16\,\mathrm{nmol\,m^{-1}\,L^{-1}\,s^{-1}}$. After coagulation initially forms on the walls in low shear regions, further coagulant production and thrombus formation is controlled by a feedback loop, which drives thrombus formation in areas of low shear, and high RRT, AP and C ($\varphi_{\mathrm{BP\,control}}$). This then feeds back to further coagulant production which occurs in areas of high BP ($\varphi_{C\,\mathrm{control}}$). Mathematically, this is implemented through switching functions, seen in Eqs. (4) and $(8)_1$, which vary between 0 and 1, depending on a variable's concentration relative to threshold values ($\mathrm{RRT}_t = 0.85$, $C_t = 10\,\mathrm{nmol/L}$, $\mathrm{BP}_t = 20\,\mathrm{nmol/L}$, $\mathrm{AP}_t = 15$). Thus,

$$\varphi_i = \frac{X_i^2}{X_i^2 + X_{i_t}^2}. \tag{10}$$

Finally, in order to account for the influence of thrombus formation on the flow field the momentum equation is modified to include the porosity ε of the domain (ranging between 0.75 and 1, depending on the BP concentration) and a momentum source proportional to the BP concentration to represent the resistance to flow produced by the presence of thrombus, i.e.

$$\rho \frac{\partial \varepsilon \mathbf{u}}{\partial t} + \nabla \cdot (\rho \varepsilon \mathbf{u} \times \mathbf{u}) = -\varepsilon \nabla p + \nabla \cdot [\mu \varepsilon (\nabla^2 \mathbf{u})] - \varepsilon S_M, \tag{11}$$

$$\varepsilon = \max \left(1 - \frac{\mathrm{BP}^2}{\mathrm{BP}^2 + \mathrm{BP}_t^2}, 0.75 \right), \tag{12}$$

$$S_M = k_M \frac{\mathrm{BP}^2}{\mathrm{BP}^2 + \mathrm{BP}_t^2} \mathbf{u}, \tag{13}$$

where $k_M = 10^7\,\mathrm{kg\,m^3\,s^{-1}}$. The blood viscosity μ was modeled as non-Newtonian using the Quemada model, with constants taken from the literature, i.e.

$$\mu = \mu_\mathrm{p}\left(1 - \frac{1}{2}\frac{k_0 - k_\infty\sqrt{\dot{\gamma}/\gamma_\mathrm{c}}}{1 + \sqrt{\dot{\gamma}/\gamma_\mathrm{c}}}H_\mathrm{t}\right)^{-2}, \qquad (14)$$

where μ_p is the plasma viscosity (1.2 mPa s), k_0 and k_∞ are intrinsic viscosities at zero and infinite shear, respectively ($k_0 = 4.33$; $k_\infty = 2.07$), γ_c is the critical shear rate ($1.88\,\mathrm{s}^{-1}$), and H_t is the hematocrit (0.45) (Neofytou 2004).

2.2 *Model Geometry*

Images of two patients from the ADSORB trial (Brunkwall et al. 2014), one post-TEVAR (P1) and one pre-TEVAR (P2), were used in this study. Geometries were segmented and reconstructed to 3D models from computed-tomography (CT) scans in Mimics (Materialise, Leuven, Belgium), using a range of semi-automatic and manual tools. First, both patient-specific models were built from the ascending aorta distal to the coronary arteries down to the aortic bifurcation, including the 9 main vessels that protrude from the aorta: brachiocephalic artery (BRAC), left common carotid artery (LCCA), left subclavian artery (not included in the P1 TEVAR model due to occlusion by the stent graft), celiac trunk (CEL), superior mesenteric artery, left and right renal arteries, and left and right iliac arteries. Second, an additional model for each patient was created which included additional false lumen fed branches. The inferior mesenteric artery (IMA) in the abdominal aorta was identified to branch from the FL in P1 (creating model P1+IMA), and 11 FL fed intercostal arteries (ICA) were identified in P2 (creating model P2+ICA). All four models are shown in Fig. 1. As both the IMA and ICA are small branches, segmenting them straight from the CT images can be challenging due to the relatively low pixel quality. Therefore, after accurately locating the branches from the CT scans their radius was manually set to match literature data—about 1.6 mm for the IMA; about 2 mm for each ICA (Du et al. 2015).

Each model was meshed in ICEM (Ansys Inc, Canonsburg, PA, USA) into unstructured meshes that consisted of a hexahedral core with 10 prismatic layers at the wall, with local refinement in areas such as tears and small branches. Mesh sensitivity tests were conducted to ensure mesh-independent solutions, and the final chosen meshes contained the following number of elements: P1—5.47 million; P1+IMA—5.49 million; P2—5 million; P2+ICA—6.2 million.

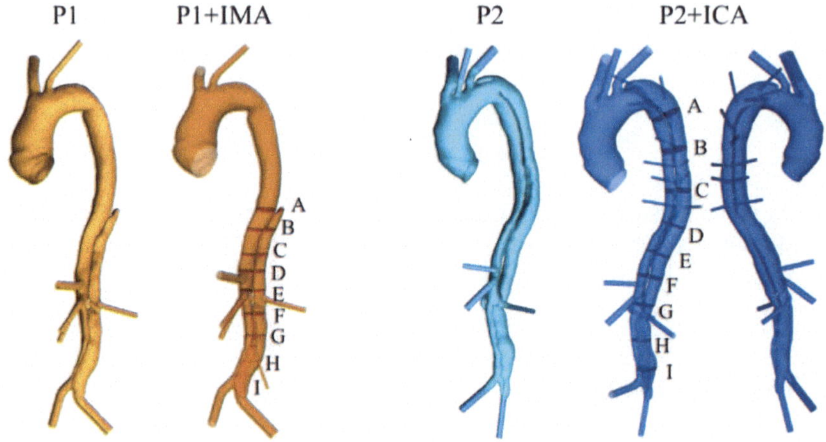

Fig. 1 Geometric models used for simulations: (left) P1 geometry without and with the false lumen (FL) fed inferior mesenteric artery (IMA); (right) P2 geometry without and with 11 FL fed intercostal arteries (ICA)

2.3 Boundary Conditions and Computational Details

As no patient-specific flow information was available for either patient, inlet and outlet boundary conditions were developed based on data from the literature. In all simulations, a flat inlet velocity profile, of period 1.3 s, was applied in the ascending aorta, distal to the coronary arteries (Dillon-Murphy et al. 2015). Three-element Windkessel models (3EWK), identified as the most physiologically accurate outlet boundary conditions (Pirola et al. 2017), were tuned for each outlet, expect the intercostal arteries. Total resistance R_T and compliance for each branch were taken from a previous study (Dillon-Murphy et al. 2015), and the individual proximal R_1 and distal R_2 resistances were calculated through the methodology reported by Pirola et al. (2019). For the intercostal arteries in model P2+ICA, a fixed flowrate boundary condition was applied, with 5% of the inlet flow was directed to the 11 intercostals in total (Lewis and McKenna 2010).

The walls were assumed to be rigid with a no-slip boundary condition, with zero flux for all species in the thrombus model except coagulant which is controlled by a variable flux condition described above, Eq. (9). Flow was assumed to be laminar in all simulations presented in this study. The thrombosis model described above was implemented in Ansys CFX (Ansys Inc, Canonsburg, PA, USA), as reported previously (Menichini and Xu 2016). Initial simulations were run for flow only without using the thrombosis model; this required a minimum of three cardiac cycles in order to reach a periodic solution. The model for thrombus formation and growth was then switched on, and the simulations were run for 16 cycles for P1 and P1+IMA. A uniform time step of 0.005 s was set in all simulations. Post-processing of numerical results was performed using EnSight (Ansys Inc, Canonsburg, PA, USA).

3 Results

3.1 *Flow Patterns*

Analysis of flow patterns, wall shear stress and pressure was carried out using the simulation results obtained before the thrombosis model was switched on. Figure 2 shows the velocity streamlines at mid-systolic acceleration, peak systole, and mid-systolic deceleration for all models. Comparing P1 and P1+IMA (Fig. 2), there was little difference in the thoracic aorta at any time point. However, in the abdominal aorta, the inclusion of the IMA altered the local flow in the FL, particularly at peak systole. About 2.2% of the total inlet flow reported to the IMA, which caused a small increase (3.9%) in the percentage of abdominal flow (below the renal arteries) recorded in the FL. As the same inlet flowrate was adopted in all simulations, the IMA flow came from redistribution of flow through other branches. This resulted in small changes in flow through the BRAC (0.8%), LCCA (0.5%) and CEL (0.5%).

Figure 3 shows the velocity streamlines for P2 and P2+ICA at the selected systolic time points. Comparing P2 and P2+ICA, there was hardly any difference globally, although quantitative variations were noted. Including the ICA caused a 13.4% increase in peak systole maximum velocity in the jet through the primary entry tear distal to the LSA (about 0.73 m/s in P2; about 0.82 m/s in P2+ICA), and a smaller yet noticeable decrease in velocity in the TL approximately two-thirds down the thoracic aorta. These local changes are due to the inclusion of the ICA drawing more blood into the FL—the percentage of descending aorta flow entering the FL increased from 30 to 32% when the ICA were included.

3.2 *Wall Shear Stress*

Figure 4 shows the time-averaged wall shear stress (TAWSS) maps for the four models. As with the velocity results, for both patients the inclusion of the additional FL branches did not induce any qualitative changes in TAWSS patterns. For P1/P1+IMA, the most notable difference can be observed in the abdominal aorta where including the IMA substantially increased the local TAWSS in the abdominal FL. As a result, the volume-averaged TAWSS in the abdominal FL increased by 25.1%, from about 0.68 to about 0.85 Pa. Considering P2 and P2+ICA, the presence of the ICA did induce local increases of TAWSS around the branch ostia, but the volume-averaged TAWSS in the segment of thoracic FL containing the ICA only increased by 7.8%, from about 0.5 to about 0.53 Pa.

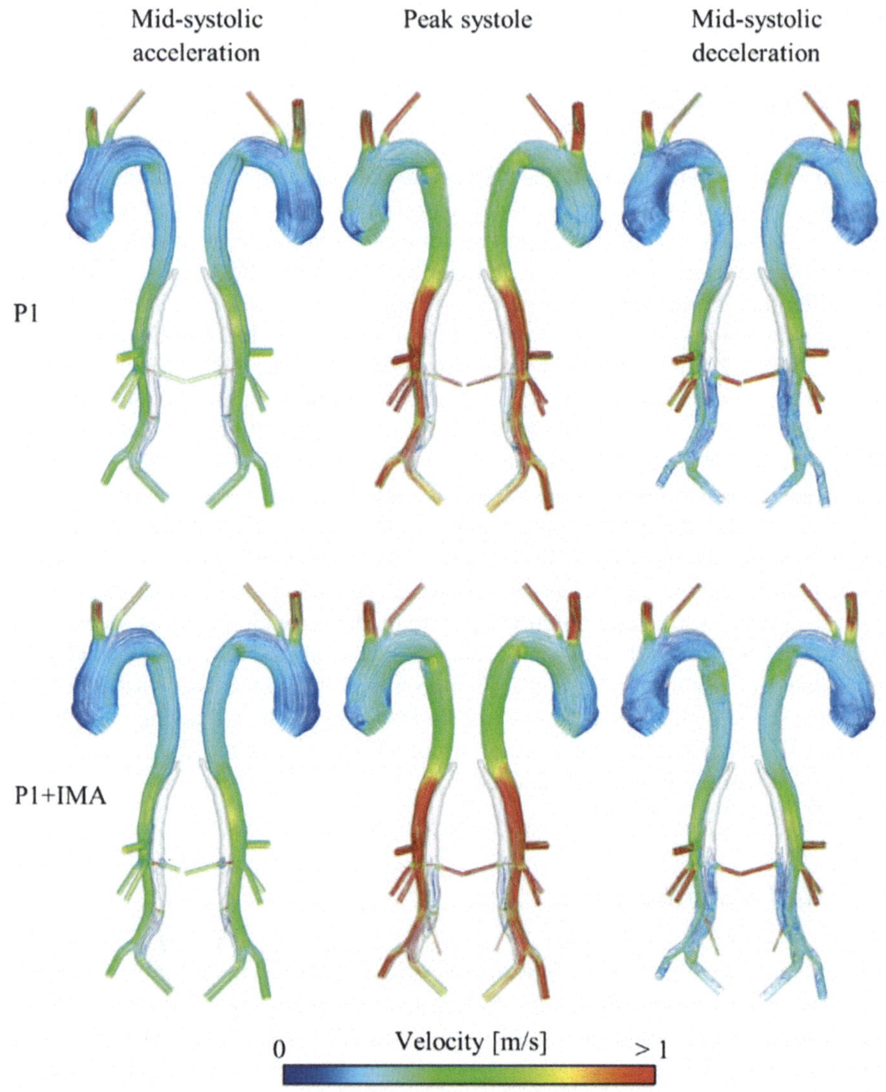

Fig. 2 Velocity streamlines throughout the systolic phase for P1 and P1+ICA

3.3 *Pressure*

The pressure difference between the TL and FL was evaluated on planes in each model (A-I, shown in Fig. 1) and the peak systolic results are presented in Fig. 5. For P1 where the TL pressure was higher than the FL throughout the aorta, the inclusion of the IMA resulted in a reduction in FL pressure, hence an increase in

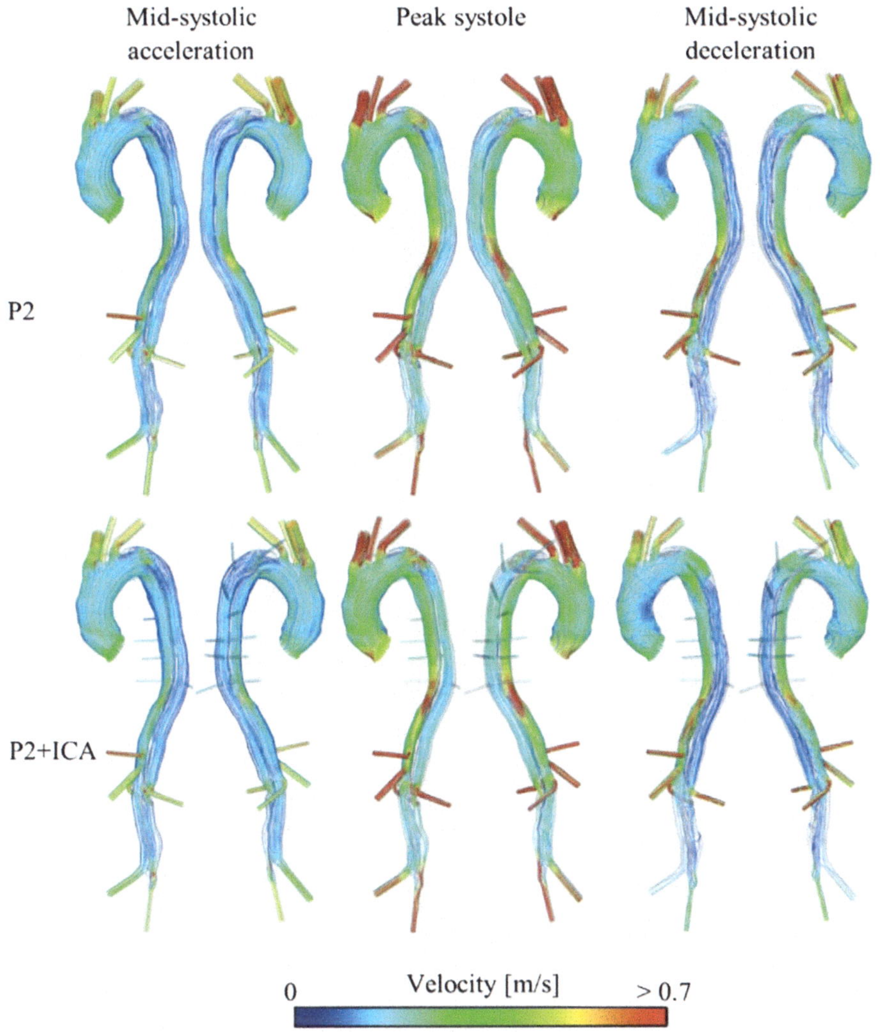

Fig. 3 Velocity streamlines throughout the systolic phase for P2 and P2+ICA

TL-FL pressure difference. For P2 where the FL pressure was higher than the TL except in the upper thoracic aorta (planes A and B), the inclusion of the ICA decreased the cross-lumen pressure difference. Generally, for both patients, the change in pressure difference was small, with a maximum shift in pressure difference observed across all simulations being just over 1 mmHg on plane A in P1/P1+IMA. Furthermore, cross-lumen difference in mean pressure over the entire cardiac cycle varied by <0.13 mmHg for all models at all locations.

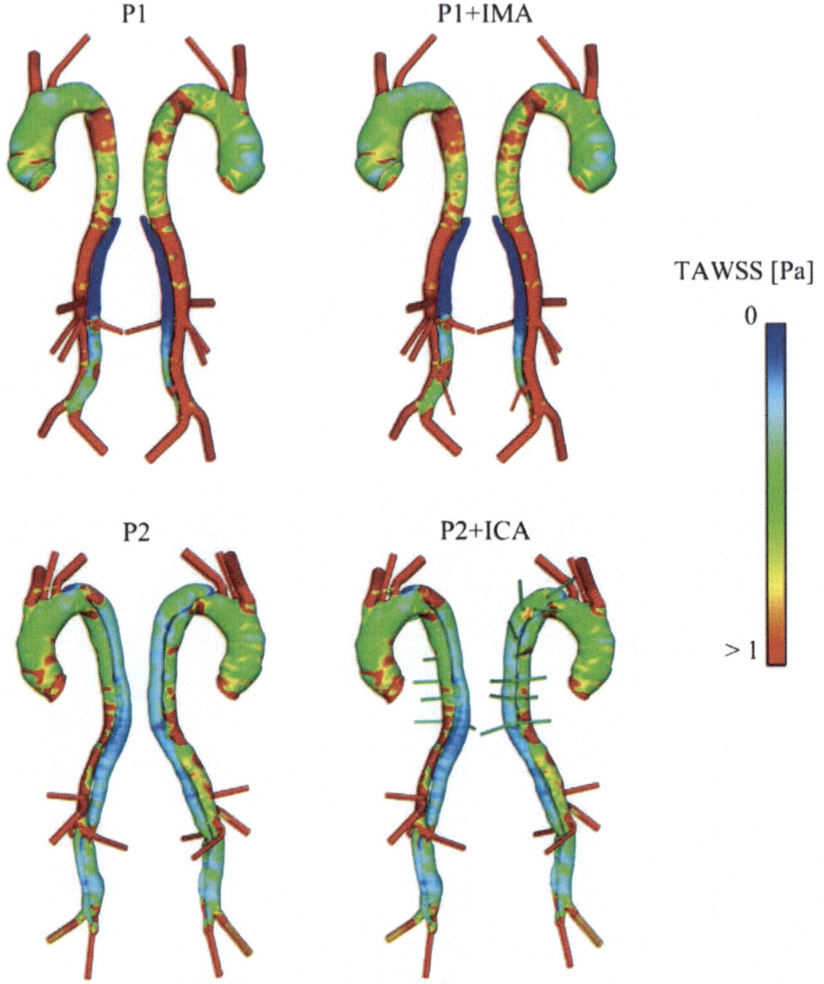

Fig. 4 Time-averaged wall shear stress (TAWSS) for models P1, P1+IMA, P2 and P2+ICA

3.4 Thrombus Formation

Predicted thrombus formation, alongside actual observed thrombus formation from the 1-year follow-up CT scan can be seen in Fig. 6 for P1 and P1+IMA. P1 correctly predicted complete thrombosis in the thoracic FL above the CEL. However, there was overprediction of thrombus growth in the abdominal aorta, as seen in the follow-up scan there was virtually no thrombus formation in the abdominal FL. With the inclusion of the IMA, P1+IMA saw a reduction of this overprediction in the abdominal FL, with 15.3% less abdominal thrombus formation compared to P1, while still correctly predicting the thoracic FLT.

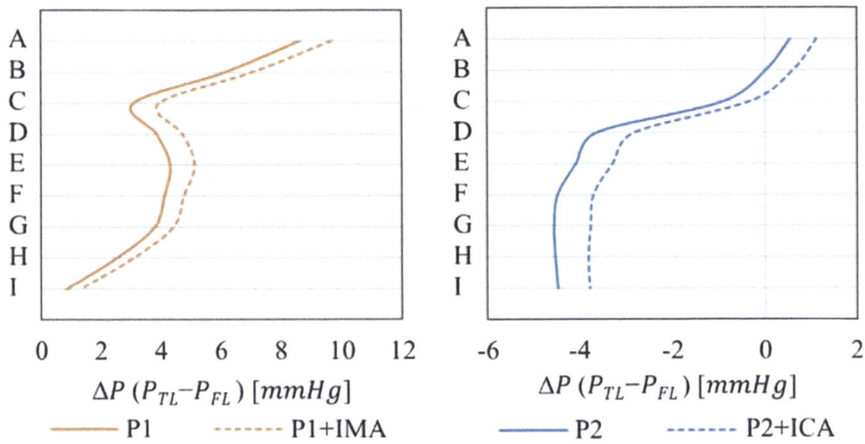

Fig. 5 Pressure difference ΔP between the true (TL) and false lumen (FL) on planes A-I (shown in Fig. 1) throughout the aorta for each model

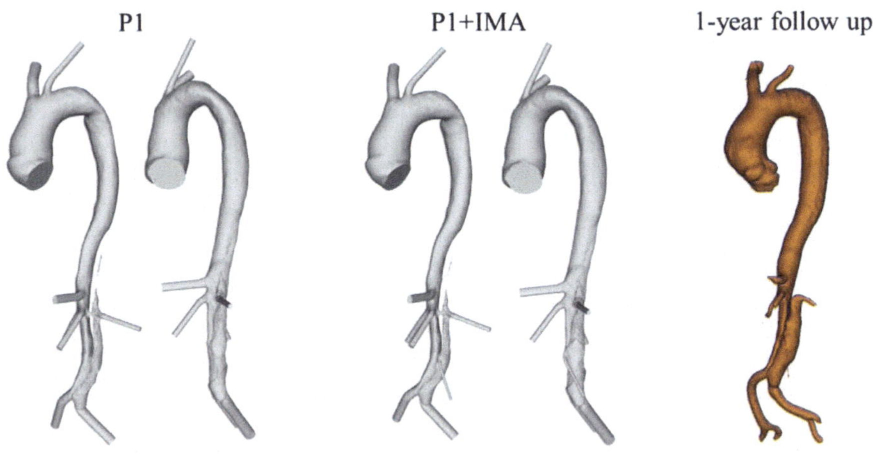

Fig. 6 Predicted thrombus formation after 16 cycles for P1 and P1+IMA alongside actual observed thrombus formation from 1-year follow up CT scan

4 Discussion

Type B aortic dissection is a serious cardiovascular disease, which can present in a wide variety of morphologies, can often be complex, and can have high mortality rates. While treatment is typically assigned based on diagnostic status (uncomplicated patients treated medically, and complicated patients treated with TEVAR), with accurate predictive tools, there is a potential for early intervention in uncomplicated patients that go on to develop complications.

The presence of FL perfused side branches has been identified as a predictor of aortic growth in medically managed patients (Kamman et al. 2017) and incomplete thrombosis in TEVAR patients (Qin et al. 2012; Tolenaar et al. 2014; Ge et al. 2017; Liu et al. 2018). Sailer et al. (2017) specifically identified FL perfused intercostal arteries as a risk factor for late-onset complications, such as aneurysms, rapid aortic growth, ischemia, and aortic rupture. This study aimed to gain a mechanistic understanding of the role of such branches in dissection progression, through computational flow simulations of two TBAD cases. A post-TEVAR case was modeled without (P1) and with (P1+IMA) the FL perfused IMA, and a pre-TEVAR case was modeled without (P2) and with (P2+ICA) 11 FL perfused ICA. Thrombus formation was also simulated in the post-TEVAR models, to address the connection between FL branches and post-TEVAR thrombosis.

The presence of FL perfused branches was shown to locally alter flow patterns, with increased flow and flow disturbance around the IMA in P1+IMA compared to P1, and increased flow and velocities throughout the thoracic region in P2+ICA compared to P2. The additional outlet(s) in these models provide an additional low pressure exit for the blood, thus pulling more flow to the local area.

This increase in flow and velocities translated to an increase in TAWSS in the local branch areas. The inclusion of the IMA in P1+IMA resulted in a nearly 10% increase in average TAWSS in the abdominal aorta, and the area of high TAWSS in this region can be seen to have expanded (Fig. 4). Including the ICA in P2+ICA caused a relatively small increase in TAWSS at about 2%. This is likely due to the fact that while the intercostals took 5% of inlet flow compared to the IMA's 2.2%, the 5% was split between 11 branches, meaning each branch received <0.5% of inlet flow, resulting in the increase in TAWSS being confined to the branch entry points.

Due to the local increase in TAWSS, P1+IMA predicted less thrombus growth in the abdominal FL compared to P1 (Fig. 6). The result of P1+IMA matched more closely to the follow-up CT scan; the latter showed virtually no thrombus formation in the abdominal aorta. This suggests that FL perfused side branches do inhibit thrombus formation, as identified in anatomical studies (Qin et al. 2012; Tolenaar et al. 2014; Ge et al. 2017; Liu et al. 2018), due to the increased flow thereby reducing stasis and the likelihood for activated platelets to aggregate and adhere to the wall. However, there were still small areas of thrombus formation in P1+IMA that deviate from the follow-up scan. Armour et al. (2021) showed that the use of a non-patient-specific stroke volume (as in this study) could significantly alter TAWSS results, with a 25% reduction in stroke volume leading to a 35% reduction in TAWSS. While the applied stroke volume of 93 mL falls in the typical range of 94 ± 15 mL (Maceira et al. 2006), it was not patient-specific, which might account for the slight difference between the predicted thrombus formation in P1+IMA and the in vivo results.

Cross-lumen pressure difference is another parameter of interest, as it has been implicated in FL expansion. Armour et al. (2021) showed that the use of non-patient-specific flow waveform and stroke volume can affect TL and FL pressures and their difference. There is therefore uncertainty in the absolute values of these parameters. However, this should not affect our comparison between two models (P1 vs. P1+IMA or P2 vs. P2+ICA) which used a consistent inlet velocity profile.

Table 1 Percentage change in FL volume by region between the baseline and 1-year follow-up CT scan for P1 and P2. Thoracic: above celiac trunk; abdominal: below celiac trunk

	Thoracic FL (%)	Abdominal FL (%)
P1	−100	4.9
P2	17.6	−100

The inclusion of the additional FL branches in P1+IMA and P2+ICA had a similar effect on reducing FL pressure. This is logical when considering the fact that these additional branches provide low-pressure exit points for the flow that help to alleviate FL pressure. Regarding cross-lumen pressure difference, it was increased in P1+IMA versus P1, but decreased in P2+ICA versus P2. This is because in the case of P1, the TL pressure was higher than FL pressure, hence a reduction in FL pressure would widen the TL and FL pressure difference, whereas the opposite was true for P2. Nevertheless, the predicted cross-lumen pressure differences were small in both patients, and the effect of including IMA in P1 and ICA in P2 on pressure difference was insignificant.

In this study, false lumen expansion could not be modeled, but the 1-year follow-up CT scan of P1 revealed little change in abdominal FL volume from its baseline scan—Table 1 reports FL volume percentage changes for both P1 and P2. P2 did show FL expansion in the thoracic region at the 1-year follow-up CT scan, while complete thrombosis was achieved in the abdominal FL. The increased velocity in the jet through the entry tear due to the presence of the ICA led to an 11.6% increase in maximum TAWSS on the FL wall hit by the jet. This would likely cause further degradation of the already weakened FL wall and could contribute to FL expansion, potentially explaining the results of Kamman et al. (2017), which showed FL branches to increase the risk of FL growth in medically managed patients. The higher-pressure FL in P2 may also promote FL expansion, however a larger cohort study would be required to assess the relative importance of the impact of FL perfused branches on TAWSS and cross-lumen pressure difference.

Sailer et al. (2017) identified ICA to be a predictor of late-onset complications. This cannot be explained by the results presented in this study as late-onset complications cover a variety of potential outcomes, such as rapid aortic growth, aortic rupture, limb or organ ischemia, and new refractory hypertension or pain. However, our results do show that the inclusion of ICA can significantly alter velocity magnitudes in critical areas such as the entry tear, and the presence of ICA should not be simply ignored, as has been done commonly.

While the progression of TBAD in all cases is a multi-factor process, the results of this study help to elucidate the role of FL perfused side branches in flow distribution and thrombus formation in TBAD patients. It is hoped that detailed studies, including the work presented here, into the key morphological parameters known to be linked to various progressions of TBAD will aid in treatment planning to improve patient prognosis.

There are limitations to this study. The inlet and outlet boundary conditions were taken from the literature, when ideally these would be tuned with patient-specific data. As discussed, the non-patient-specific stroke volume likely had an impact on TAWSS and therefore thrombus results, and a more accurate prediction can be made with a patient-specific tuned boundary condition. Additionally, all models were assumed to have a rigid wall. This assumption means the aortic wall motion and more importantly the intimal flap motion were neglected. The intimal flap is known to be flexible in early dissection stages, with increasing stiffness as the dissection ages (Peterss et al. 2016). A recent study in an idealized geometry by Chong et al. (2020) showed that flap motion increased FL flow and diastolic wall shear stress values, in turn significantly reducing the total area of low TAWSS—a parameter shown to be crucial for thrombus formation predictions. Patient-specific studies (Alimohammadi et al. 2015; Qiao et al. 2019b; Bäumler et al. 2020) have also highlighted the effect of wall motion on regions of low TAWSS. Furthermore, in relation to P1 and P1+IMA (post-TEVAR models) the compliance of the stent-graft (the focus of recent studies by Qiao et al. 2019a, 2020) will be significantly different to the aortic wall. The incorporation of a fluid-structure interaction model would allow for the effect of wall, flap and stent mechanical behaviors on thrombus formation to be assessed.

5 Conclusion

This study demonstrates the impact of FL perfused branches on TBAD hemodynamics and thrombus formation. From a methodological standpoint, our results show that FL branches, even low flow non-visceral branches such as intercostals, should be included to accurately assess FL hemodynamics. Our results provide a mechanistic understanding of the previously reported connection between FL branches and lack of post-TEVAR FLT, with increased FL flow and elevated wall shear stress due to FL perfused branches, which inhibit platelet aggregation and deposition. Further studies on a larger cohort of both pre- and post-TEVAR patients would allow for a deeper understanding of the influence of FL branches in various aortic morphologies and may help clinicians tailor treatment to individual patient needs.

Acknowledgements This work was funded by Ph.D. scholarships awarded to CHA by EPSRC Doctoral Training Partnership Grant (EP/R512540/1) and to Claudia Menichini by the Department of Chemical Engineering at Imperial College London and Imperial College Healthcare Charity.

References

Alimohammadi, M., Sherwood, J.M., Karimpour, M., Agu, O., Balabani, S., Díaz-Zuccarini, V.: Aortic dissection simulation models for clinical support: fluid-structure interaction vs. rigid wall models. Biomed. Eng. Online **14**, 34 (2015)

Anand, M., Rajagopal, K., Rajagopal, K.R.: A model for the formation and lysis of blood clots. Pathophysiol. Haemost. Thromb. **34**, 109–120 (2005)

Armour, C.H., Guo, B., Pirola, S., Saitta, S., Liu, Y., Dong, Z., Xu, X.Y.: The influence of inlet velocity profile on predicted flow in type B aortic dissection. Biomech. Model. Mechanobiol., in press

Armour, C.H., Menichini, C., Milinis, K., Gibbs, R.G.J., Xu, X.Y.: Location of reentry tears affects false lumen thrombosis in aortic dissection following TEVAR. J. Endovasc. Ther. **27**, 396–404 (2020)

Bäumler, K., Vedula, V., Sailer, A.M., Seo, J., Chiu, P., Mistelbauer, G., Chan, F.P., Fischbein, M.P., Marsden, A.L., Fleischmann, D.: Fluid-structure interaction simulations of patient-specific aortic dissection. Biomech. Model. Mechanobiol. **19**, 1607–1628 (2020)

Brunkwall, J., Kasprzak, P., Verhoeven, E., Heijmen, R., Taylor, P., ADSORB Trialists; Alric, P., Canaud, L., Janotta, M., Raithel, D., Malina, W., Resch, T., Eckstein, H.H., Ockert, S., Larzon, T., Carlsson, F., Schumacher, H., Classen, S., Schaub, P., Lammer, J., Lönn, L., Clough, R.E., Rampoldi, V., Trimarchi, S., Fabiani, J.N., Böckler, D., Kotelis, D., Böckler, D., Kotelis, D., von Tenng-Kobligk, H., Mangialardi, N., Ronchey, S., Dialetto, G., Matoussevitch, V.: Endovascular repair of acute uncomplicated aortic type B dissection promotes aortic remodelling: 1 year results of the ADSORB trial. Eur. J. Vasc. Endovasc. Surg. **48**, 285–291 (2014)

Buth, J., Harris, P.L., Hobo, R., Cuypers, P., Duijm, L., Tielbeek, X.: Neurologic complications associated with endovascular repair of thoracic aortic pathology: incidence and risk factors. A study from the European Collaborators on Stent/Graft Techniques for Aortic Aneurysm Repair (EUROSTAR) Registry. J. Vasc. Surg. **46**, 1103–1111 (2007)

Chen, I.M., Chen, P.L., Huang, C.Y., Weng, S.H., Chen, W.Y., Chih, C.C.: Factors affecting optimal aortic remodeling after thoracic endovascular aortic repair of type B (IIIb) aortic dissection. Cardiovasc. Interv. Radiol. **40**, 671–681 (2017)

Chong, M.Y., Gu, B., Chong, B.T., Gu, B., Chan, B.T., Ong, Z.C., Xu, X.Y., Lim, E.: Effect of intimal flap motion on flow in acute type B aortic dissection by using fluid-structure interaction. Int. J. Numer. Method Biomed. Eng. e3399 (2020)

Clough, R.E., Nienaber, C.A.: Management of acute aortic syndrome. Nat. Rev. Cardiol. **12**, 103–114 (2015)

Dillon-Murphy, D., Noorani, A., Nordsletten, D., Figueroa, C.A.: Multi-modality image-based computational analysis of haemodynamics in aortic dissection. Biomech. Model. Mechanobiol. **15**, 857–876 (2015)

Du, T., Hu, D., Cai, D.: Outflow boundary conditions for blood flow in arterial trees. PLoS ONE **10**, 1–22 (2015)

Evangelista, A., Salas, A., Ribera, A., Ferreira-González, I., Cuellar, H., Pineda, V., González-Alujas, T., Bijnens, B., Permanyer-Miralda, G., Garcia-Dorado, D.: Long-term outcome of aortic dissection with patent false lumen: predictive role of entry tear size and location. Circulation **125**, 3133–3141 (2012)

Fattori, R., Tsai, T.T., Myrmel, T., Evangelista, A., Cooper, J.V., Trimarchi, S., Li, J., Lovato, L., Kische, S., Eagle, K.A., Isselbacher, E.M., Nienaber, C.A.: Complicated acute type B dissection: is surgery still the best option? A report from the International Registry of Acute Aortic Dissection. JACC Cardiovasc. Interv. **1**, 395–402 (2008)

Ge, Y.Y., Guo, W., Cheshire, N., Liu, X.P., Jia, X., Xiong, J., Ma, X.H., Zhang, H.P.: Preoperative thoracic false lumen branches relate to aortic remodeling after thoracic endovascular aortic repair for DeBakey IIIb aortic dissection. J. Vasc. Surg. **65**, 659–668 (2017)

Girish, A., Padala, M., Kalra, K., McIver, B.V., Veeraswamy, R.K., Chen, E.P., Leshnower, B.G.: The impact of intimal tear location and partial false lumen thrombosis in acute type B aortic dissection. Ann. Thorac. Surg. **102**, 1925–1932 (2016)

Goodman, P., Barlow, E.T., Crapo, P.M., Mohammad, S.F., Solen, K.A.: Computational model of device-induced thrombosis and thromboembolism. Ann. Biomed. Eng. **33**, 780–797 (2005)

Kamman, A.V., Brunkwall, J., Verhoeven, E.L., Heijmen, R.H., Trimarchi, S.: ADSORB trialists: predictors of aortic growth in uncomplicated type B aortic dissection from the Acute Dissection Stent Grafting or Best Medical Treatment (ADSORB) database. J. Vasc. Surg. **65**, 964–971 (2017)
Khoynezhad, A., Walot, I., Kruse, M.J., Rapae, T., Donayre, C.E., White, R.A.: Distribution of intimomedial tears in patients with type B aortic dissection. J. Vasc. Surg. **52**, 562–568 (2010)
Kim, T.H., Ko, Y.G., Kwon, S.W., Choi, D., Lee, D.Y., Shim, W.H., Hyon, M.S.: Large false lumen area is a predictor of failed false lumen volume reduction after stent-graft repair in type B aortic dissection. J. Endovasc. Ther. **21**, 697–706 (2014)
Kitamura, T., Torii, S., Oka, N., Horai, T., Itatani, K., Yoshii, T., Nakamura, Y., Shibata, M., Tamura, T., Araki, H., Matsunaga, Y., Sato, H., Miyaji, K.: Impact of the entry site on late outcome in acute Stanford type B aortic dissection. Eur. J. Cardiothorac. Surg. **48**, 655–661 (2015)
Kotelis, D., Grebe, G., Kraus, P., Müller-Eschner, M., Bischoff, M., von Tengg-Kobligk, H., Böckler, D.: Morphologic predictors of aortic expansion in chronic type B aortic dissection. Vascular **24**, 187–193 (2016)
Leiderman, K., Fogelson, A.L.: Grow with the flow: a spatial-temporal model of platelet deposition and blood coagulation under flow. Math. Med. Biol. **28**, 47–84 (2011)
Lewis, M.I., McKenna, R.J.: Disorders of the lung. In: Falk, J.A., Chaux, G.E. (eds.) Medical Management of the Thoracic Surgery Patient, pp. 139–146. Sanders, Philadelphia (2010)
Liu, F., Ge, Y.Y., Guo, W., Liu, X.P., Jia, X., Xiong, J., Ma, X.H.: Preoperative thoracic false lumen branches are predictors of aortic enlargement after stent grafting for DeBakey IIIb aortic dissection. J. Thorac. Cardiovasc. Surg. **155**, 21–29 (2018)
Maceira, A.M., Prasad, S.K., Khan, M., Pennell, D.J.: Reference right ventricular systolic and diastolic function normalized to age, gender and body surface area from steady-state free precession cardiovascular magnetic resonance. Eur. Heart J. **27**, 2879–2888 (2006)
Marui, A., Mochizuki, T., Koyama, T., Mitsui, N.: Degree of fusiform dilatation of the proximal descending aorta in type B acute aortic dissection can predict late aortic events. J. Thorac. Cardiovasc. Surg. **134**, 1163–1170 (2007)
McMahon, M., Squirrell, C.: Multidetector CT of aortic dissection: a pictorial review. Radiographics **30**, 445–460 (2010)
Menichini, C., Xu, X.Y.: Mathematical modeling of thrombus formation in idealized models of aortic dissection: initial findings and potential applications. J. Math. Biol. **73**, 1205–1226 (2016)
Menichini, C., Cheng, Z., Gibbs, R.G.J., Xu, X.Y.: Predicting false lumen thrombosis in patient-specific models of aortic dissection. J. R. Soc. Interface **13**, 20160759 (2016)
Menichini, C., Cheng, Z., Gibbs, R.G.J., Xu, X.Y.: A computational model for false lumen thrombosis in type B aortic dissection following thoracic endovascular repair. J. Biomech. **66**, 36–43 (2018)
Neofytou, P.: Comparison of blood rheological models for physiological flow simulation. Biorheology **41**, 693–714 (2004)
Nienaber, C.A., Divchev, D., Palisch, H., Clough, R.E., Richartz, B.: Early and late management of type B aortic dissection. Heart **100**, 1491–1497 (2014)
Nienaber, C.A., Clough, R.E., Sakalihasan, N., Suzuki, T., Gibbs, R., Mussa, F., Jenkins, M.P., Thompson, M.M., Evangelista, A., Yeh, J.S.M., Cheshire, N., Rosendahl, U., Pepper, J.: Aortic dissection. Nat. Rev. Dis. Primers **2**, 16071 (2016)
Peterss, S., Mansour, A.M., Ross, J.A., Vaitkeviciute, I., Charilaou, P., Dumfarth, J., Fang, H., Ziganshin, B.A., Rizzo, J.A., Adeniran, A.J., Elefteriades, J.A.: Changing pathology of the thoracic aorta from acute to chronic dissection: literature review and insights. J. Am. Coll. Cardiol. **68**, 1054–1065 (2016)
Pirola, S., Cheng, Z., Jarral, O.A., O'Regan, D.P., Pepper, J.R., Athanasiou, T., Xu, X.Y.: On the choice of outlet boundary conditions for patient-specific analysis of aortic flow using computational fluid dynamics. J. Biomech. **60**, 15–21 (2017)
Pirola, S., Guo, B., Menichini, C., Saitta, S., Fu, W., Dong, Z., Xu, X.Y.: 4-D flow MRI-based computational analysis of blood flow in patient-specific aortic dissection. IEEE Trans. Biomed. Eng. **66**, 3411–3419 (2019)

Qiao, Y., Fan, J., Ding, Y., Zhu, T., Luo, K.: A primary computational fluid dynamics study of pre- and post-TEVAR with intentional left subclavian artery coverage in a type B aortic dissection. J. Biomech. Eng. **141**, 111002 (2019a)

Qiao, Y., Zeng, Y., Ding, Y., Fan, J., Luo, K., Zhu, T.: Numerical simulation of two-phase non-Newtonian blood flow with fluid-structure interaction in aortic dissection. Comput. Methods Biomech. Biomed. Eng. **22**, 620–630 (2019b)

Qiao, Y., Mao, L., Zhu, T., Fan, J., Luo, K.: Biomechanical implications of the fenestration structure after thoracic endovascular aortic repair. J. Biomech. **99**, 109478 (2020)

Qin, Y.L., Deng, G., Li, T.X., Jing, R.W., Teng, G.J.: Risk factors of incomplete thrombosis in the false lumen after endovascular treatment of extensive acute type B aortic dissection. J. Vasc. Surg. **56**, 1232–1238 (2012)

Ray, H.M., Durham, C.A., Ocazionez, D., Charlton-Ouw, K.M., Estrera, A.L., Miller, C.C., 3rd., Safi, H.J., Azizzadeh, A.: Predictors of intervention and mortality in patients with uncomplicated acute type B aortic dissection. J. Vasc. Surg. **64**, 1560–1568 (2017)

Riambau, V., Böckler, D., Brunkwall, J., Cao, P., Chiesa, R., Coppi, G., Czerny, M., Fraedrich, G., Haulon, S., Jacobs, M.J., Lachat, M.L., Moll, F.L., Setacci, C., Taylor, P.R., Thompson, M., Trimarchi, S., Verhagen, H.J., Verhoeven, E.L., Esvs Guidelines Committee, Kolh, P., de Borst, G.J., Chakfé, N., Debus, E.S., Hinchliffe, R.J., Kakkos, S., Koncar, I., Lindholt, J.S., de Ceniga, M.V., Vermassen, F., Verzini, F., Document Reviewers P. Kolh, Black 3rd, J.H., Busund, R., Björck, M., Dake, M., Dick, F., Eggebrecht, H., Evangelista, A., Grabenwöger, M., Milner, R., Naylor, A.R., Ricco, J.B., Rousseau, H., Schmidli, J.: Editor's choice – Management of descending thoracic aorta diseases: clinical practice guidelines of the European Society for Vascular Surgery (ESVS). Eur. J. Vasc. Endovasc. Surg. **53**, 4–52 (2017)

Rizvi, A.Z., Murad, M.H., Fairman, R.M., Erwin, P.J., Montori, V.M.: The effect of left subclavian artery coverage on morbidity and mortality in patients undergoing endovascular thoracic aortic interventions: a systemic review and meta-analysis. J. Vasc. Surg. **50**, 1159–1169 (2009)

Sailer, A.M., van Kuijk, P.J., Nelemans, S.M.J., Chin, A.S., Kino, A., Huininga, M., Schmidt, J., Mistelbauer, G., Bäumler, K., Chiu, P., Fischbein, M.P., Dake, M.D., Miller, D.C., Schurink, G.W.H., Fleischmann, D.: Computed tomography imaging features in acute uncomplicated Stanford type-B aortic dissection predict late adverse events. Circ. Cardiovasc. Imaging **10**, e005709 (2017)

Salameh, M.J., Ratchford, E.V.: Aortic dissection. Vasc. Med. **21**, 276–280 (2016)

Sorensen, E.N., Burgreen, G.W., Wagner, W.R., Antak, J.F.: Computational simulation of platelet deposition and activation: I. Model development and properties. Ann. Biomed. Eng. **27**, 449–458 (1999)

Steuer, J., Eriksson, M.O., Nyman, R.: Early and long-term outcome after Thoracic Endovascular Aortic Repair (TEVAR) for acute complicated type B aortic dissection. Eur. J. Vasc. Endovasc. Surg. **41**, 318–323 (2011)

Tolenaar, J.L., Van Keulen, J.W., Jonker, F.H.W., van Herwaarden, J.A., Verhagen, H.J., Moll, F.L., Muhs, B.E., Trimarchi, S.: Morphologic predictors of aortic dilatation in type B aortic dissection. J. Vasc. Surg. **58**, 1220–1225 (2013a)

Tolenaar, J.L., Van Keulen, J.W., Trimarchi, S., Jonker, F.H.W., van Herwaarden, J.A., Verhagen, H.J.M., Moll, F.L., Muhs, B.E.: Number of entry tears is associated with aortic growth in type B dissections. Ann. Thorac. Surg. **96**, 39–42 (2013b)

Tolenaar, J.L., Kern, J.A., Jonker, F.H.W., Cherry, K.J., Tracci, M.C., Angle, J.F., Sabri, S., Trimarchi, S., Strider, D., Alaiwaidi, G., Upchurch, G.R., Jr.: Predictors of false lumen thrombosis in type B aortic dissection treated with TEVAR. Ann. Cardiothorac. Surg. **3**, 255–263 (2014)

Trimarchi, S., Tolenaar, J.L., Jonker, F.H.W., Murray, B., Tsai, T.T., Eagle, K.A., Rampoldi, V., Verhagen, H.J.M., van Herwaarden, J.A., Moll, F.L., Muhs, B.E., Elefteriades, J.A.: Importance of false lumen thrombosis in type B aortic dissection prognosis. J. Thorac. Cardiovasc. Surg. **145**, S208–S212 (2013)

Tsai, T.T., Evangelista, A., Nienaber, C.A., Myrmel, T., Meinhardt, G., Cooper, J.V., Smith, D.E., Suzuki, T., Fattori, R., Llovet, A., Froehlich, J., Hutchison, S., Distante, A., Sundt, T., Beckman,

J., Januzzi, J.L., Jr., Isselbacher, E.M., Eagle, K.A.: International Registry of Acute Aortic Dissection: partial thrombosis of the false lumen in patients with acute type B aortic dissection. N. Engl. J. Med. **357**, 349–359 (2007)

Wootton, D.M., Markou, C.P., Hanson, S.R., Ku, D.N.: A mechanistic model of acute platelet accumulation in thrombogenic stenoses. Ann. Biomed. Eng. **29**, 321–329 (2001)

Zeeshan, A., Woo, E.Y., Bavaria, J.E., Fairman, R.M., Desai, N.D., Pochettino, A., Szeto, W.Y.: Thoracic endovascular aortic repair complicated type B aortic dissection: superiority relative to conventional open surgical and medical therapy. J. Thorac. Cardiovasc. Surg. **140**, S109–S115 (2010)

Zipfel, B., Buz, S., Redlin, M., Hullmeine, D., Hammerschmidt, R., Hetzer, R.: Spinal cord ischemia after thoracic stent-grafting: causes apart from intercostal artery coverage. Ann. Thorac. Surg. **96**, 31–38 (2013)

Structural and Mechanical Inhomogeneity in Arterial ECM: Implications for Physiology and Disease

Yanhang Zhang

It is a great honor and privilege to contribute a chapter to this book, dedicated to Dr. Holzapfel's 60th birthday. I first became interested in vascular biomechanics during my postdoctoral research. Dr. Holzapfel's work enriched me with knowledge in this field and inspired my research. Over the past 15 years of my academic career, I have learned tremendously from Dr. Holzapfel's insightful work. I am very fortunate to have the opportunity to know Dr. Holzapfel, a role model who I aspire to be like.

Katherine Zhang

Abstract The extracellular matrix (ECM) of an artery endows the tissue its load bearing and damage resistance capacities. This chapter presents a review on the complex interplay between the multiscale ECM structural inhomogeneity and mechanics of large elastic arteries. Our recent studies integrating multiphoton imaging and quantification, biomechanical characterization, and computational modeling showed that ECM structural inhomogeneity exists at multiple structural levels of the arterial wall. At the intralamellar level, varying fiber orientation distribution and undulation contributes to local ECM mechanical properties. At the interlamellar level, transmural variation in in-plane fiber orientation distribution determines the anisotropic mechanical behavior of the elastin network. Furthermore, the waviness gradient among the elastic lamellar layers plays an important role in maintaining tissue homeostasis. Finally, structural inhomogeneity in transmural interlamellar fibers, and the development and propagation of aortic dissection will be discussed.

Y. Zhang (✉)
Department of Mechanical Engineering, Department of Biomedical Engineering, Division of Materials Science and Engineering, Boston University, 110 Cummington Mall, Boston, MA 02215, USA
e-mail: yanhang@bu.edu

G. Sommer et al. (eds.), *Solid (Bio)mechanics: Challenges of the Next Decade*,
Studies in Mechanobiology, Tissue Engineering and Biomaterials 24,
https://doi.org/10.1007/978-3-030-92339-6_3

1 Introduction

The extracellular matrix (ECM) in the arterial wall, composed of elastic fibers, collagen fibers, and ground substances including glycosaminoglycans/proteoglycans (GAGs/PGs), largely determines the passive mechanical properties of large elastic arteries. Elastic fibers, predominantly existing in the medial layer of the arterial wall (Wolinsky and Glagov 1967), imparts elastic property to an artery in order to accommodate cyclic physiological deformation. Large bundles of collagen fibers in the adventitia layer play an important role in preventing artery rupture at high pressures (Fung 1993).

In elastic arteries such as aorta, elastic fibers form concentric layers of elastic lamellae that are subjected to billions of stretch cycles during a lifetime (Kielty et al. 2002) (Fig. 1a, b). The elastic lamellae layers are interconnected through interwoven elastic and collagen fibers in the narrow interlamellar space (Davis 1993) (Fig. 1c, d). Alternating layers of smooth muscle cells anchor on either side to the adjacent lamellar layers through elastin extensions (Davis 1993). The ECM components, together with smooth muscle cells, are organized into lamellar units, which are considered as the functional units of the arterial wall (Wolinsky and Glagov 1964).

The ECM fiber networks in the arterial wall are highly inhomogeneous in structure with varying fiber lengths, diameters, density, orientation distribution, etc. Structural inhomogeneity can lead to large variations in local fiber stresses (Zhang et al. 2013), as well as in local ECM mechanical properties (Li et al. 2019). Wall stress distribution can alter the local permeability and pressure gradient, which governs many important physiological events (Chuong and Fung 1986). Microstructural inhomogeneity also plays an important role in regulating the failure behavior in laminated biological tissues (Iatridis and ap Gwynn 2004; Katsamenis et al. 2015). Arterial failure has been reported to initiate locally and may involve a combination of both in-plane fracture and delamination of the elastin lamellae (Rajagopal et al. 2007; Brunet et al. 2019).

This chapter presents our recent studies that reveal the existence of structural inhomogeneity in arterial ECM at the intra- and interlamellar structural levels. By integrating experimental and modeling approaches, these studies provide new understandings of the underlying microstructural mechanisms of ECM inhomogeneity in contributing to tissue homeostasis, vascular mechanics, and arterial failure such as the development and propagation of aortic dissection (AD).

Fig. 1 Cross section view of a porcine thoracic aorta showing layers of elastic lamellae. (a) Movat's stain of the aorta from adventitial (left) to intimal (right) side. Collagen (yellow), elastin (black), smooth muscle (red), and GAGs (blue). Scale bar is 100 μm (Mattson et al. 2017). (b) Multiphoton image of the aorta showing the lamellar layers in the media and the presence of elastin (green) and collagen (blue) in the arterial wall (Turcotte and Zhang 2020) with slight modifications. Inset picture shows the zoom in view of the interconnected lamellar layers. The thickness of the aorta is about 1 mm. (c), (d) Transmission electron microscopy images of the aorta at two magnifications with the intersect square image zoomed in Wang et al. (2015). EL = elastic laminae; vSMC = vascular smooth muscle cells; Coll (LS) = collagen fibers in longitudinal section; Coll (XS) = collagen fibers in cross-section. Arrows indicate inter-laminar elastin deposits

2 Multiscale ECM Inhomogeneity

2.1 Interlamellar Transmural Variation in Elastic Fiber Orientation Distribution and the Anisotropic Tissue Mechanical Behavior

2.1.1 Interlamellar Transmural Variation in Elastic Fiber Orientation Distribution

Natural biological tissue and tissue-derived soft biomaterial often exhibit pronounced mechanical anisotropy due to preferred fibrous orientation (Chuong et al. 1991; Hoffmeister et al. 1996; Ono et al. 1990; Demer et al. 1983). The anisotropic mechanical response for vascular tissue was suggested to directly result from fiber orientation (Timmins et al. 2010). Elastic fiber network has been shown to have anisotropic mechanical behavior (Sherebrin et al. 1983; Lillie et al. 2010; Zou et al. 2011). Previous studies attributed the anisotropy of elastic fiber network to fiber reorientation (Ronchetti et al. 1998; Rezakhaniha and Stergiopulos 2008) and that the elastic fibers being orientated in the circumferential direction (Rezakhaniha et al. 2010). However, several studies including ours reported that elastic fibers in large arteries have a relative uniform distribution, and the elastic fiber network does not show obvious structural changes under deformation (Fata et al. 2013; Chow et al. 2014). In these studies, multiphoton microscopy was used to image the elastic fibers in the medial layer and the arteries were imaged from the intimal surface. Due to limited laser penetration depth, however, imaging of the elastic fiber network was achieved from the intimal side of the arterial wall with a maximum depth of about 200 μm, which is very small compared to the arterial wall thickness ($\sim$1.5 mm). Thus, the fiber distribution obtained from such a limited imaging depth from the intimal surface may not be representative of the tissue structure. Yu et al. (2018b) examined the transmural variation in elastic fiber orientation distribution in porcine thoracic aorta using multiphoton microscopy and its association with tissue anisotropy. The elastic fibers orientation distribution was found to change from a relatively uniform distribution in regions close to the luminal surface to a more circumferential distribution in regions that dominates the media, then to a longitudinal distribution in regions close to the outer media (Figs. 2, 3a, b). The middle media appears to be much thicker than the inner and outer media.

The transmural structural inhomogeneity is important for the understanding of the structure and function of elastic fiber network in the arterial wall. Transmural alternations in elastic fiber orientation in the arterial wall was reported earlier from studies using electron microscopy (Clark et al. 1985; Farand et al. 2007). It was also observed that the smooth muscle cells follow a similar transmural distribution as the elastic fibers (Clark et al. 1985). The distinct transmural variation in fiber orientation helps to accommodate the complex loading in the artery. The axially oriented fibers from the intima side support the shear stress from blood flow in the longitudinal

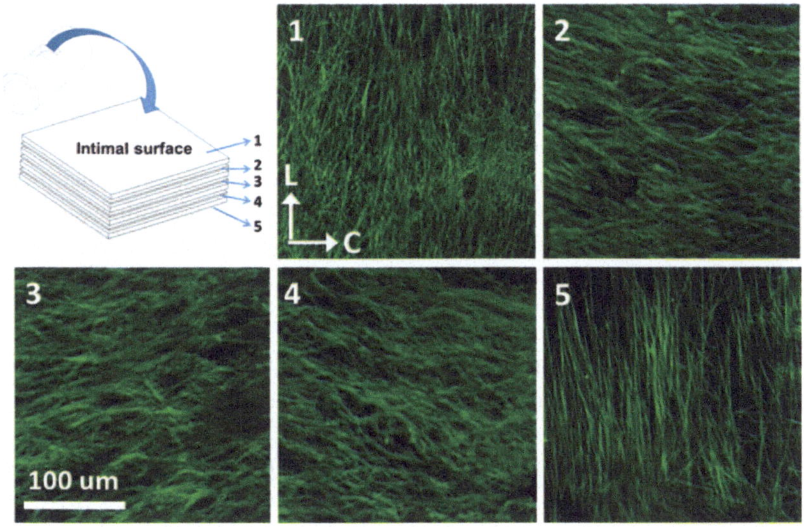

Fig. 2 Multiphoton images showing the distribution of elastic fibers in the inner media (1), middle media (2, 3, 4), and outer media (5). Here L refers to longitude; and C refers to circumference (Yu et al. 2018b)

direction, while the circumferentially oriented elastic fibers in the middle media bears the pulsatile load in the circumferential direction (Clark et al. 1985; Farand et al. 2007). Axially aligned elastic fibers in the outer media, however, play a role in transmitting axial shear stress from organ movement and tethering by adventitia and branches to media (Clark et al. 1985).

2.1.2 Structural Origin for the Anisotropic Tissue Mechanical Behavior in Constitutive Models

In structure-based constitutive models, fiber orientation distribution usually is the primary source of tissue anisotropy. Experimentally measured structural parameters from imaging, such as fiber orientation distribution, fiber crimping, and constituent volume fractions, can be incorporated directly into structure-based constitutive models to reduce the number of parameters and improve prediction accuracy (Wang et al. 2016). The transmural fiber orientation distribution of collagen, another major ECM constituent, has been found to change from nonuniform distribution in subendothelial layer to circumferentially oriented in the media, and to helically arranged fiber families in the adventitia (Gasser et al. 2006). The experimentally measured collagen fiber orientations through the thickness of mouse carotid arteries was considered by using a four-fiber family model with improvement in predicting capability (Wan et al. 2012).

While there are wide studies on structure-based models considering collagen fiber microstructural organization, only a few studies attempted to incorporate elastin network anisotropy with reported better description of the biomechanical behavior of the arterial wall (Rezakhaniha et al. 2010). Study by Yu et al. (2018b) provides further understandings on the origin of the anisotropic behavior of elastin network. Besides fiber orientation distributions, transmural variation in fiber orientation distributions, and the proportions of the inner, middle and outer media regions with distinct elastic fiber orientation distributions play an important role in determining tissue anisotropy (Yu et al. 2018b). The dominated circumferentially oriented elastic fibers explain well the generally stiffer behavior in the circumferential direction of the elastic fiber network.

To account for transmural variation in elastin fiber orientation distribution, the elastic fiber network was modeled using a constitutive model that incorporates the orientation distributions of elastic fibers at different transmural depths (Yu et al. 2018b). The total strain-energy function was assumed to be the sum of strain-energy function at the inner, middle, and outer media as: $W = \sum_{i=1}^{3} n_i \int_{-\pi/2}^{\pi/2} w(\rho) R_i(\theta) \mathrm{d}\theta$, where $w(\rho)$ is the strain-energy function at the fiber level, n_i is the elastin content that defines the proportions of the inner, middle and outer media regions. $R_i(\theta)$ is the elastic fiber orientation distribution function from the inner, middle, and outer media from Fig. 3a. For comparison, elastin network was also modeled using a constitutive model that only incorporates the fiber orientation distribution from the inner media, which does not consider the transmural variation in fiber orientation distribution. The anisotropic mechanical behavior of elastic fiber network is well described by the structure-based model that considers the transmural variation in elastin fiber distribution (Fig. 3c). The new model well captures the anisotropic behavior of elastin network during equi- and nonequi-biaxial mechanical loading. However only accounting for elastin fiber orientation distribution from the inner media results in poor fitting capabilities (Fig. 3d). Therefore, by considering the transmural elastin fiber distribution, the structure-based model showed improved fitting and predicting capability.

These results suggest fiber orientation is the main source of the anisotropy in elastic fiber network, which arises from the dominated circumferentially orientated elastin fibers in the middle media. The transmural variation in fiber orientation distributions, and the proportions of the inner, middle and outer media regions in elastin needs to be considered in structure-based constitutive models for improved fitting and predicting capabilities. Furthermore, the transmural structural inhomogeneity in elastic fiber distribution is essential in accommodating the complex loading in the artery, and should be included in constitutive modeling to study the mechanical properties of arteries in health and diseases.

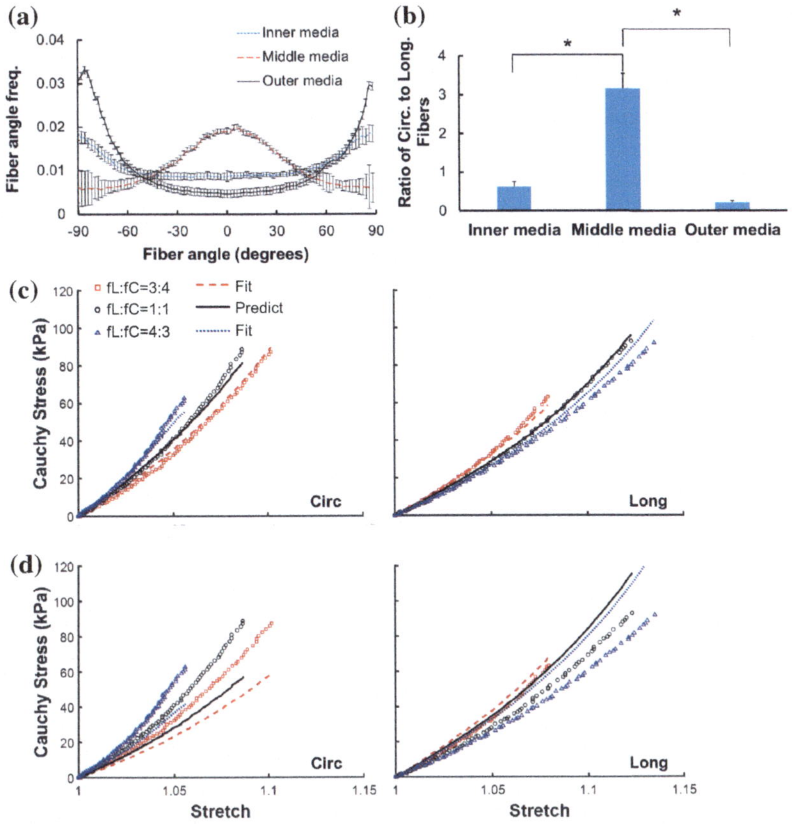

Fig. 3 (a) Transmural variation in elastic fiber orientation distribution in the inner, middle, and outer media with 0° being the circumferential and 90° being the longitudinal direction. (b) Ratio of circumferentially to longitudinally distributed elastic fibers in the inner, middle, and outer media shows that the middle media has significantly more circumferentially oriented elastic fibers than the inner and outer media ($*p < 0.01$). (c), (d) Representative results of Cauchy stress versus stretch in the circumferential (Circ) and longitudinal (Long) directions when fitting two sets of nonequi-biaxial testing data fL : fC = 75 : 100 and 100 : 75 N/m, and predicting equi-biaxial testing data fL : fC = 100 : 100 N/m using (c) a structural based constitutive model that considers transmural variation in elastin fiber orientation distribution, and (b) a structural based constitutive model that considers only fiber orientation distribution in inner media. Symbols represent experimental measurements and curves represent modeling results (Yu et al. 2018b)

2.2 Transmural Waviness Gradient in Elastic Lamellar Layers and Implications for Tissue Homeostasis

2.2.1 Transmural Waviness Gradient in Elastic Lamellar Layers

The lamellar units in large elastic arteries support and evenly distribute the stresses in the arterial wall caused by intraluminal distending pressure and plays an important role in maintaining tissue homeostasis (Wolinsky and Glagov 1964). Early studies found that the number of aortic lamellae layers is linearly proportional to the artery diameter, with a few layers for a mouse aorta while more than 50 layers for human (Wolinsky and Glagov 1967). Considering the law of Laplace and the relatively small variation in mean blood pressure, the average tension in each lamellae layer was suggested to be about the same regardless of species (Wolinsky and Glagov 1967), although it was unclear at the time as how the highly ordered concentric lamellar layers in the arterial wall are able to evenly distribute the stress radially through the arterial wall. Nevertheless, these earlier insightful studies pointed out the important role of elastic lamella as a fundamental functional unit in physiology and in maintaining vascular homeostasis. Waviness of elastic lamellae was noticed in transverse histological section of arterial tissue (Wolinsky and Glagov 1964). Several studies on the structure of elastic lamellae relying on histological analysis of biological tissue found that the elastic lamellar layers uncoil when subjected to deformation (Dobrin 1999; Sokolis et al. 2006). However, histological preparation and tissue retraction upon removal of mechanical loading may alter the structure of ECM.

To capture the three-dimensional (3D) architecture of elastic lamellae, multi-photon microscopy was used to visualize the microstructural deformation of elastic lamellae in mouse carotid arteries under biaxial pressurization and axial stretching (Yu et al. 2018a) (Fig. 4a, b). The 3D structure of the elastic lamellae was reconstructed to analyze its spatial distribution and unfolding under biaxial mechanical loading (Fig. 4c–f). The waviness of elastic lamellar layers decreases as intraluminal pressure increases (Fig. 5a). Further structural quantification at lamellar level were measured directly using the reconstructed 3D structure showed that the waviness of elastic lamellae varies transmurally with wavier elastic lamella towards the inner surface of the arterial wall. Lamellar layer closer to the inner wall are more undulated than lamellar layer closer to the outer wall (Fig. 5b). When the artery is pressurized, the more undulated inner lamellar layer also undergoes larger unfolding (Fig. 5c).

2.2.2 Structural Mechanisms That Enable the Elastic Lamellar Layers to Evenly Distribute the Stress Across the Arterial Wall and Maintain Tissue Homeostasis

To understand the structural mechanisms that enable the elastic lamellar layers to evenly distribute the stress across the arterial wall, Yu et al. (2018a) studied the stretching and unfolding of elastic lamellae in the arteries and its relationship with tissue-level deformation. At the tissue level, stress analysis of a homogeneous thick cylindrical wall subjected to intraluminal pressure also showed that the circumferential stretch decreases from the inner to the outer wall, however the tissue-level circumferential stretch (Fig. 5d) is much greater than the lamellar unfolding stretch (Fig. 5c). Study by Yu et al. (2018a) suggested that lamellar unfolding and stretching occurs simultaneously and both contribute to tissue-level circumferential stretch. The inner lamellar layers are wavier and unfold more than the outer layers. As shown in the schematic drawings in Fig. 6, the transmural waviness gradient in the lamellar

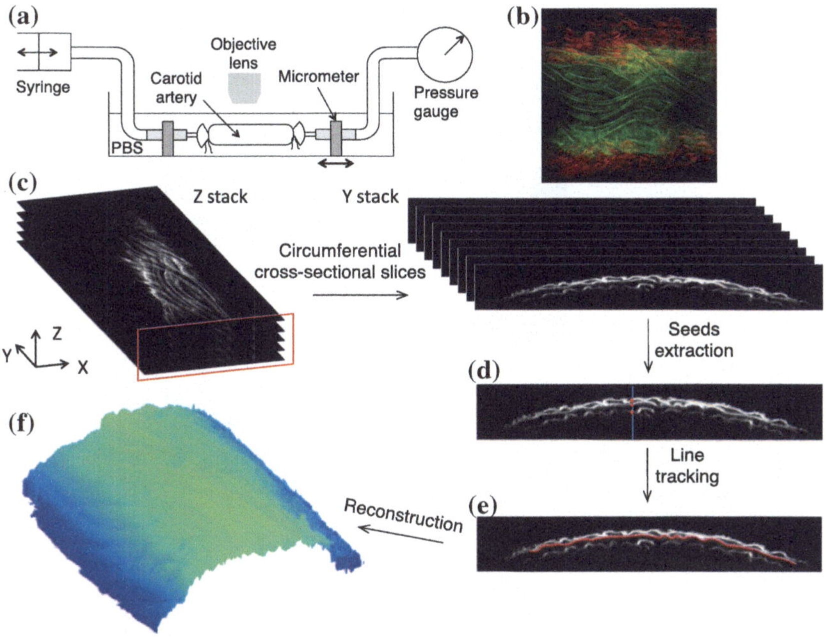

Fig. 4 (a), (b) Multiphoton imaging of a mouse carotid artery while the artery was subjected to mechanical loading. (a) Schematic of the experimental setup. A multiphoton microscope was used to image the elastic lamellar layers in the artery while it was subjected to biaxial extension-inflation. (b) Representative multiphoton images of a carotid artery showing the wavy collagen fibers in the adventitial layer and the longitudinally undulated elastic lamella. Red: second harmonic generation from collagen; green: fluorescence from elastin; image is 400 × 400 μm. (c)–(f) 3D reconstruction of elastic lamellae and imaging analysis. (c) The original z-stack multiphoton images were resliced along the longitudinal direction (y-axis) to show the transverse cross-sectional view of the elastic lamellae. An image processing algorithm was performed to reconstruct the lamellar layers and consisted primarily of (d) extraction of seed points and (e) line tracking from the seed points to identify the three layers. (f) Representative image of a reconstructed elastic lamella (Yu et al. 2018a)

layers compensates the larger tissue circumferential stretch experienced at the inner surface, thus equalizing lamellar layer extension through the arterial wall (Fig. 5e).

Research in the 1960s suggest the existence of residual stress in an artery even when there is no distending pressure (Bergel 1961), although the origin of residual stress is still not well understood. Since then, residual stress was considered in numerous computational models by including a stress-free configuration, which induces a negative stress gradient in the arterial wall with negative stress at the lumen surface at the load-free configuration (Chuong et al. 1986). With such compensation, a more evenly distributed wall stress can be achieved. Study by Yu et al. (2018a) reveals the importance of structural inhomogeneity in maintaining tissue homeostasis through the arterial wall. By combining the deformation of elastic lamellae and constitutive modeling, a new approach was developed to quantify the local deformation of elastic

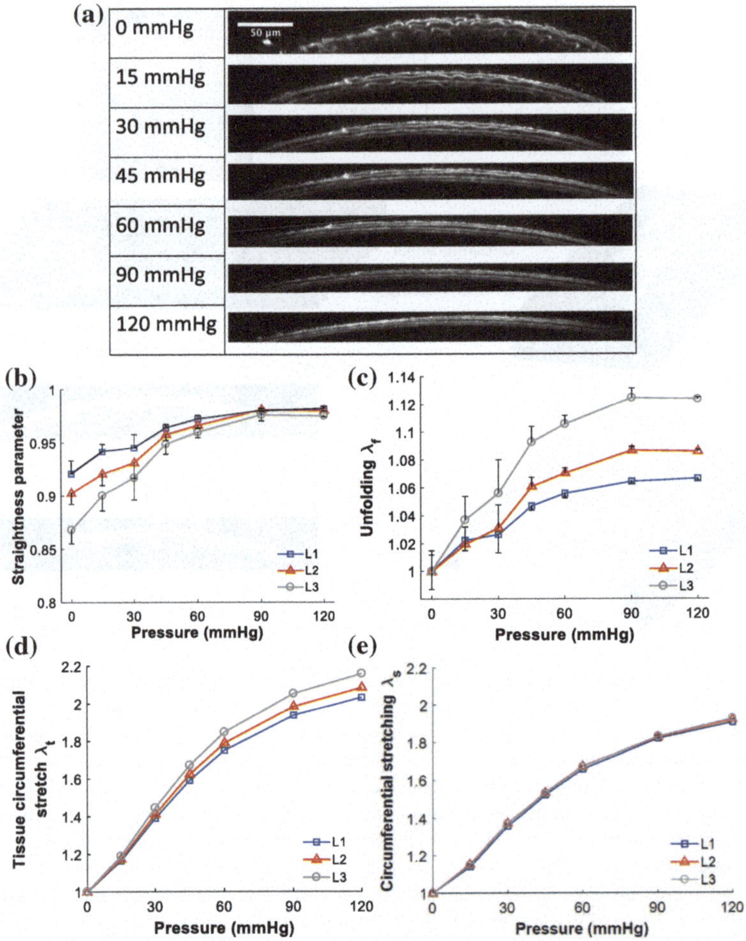

Fig. 5 (a) Reconstructed transverse cross-sectional view of the three lamellar layers in the mouse carotid artery when pressure increases from 0 to 120 mmHg at a physiological axial stretch ratio of 1.6. (b) Straightness parameter as a function of pressure for the three lamellar layers L1, L2 and L3 with L1 being the outermost layer. (c) Unfolding of the three elastic lamellar layers when pressure increases from 0 to 120 mmHg at an axial stretch of 1.6. (d) Tissue circumferential stretch as a function of pressure at the positions of the three lamellar layers (a). (e) Lamellar stretching as a function of pressure, determined from the tissue level deformation (d) and lamellar unfolding (c), was equivalent in all layers (Yu et al. 2018a)

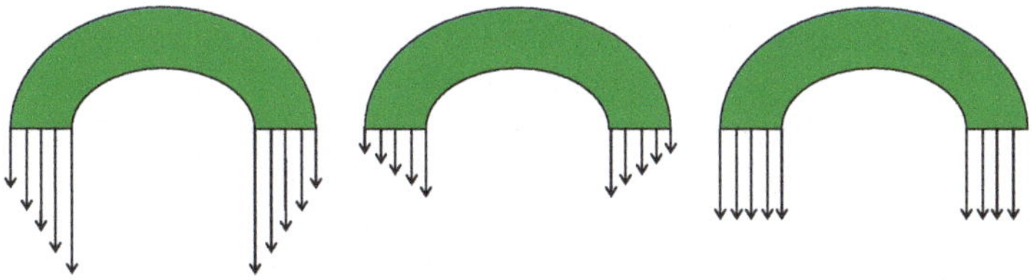

Fig. 6 Schematic drawings show that both lamellar unfolding and stretching contribute to tissue-level deformation. The transmural waviness gradient of elastic lamellar layers plays a critical role in equalizing the local circumferential stresses throughout the arterial wall (Yu et al. 2018a)

lamellae and to uncover the role of structural inhomogeneity in multiscale tissue mechanics. Specifically, the higher lamellae unfolding in the inner lamellae layer compensates the tissue circumferential stretch gradient, and thus plays an important role in maintaining a more evenly distributed stretching in the lamellar layers through the arterial wall (Fig. 6). This is important as lamellar stretching/elongation is directly related to stress development in the lamellar layers. These findings suggest a fundamental mechanism, originated from structural inhomogeneity, of how the arterial wall behaves mechanically as a homogeneous material, despite its structural inhomogeneity.

2.3 *Contribution of Structural Inhomogeneity to ECM Local Mechanical Properties*

The mechanical properties of ECM, known to have a 3D hierarchical structure, are highly dependent on the scale of measurement. Significant differences in the mechanical properties of ECM were shown ranging from the macro to nano scales as $E_{tendon} < E_{fiber} < E_{fibril}$ (Aifantis et al. 2011). In microscopic mechanical measurements, controlled mechanical forces were applied using a variety of techniques including indentation (Wenger et al. 2007; Aifantis et al. 2011; McKee et al. 2011), laser tracking microrheology (Velegol and Lanni 2001; Sun et al. 2004; Shayegan and Forde 2013), magnetic twisting cytometry (Leung et al. 2007), and optical magnetic twisting cytometry (OMTC) (Li et al. 2019). Studies on the mechanical properties of ECM usually focus on a single measurement scale. Advances in imaging techniques promote the integration of multiscale mechanical loading and imaging modalities to provide perspectives on load induced changes in ECM microstructure, cellular morphology and alignment (Roeder et al. 2009; Bell et al. 2012; Quinlan and Biliar 2012). To obtain local mechanical properties, local strains were measured, and the macro-scale mechanical properties of a substrate are usually used in the calculation of local traction forces (Franck et al. 2011; Toyjanova et al. 2014).

To understand the role of the inhomogeneous ECM fiber network structure in determining the local ECM mechanical properties, Li et al. (2019) created a multiscale experimental approach that combines OMTC and biaxial tensile testing approaches to study the multiscale ECM mechanics of the collagen network in the adventitial layer of the arterial wall (Fig. 7a, b). By oscillating microscale ferromagnetic beads bound to the collagen fibers, OMTC was used to measure the local mechanical properties of a collagen matrix that is subjected to controlled global biaxial mechanical loading by tracking the lateral bead displacement in response to oscillatory magnetic twisting torque (Fig. 7c). The orientation distribution of bead displacement in Fig. 7c shows a trimodal feature with a central major peak and two minor peaks (Fig. 7d), which resemble the multiple fiber family distribution in adventitial collagen.

Study by Li et al. (2019) shows that the ECM structural inhomogeneity, fiber orientation and fiber recruitment, play important roles in determining the local ECM mechanics. The wide range distribution of local stiffness (Fig. 7e–g) is associated with the inhomogeneous distribution of thick and wavy collagen fiber bundles in the adventitia. At low equal biaxial stretch levels, the wavy collagen fibers do not take much load, and have little restriction to the rotation of the magnetic beads. As the tissue is deformed, engagement/straightening of collagen fibers results in a redistribution of local ECM stiffness. The heterogeneity in local mechanical property measurements increases with global tissue deformation, as shown in the increasing median absolute deviation (MAD), which characterizes the variability of the dispersed data with tissue stretch. At higher stretch levels, the proportion low stiffness in the OMTC measurements gradually decreases, but still exists. In the meantime, the higher stiffness region starts to emerge (Fig. 7f, g) and result in a more spread distribution of the local modulus and increased MAD (Fig. 7h).

Fiber orientation also plays an important role in dictating the local ECM properties. In OMTC measurements, when the collagen fiber orientation coincides with the bead magnetizing direction, bead rotation tends to stretch and straighten the fibers. On the other hand, when the collagen fiber orientation is not aligned with the bead magnetization direction, bead rotation tends to bend and twist the fibers. This latter scenario often results in a relatively large lateral bead displacement, which was interpreted as lower stiffness. However, in most cases, it is likely that both fiber orientation and fiber recruitment play a role in determining local ECM mechanical properties, and thus contribute to the heterogeneity in local stiffness measurements. Furthermore, measurements from beads moving along the directions under the major peaks show a stronger linear dependency on tissue stretch (Li et al. 2019). Such results further suggest that the complex interaction between ferromagnetic beads and collagen fiber orientation distribution contribute to the heterogeneity of local ECM mechanical property measurements.

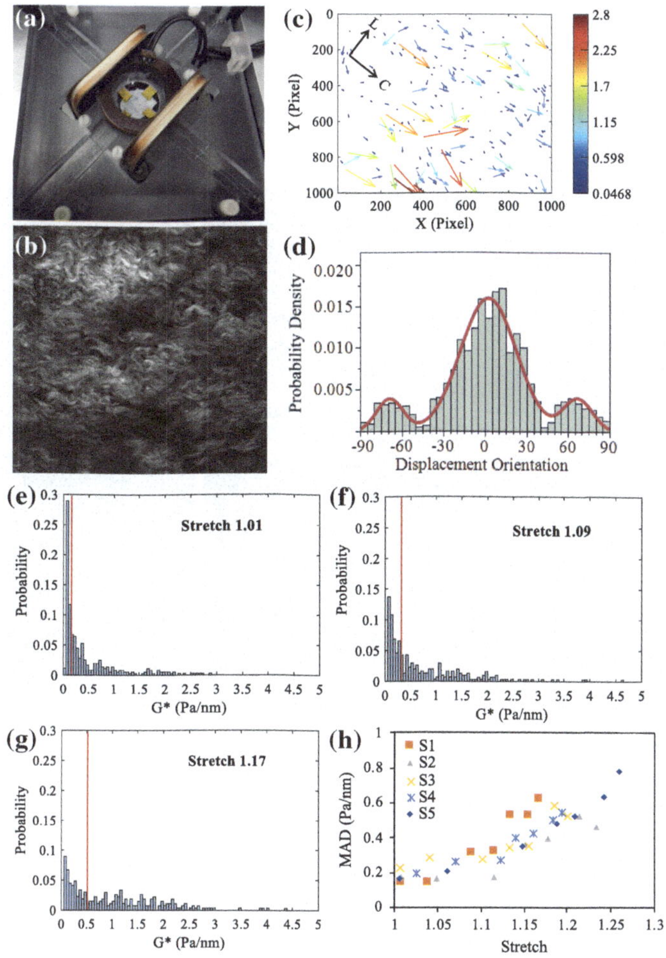

Fig. 7 (a) Picture showing a tissue sample that is centered within the magnetic twisting coils of optical magnetic twisting cytometry (OMTC) and loaded biaxially to allow the measurement of local mechanical properties while the sample is subjected to biaxial tissue loading. (b) Multiphoton image of unloaded adventitial collagen. Images are $425 \times 425\,\mu m^2$. (c) The magnitude of bead lateral displacement (μm) under biaxial loading of 20 g (1.15 stretch). The bead lateral displacement is oriented in the arrow direction and the magnitude of the displacement is represented by the length of the arrow ($\times 100$). C and L represent the circumferential and longitudinal direction of the sample. Note that the beads are magnetized along the circumferential direction. (d) Histogram of displacement orientation distribution of beads in (c) with trimodal distribution fitting (solid curve). Representative distributions of apparent complex modulus of sample 1 at biaxial stretch levels of (e) 1.01, (f) 1.09, and (g) 1.17. The position of the median, used as a measure of the local mechanical properties in the OMTC experiment, is marked with a red curve. (h) The median absolute deviation of locally measured apparent complex modulus with increasing biaxial stretch for all adventitia samples (Li et al. 2019)

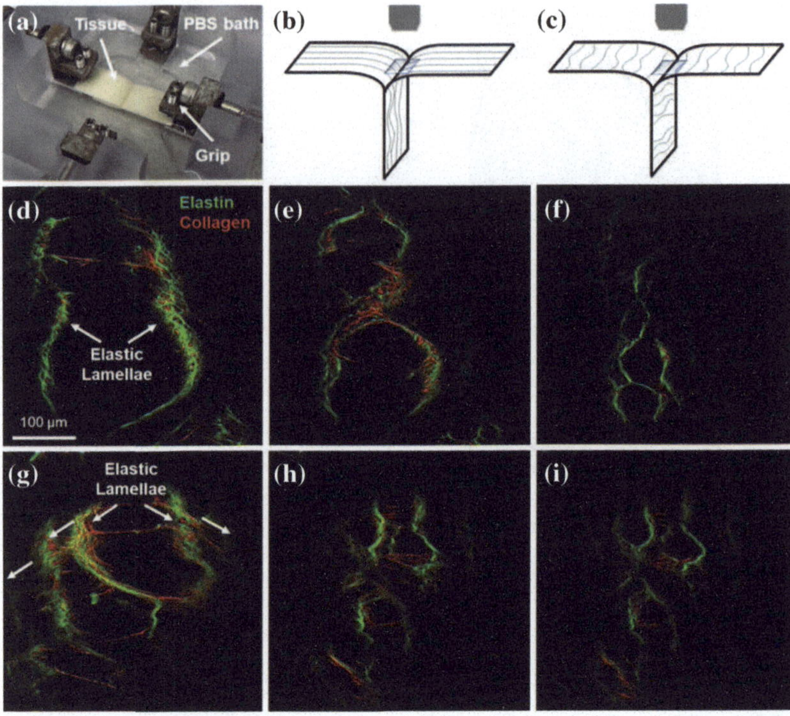

Fig. 8 (a) Custom-built device for multiphoton imaging of the peeling forefront; (b), (c) schematic drawings of multiphoton imaging of the peeling forefront of aortic media cut along the circumferential and longitudinal directions, respectively, with preferred orientation of in-plane fiber distribution; multiphoton images of different depth of the peeling site of porcine aortic media obtained from a circumferential tissue strip (d)–(f) and (g)–(i) a longitudinal tissue strip. Imaging depth increases from (d) to (f) and from (g) to (i) (Wang et al. 2021)

2.4 *Structural Inhomogeneity of Interlamellar ECM Fibers and Propagation of Aortic Dissection*

2.4.1 Structural Inhomogeneity of Interlamellar ECM Fibers

The concentric layers of elastic lamellae in the medial layer of the arterial wall are interconnected through interlamellar elastic and collagen fibers and smooth muscle cells (Wolinsky and Glagov 1967; Davies 1993). The elastic and collagen fibers are distributed predominantly parallel to the lamellar planes, with minor amounts of interlamellar fibers. The spatial distribution of interlamellar fibers are highly nonuniform in terms of the fiber density as well as fiber orientation with respect to lamellar layers. Multiphoton images taken at the separation point of adjacent lamellar layers (Fig. 8a–c) show the interlamellar elastic and collagen fibers connecting the two elastic lamellar layers (Fig. 8d–i) (Wang et al. 2021).

Compared with the dense elastic and collagen fibers that form the lamellar layers (Figs. 2 and 7b), interlamellar fibers are more sparsely arranged, and their distributions are highly nonuniform in density and orientation. This suggests that the local interlamellar resistance to layer separation could vary dramatically from one location to another.

2.4.2 Avalanches and Power Law Behavior in Aortic Dissection Propagation

Due to the laminated structure of the aortic media, a tearing of the intima may propagate parallel to the lumen, forcing the layers apart and resulted in aortic dissection (Tam et al. 1998). Arterial dissection is a catastrophic event that may occur spontaneously or as a result of traumatic injuries in many arterial branches, including the aorta (Nienaber et al. 2016). It is characterized by the tearing of the intimal layer which allows blood to enter between the lamellar layers of the aortic wall, resulting in the separation of the layers. A false lumen is formed in the separated media, which may narrow or occlude the true lumen. Due to diagnostic challenges, initial presentation of aortic dissection is missed in >30% of the cases (Nienaber et al. 2016). In the absence of intervention, acute aortic dissections result in mortality up to 90%, with the majority of deaths occurring with 48 h of the onset of aortic dissection (Kouchoukos and Dougenis 1997). Thus, acute aortic dissection is one of the most devastating cardiovascular diseases. Once initiated, the dissection propagates rapidly, leading to high morbidity and mortality. Prior studies of the biomechanics of aortic dissections, both experimental and computational, focused on the blood pressure and energy required for the propagation of an existing dissection (Carson and Roach 1990; Sommer et al. 2008).

As an acute cardiovascular disease, aortic dissection is often associated with its rapid development, which presents substantial challenges in diagnosis and treatment. Recent study by Yu et al. (2020) reveals the mechanisms underlying the propagation of aortic dissection within the arterial wall, a highly inhomogeneous biological material. In the study by Yu et al. (2020), aorta media samples were subjected to a slow peeling test that allowed us to mimic the propagation of aortic dissection. The peeling force F (normalized by the width of the sample) showed significant fluctuations in the plateau region with many sudden force drops of various magnitudes (Fig. 9a–c). The resulting probability density distribution of force drops showed a linear decrease implying a power law distribution over almost two orders of magnitude from 0.1 to 10 N/m. The power law distribution in the circumferential and longitudinal direction has an exponent, defined as the negative slope of a straight line fit, of 1.51 and 1.33 ($p < 0.05$), respectively (Fig. 9d, e).

The sudden catastrophic nature of aortic dissections is largely similar to an avalanching behavior. Avalanches are events that are accompanied by a sudden transfer of energy or material with a wide range of magnitudes without a characteristic length scale (Ludington et al. 2013). In an avalanching system, the energy or material accumulates gradually driving the system to an unstable state. When a critical

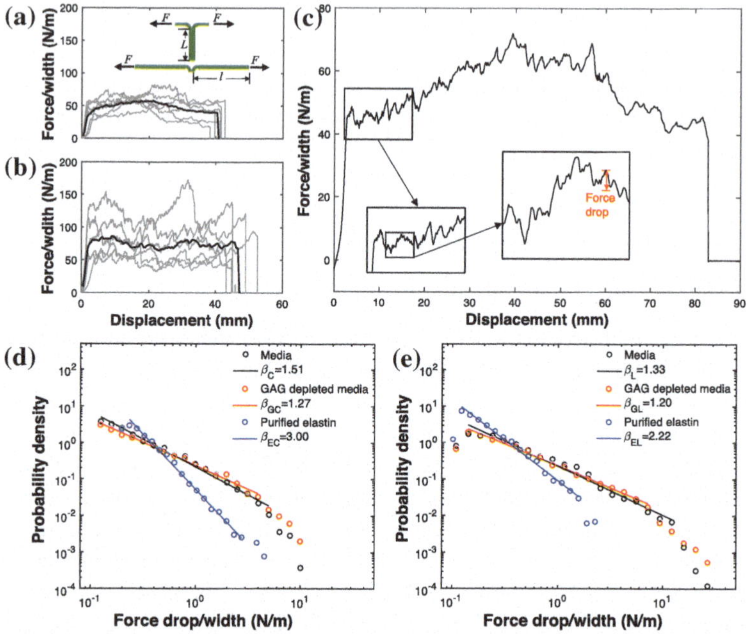

Fig. 9 Power law behavior in dissection propagation within the aortic media. Peeling force (force per unit width) versus displacement for (a) circumferentially oriented aortic media, and (b) longitudinally oriented aortic media ($n = 6$). The thick curve characterizes the arithmetic mean response. Inset figure shows schematic drawings of the peeling test experimental set up before loading and right before full separation. L is the effective length of the tissue and l is the length of the tissue right before separation. (c) Representative peeling force (force per unit width) versus displacement curve showing the fluctuating force profile during peeling test. Force drop was calculated as the difference between the peak and following valley. Probability density distribution of the force drops in dissection propagation within the media, GAG depleted media, and purified elastin samples in (d) the circumferentially oriented samples, and (e) the longitudinally oriented samples. Straight lines in (d) and (e) represent linear fits to the experimental data in double logarithmical graphs, and β is defined as the negative slope of a straight line fit. Here subscripts C, GC, and EC represent the media, GAG depleted media, and purified elastin, respectively, in the circumferential direction; subscripts L, GL, and EL represent the media, GAG depleted media, and purified elastin, respectively, in the longitudinal direction (Yu et al. 2020)

threshold is reached, the energy is released through avalanches of all sizes until the system returns back to a more stable state. Avalanching behavior is characterized by a power law distribution of avalanche sizes, exhibiting self-similar behavior over a wide range of scales. The power law behavior implies that the force drops have no characteristic scale and hence the dissection process is not dominated by any specific length scale other than the system size, which is consistent with the force drops originating from cascade failures of structural units in the arterial wall.

The adjacent elastic lamellar layers are connected by interlamellar elastic and collagen fibers (Fig. 8), which are the key elastic elements contributing to the force development during lamellar separation. The interlamellar fibers also play an important role in transferring the transmural stress across the wall (Wagenseil and Mecham 2009). In the force displacement curve, the fluctuating force profile of individual samples in Fig. 9c originates from the failure of nonuniformly distributed interlamellar fibers. Each force drop corresponds to one small avalanche or microfailure that may be comprised of multiple fibers rupturing simultaneously. Indeed, from the estimation of inter-lamellar fiber strength, it's unlikely that a single microfailure corresponds to the failure of an individual interlamellar fiber (Miyazaki and Hayashi 1999). Rather, the microfailure is likely due to a group of inhomogeneously arranged interlamellar collagen fibers rupturing in a single avalanche. When dissection advances between the lamellar layers, the interlamellar fibers are stretched and store more elastic energy. Consequently, stress concentration gradually builds up at the front of the dissection. When the strain in an interlamellar fiber reaches its failure limit, the fiber fails and the load it carried before rupture is redistributed among the neighboring intact fibers. Since the failure threshold of the fibers is similar, failure of a single fiber can immediately stretch more fibers beyond their failure threshold. Consequently, these neighboring fibers will also fail in a rapid succession generating an avalanche. Various digestion experiments and systematic sensitivity analyses further suggested that the avalanche failure in AD is closely associated with the failure of the inhomogeneously distributed inter-lamellar collagen fibers (Yu et al. 2020).

2.4.3 Discrete Finite Element Modeling of Aortic Dissection Propagation

To further determine the mechanism that governs the dissection process, a finite element model capturing the dissection propagation in the peeling test was created that considers the contribution of strain-induced failure of discrete inter-lamellar collagen fibers to layer separation (Fig. 3a). The inhomogeneity of the interlamellar fibers was included by introducing the randomness in the interlamellar fiber orientation and density, as well as the failure strain. Compared with earlier statistical models of parallel fiber bundles, this inhomogeneity was introduced along the propagation direction of the dissection. The model simulated the controlled peeling process in the experiment by separating the media from the inter-lamellar space (Fig. 10a). When the dissection propagates through the region where the interlamellar fibers are distributed sparsely, the fibers break, generating discrete force drops. The resistance against the advance of a crack in the material gradually builds up when the dissection front reaches a region where the interlamellar fibers are more densely distributed. Moreover, the angled interlamellar fibers can reorient and stretch during the separation before the deformation exceeds their failure strain (Fig. 10b), which increases the force required to propagate the dissection. When the fibers at the dissection front fail, the dissection advances easily through the neighboring scattered fibers, creating successive avalanches.

During layer separation, avalanches were generated when interlamellar fibers were gradually stretched until failure (Fig. 10c). The corresponding distribution of

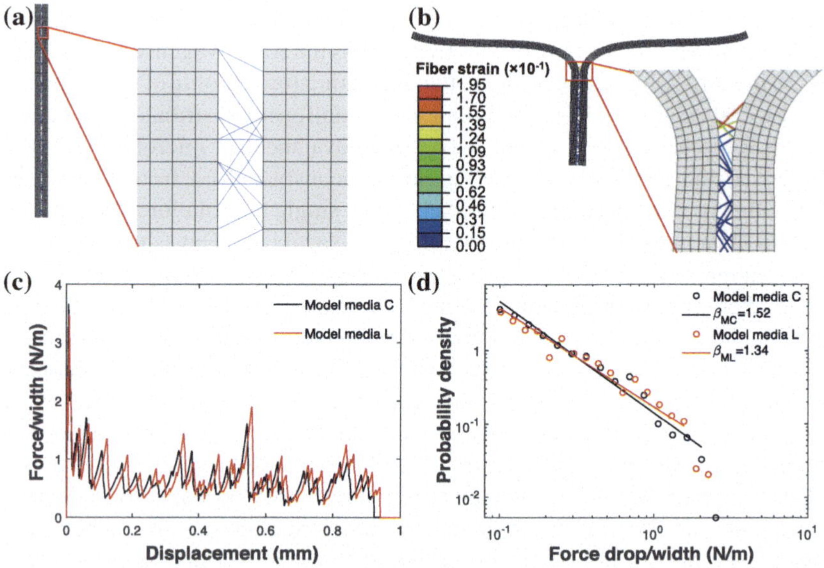

Fig. 10 Power law behavior of dissection propagation simulated using a finite element model considering the discrete interlamellar collagen fibers. (a) Finite element model showing the two media strips connected via discrete interlamellar collagen fibers. (b) Engineering strain in the interlamellar fibers during dissection propagation. Fibers reached failure criteria were automatically removed from the model. It can be seen that the fibers exactly at the front of the separation are the most stretched. (c) Peeling force (force per unit width) versus displacement from the finite element simulation. (d) Probability density distribution of the force drops in dissection propagation within the aortic media from simulations. Straight lines represent linear fits to the model results in double logarithmical graphs, and β is defined as the negative slope of a straight line fit. Here subscripts MC and ML represent the model results in the circumferential and longitudinal directions, respectively (Yu et al. 2020)

the force drops from the model (Fig. 10d) is also consistent with a power law in both circumferential and longitudinal directions, with exponents that agree very well with those from the experiments (1.51 and 1.33 in the circumferential and longitudinal directions, respectively, as shown in Fig. 9d, e). ECM fibers in the arterial wall are structurally and mechanically inhomogeneous. In this model, the mechanical inhomogeneity is included in the random failure strain of interlamellar fibers. The structural inhomogeneity is present as a nonuniform spatial distribution of interlamellar fibers, in terms of the fiber density along the dissection direction, as well as their orientation with respect to lamellar layers. The variance of failure strain of elastic and collagen fibers have been observed in tensile tests of a single fiber (Miyazaki and Hayashi 1999). Yu et al. (2020) showed that the power law distribution doesn't depend on the mechanical inhomogeneity, however structural inhomogeneity is the underlying mechanism of the avalanche behavior in the propagation of aortic dissection. Simulation results showed that when structural inhomogeneity was eliminated by having evenly spaced parallel interlamellar fibers between the adjacent elastic lamellae, the power law behavior completely disappeared, indicating that structural inhomogeneity is the ultimate source of the power law behavior in delaminating the lamellar layers (Yu et al. 2020). In other words, the random structural arrangement in interlamellar fibers is the source of the generated avalanches of energy release, or force drop, on a wide range of scales.

3 Conclusion and Future Outlook

ECM structural inhomogeneity exists at multiple structural levels. This chapter summarizes our recent findings on the ECM structural inhomogeneities and the important roles they play in contributing to tissue mechanics and homeostasis. Considering ECM structural inhomogeneity is important when studying the physiological function and failure behavior of the arterial wall. Understanding the complex interplay between structural inhomogeneity and mechanical function calls for new approaches that integrate advanced imaging, mechanical characterization, and computational modeling.

The transmural variations in elastic fiber orientation distribution is essential in accommodating the complex loading in the artery. In elastic fiber network, fiber orientation is the main source of elastin anisotropy, which arises from the dominated circumferentially orientated elastic fibers in the middle media. The transmural variation in fiber orientation distributions, and the proportions of the inner, middle and outer media regions need to be considered in structure-based constitutive modeling of elastic fiber network for improved fitting and predicting capabilities. Furthermore, fiber orientation and recruitment both contribute to the wide distribution in local ECM mechanical properties, thus ECM structural inhomogeneity can have a large impact on cellular-level deformation and stresses.

Stretching and unfolding of elastic lamellae occurs simultaneously during deformation and both contribute to tissue-level circumferential stretch. While the lamellar stretching dominates the tissue deformation, the radial waviness gradient of elastic lamellar layers plays a critical role in equalizing the local circumferential stresses throughout the arterial wall. These findings, by looking at micromechanics of individual elastic lamellar layers, reveal the importance of structural inhomogeneity in maintaining tissue homeostasis and the underlying structural origin that enables elastic lamellar layers to evenly distribute the stresses through the arterial wall, a fundamental requirement for tissue and cellular function.

Interlamellar fiber inhomogeneity was found to govern the rapid evolution of aortic dissection by avalanches of various sizes that follow a power law distribution. The avalanche behavior results from the local build-up of strain-energy followed by a cascade of mechanical failures of interlamellar fibers, a key component of which is the inhomogeneously distributed interlamellar collagen fibers in the aortic media. Thus, characterizing the ECM structural inhomogeneity *in vivo* with future imaging modalities may provide a useful index of the risk of developing dissections.

Aging and diseases could have profound impact on the composition and structure of the arterial wall. Findings from these studies may have profound implications on vascular remodeling in aging and diseases, as well as in tissue engineering of functional blood vessels. It is thus necessary to develop microstructure-inspired models and to incorporate such structural inhomogeneities within the ECM in the existing theories on growth and remodeling in future studies. The structural and mechanical interplays need to be considered in order to advance the current understanding of multiscale ECM mechanics and to establish the connections between the macro- and microscopic mechanical properties of the ECM. In summary, our study emphasized the importance of adopting a multiscale and multi-modal perspective when dissecting the role of ECM components in tissue mechanics.

References

Aifantis, K., Shrivastava, S., Odegard, G.: Transverse mechanical properties of collagen fibers from nanoindentation. J. Mater. Sci. Mater. Med. **22**, 1375–1381 (2011)

Bell, B., Nauman, E., Voytik-Harbin, S.: Multiscale strain analysis of tissue equivalents using a custom-designed biaxial testing device. Biophys. J. **102**, 1303–1312 (2012)

Bergel, D.H.: The static elastic properties of the arterial wall. J. Physiol. **156**, 445–457 (1961)

Brunet, J., Pierrat, B., Maire, E., Adrien, J., Badel, P.: A combined experimental-numerical lamellar-scale approach of tensile rupture in arterial medial tissue using X-ray tomography. J. Mech. Behav. Biomed. Mater. **95**, 116–123 (2019)

Carson, M.W., Roach, M.R.: The strength of the aortic media and its role in the propagation of aortic dissection. J. Biomech. **23**, 579–588 (1990)

Chow, M.J., Turcotte, R., Lin, C.P., Zhang, Y.: Arterial extracellular matrix: a mechanobiological study of the contributions and interactions of elastin and collagen. Biophys. J. **106**, 2684–2692 (2014)

Chuong, C., Sacks, M., Johnson, R., Reynolds, R.: On the anisotropy of the canine diaphragmatic central tendon. J. Biomech. **24**, 563–576 (1991)

Chuong, C.J., Fung, Y.C.: On residual stress in arteries. J. Biomech. Eng. **108**, 189–192 (1986)

Clark, J.M., Glagov, S.: Transmural organization of the arterial media: the lamellar unit revisited. Arteriosclerosis **5**, 19–34 (1985)

Davies, M.J., Richardson, P.D., Woolf, N., Katz, D.R., Mann, J.: Risk of thrombosis in human atherosclerotic plaques: role of extracellular lipid, macrophage, and smooth muscle cell content. Br. Heart J. **69**, 377–381 (1993)

Demer, L.L., Yin, F.C.P.: Passive biaxial mechanical properties of isolated canine myocardium. J. Physiol. Lond. **339**, 615–630 (1983)

Dobrin, P.B.: Distribution of lamellar deformations: implications for properties of the arterial media. Hypertension **33**, 806–810 (1999)

Farand, P., Garon, A., Plante, G.: Structure of large arteries: orientation of elastin in rabbit aortic internal elastic lamina and in the elastic lamellae of aortic media. Microvasc. Res. **73**, 95–99 (2007)

Fata, B., Carruthers, C.A., Gibson, G., Watkins, S.C., Gottlieb, D., Mayer, J.E., Sacks, M.S.: Regional structural and biomechanical alterations of the ovine main pulmonary artery during postnatal growth. J. Biomech. Eng. **135**, 021022 (2013)

Franck, C., Maskarinec, S., Moore, D., Ravichandran, G.: Three-dimensional traction force microscopy: a new tool for quantifying cell-matrix interactions. PLoS ONE **6**, e17833 (2011)

Fung, Y.C.: Biomechanics. Mechanical Properties of Living Tissues, 2nd edn. Springer, New York (1993)

Gasser, T.C., Ogden, R.W., Holzapfel, G.A.: Hyperelastic modelling of arterial layers with distributed collagen fibre orientations. J. R. Soc. Interface **3**, 15–35 (2006)

Hoffmeister, B., Handley, S., Wickline, S., Miller, J.: Ultrasonic determination of the anisotropy of Young's modulus of fixed tendon and fixed myocardium. J. Acoust. Soc. Am. **100**, 3933–3940 (1996)

Iatridis, J., ap Gwynn, I.: Mechanisms for mechanical damage in the intervertebral disc annulus fibrosus. J. Biomech. **37**, 1165–1175 (2004)

Katsamenis, O., Jenkins, T., Thurner, P.: Toughness and damage susceptibility in human cortical bone is proportional to mechanical inhomogeneity at the osteonal-level. Bone **76**, 158–168 (2015)

Kielty, C.M., Sherratt, M.J., Shuttleworth, C.A.: Elastic fibres. J. Cell Sci. **115**, 2817–2828 (2002)

Kouchoukos, N.T., Dougenis, D.: Surgery of the thoracic aorta. N. Engl. J. Med. **336**, 1876–1888 (1997)

Leung, L., Tian, D., Brangwynne, C., Weitz, D., Tschumperlin, D.: A new microrheometric approach reveals individual and cooperative roles for TGF-β1 and IL-1β in fibroblast-mediated stiffening of collagen gels. FASEB J. **21**, 2064–2073 (2007)

Li, H., Mattson, J., Zhang, Y.: Integrating structural heterogeneity, fiber orientation, and recruitment in multiscale ECM mechanics. J. Mech. Behav. Biomed. Mater. **92**, 1–10 (2019)

Lillie, M., Shadwick, R., Gosline, J.: Mechanical anisotropy of inflated elastic tissue from the pig aorta. J. Biomech. **43**, 2070–2078 (2010)

Ludington, W., Wemmer, K., Lechtreck, K., Witman, G., Marshall, W.: Avalanche-like behavior in ciliary import. Proc. Natl. Acad. Sci. USA **110**, 3925–3930 (2013)

Mattson, J.M., Turcotte, R., Zhang, Y.: Glycosaminoglycans contribute to extracellular matrix fiber recruitment and arterial wall mechanics. Biomech. Model. Mechanobiol. **16**, 213–225 (2017)

Mckee, C., Last, J., Russell, P., Murphy, C.: Indentation versus tensile measurements of Young's modulus for soft biological tissues. Tissue Eng. Part B Rev. **17**, 155–164 (2011)

Miyazaki, H., Hayashi, K.: Tensile tests of collagen fibers obtained from the rabbit patellar tendon. Biomed. Microdevices **2**, 151–157 (1999)

Nienaber, C.A., Clough, R.E., Sakalihasan, N., Suzuki, T., Gibbs, R., Mussa, F., Jenkins, M.P., Thompson, M.M., Evangelista, A., Yeh, J.S.M., Cheshire, N., Rosendahl, U., Pepper, J.: Aortic dissection. Nat. Rev. Dis. Primers **2**, 16053 (2016)

Ono, K., Kikuch, Y., Higashi, K., Tamiya, N., Yasuoka, N.: Elastic anisotropy of bivalve hinge-ligament. J. Biomech. **23**, 307–312 (1990)

Quinlan, A., Billiar, K.: Investigating the role of substrate stiffness in the persistence of valvular interstitial cell activation. J. Biomed. Mater. Res. A **100**, 2474–2482 (2012)

Rajagopal, K., Bridges, C., Rajagopal, K.R.: Towards an understanding of the mechanics underlying aortic dissection. Biomech. Model. Mechanobiol. **6**, 345–359 (2007)

Rezakhaniha, R., Stergiopulos, N.: A structural model of the venous wall considering elastin anisotropy. J. Biomech. Eng. **130**, 031017 (2008)

Rezakhaniha, R., Fonck, E., Genoud, C., Stergiopulos, N.: Role of elastin anisotropy in structural strain energy functions of arterial tissue. Biomech. Model. Mechanobiol. **10**, 599–611 (2010)

Roeder, B., Kokini, K., Voytik-Harbin, S.: Fibril microstructure affects strain transmission within collagen extracellular matrices. J. Biomech. Eng. **131**, 031004 (2009)

Ronchetti, I., Alessandrini, A., Baccarani Contri, M., Fornieri, C., Mori, G., Quaglino, D., Valdré, U.: Study of elastic fiber organization by scanning force microscopy. Matrix Biol. **17**, 75–83 (1998)

Shayegan, M., Forde, N.: Microrheological characterization of collagen systems: from molecular solutions to fibrillar gels. PLoS ONE **8**, 1–12 (2013)

Sherebrin, M., Song, S., Roach, M.: Mechanical anisotropy of purified elastin from the thoracic aorta of dog and sheep. Can. J. Physiol. Pharmacol. **61**, 539–545 (1983)

Sokolis, D., Kefaloyannis, E., Kouloukoussa, M., Marinos, E., Boudoulas, H., Karayannacos, P.: A structural basis for the aortic stress-strain relation in uniaxial tension. J. Biomech. **39**, 1651–1662 (2006)

Sommer, G., Gasser, T.C., Regitnig, P., Auer, M., Holzapfel, G.A.: Dissection properties of the human aortic media: an experimental study. J. Biomech. Eng. **130**, 021007–1–12 (2008)

Sun, Y.L., Luo, Z.P., Fertala, A., An, K.N.: Stretching type II collagen with optical tweezers. J. Biomech. **37**, 1665–1669 (2004)

Tam, A.S.M., Sapp, M.C., Roach, M.R.: The effect of tear depth on the propagation of aortic dissections in isolated porcine thoracic aorta. J. Biomech. **31**, 673–676 (1998)

Timmins, L.H., Wu, Q., Yeh, A.T., Moore, J.E., Jr., Greenwald, S.E.: Structural inhomogeneity and fiber orientation in the inner arterial media. Am. J. Physiol. Heart Circ. Physiol. **298**, 1537–1545 (2010)

Toyjanova, J., Bar-Kochba, E., López-Fagundo, C., Reichner, J., Hoffman-Kim, D., Franck, C.: High resolution, large deformation 3D traction force microscopy. PLoS ONE **9**, e90976 (2014)

Turcotte, R., Zhang, Y.: Intrinsic optical imaging of ECM mechanics. In: Zhang, Y. (ed.) Multi-scale Extracellular Matrix Mechanics and Mechanobiology, pp. 165–202. Springer, Heidelberg (2020)

Velegoll, D., Lanni, F.: Cell traction forces on soft biomaterials. I. Microrheology of type I collagen gels. Biophys. J. **81**, 1786–1792 (2001)

Wagenseil, J.E., Mecham, R.P.: Vascular extracellular matrix and arterial mechanics. Physiol. Rev. **89**, 957–989 (2009)

Wan, W., Dixon, J., Gleason, R.: Constitutive modeling of mouse carotid arteries using experimentally measured microstructural parameters. Biophys. J. **102**, 2916–2925 (2012)

Wang, R., Yu, X., Gkousioudi, A., Zhang, Y.: Effect of glycation on interlamellar bonding of arterial elastin. Exp. Mech. **61**, 81–94 (2021)

Wang, Y., Zeinali-Davarani, S., Davis, E., Zhang, Y.: Effect of glucose on the biomechanical function of arterial elastin. J. Mech. Behav. Biomed. Mater. **49**, 244–254 (2015)

Wang, Y., Zeinali-Davarani, S., Zhang, Y.: Arterial mechanics considering the structural and mechanical contributions of ECM constituents. J. Biomech. **49**, 2358–2365 (2016)

Wenger, M., Bozec, L., Horton, M., Mesquida, P.: Mechanical properties of collagen fibrils. Biophys. J. **93**, 1255–1263 (2007)

Wolinsky, H., Glagov, S.: Structural basis for the static mechanical properties of the aortic media. Circ. Res. **14**, 400–413 (1964)

Wolinsky, H., Glagov, S.: A lamellar unit of aortic medial structure and function in mammals. Circ. Res. **20**, 90–111 (1967)

Yu, X., Turcotte, R., Seta, F., Zhang, Y.: Micromechanics of elastic lamellae: unravelling the role of structural inhomogeneity in multi-scale arterial mechanics. J. R. Soc. Interface **15**, 20180492 (2018a)

Yu, X., Wang, Y., Zhang, Y.: Transmural variation in elastin fiber orientation distribution in the arterial wall. J. Mech. Behav. Biomed. Mater. **77**, 745–753 (2018b)

Yu, X., Suki, B., Zhang, Y.: Avalanches and power law behavior in aortic dissection propagation. Sci. Adv. **6**, eaaz1173 (2020)

Zhang, L., Lake, S., Lai, V., Picu, C., Barocas, V., Shephard, M.: A coupled fiber-matrix model demonstrates highly inhomogeneous microstructural interactions in soft tissues under tensile load. J. Biomech. Eng. **135**, 011008 (2013)

Zou, Y., Zhang, Y.: The orthotropic viscoelastic behavior of aortic elastin. Biomech. Model. Mechanobiol. **10**, 613–625 (2011)

Cohesive Zone Model Analysis, Development, and Application in Mixed-Mode Arterial Dissection

Brian FitzGibbon, Behrooz Fereidoonnezhad, and Patrick McGarry

From an early stage in my career I have closely followed Gerhard's seminal work on anisotropic soft tissue behavior. His rigorous and fundamental mechanistic advances have been key to the establishment of the field of computational biomechanics. Gerhard has been a generous and inspirational mentor and friend, and there have been many highlights over the years. He invited me to present at his special symposium on soft tissue modeling at ESMC in Madrid, after which we had some truly excellent discussions. I was particularly honored when Gerhard traveled to Galway to deliver a wonderful keynote lecture when I chaired the National Irish Bioengineering Conference in 2016. I was truly delighted when Gerhard appointed me as Visiting Professor at TU Graz in 2017, allowing me to spend two wonderful weeks at his Institute for Biomechanics. This was a memorable experience; in addition to delivering a graduate module, I had the pleasure of working with Gerhard on the development of a micromechanical model of the myocardium. Gerhard also gave me fantastic experience of Austrian culture and cuisine, in addition to a unique St. Patrick's day celebration in Graz! Gerhard, thank you always for your generosity, support and friendship. Patrick

Gerhard has been my role model since I started my masters in the field of solid mechanics in 2009. Five years later, I first met Gerhard in Graz when I had the honor of a six month visit to his institute as a visiting researcher. I had the pleasure of working with Gerhard on the development of a mechanobiological model for growth of arterial tissue. He gave me a deep insight into the field of continuum thermodynamics. I was particularly honored when Gerhard accepted to be my PhD co-supervisor. Gerhard, I appreciate all of your supports and kindness. I look forward to continue collaboration in the future. Behrooz

I attended Professor Holzapfel's Summer School at the Institute of Biomechanics in Graz in 2016. It was a truly inspirational and memorable experience. Thank you for providing this wonderful opportunity. Brian

B. FitzGibbon · B. Fereidoonnezhad · P. McGarry (✉)
School of Engineering, National University of Ireland Galway, Galway, Ireland
e-mail: patrick.mcgarry@nuigalway.ie

G. Sommer et al. (eds.), *Solid (Bio)mechanics: Challenges of the Next Decade*, Studies in Mechanobiology, Tissue Engineering and Biomaterials 24,
https://doi.org/10.1007/978-3-030-92339-6_4

Abstract Recent studies have presented evidence that aortic dissection entails mode II and mixed-mode crack growth. In the present chapter we analyze the mixed-mode behavior of two exponentially softening path-dependent cohesive zone models (CZMs). We demonstrate that these models provide physically realistic tractions and positive incremental dissipation during proportional and non-proportional loading paths. In contrast, we demonstrate that potential-based path-independent CZMs can potentially result in the calculation of physically unrealistic repulsive normal tractions and negative incremental dissipation. Our recent experiments suggest that mode II fracture strength of the aorta in the circumferential-axial plane is significantly higher than the mode I strength in the same plane. We demonstrate that such anisotropic interface behavior renders potential-based CZMs unsuitable for simulation of aortic dissection. Using our new non-potential path-dependent CZM we simulate aortic dissection due to the presence of a notch in the radial-axial plane of the aorta. Simulations suggest that significant dissection propagation only occurs in cases of extreme hypertension. Additionally, we demonstrate that blood pressure loading in a false lumen will result in a mixed-mode traction state at the crack tip, with significant crack-tip blunting acting as a toughening mechanism against significant propagation of the false lumen.

1 Introduction

The progression and pathogenesis of aortic dissection (AD) is relatively poorly understood compared to other cardiovascular diseases (Vilacosta et al. 2009; Criado 2011). For example, several clinical studies suggest that AD occurs as a result of initial damage to the intima, with such damage referred to in clinical literature as an 'intimal tear' or 'entry tear' (Larson and Edwards 1984; Criado 2011; Lemaire and Russel 2011; Kim et al. 2014). However, a number of clinical studies report AD without an intimal tear (Lui et al. 1992; Eichelberger 1994; Utoh et al. 1997; Colli et al. 2018). Conversely, an intimal tear may present without extensive AD, indicating early arrest of tear propagation (Svensson et al. 1999). AD is typically characterized in the clinical literature as the occurrence of a true and false lumen separated by an intraluminal septum (Hasleton and Leonard 1979; Sayer et al. 2008; Nienaber et al. 2010; Huang et al. 2015). The end point of the false lumen is typically used to indicate the extent of the AD, which may be as short as 1 cm from the original entry tear (Svensson et al. 1999) or as long as the entire aorta and iliac vessels (>1 m) (Dotter et al. 1950; Hagan et al. 2000; Dake et al. 2013; Qiao et al. 2015; Gambardella et al. 2017).

In recent years there have been a myriad of numerical studies examining the fluid dynamics of AD, and more recently, studies examining AD from a fluid-structure interaction standpoint (Cheng et al. 2013, 2014; Alimohammadi et al. 2015; Doyle and Norman 2016; Ryzhakov et al. 2019; Bäumler et al. 2020; Xiong et al. 2020; Zorrilla et al. 2020). Many of these studies explore subject matter such as the effect of bypassing the false lumen on fluid flow using a bypass graft (Qiao et al. 2015), development of patient-specific models (Alimohammadi et al. 2015; Xiong et al.

2020), and analysis of intraluminal septum rigidity (Bonfanti et al. 2018). However, to the authors' knowledge, no study to date has examined the risk of false lumen propagation from a fracture mechanics standpoint. In the present chapter we investigate the application of mixed-mode cohesive zone models (CZMs) to analyze AD risk. We begin the chapter with an analysis of mixed-mode cohesive zone formulations. Cohesive zone models have been extensively used to describe an interface undergoing separation (Barenblatt 1959; Xu and Needleman 1993; van den Bosch et al. 2006; McGarry et al. 2014). CZMs have been used to model crack propagation in ductile metals (Nielsen 2012), polyethylene (Ivankovic et al. 2004), porous materials (Nakamura and Wang 2001) and concrete beams (Aure and Ioannides 2010). They have also been used to model the delamination of cells from substrates (McGarry and McHugh 2008), polymer coatings from stents (Hopkins et al. 2010; McGarry et al. 2014), inter-laminar failure in carbon-fiber laminates (Gallagher et al. (2018, 2019)), failure of coatings of diamond-coated cutting tools (Hu et al. 2008), and separation of arterial layers (Gasser and Holzapfel (2006, 2007); Ferrara and Pandolfi 2010; Wang et al. 2017; Noble et al. 2017; Gültekin et al. 2019; FitzGibbon and McGarry 2020). The calibration of CZM parameters requires experimental testing data (Ivankovic et al. 2004; Di Leo et al. 2014; Wu et al. 2016). The primary outputs of these tests are interface strength and fracture energy. It is extremely difficult to measure the stiffness of interfaces in highly elastic materials. The characteristic length (δ) or the modulus (K) of the initial elastic region of some CZMs often dictates the fracture energy (G) and vice versa. This is common in exponential CZM formulations as seen in Xu and Needleman (1993), van den Bosch et al. (2006) and Dimitri et al. (2015). Therefore, the choice of fracture energy may anomalously influence the stiffness and characteristic length of the interface.

Piecewise CZMs allow for the specification of K (and by extension δ) and G independently (Camanho et al. 2003; Park et al. 2009). Choice of CZM should be motivated by experimental data and boundary conditions. CZMs may be coupled or uncoupled. In a coupled CZM the traction depends on the separation vector which will include both normal and tangential opening (Δ_n, Δ_t), whereas in an uncoupled CZM the normal traction (T_n) depends only on the normal opening (Δ_n). Most engineering applications involve mixed separation and therefore should implement mixed-mode (coupled) CZMs. Achieving a physically realistic mixed-mode response is important in such applications. Many constitutive relationships of traction-separation laws have been proposed, these include but are not limited to, linear softening (Camacho and Ortiz 1996), bilinear softening, cubic polynomial, exponential (Xu and Needleman 1993; van den Bosch et al. 2006; McGarry et al. 2014), exponential softening, and trapezoidal (Tvergaard and Hutchinson 1992).

In the first part of this chapter we investigate the suitability of several cohesive zone formulations for mixed-mode dissection. We firstly consider exponential softening model and introduce a feature whereby tangential interface strength can be augmented as a function of compressive normal traction. We also explore the extension of such exponential softening models to a potential-based formulation. We demonstrate that the derivation of path-independent tractions separation relationships from a potential function results in non-physical behavior. In the second part of this chapter we apply our mixed-mode exponential softening CZM to the analysis of AD. We consider the evolution of AD due to pre-existing intraluminal septum and a patent false lumen. Simulations suggest that extensive propagation of a false lumen will not occur at a slightly hypertensive systolic pressure of 140 mmHg in a healthy aorta. Even in extreme hypertensive loading conditions AD propagation is arrested due to blunting of the crack tip and an increase in the mode angle towards mode II. We also investigate if an intimal tear (radial notch) will result in extensive AD under physiological and super-physiological lumen blood pressure. Simulations suggest that propagation of an intimal tear is not predicted for pressures less than $p = 275$ mmHg in a healthy aorta.

2 Analysis of CZMs in Mixed-Mode Dissection

2.1 Development of Non-Potential Based Mixed-Mode CZMs with Exponential Damage and Overclosure Penalization (CZM1 and CZM2)

For a given interface displacement vector $\mathbf{\Delta}$ with normal and tangential (shear) components Δ_n and Δ_t, respectively, the corresponding displacement magnitude is given as $\Delta_m = \left(\Delta_n{}^2 + \Delta_t{}^2\right)^{1/2}$ and the mode angle is given as $\varphi = \tan^{-1}(\Delta_t/\Delta_n)$. The magnitude of the interface traction is expressed as a function of Δ_n and φ by the following formulation:

$$T_m(\Delta_m, \varphi) = \begin{cases} K_m \Delta_m, & \Delta_m < T_m^{max}(\varphi)/K_m. \\ K_m T_m^{max}(\varphi)/K_m \Psi(\varphi), & \Delta_m \geq T_m^{max}(\varphi)/K_m. \end{cases} \tag{1}$$

We refer to Ψ as the integrity of the interface (i.e. $\Psi = (1 - D)$, where D is referred to as the interface damage). Ψ monotonically decreases from 1 to 0 with increasing interface separation, such that

$$\Psi(\varphi) = \exp\left(-\frac{\Delta_m^{max} - T_m^{max}(\varphi)/K_m}{\delta_m^*(\varphi)}\right)\frac{\Delta_m}{\Delta_m^{max}}, \tag{2}$$

where K_m is the intrinsic elastic stiffness of the interface. K_m is assumed to be mode-independent. $T_m^{max}(\varphi)$ is the specified mode-dependent interface strength; the corresponding displacement ($\delta_m^{el}(\varphi) = T_m^{max}(\varphi)/K_m$) represents the elastic limit of interface separation at the point of damage initiation. The mode-dependent parameter $\delta_m^*(\varphi)$ governs the rate of softening in the damage region $\Delta_m \geq T_m^{max}(\varphi)/K_m$. The interface strength is defined as a function of the mode as follows:

$$T_m^{max}(\varphi) = \tau_{max} - \left[\frac{\tau_{max} - \sigma_{max}}{1 - \exp\left(-\frac{\pi/2}{\Omega^T}\right)}\right]\left[1 - \exp\left(-\frac{\varphi}{\Omega^T}\right)\right], \tag{3}$$

where $\tau_{\max}$ is the mode II interface strength, Ω^{T} sets the nonlinearity of the transition from mode II to mode I, and $\sigma_{\max}$ is the mode I interface strength. The mode mixity of the initial interface damage parameter $\delta_{\mathrm{m}}^{*}(\varphi)$ is obtained from

$$\delta_{\mathrm{m}}^{*}(\varphi)=\frac{G_{\mathrm{m}}(\varphi)}{T_{\mathrm{m}}^{\max}(\varphi)}-\frac{T_{\mathrm{m}}^{\max}(\varphi)}{2K_{\mathrm{m}}}, \tag{4}$$

where $G_{\mathrm{m}}(\varphi)$ is the mode-dependent fracture energy

$$G_{\mathrm{m}}(\varphi)=\frac{1}{2}K_{\mathrm{m}}\delta_{\mathrm{m}}^{\mathrm{el}}(\varphi)^{2}+\int_{T_{\mathrm{m}}^{\max}(\varphi)/K_{\mathrm{m}}}^{\infty}T_{\mathrm{m}}^{\max}(\varphi)\exp\left(-\frac{\Delta_{\mathrm{m}}^{\max}-T_{\mathrm{m}}^{\max}(\varphi)/K_{\mathrm{m}}}{\delta_{\mathrm{m}}^{*}}\right)\mathrm{d}\Delta_{\mathrm{m}}. \tag{5}$$

We may specify the mode-dependence of $G_{\mathrm{m}}(\varphi)$ using the following function:

$$G_{\mathrm{m}}(\varphi)=G_{\mathrm{t}}^{0}-\left[\frac{G_{\mathrm{t}}^{0}-G_{\mathrm{n}}^{0}}{1-\exp\left(-\frac{\pi/2}{\Omega^{\mathrm{G}}}\right)}\right]\left[1-\exp\left(-\frac{\varphi}{\Omega^{\mathrm{G}}}\right)\right], \tag{6}$$

where G_{t}^{0} is the mode II fracture energy and G_{n}^{0} is the mode I fracture energy. Ω^{G} sets the nonlinearity of the transition from mode II to mode I. Finally, we complete the description of the cohesive zone formulation by decomposing T_{m} into the normal and tangential components, T_{n} and T_{t}, respectively, such that

$$T_{\mathrm{n}}=\begin{cases}K_{\mathrm{n_{oc}}}\Delta_{\mathrm{n}}, & \Delta_{\mathrm{n}}<0,\\ T_{\mathrm{m}}\sin\varphi, & \Delta_{\mathrm{n}}\geq 0,\end{cases}\qquad T_{\mathrm{t}}=T_{\mathrm{m}}\cos(\varphi), \tag{7}$$

where $K_{\mathrm{n_{oc}}}$ is the overclosure penalty stiffness. We include a dependence of the mode II interface strength, $\tau_{\max}$, on normal compression at the interface, such that

$$\tau_{\max}(T_{\mathrm{n}})=\begin{cases}\tau_{\max}^{\mathrm{nc}}(T_{\mathrm{n}}), & \Delta_{\mathrm{n}}<0,\\ \tau_{\max}^{0}, & \Delta_{\mathrm{n}}\geq 0,\end{cases} \tag{8}$$

as described in Eq. (9). The mode II interface strength increases with increasing (negative) normal traction, such that

$$\tau_{\max}^{\mathrm{nc}}(T_{\mathrm{n}})=\tau_{\max}^{0}+\tau_{\max}^{0}(F_{\mathrm{oc}}-1)\left[1-\exp\left(\frac{T_{\mathrm{n}}K_{\mathrm{n_{oc}}}^{*}/\sigma_{\max}}{K_{\mathrm{m}}}\right)\right], \tag{9}$$

where $\tau_{\max}^{\mathrm{nc}}(\Delta_{\mathrm{n}})$ is the increased value of tangential strength due to a normal compression at the interface. $\tau_{\max}^{0}$ is the maximum tangential strength, as encountered during a pure mode II separation when $\Delta_{\mathrm{n}}=0$ and $T_{\mathrm{n}}=0$. The parameter F_{oc} prescribes the maximum (plateau) value of increased shear strength due to compressive normal tractions at the interface. The parameter $K_{\mathrm{n_{oc}}}^{*}$ governs the sensitivity of maximum (plateau) shear stress to compressive normal tractions. Unless otherwise stated, we assume that $\tau_{\max} = \tau^{0}_{\max}$, i.e. maximum tangential strength is not increased due to normal compression at the interface.

2.2 Alternative Form of Damage and Softening (CZM2)

To demonstrate that alternative forms of damage softening laws can be readily incorporated into our CZM framework, we next present a small modification of Eq. (2) whereby in which exponential damage is assumed to depend on the square of the interface separation. Again, starting with Eq. (1), and the integrity of the interface, Ψ is now defined as

$$\Psi(\varphi) = \exp\left[-\left(\frac{\Delta_{\mathrm{m}}^{\max} - T_{\mathrm{m}}^{\max}(\varphi)/K_{\mathrm{m}}}{\delta_{\mathrm{m}}^{*}(\varphi)}\right)^{2}\right]\frac{\Delta_{\mathrm{m}}}{\Delta_{\mathrm{m}}^{\max}}. \tag{10}$$

We make use of the same relation of $G(\varphi) = \int_0^{\infty} T_{\mathrm{m}}(\Delta_{\mathrm{m}}, \varphi)\,\mathrm{d}\Delta_{\mathrm{m}}$ in order to obtain an expression for δ_{m}^{*}. The description of $G_{\mathrm{m}}(\varphi)$ is given as follows

$$\begin{aligned} G_{\mathrm{m}}(\varphi) &= \frac{1}{2}K_{\mathrm{m}}\delta_{\mathrm{m}}^{\mathrm{el}}(\varphi)^{2} \\ &+ \int_{T_{\mathrm{m}}^{\max}(\varphi)/K_{\mathrm{m}}}^{\infty} T_{\mathrm{m}}^{\max}(\varphi)\exp\left[-\left(\frac{\Delta_{\mathrm{m}}^{\max} - T_{\mathrm{m}}^{\max}(\varphi)/K_{\mathrm{m}}}{\delta_{\mathrm{m}}^{*}(\varphi)}\right)^{2}\right]\mathrm{d}\Delta_{\mathrm{m}}. \end{aligned} \tag{11}$$

Integrating we get the following expression which includes the error function

$$G_{\mathrm{m}} = \frac{1}{2}K_{\mathrm{m}}\delta_{\mathrm{m}}^{\mathrm{el}}(\varphi)^{2} + T_{\mathrm{m}}^{\max}(\varphi)\left\{\lim_{\Delta_{\mathrm{m}}\to\infty}\left[\frac{1}{2}\sqrt{\pi}\,\delta_{\mathrm{m}}^{*}(\varphi)\left(\frac{\Delta_{\mathrm{m}} - T_{\mathrm{m}}^{\max}(\varphi)/K_{\mathrm{m}}}{\delta_{\mathrm{m}}^{*}(\varphi)}\right)\right]\right\}. \tag{12}$$

As $\Delta_{\mathrm{m}} \to \infty$ we arrive at the final expression for $\delta_{\mathrm{m}}^{*}(\varphi)$, i.e.

$$\delta_{\mathrm{m}}^{*}(\varphi) = -\frac{T_{\mathrm{m}}^{\max}(\varphi)}{K_{\mathrm{m}}\sqrt{\pi}} + \frac{2G_{\mathrm{m}}(\varphi)}{\sqrt{\pi}\,T_{\mathrm{m}}^{\max}(\varphi)}. \tag{13}$$

2.3 Exploration of CZM1 and CZM2 Behavior

We firstly present the response of CZM1 and CZM2 to pure mode II loading, as shown in Fig. 1a. Normalized tangential traction $T_{\mathrm{t}}/\tau_{\max}$ is presented as a function of normalized tangential separation $\Delta_{\mathrm{t}}/\delta_{\mathrm{t}}^{\mathrm{el}}$. Traction increases according to the interfacial stiffness K_{m} until the prescribed mode II strength $\tau_{\max}$ is reached. Further deformation results in damage and softening. Due to the form of the respective exponential

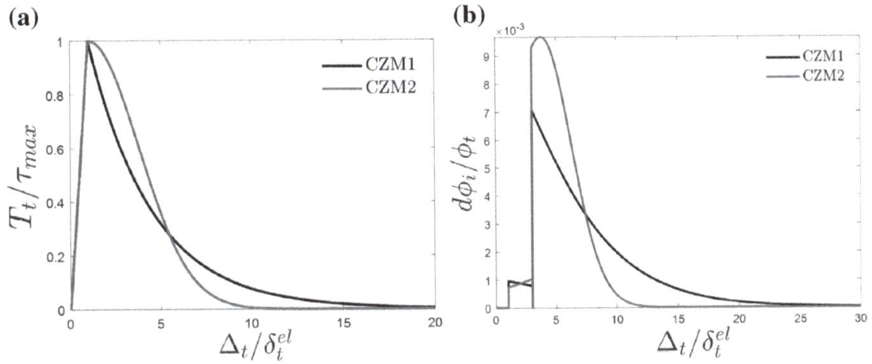

Fig. 1 (a) Normalized traction separation curves for a mode II separation for both formulations of the proposed model ($K_m = 10$ MPa, $\tau_{max} = 5$ MPa, $G_t^0 = 10$ N/mm); (b) instantaneous incremental energy dissipation $d\phi_i/\phi_t$ as a function of normalized tangential separation for CZM1 and CZM2 during a mode II displacement

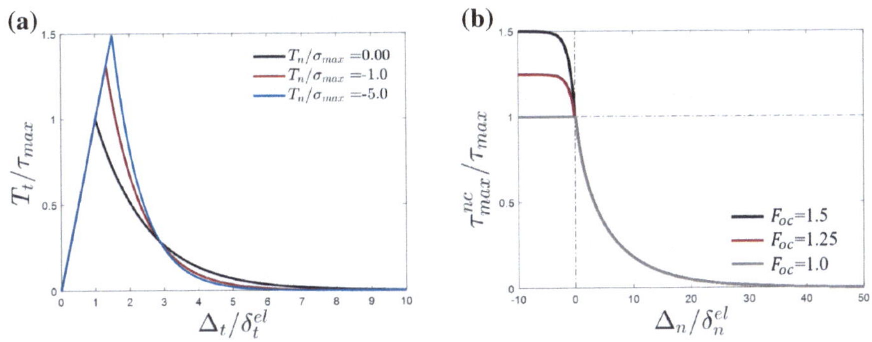

Fig. 2 (a) Influence of compressive normal tractions on tangential fracture; (b) maximum tangential traction shown as a function of the F_{oc} parameter

damage laws, CZM1 initially softens at a faster rate than CZM2. An expression for the instantaneous incremental energy dissipation (Cazes et al. 2009) is given as $d\phi_i = 0.5(\mathbf{T}d\mathbf{\Delta} - \mathbf{\Delta}d\mathbf{T})$. As shown in Fig. 1b, positive instantaneous incremental energy dissipation is observed throughout the mode II separation for both CZM1 and CZM2.

In Fig. 2a we explore the influence of normal compressive tractions on tangential fracture. An increase in magnitude of a compressive normal traction T_n (with $T_n < 0$) results in an increase in $\tau_{max}^{nc}(T_n)$, up to a plateau value of $\tau_{max}^{0} F_{oc}$ as $T_n \rightarrow -\infty$. Figure 2b further explores the influence of a fixed applied normal separation on the maximum tangential traction, T_{max}, encountered during a subsequent tangential separation. As expected, normal interface separation $\Delta_n > 0$ results in a reduction in T_{max}. In contrast, if $\Delta_n < 0$ a negative normal traction $T_n < 0$ is obtained through

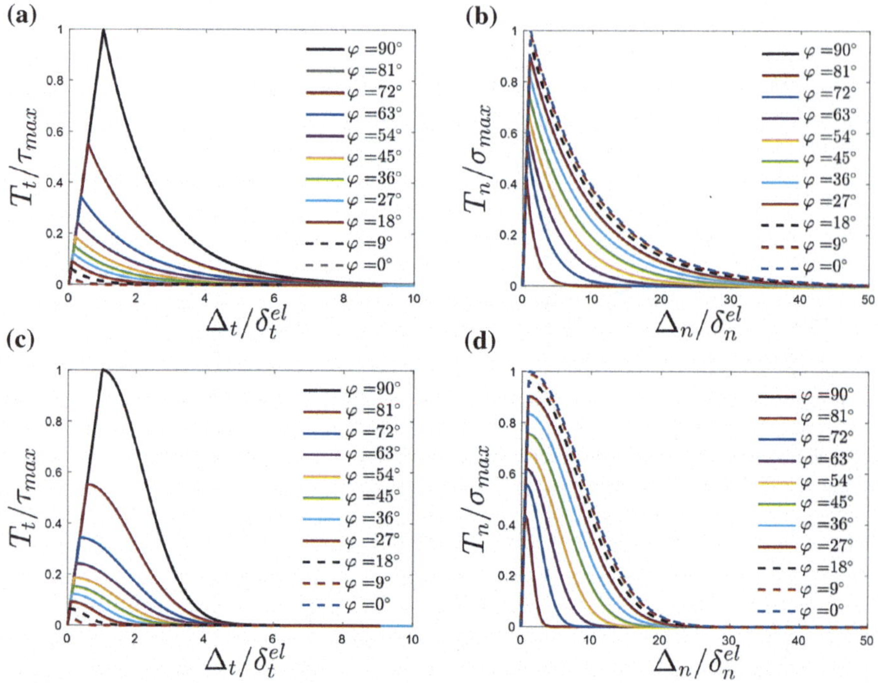

Fig. 3 Mixed-mode response of each CZM undergoing proportional loading ranging from mode II to mode I. Normalized traction versus separation in the normal direction (b), (d) and tangential direction (a), (c) for CZM1 (a), (b), CZM2 (c), (d) undergoing proportional loading. $K_{\rm m} = 10\,{\rm MPa}$, $\tau_{\rm max} = 5\sigma_{\rm max}$, $G_{\rm t} = 5G_{\rm n}$

the overclosure penalty term (Eq. (8)). This results in an increase in $\tau_{\rm max}^{\rm nc}(T_{\rm n})$ up to a plateau value with increasingly negative normal traction.

Proportional loading paths: CZM1 and CZM2 are examined under mixed-mode proportional loading conditions whereby $\Delta_{\rm m}$ increases at a constant mode angle φ. Computed traction-separation responses are presented in Fig. 3 for the case of $K_{\rm m} = 10\,{\rm MPa}$, $\tau_{\rm max} = 5\sigma_{\rm max}$, $G_{\rm t} = 5G_{\rm n}$, $\Omega^{\rm T} = \pi/16$, $\Omega^{\rm G} = \Omega^{\rm T}$ for both models. CZM1 is shown in Fig. 3a, b and CZM2 in Fig. 3c, d. Consistent mixed-mode behavior is observed for both formulations, with a gradual transition from mode II behavior to mode I behavior. In accordance with Eqs. (3) and (6), identical fracture energy is obtained for both formulations for such proportional loading paths, as shown in Fig. 4. The computed total fracture energy $G_{\rm total}$ monotonically increases from $G_{\rm n}^0$ to $G_{\rm t}^0$.

Non-proportional loading paths: We next consider non-proportional loading path whereby the interface undergoes an initial mode II separation to a prescribed value of $\Delta_{\rm t} = \Delta_{\rm t}^{\rm max}$, followed by a subsequent normal separation to complete failure. Figure 5 shows the traction-separation response, again for the case of $K_{\rm m} = 10\,{\rm MPa}$, $\tau_{\rm max} = 5\sigma_{\rm max}$, $G_{\rm t} = 5G_{\rm n}$, $\Omega^{\rm T} = \pi/16$, $\Omega^{\rm G} = \Omega^{\rm T}$ for both models. CZM1 is shown

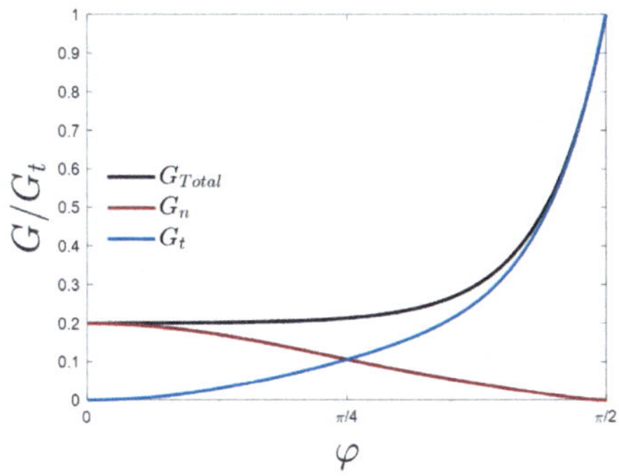

Fig. 4 Computed fracture energy as a function of mode angle φ for CZM1 and CZM2 under proportional loading in the case of $K_m = 10\,\text{MPa}$, $\tau_{max} = 5\sigma_{max}$, $G_t = 5G_n$, $\Omega^T = \pi/16$, $\Omega^G = \Omega^T$

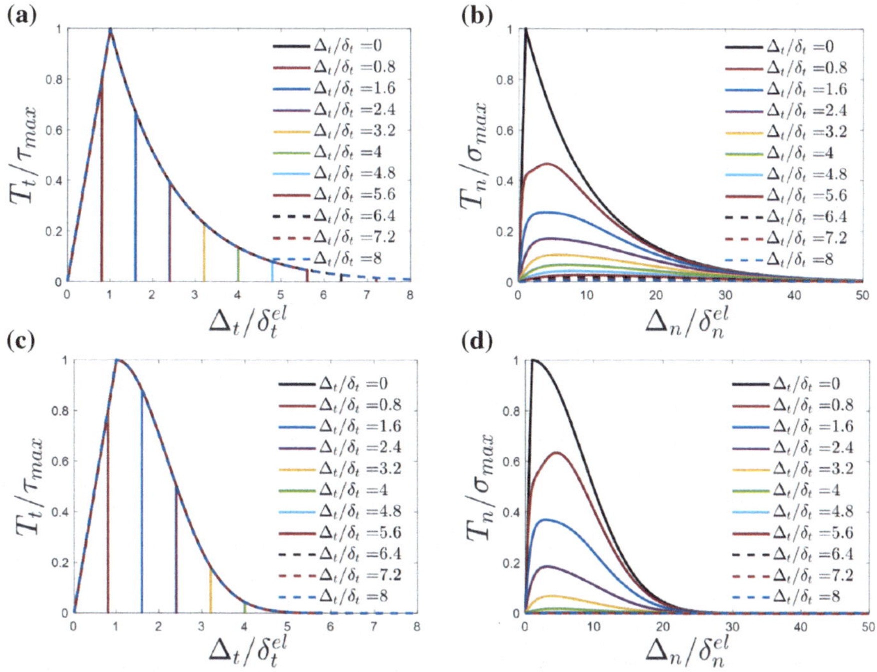

Fig. 5 Normalized traction versus separation in the normal direction (b), (d) and tangential direction (a), (c) for CZM1 (a), (b) and CZM2 (c), (d) undergoing non-proportional loading. Each simulation involves incrementally increasing tangential displacements beginning at $\Delta_t/\delta_t = 0$ and increasing until the interface is fully debonded in the tangential direction followed by a complete normal separation ($\Delta_n/\delta_n = 50$. $K_m = 10\,\text{MPa}$, $\tau_{max} = 5\sigma_{max}$, $G_t = 5G_n$)

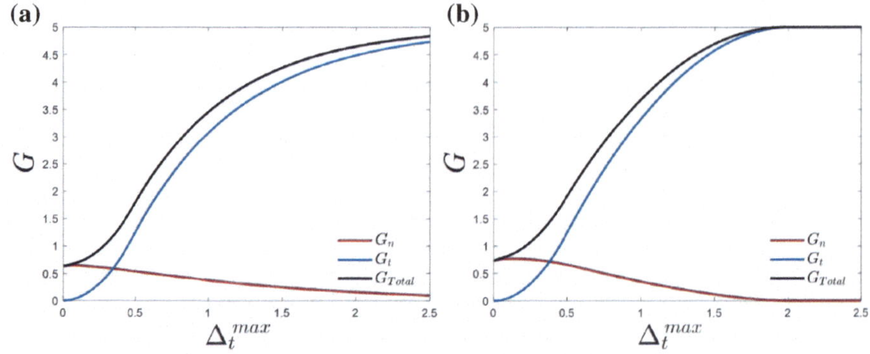

Fig. 6 Fracture energy (G) for CZM1 (a) and CZM2 (b) undergoing non-proportional loading paths of varying initial tangential separation Δ_t followed by normal separation Δ_n until complete failure. Parameters are the same as those implemented in Fig. 5

in Fig. 5a, b and CZM2 in Fig. 5c, d. Consistent behavior is again observed for both models. If significant damage is computed during the initial mode II separation, the maximum normal traction computed during subsequent normal separation is reduced. This demonstrates that sensible mixed-mode coupling is provided by both CZM1 and CZM2. Computed fracture energy is presented in Fig. 6 for the same non-proportional loading paths as those presented in Fig. 5. While identical fracture energy is not obtained for both formulations, a monotonic increase as a function of increasing Δ_t^{max} is computed both formulations. Additionally, monotonically increasing/decreasing tangential/normal fracture energy contributions are also computed.

The dissipation of CZM1 is presented in Fig. 7a, c and CZM2 is presented in Fig. 7b, d for the proportional loading (Fig. 7a, b) and non-proportional loading (Fig. 7c, d). Positive instantaneous incremental dissipation is computed throughout each of the analyses presented. The response of CZM1 and CZM2 to load-unload boundary conditions is demonstrated in Fig. 8. Specifically, as shown in Fig. 8a, a mixed-mode proportional loading path is followed so that partial damage is computed. The interface is then unloaded along the same mode angle, until it returns to its original configuration ($\Delta_n = \Delta_t = 0$). The interface is then subjected to the same mode angle in the reverse direction ($\Delta_t < 0$) to failure. As shown in Fig. 8b–d, consistent behavior is obtained for all mode angles. In all cases, the prescribed strength for the given mode angle is exceeded and softening/damage is computed. Reversal of loading results in elastic unloading at a reduced/damaged interface stiffness. Reversal of the direction of tangential loading direction for the final mixed-mode component of the loading path results in a continuation of elastic deformation at the reduced damaged interface stiffness. Eventually, the tangential traction reaches the reduced/damaged tangential strength and further interface softening proceeds during ultimate mode II separation and failure. Similar consistent trends are computed for both CZM1 and CZM2. Figure 8f, g demonstrate that instantaneous incremental dissipation is positive throughout the entire mixed-mode loading histories shown in Fig. 8b–e.

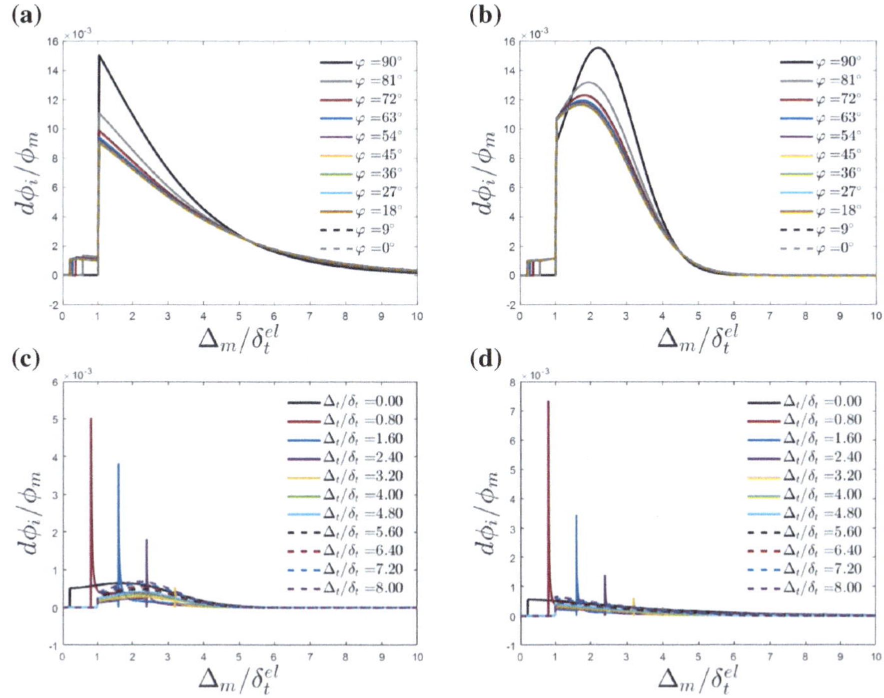

Fig. 7 Instantaneous incremental energy dissipation of CZM1 (a), (c) and CZM2 (b), (d) under proportional loading (a), (b) and non-proportional loading (c), (d)

2.4 *Comparison of CZM1 and CZM2 with Abaqus Exponential Softening Formulation (CZM3)*

The commercial finite element software Abaqus provides a cohesive zone functionality in which damage evolution may be specified in terms of mixed-mode fracture energy. In this section we describe this formulation, which we refer to as CZM3, and we provide critical comparisons with CZM1 and CZM2. The constitutive law for the traction is specified as follows

$$\mathbf{T} = \begin{pmatrix} T_\mathrm{n} \\ T_\mathrm{t} \end{pmatrix} = \begin{bmatrix} K_\mathrm{nn} & 0 \\ 0 & K_\mathrm{tt} \end{bmatrix} \begin{pmatrix} \Delta_\mathrm{n} \\ \Delta_\mathrm{t} \end{pmatrix} = \mathbf{K}\boldsymbol{\Delta}, \tag{14}$$

where K_nn is a prescribed mode I stiffness and K_tt is a prescribed mode II stiffness. The traction increases elastically in accordance with Eq. (14) until the chosen damage initiation criterion is met. As an example, the quadratic failure criterion (Tsai and Wu 1971) can readily be chosen, such that damage initiates when

$$\left(\frac{\langle T_\mathrm{n} \rangle}{\sigma_\mathrm{max}}\right)^2 + \left(\frac{T_\mathrm{t}}{\tau_\mathrm{tmax}}\right)^2 = 1, \tag{15}$$

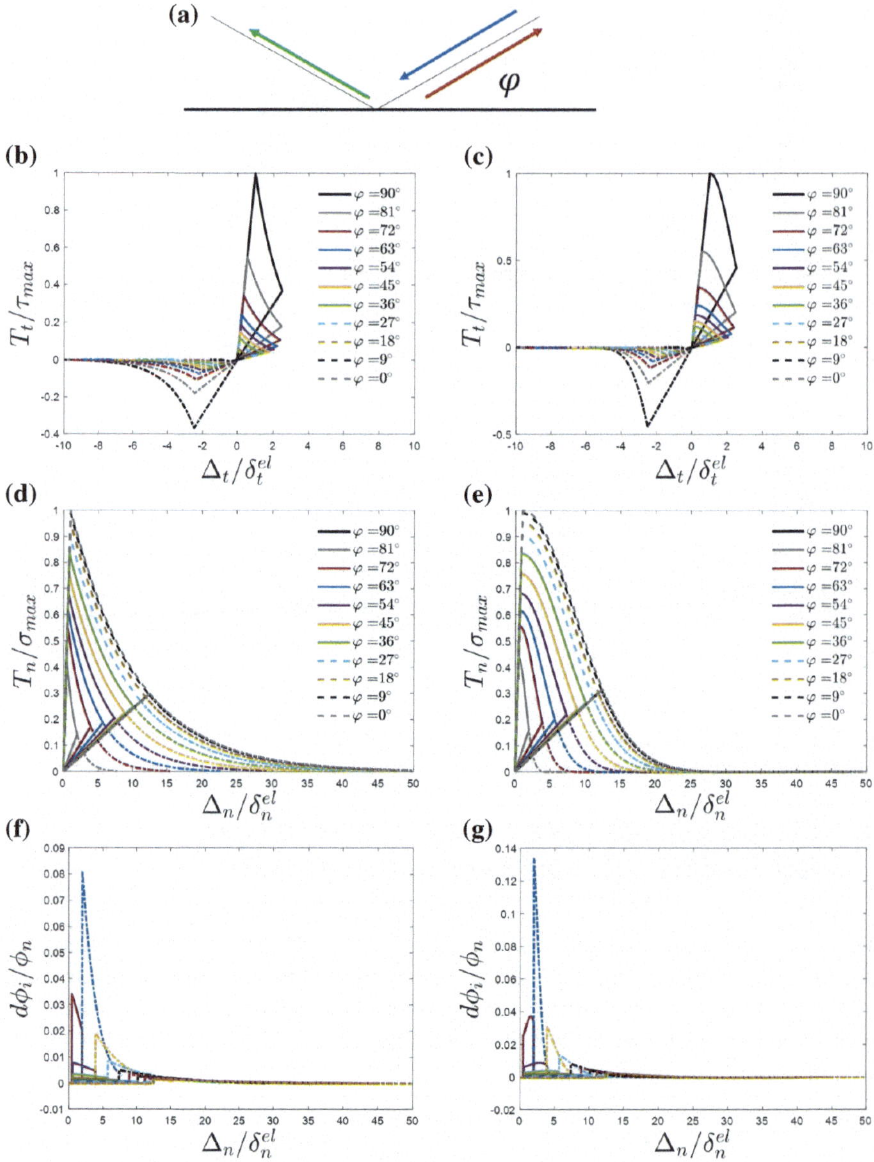

Fig. 8 Loading-unloading path (a); demonstration of the interface integrity variable in CZM1 (b), (d) and CZM2 (c), (e) for a series of proportional loading modes ranging from mode II to mode I ($K_m = 10\,\text{MPa}$, $\tau_{max} = 5\sigma_{max}$, $G_t = 5G_n$). Instantaneous incremental dissipation is shown in (f), (g) for CZM1 and CZM2, respectively

where brackets describe the ramp function

$$\langle T_{\mathrm{n}} \rangle = \begin{cases} 0, & T_{\mathrm{n}} < 0, \\ T_{\mathrm{n}}, & T_{\mathrm{n}} \geq 0. \end{cases} \tag{16}$$

Once the damage initiation criterion is satisfied, the scalar damage variable D increases monotonically from 0 to 1 with increasing separation until complete failure of the interface occurs. The traction during damage evolution is given as

$$\mathbf{T} = (1 - D)\,\overline{\mathbf{T}}, \tag{17}$$

where $\overline{\mathbf{T}}$ is the peak elastic traction calculated in the absence of damage according to Eq. (14). This applies in both tangential directions, however, in normal displacement the traction is given as follows

$$T_{\mathrm{n}} = \begin{cases} (1 - D)\overline{T}_{\mathrm{n}}, & \overline{T}_{\mathrm{n}} \geq 0, \\ \overline{T}_{\mathrm{n}}, & \overline{T}_{\mathrm{n}} < 0. \end{cases} \tag{18}$$

This ensures the stiffness of the interface in overclosure cannot be affected by the damage variable D whilst allowing stiffness of the interface in normal opening K_{nn} to be affected in the same manner as Eq. (17). For exponential damage evolution specified according to energy the damage variable D reduces to

$$D = \int_{\delta_{\mathrm{m}}^{\mathrm{el}}}^{\delta_{\mathrm{m}}^{*}} \frac{T_{\mathrm{m}} \mathrm{d}\Delta_{\mathrm{m}}}{G_{\mathrm{m}} - G_{0}}, \tag{19}$$

where T_{m} is the effective traction and Δ_{m} is the effective separation, G_0 is the stored-elastic energy at damage initiation and G_{m} is the mixed-mode fracture energy. In Fig. 9a we show the response of CZM3 to a simple mode II separation. The computed tangential traction exceeds the prescribed mode II fracture strength of the interface, providing a non-physical prediction. In contrast, CZM1 and CZM2 correctly reproduce the prescribed mode II fracture strength, in addition to the prescribed mode II fracture energy. Figure 9b shows the computed tangential traction for CZM3 as a function of the damage ($D = (1 - \Psi)$, where Ψ is the interface integrity) during a mode II separation. The relationship is non-monotonic, whereby the mode II interface strength incorrectly increases with increasing interface damage. Peak tangential traction occurs when the scalar damage variable $D \approx 0.3$. Furthermore, the tangential traction does not decrease to the specified mode II fracture strength until $D \approx 0.7$, i.e. when the interface is 70% damaged. In contrast, tangential tractions are correctly computed to decrease monotonically with increasing interface damage for CZM1 and CZM2.

Such anomalous behavior in CZM3 is further exposed in Fig. 10. Figure 10a shows the computed maximum normal traction as a function of mode angle during proportional mixed-mode separation. In the case of CZM1 and CZM2 the prescribed mode I strength is reached only for a pure mode I, as expected. However, for CZM3 the

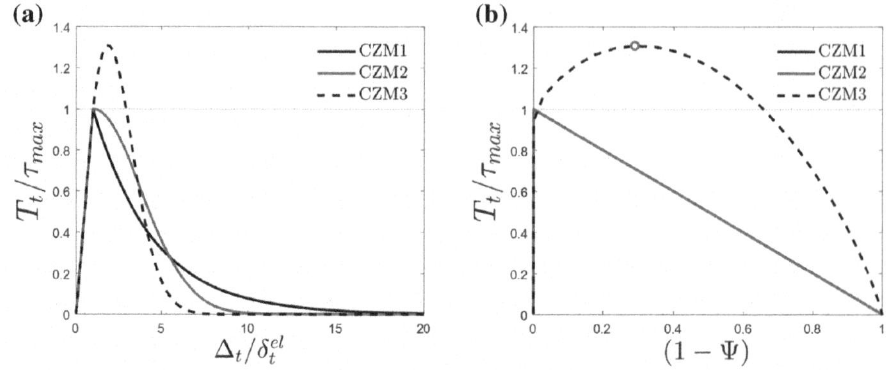

Fig. 9 (a) Normalized traction-separation response of CZM3 (Abaqus exponential softening) formulation presented with CZM1 and CZM2 for a pure mode II separation. Note $T_t > \tau_{max}$ for the Abaqus exponential softening ($K_m = 10\,\text{MPa}$, $\tau_{max} = 5\,\text{MPa}$, $G = 10\,\text{N/mm}$); (b) normalized tangential traction as a function of damage D (where $D = (1 - \Psi)$). Monotonic reduction is mode II interface strength with increasing damage for CZM1 and CZM2, but not for CZM3

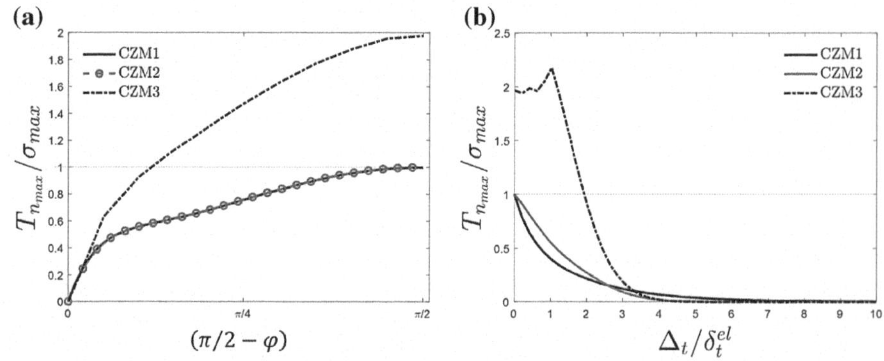

Fig. 10 (a) Normalized normal traction $T_{n\max}/\sigma_{max}$ as a function of mode angle φ for the three models in proportional loading scenarios (as seen in Fig. 3); (b) normalized traction as a function of maximum normalized tangential displacement prior to complete normal separation (as seen in Fig. 5). The dotted grey curve indicates σ_{max} in both (a) and (b). Note the non-monotonic behavior of the Abaqus exponential softening model

prescribed mode I strength is incorrectly exceeded for a wide range of applied mode angles. Similar results are presented in Fig. 10b for non-proportionate mixed-mode loading paths.

2.5 *Construction of a Potential-Based CZM*

Several previous studies have used traction-separation laws derived from a potential function (Xu et al. 1993; Park et al. 2009), so that the work of separation associated with a given interface separation vector should be path independent. A study by

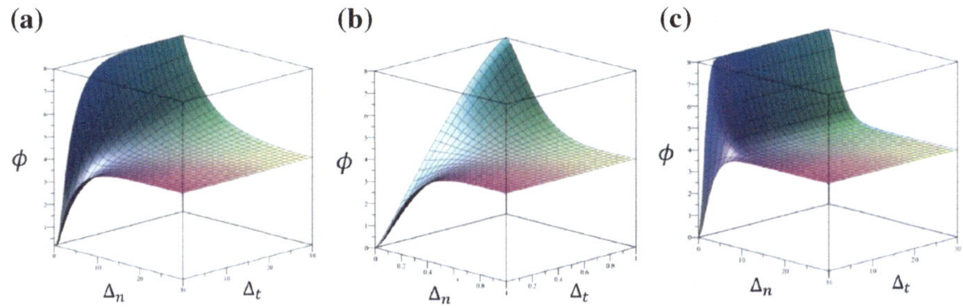

Fig. 11 Potential surfaces shown as a function of normal separation and tangential separation: (a) PF-CZM1 ($K = 1, \varphi_t = 2\varphi_n, \tau_{max} = \sigma_{max}, \varphi^0 = \pi/8$); (b) PPR ($\sigma_{max} = \tau_{max}, \varphi_t = 2\varphi_n, \delta_n = \delta_t$); (c) Xu-Needleman ($r = 0, q = 2$)

McGarry et al. (2014) uncovered problematic behavior associated with potential-based CZMs, such as repulsive tractions during mixed-mode loading paths. To further demonstrate this limitation of potential-based CZMs, we construct a potential surface by integrating the traction magnitude-separation magnitude relationships specified for CZM1 along proportional loading paths for $0 \leq \varphi \leq \pi/2$, giving the following piece-wise expression potential function

$$\phi(\Delta_m, \varphi) = \begin{cases} \frac{1}{2} K_m \Delta_m^2, & \Delta_m < T_m^{max}(\varphi)/K_m, \\ \frac{1}{2} K_m \delta_m^{el}(\varphi)^2 - \\ T_m^{max}(\varphi)\delta_m^*(\varphi)\left[\exp\left(-\frac{\Delta_m - \delta_m^{el}(\varphi)}{\delta_m^*(\varphi)}\right) - 1\right], & \Delta_m \geq T_m^{max}(\varphi)/K_m. \end{cases} \tag{20}$$

We refer to this formulation as PF-CZM1 (potential function extension of CZM1). The traction separation relationships are derived from $T_{(i=n,t)} = \partial\varphi/\partial\Delta_i$. Figure 11a shows φ for PF-CZM1 for the case where the specified mode II fracture energy is two times higher than the mode I fracture energy. The potential function (φ) of the PPR model (2009) and the Xu-Needleman model (1993) are also shown for comparison. Figure 12a shows the normal traction $\partial\phi/\partial\Delta_n$ derived from the PF-CZM1 potential function ϕ as a function of normal separation Δ_n and tangential separation Δ_t. As clearly shown in Fig. 12b repulsive tractions are computed for increasing tangential separation, even when no normal separation exists ($\Delta_n = 0$).

Figure 13 illustrates the response of the PF-CZM1, XN and PPR potential-based formulations when subjected to a mixed-mode monotonic loading path with $\varphi = 45°$. In all cases tangential strength is higher than the normal strength ($\tau_{max} = 5\sigma_{max}$) and the tangential work of separation is higher than the corresponding normal value ($\phi_t = 5\phi_n$). In all cases significant repulsive normal tractions are computed from the potential function during the monotonic deformation path, while tangential tractions remain positive (non-repulsive). Negative instantaneous incremental dissipation is computed for the XN model, whereas such a violation of the second law is avoided for the PF-CZM1 and the PPR model simply due to the dominance of the positive dissipation of the tangential contribution over the negative normal contributions.

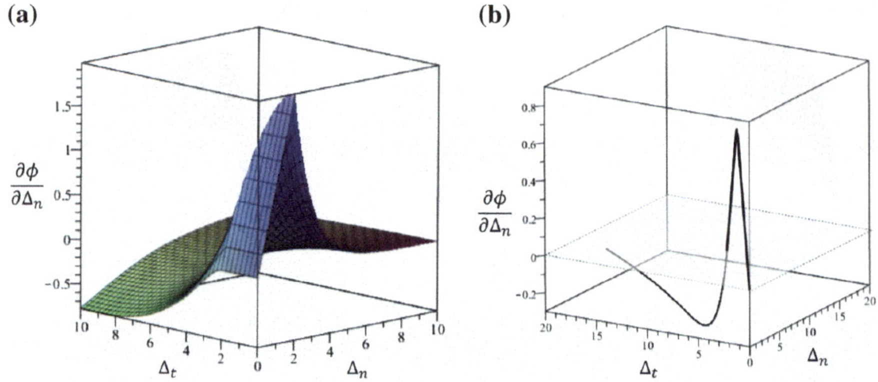

Fig. 12 (a) Normal traction derived from the potential function $\varphi(\Delta_\mathrm{n}, \Delta_\mathrm{t})$ seen in Fig. 11. Note repulsive normal tractions exist for increasing tangential opening even for $\Delta_\mathrm{n} = 0$; (b) repulsive normal tractions are further demonstrated for a proportional loading of $\Delta_\mathrm{t} = 2\Delta_\mathrm{n}$. The transparent surface indicates $\partial\phi/\partial\Delta_\mathrm{n} = 0$

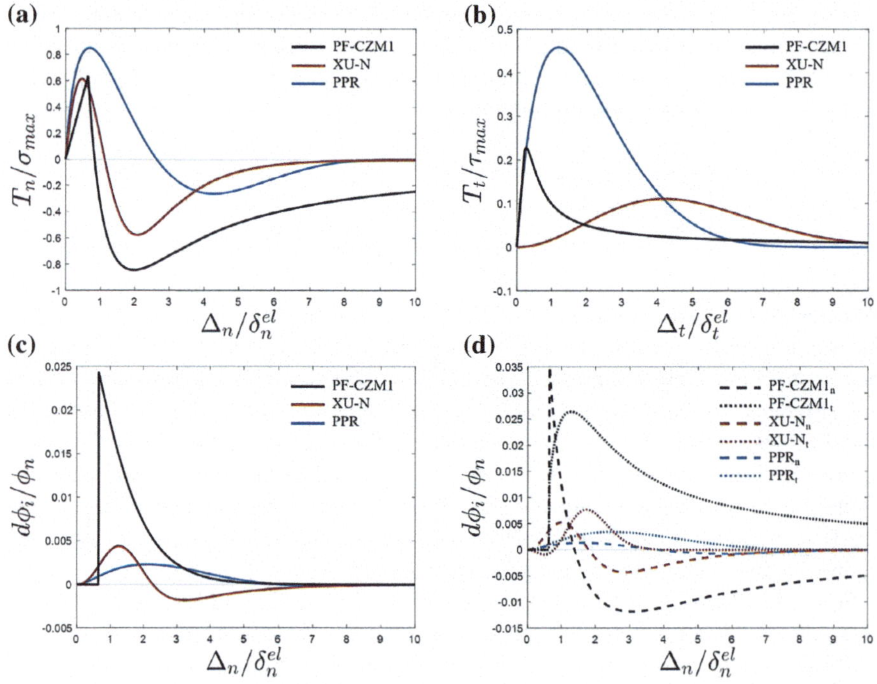

Fig. 13 PPR model subject to a proportional loading path where $\varphi = 45°$. Model parameters are the same as those seen in Fig. 5 ($\tau_{\max} = 5\sigma_{\max}$, $\phi_\mathrm{t} = 5\phi_\mathrm{n}$, $\delta_\mathrm{t} = 0.5$, $\delta_\mathrm{n} = \delta_\mathrm{t}/5$): (a) normalized normal traction $T_\mathrm{n}/\sigma_{\max}$ as a function of normal separation Δ_n. Repulsive (negative) tractions are computed; (b) normalized tangential traction $T_\mathrm{t}/\tau_{\max}$ as a function of tangential separation Δ_t. Tractions remain positive throughout; (c) instantaneous incremental energy dissipation $\mathrm{d}\phi_i/\phi_\mathrm{n}$ as a function of normal separation Δ_n. Negative instantaneous incremental energy dissipation is observed during the loading history; (d) decomposition of the instantaneous incremental energy dissipation (c) into normal and tangential components presented as a function of normal separation Δ_n

However, Fig. 13 illustrates that potential-based models can result in repulsive normal tractions and negative instantaneous incremental dissipation during mixed-mode loading paths.

3 CZM Simulation of Mixed-Mode Aortic Dissection

In the previous section we demonstrated the advantages of path-dependent exponential softening CZM1 and CZM2 formulations. In this section we use CZM1 to investigate mixed-mode AD.

3.1 Examination of an Artery with an Intimal Tear

We first examine the risk of dissection propagation in an artery with a pre-existing intimal tear in the form of a radial notch. Figure 14e shows an illustration of the intimal tear highlighted in red on the arterial geometry. In order to assess the risk of the intimal tear propagating under a luminal pressure load a cohesive zone is prescribed in the circumferential plane around the intimal tear, as shown by the red surface in Fig. 14e. In order to determine an upper limit of allowable lumen pressure a super-physiological blood pressure of 500 mmHg is applied in the lumen. A 10% axial stretch is applied to the artery prior to lumen pressurization. Hyperelastic anisotropic fiber-reinforced material behavior is prescribed to the artery according to the bilinear fiber model proposed by Fereidoonnezhad et al. (2020). The shear strength of the CZM interface is taken as $\tau_{\text{max}} = 1.6\,\text{MPa}$, as determined in the experimental-computational study of mode II dissection (Fig. 15) by FitzGibbon and McGarry (2020). The normal strength of the interface is taken as $\sigma_{\text{max}} = 0.2\,\text{MPa}$, based on standard peel test experiments (FitzGibbon and McGarry 2020).

Figure 14f shows the max principal stress in the aorta at a lumen pressure of 500 mmHg. The crack initiates at a pressure of 275 mmHg and continues to propagate with increasing applied pressure. The final crack growth is ~5.75 mm at a pressure of 500 mmHg. No crack propagation is recorded at lumen pressures below 275 mmHg. The crack initiates and propagates in a pure mode II due to compressive tractions at the medial interface caused by the hypertensive blood pressure. Figure 14g shows the interface strength as a function of the lumen pressure at the point of intimal tear propagation. $\tau^{\text{max}}/\tau^{\text{exp}} = 1$ indicates the interface strength is the experimentally recorded value of $\tau^{\text{max}} = 1.6\,\text{MPa}$ (FitzGibbon and McGarry 2020). As shown, a 50% reduction in interface strength will lead to an intimal tear propagation under a typical hypertensive blood pressure load of $p \approx 190\,\text{mmHg}$. A further reduction in interface strength ($\tau^{\text{max}}/\tau^{\text{exp}} = 0.25$) results in tear propagation in the normotensive blood pressure range. The highest recorded blood pressure in humans is 480/350 mmHg during heavy-resistance exercise (MacDougall et al. 1985), these are referred to

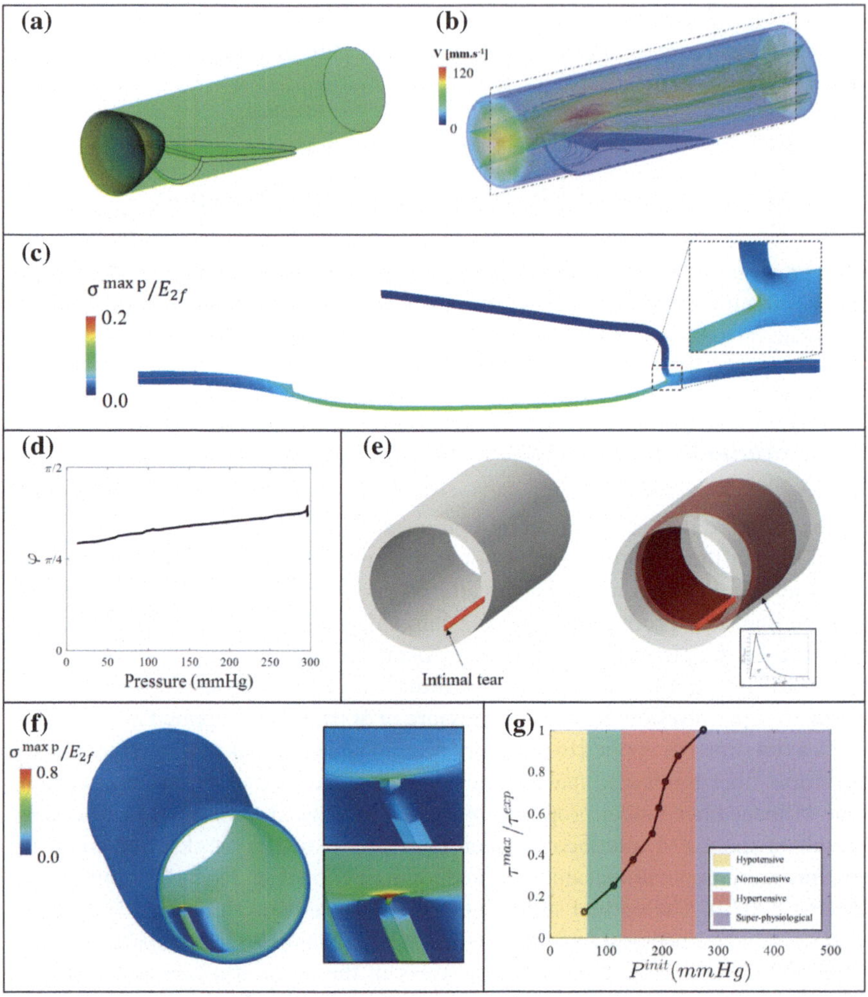

Fig. 14 (a) Illustration of the dissected geometry with the paraboloid flow profile. The intraluminal septum is shown; (b) velocity magnitude (mm s^{-1}) streamline plot of the dissected geometry; (c) blunting of the crack tip in the axisymmetric cohesive zone finite element analysis; (d) mode at the crack tip in (c) is shown as a function of the lumen pressure; (e) illustration of the intimal tear in the artery. The surface of the CZM is shown in red; (f) maximum principal stress in the aorta subject to 500 mmHg of lumen pressure; (g) interface strength ($\tau^{\max}$) as a function of pressure at crack propagation (p^{init}). Four regions of blood pressure are shown: hypotensive, normotensive, hypertensive, and super-physiological. In summary, this analysis suggests that propagation of an intimal tear of the aorta will occur only in cases of extreme hypertensive pressures ($\geq$275 mmHg). Extensive propagation of a false lumen is not predicted at a slightly hypertensive blood pressure of 140 mmHg. Even in extreme hypertensive loading conditions propagation is arrested due to mixed-mode blunting of the crack tip and an increase in the mode angle towards mode II

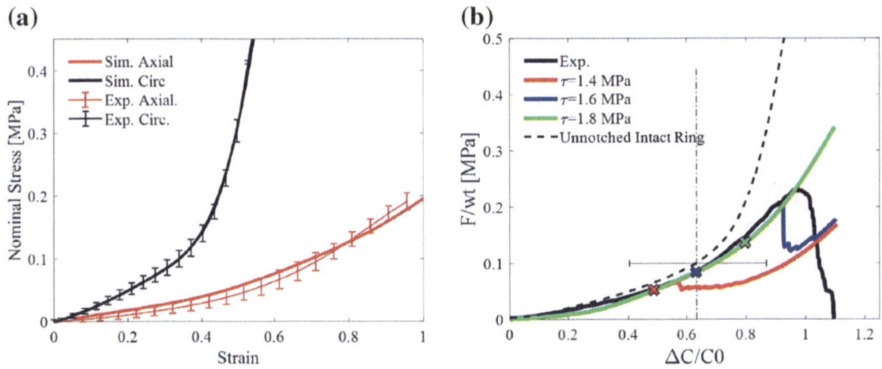

Fig. 15 (a) Computational fit of the bilinear fiber model to the experimental results (adapted from FitzGibbon and McGarry 2020); (b) calibrated cohesive zone shear interface strength parameters (adapted from FitzGibbon and McGarry 2020)

here as super-physiological. Our results seem to suggest under such blood pressure conditions an intimal tear is likely to propagate in a mode II in a healthy artery.

3.2 Examination of a Dissected Artery with a Patent False Lumen

We next explore the risk of a false lumen propagating further in an artery with a pre-existing patent false lumen, as shown in Fig. 14a. Computational fluid dynamics simulations are implemented to determine the loading on the vessel wall and on the intimal flap. A parabolic flow profile is prescribed at the inlet boundary (Fig. 14a). A zero-velocity (no-slip) condition is imposed at the wall for the fluid. Physiologically accurate transient flow profiles are implemented. A pressure boundary condition is imposed at the outlet boundary. Using a one-way fluid-structure-interaction approach, computed pressure distributions are applied to a solid finite element, including CZM interfaces, to simulate propagation of the false lumen (Fig. 14c). Velocity streamlines in the aorta are shown in Fig. 14b. Maximum velocity occurs above the intraluminal septum and minimum velocities occur in the false lumen. Figure 14c, d explores crack arrest due to blunting of the crack tip. While recent studies demonstrate that initiation of AD is a pure mode II process (Haslach et al. 2018; Gültekin et al. 2019; FitzGibbon and McGarry 2020), the simulations presented in Fig. 14c suggests that the eventual formation of a false lumen, following mode II initiation, will result in mixed-mode conditions at the crack-tip. A form of toughening of the crack tip is also evident in Fig. 14d as the mode angle at the crack tip increases (tending towards mode II propagation) with increasing pressure. As the mode II strength has been shown to be eight times higher than mode I strength (FitzGibbon and McGarry 2020), such an increase in the mode angle with increasing pressure will be accompanied by a significant increase fracture resistance. Such a toughening mechanism is incorporated into our model through our calibrated mixed-mode CZM formulation.

4 Conclusion

In the present chapter we analyze two new exponentially softening path-dependent cohesive zone formulations. An extensive exploration of the mixed-mode coupling in proportional and non-proportional loading paths is presented. Physically realistic mixed-mode behavior of a CZM is of fundamental importance for applications in materials with anisotropic fracture properties. Our recent experimental work demonstrates that the mode II fracture strength of the aorta in the circumferential-axial plane is eight times higher than the mode I strength in the same plane (FitzGibbon and McGarry 2020). Recent studies have presented evidence that initiation of aortic dissection is predominantly a mode II fracture process (Haslach et al. 2018; Gültekin et al. 2019) while the further propagation of aortic dissection is suspected to occur in a mixed-mode, as demonstrated in the finite element cohesive zone analysis presented in Fig. 14c, d.

Accurate cohesive zone simulation of these complex fractures cannot be achieved without robust analysis of the mixed-mode behavior of the cohesive zone under a range of boundary conditions and material properties. This is clearly demonstrated in the simulation presented in Fig. 14c, d. Accurate simulation of the crack tip blunting in the false lumen, as presented in Fig. 14c, d, requires the use of a cohesive zone formulation with robust mixed-mode behavior (CZM1). As previously demonstrated by McGarry et al. (2014), the use of potential-based cohesive zone formulations in mixed-mode applications results in problematic repulsive normal tractions. This is clearly demonstrated in Fig. 13 where repulsive tractions are calculated for three different potential-based models subject to a proportional separation at a mode angle of $\varphi = \pi/4$ (Xu and Needleman 1993; Park et al. 2009). The newly proposed PF-CZM1 is subject to the same limitations as previously proposed potential-based CZMs despite its robust mixed-mode basis. Furthermore, the use of potential-based models may result in the calculation of residual tractions after separation has occurred. The analysis presented in the present chapter clearly demonstrates the importance of using a non-potential-based cohesive zone formulation with robust mixed-mode behavior in the simulation of aortic dissection.

Acknowledgements Funding for this work was provided by Science Foundation Ireland grant 16/US-C2C/3291 and by the European Union Horizon 2020 Research and Innovation Program, under grant agreement No. 777072 (INSIST).

References

Alimohammadi, M., Sherwood, J.M., Karimpour, M., Agu, O., Balabani, S., Díaz-Zuccarini, V.: Aortic dissection simulation models for clinical support: fluid-structure interaction vs. rigid wall models. Biomed. Eng. Online **14**, 34 (2015)

Aure, T., Ioannides, A.: Simulation of crack propagation in concrete beams with cohesive elements in ABAQUS. Transp. Res. Rec. **2154**, 12–21 (2010)

Barenblatt, G.I.: The formation of equilibrium cracks during brittle fracture. General ideas and hypotheses. Axially-symmetric cracks. J. Appl. Math. Mech. **23**, 622–636 (1959)

Bäumler, K., Vedula, V., Sailer, A.M., Seo, J., Chiu, P., Mistelbauer, G., Chan, F.P., Fischbein, M.P., Marsden, A.L., Fleischmann, D.: Fluid-structure interaction simulations of patient-specific aortic dissection. Biomech. Model. Mechanobiol. **19**, 1607–1628 (2020)

Bonfanti, M., Balabani, S., Alimohammadi, M., Agu, O., Homer-Vanniasinkam, S., Díaz-Zuccarini, V.: A simplified method to account for wall motion in patient-specific blood flow simulations of aortic dissection: comparison with fluid-structure interaction. Med. Eng. Phys. **58**, 72–79 (2018)

van den Bosch, M.J., Schreurs, P.J.G., Geers, M.G.D.: An improved description of the exponential Xu and Needleman cohesive zone law for mixed-mode decohesion. Eng. Fract. Mech. **73**, 1220–1234 (2006)

Camacho, G.T., Ortiz, M.: Computational modelling of impact damage in brittle materials. Int. J. Solids Struct. **33**, 2899–2938 (1996)

Camanho, P.P., Dávila, C.G., De Moura, M.F.: Numerical simulation of mixed-mode progressive delamination in composite materials. J. Compos. Mater. **37**, 1415–1438 (2003)

Cazes, F., Coret, M., Combescure, A., Gravouil, A.: A thermodynamic method for the construction of a cohesive law from a nonlocal damage model. Int. J. Solids Struct. **46**, 1476–1490 (2009)

Cheng, Z., Juli, C., Wood, N.B., Gibbs, R.G.J., Xu, X.Y.: Predicting flow in aortic dissection: comparison of computational model with PC-MRI velocity measurements. Med. Eng. Phys. **36**, 1176–1184 (2014)

Cheng, Z., Riga, C., Chan, J., Hamady, M., Wood, N.B., Cheshire, N.J.W., Xu, Y., Gibbs, R.G.J.: Initial findings and potential applicability of computational simulation of the aorta in acute type B dissection. J. Vasc. Surg. **57**, 35S-43S (2013)

Colli, A., Carrozzini, M., Francescato, A., Galuppo, M., Comisso, M., Toto, F., Gregori, D., Gerosa, G.: Acute DeBakey Type I aortic dissection without intimal tear in the arch: is total arch replacement the right choice? Interact. Cardiovasc. Thorac. Surg. **26**, 84–90 (2018)

Criado, F.J.: Aortic dissection: a 250-year perspective. Tex. Heart Inst. J. **38**, 694–700 (2011)

Dake, M.D., Thompson, M., van Sambeek, M., Vermassen, F., Morales, J.P.: DISSECT: a new mnemonic-based approach to the categorization of aortic dissection. Eur. J. Vasc. Endovasc. Surg. **46**, 175–90 (2013)

Dimitri, R., Trullo, M., De Lorenzis, L., Zavarise, G.: Coupled cohesive zone models for mixed-mode fracture: a comparative study. Eng. Fract. Mech. **148**, 145–179 (2015)

Dotter, C.T., Robers, D.J., Steinberg, I.: Aortic length: angiocardiographic measurements. Circulation **2**, 915–920 (1950)

Doyle, B.J., Norman, P.E.: Computational biomechanics in thoracic aortic dissection: today's approaches and tomorrow's opportunities. Ann. Biomed. Eng. **44**, 71–83 (2016)

Eichelberger, J.P.: Aortic dissection without intimal tear: case report and findings on transesophageal echocardiography. J. Am. Soc. Echocardiogr. **7**, 82–86 (1994)

Fereidoonnezhad, B., O'Connor, C., McGarry, J.P.: A new anisotropic soft tissue model for elimination of unphysical auxetic behavior. J. Biomech. **111**, 110006 (2020)

Ferrara, A., Pandolfi, A.: A numerical study of arterial media dissection processes. Int. J. Fracture **166**, 21–33 (2010)

FitzGibbon, B., McGarry, P.: Development of a test method to investigate mode II fracture and dissection of arteries. Acta Biomater. **121**, 444–460 (2020)

Gallagher, E.A., Lamorinière, S., McGarry, P.: Finite element investigation into the use of carbon fibre reinforced PEEK laminated composites for distal radius fracture fixation implants. Med. Eng. Phys. **67**, 22–32 (2019)

Gallagher, E.A., Lamorinière, S., McGarry, P.: Multi-axial damage and failure of medical grade carbon fibre reinforced PEEK laminates: experimental testing and computational modelling. J. Mech. Behav. Biomed. Mater. **82**, 154–167 (2018)

Gambardella, I., Gaudino, M., Lau, C., Munjal, M., Elsayed, M., Girardi, L.N.: Surgical outcomes of chronic descending dissections: type I versus III DeBakey. Ann. Thorac. Surg. **104**, 593–598 (2017)

Gasser, T.C., Holzapfel, G.A.: Modeling the propagation of arterial dissection. Eur. J. Mech. - A/Solids **25**, 617–633 (2006)

Gasser, T.C., Holzapfel, G.A.: Modeling plaque fissuring and dissection during balloon angioplasty intervention. Ann. Biomed. Eng. **35**, 711–723 (2007)

Gültekin, O., Hager, S.P., Dal, H., Holzapfel, G.A.: Computational modeling of progressive damage and rupture in fibrous biological tissues: application to aortic dissection. Biomech. Model. Mechanobiol. **18**, 1607–1628 (2019)

Hagan, P.G., Nienaber, C.A., Isselbacher, E.M., Bruckman, D., Karavite, D.J., Russman, P.L., Evangelista, A., Fattori, R., Suzuki, T., Oh, J.K., Moore, A.G: The international registry of acute aortic dissection (IRAD). J. Am. Med. Assoc. **283**, 897–903 (2000)

Haslach, H.W., Jr., Siddiqui, A., Weerasooriya, A., Nguyen, R., Roshgadol, J., Monforte, N., McMahon, E.: Fracture mechanics of shear crack propagation and dissection in the healthy bovine descending aortic media. Acta Biomater. **68**, 53–66 (2018)

Hasleton, P.S., Leonard, J.C.: Dissecting aortic aneurysms: a clinicopathological study. II. Histopathology of the aorta. Q. J. Med. **48**, 63–76 (1979)

Hopkins, C.G., McHugh, P.E., McGarry, J.P.: Computational investigation of the delamination of polymer coatings during stent deployment. Ann. Biomed. Eng. **38**, 2263–2273 (2010)

Hu, J., Chou, Y.K., Thompson, R.G.: Cohesive zone effects on coating failure evaluations of diamond-coated tools. Surf. Coatings Technol. **203**, 730–735 (2008)

Huang, X., Huang, L., Sun, L., Xu, S., Xue, Y., Zeng, Q., Guo, X., Peng, M.: Endovascular repair of Stanford B aortic dissection using two stent grafts with different sizes. J. Vasc. Surg. **62**, 43–48 (2015)

Ivankovic, A., Pandya, K.C., Williams, J.G.: Crack growth predictions in polyethylene using measured traction-separation curves. Eng. Fract. Mech. **71**, 657–668 (2004)

Kim, J.B., Choo, S.J., Kim, W.K., Kim, H.J., Jung, S.H., Chung, C.H., Lee, J.W., Song, J.K.: Outcomes of acute retrograde type a aortic dissection with an entry tear in descending aorta. Circulation **130**, S39–S44 (2014)

Larson, E.W., Edwards, W.D.: Risk factors for aortic dissection: a necropsy study of 161 cases. Am. J. Cardiol. **53**, 849–855 (1984)

Lemaire, S.A., Russell, L.: Epidemiology of thoracic aortic dissection. Nat. Rev. Cardiol. **8**, 103–113 (2011)

Di Leo, C.V., Luk-Cyr, J., Liu, H., Loeffel, K., Al-Athel, K., Anand, L.: A new methodology for characterizing traction-separation relations for interfacial delamination of thermal barrier coatings. Acta Mater. **71**, 306–318 (2014)

Lui, R.C., Menkis, A.H., McKenzie, F.N.: Aortic dissection without intimal rupture: diagnosis and management. Ann. Thorac. Surg. **53**, 886–888 (1992)

MacDougall, J.D., Tuxen, D.S.D.G., Sale, D.G., Moroz, J.R., Sutton, J.R.: Arterial blood pressure response to heavy resistance exercise. J. Appl. Physiol. **58**, 785–790 (1985)

Mártín, É.Ó., Parry, G., Beltz, G.E., McGarry, J.P.: Potential-based and non-potential-based cohesive zone formulations under mixed-mode separation and over-closure-Part II: finite element applications. J. Mech. Phys. Solids **63**, 363–385 (2014)

McGarry, J.P., McHugh, P.E.: Modelling of in vitro chondrocyte detachment. J. Mech. Phys. Solids **56**, 1554–1565 (2008)

McGarry, J.P., Mártín, É.Ó., Parry, G., Beltz, G.E.: Potential-based and non-potential-based cohesive zone formulations under mixed-mode separation and over-closure. Part I: theoretical analysis. J. Mech. Phys. Solids **63**, 336–362 (2014)

Nakamura, T., Wang, Z.: Simulations of crack propagation in porous materials. J. Appl. Mech. Trans. ASME **68**, 242–251 (2001)

Nielsen, K.L., Hutchinson, J.W.: Cohesive traction-separation laws for tearing of ductile metal plates. Int. J. Impact Eng. **48**, 15–23 (2012)

Nienaber, C.A., Kische, S., Akin, I., Rousseau, H., Eggebrecht, H., Fattori, R., Rehders, T.C., Kundt, G., Scheinert, D., Czerny, M., Kleinfeldt, T.: Strategies for subacute/chronic type B aortic dissection: the investigation of stent grafts in patients with type B aortic dissection (INSTEAD) trial 1-year outcome. J. Thorac. Cardiovasc. Surg. **140**, 101–108 (2010)

Noble, C., van der Sluis, O., Voncken, R.M., Burke, O., Franklin, S.E., Lewis, R., Taylor, Z.A.: Simulation of arterial dissection by a penetrating external body using cohesive zone modelling. J. Mech. Behav. Biomed. Mater. **71**, 95–105 (2017)

Park, K., Paulino, G.H., Roesler, J.R.: A unified potential-based cohesive model of mixed-mode fracture. J. Mech. Phys. Solids **57**, 891–908 (2009)

Qiao, A., Yin, W., Chu, B.: Numerical simulation of fluid-structure interaction in bypassed DeBakey III aortic dissection. Comput. Methods Biomech. Biomed. Engin. **18**, 1173–1180 (2015)

Ryzhakov, P., Soudah, E., Dialami, N.: Computational modeling of the fluid flow and the flexible intimal flap in type B aortic dissection via a monolithic arbitrary Lagrangian/Eulerian fluid-structure interaction model. Int. J. Numer. Method. Biomed. Eng. **35**, e3239 (2019)

Sayer, D., Bratby, M., Brooks, M., Loftus, I., Morgan, R., Thompson, M.: Aortic morphology following endovascular repair of acute and chronic type B aortic dissection: implications for management. Eur. J. Vasc. Endovasc. Surg. **36**, 522–529 (2008)

Svensson, L.G., Labib, S.B., Eisenhauer, A.C., Butterly, J.R.: Intimal tear without hematoma: an important variant of aortic dissection that can elude current imaging techniques. Circulation **99**, 1331–1336 (1999)

Tsai, S.W., Wu, E.M.: A general theory of strength for anisotropic materials. J. Compos. Mater. **5**, 58–80 (1971)

Tvergaard, V., Hutchinson, J.W.: The relation between crack growth resistance and fracture process parameters in elastic-plastic solids. J. Mech. Phys. Solids **40**, 1377–1397 (1992)

Utoh, J., Goto, H., Hirata, T., Hara, M., Yamamoto, S., Kitamura, N.: Acute aortic dissection without intimal tear. J. Cardiovasc. Surg. **38**, 419–420 (1997)

Vilacosta, I., Aragoncillo, P., Cañadas, V., San Román, J.A., Ferreirós, J., Rodríguez, E.: Acute aortic syndrome: a new look at an old conundrum. Heart **95**, 1130–1139 (2009)

Wang, L., Roper, S.M., Hill, N.A., Luo, X.: Propagation of dissection in a residually-stressed artery model. Biomech. Model. Mechanobiol. **16**, 139–149 (2017)

Wu, C., Gowrishankar, S., Huang, R., Liechti, K.M.: On determining mixed-mode traction-separation relations for interfaces. Int. J. Fract. **202**, 1–19 (2016)

Xiong, Z., Yang, P., Li, D., Qiu, Y., Zheng, T., Hu, J.: A computational fluid dynamics analysis of a patient with acute non-A-non-B aortic dissection after type I hybrid arch repair. Med. Eng. Phys. **77**, 43–52 (2020)

Xu, X.P., Needleman, A.: Void nucleation by inclusion debonding in a crystal matrix. Model. Simul. Mater. Sci. Eng. **1**, 111–132 (1993)

Zorrilla, R., Soudah, E., Rossi, R., Computational modeling of the fluid flow in type B aortic dissection using a modified finite element embedded formulation. Biomech. Model. Mechanobiol. 1–19 (2020)

Bio-Chemo-Mechanical Role of Intraluminal Thrombus Deposition on Arterial Tissue Growth and Remodeling

Lana Virag, Nino Horvat, and Igor Karšaj

I met Gerhard for the first time when I was attending 2006 Summer School in Udine, Italy, as a postdoc. At that time I was introducing myself to the area of biomechanics. All the talks were very inspiring and I was easily convinced that my decision of entering a completely new area for me was the right one. In 2012 Gerhard invited me to organize minisymposium within ESMC 2012 Congress in Graz. During the meeting we agreed that I will apply for prestigious Lise Meitner fellowship. As a Lise Meitner fellow, I spent one year at Gerhard's Institute of Biomechanics, Graz University of Technology, Austria. During my stay there, except broadening my knowledge and gaining experience within the renowned laboratory accomplishing scientific achievements, I also had the opportunity to enjoy the company of coworkers from all over the world. Besides, I cherished good relationship with Gerhard's family. After my return to Zagreb, we continued close cooperation with mutual visits. Gerhard really enjoyed, after exhausting days, to visit Zagreb during Advent period. Personally I am glad that Gerhard's knowledge and leadership strongly enriched and influenced my scientific world of ideas, which was confirmed throughout several publications in top scientific journals.

Igor

Abstract Intraluminal thrombus (ILT) is present in various arterial lesions, such as aneurysms or dissections. The role a thrombus plays during the development of these diseases is still controversial and unclear. The extraction of ILT is a standard procedure only during an open abdominal aortic aneurysm (AAA) repair. Thus a wide variety of histological analyses and mechanical testing is available mostly for AAA thrombus, whereas much less is known about other thrombi. This chapter investigates the possible, sometimes opposing, roles ILT has in arterial growth and remodeling (G&R), with a focus on AAAs, using computational methods. Based on available

L. Virag · N. Horvat · I. Karšaj (✉)
Faculty of Mechanical Engineering and Naval Architecture, University of Zagreb, Ivana Lučića 5, Zagreb, Croatia
e-mail: igor.karsaj@fsb.hr

G. Sommer et al. (eds.), *Solid (Bio)mechanics: Challenges of the Next Decade*, Studies in Mechanobiology, Tissue Engineering and Biomaterials 24,
https://doi.org/10.1007/978-3-030-92339-6_5

experimental data, a bio-chemo-mechanical ILT model was developed. The results suggest that an ILT can significantly influence AAA expansion and its rupture risk. When a luminal layer with high proteolytic activity is in proximity to the remodeling wall, it may increase AAA expansion rate. ILT may also contribute to clinically-observed discontinuous expansion patterns. Mechanically, ILT shields the wall and reduces the AAA axial growth. Additionally, thrombus deposition is believed to be driven by hemodynamic factors, meaning that the typically used G&R model of soft tissue might not suffice for a complete understanding of ILT influences; for more comprehensive modeling, a fluid-solid-growth (FSG) model is needed. Using the FSG model, asymmetrical thrombus deposition is predicted, potentially increasing the area covered by thin thrombus.

1 Introduction

Intraluminal thrombus (ILT) is a fibrin-rich structure with intertwined red and white blood cells, platelets, and other cells and molecules from the blood. It is present in many cardiovascular diseases; it causes myocardial infarction and stroke by obstruction of blood flow; whereas in aneurysms or dissections it simultaneously degrades and shields the underlying wall. The thrombus is known to significantly influence the outcome of the diseases: e.g., a study by Tsai et al. (2007) proposed that a fully thrombosed false lumen of aortic dissection (AD) and dissection without thrombus have much lower rupture risk compared to a partially thrombosed AD. Rupture of intracranial aneurysms (IA) was also related to a thin thrombus (Frösen et al. 2004). However, the treatment of IAs with flow diverters is based on disruption of blood flow within the aneurysm to induce pro-thrombotic conditions (Paliwal et al. 2017). In the thrombus-laden abdominal aortic aneurysms (AAAs), a thicker thrombus was shown to slow down the AAA growth; however, a smaller relative size of ILT was linked to a higher growth rate (Domonkos et al. 2019). In conclusion, ILT has a very complex, but undoubtedly important role in the evolution of various vascular diseases. Noting the high incidence of thrombus presence (around 70% in IA (Frösen et al. 2004), 45% in AD (Tsai et al. 2007), and 80% in AAA (Simão et al. 2000; Behr-Rasmussen et al. 2014)), it should not be neglected, neither during clinical decision making nor during numerical modeling.

Treatment of these diseases is mostly endovascular and the thrombus is not extracted during the procedure. The only exception is the open surgical repair of AAAs, during which the ILT can be removed, histologically analyzed, and mechanically tested. Because of that, there is a vast amount of data about ILT from AAA, while much less is known about other thrombi that can be examined only after autopsy. The work presented in this chapter will, consequently, be focused mostly on AAAs, but with appropriate adjustments can be applied to other thrombus-laden arterial lesions.

Formation and further growth of ILT are believed to be hemodynamically driven. It is proposed that for the formation of ILT within aneurysms two main conditions need to be fulfilled—platelets must be activated within a shear flow and then be presented to a susceptible endothelium (Di Achille et al. 2014). Besides the wall shear stress (WSS), there were several other attempts to define a measure of the likelihood of thrombus formation: Shadden et al. (2013) proposed a non-dimensional scalar index that represents the magnitude of shear rates that a fluid particle accumulates while traveling throughout the fluid domain named platelet activation potential (PLAP). Furthermore, an index referred to as thrombus formation potential (TFP), which identifies local regions of the wall that were simultaneously exposed to prothrombotic WSS stimuli and a flow rich in activated platelets, showed promising results. Comparisons between the computational prediction of ILT deposition within AAA using TFP and medical data from follow-ups on six patients gave a satisfying first estimate of thrombus deposition (Di Achille et al. 2017). Similarly, Naim et al. (2016) performed computational fluid dynamics analysis of AD to illustrate the potential mechanism behind the formation of thrombus and predict possible locations based on WSS.

An ILT, unlike other blood clots, shows a few signs of healing, i.e. fibrin fibers are not replaced with collagen during the disease evolution, (Schriefl et al. 2012), at least before treatment. Yet, follow-up imaging after treatment of aneurysms and dissections typically show shrinkage of the lesion and, in some cases, even complete disappearance of ILT (weather by replacement by collagen or by the degradation of fibrin). Thus, for modeling the ILT after treatment, herein the presented model could be expanded with a model describing the longitudinal changes in the structure and mechanical properties during the transformation of a clot from an initially fibrin-based mesh to a predominantly collagenous mass, e.g., see Karšaj and Humphrey (2009).

Depending on the thrombus size and age, ILT can consist of up to three structurally and proteolytical different layers. The portion closest to the flowing blood (i.e., the luminal layer) is characterized by an aggregation of activated platelets as well as the entrapment of erythrocytes and leukocytes within a continuously evolving fibrin mesh. The luminal layer thickness is possibly limited by the depth to which blood components can penetrate the fibrin mesh, as it is usually 2 mm thick (Vande Geest et al. 2006). Leukocytes within the luminal layer are responsible for producing enzymes that degrade the arterial wall (matrix metalloproteinases (MMPs); collagenases and elastases) or indirectly activating them. For example, activation of urokinase plasminogen activator (uPA) turns plasminogen to plasmin, which not only degrades fibrin but also activates latent MMPs. In deeper (medial and abluminal) layers, due to lack of platelets and blood cells (Houard et al. 2007), fibrin is not produced anymore, and the fibrin mesh becomes more degraded and disorganized. The abluminal layer, adjacent to the wall, can be disrupted to the point it cannot be mechanically tested (Tong et al. 2011).

Despite overwhelming evidence that thin ILT biochemically degrades the arterial wall (Houard et al. 2007; Fontaine et al. 2004), and thick mechanically shields it (Inzoli et al. 1993), thrombus has been either modeled as an inert, homogeneous material in static models that focus on the stress state in the wall (e.g., Wang et al. 2002) or neglected in most G&R studies (e.g., Watton and Hill 2009 and Wilson et al. 2012). There is, therefore, a pressing need for a greater understanding of the impact ILT has on the arterial wall and lesion evolution.

The goal of the presented research was to develop a G&R model of enlargement and possible rupture of thrombus-laden AAAs. Additionally, potential differences in model predictions for physiological ranges of parameter values related to the risk of rupture were examined. Furthermore, by coupling computational fluid dynamics analysis to predict time and position of thrombus deposition with G&R model, a fluid-solid-growth (FSG) model was formulated and used to predict natural history of AAAs.

2 Arterial Growth and Remodeling

The arterial wall and the ILT interact with each other (e.g., ILT introduces increased proteases to the wall and the wall defines thrombus deformation) and thus they cannot be modeled separately. Thus, several alterations of the traditional growth and remodeling model were necessary.

2.1 G&R of Healthy Arteries

The growth and remodeling (G&R) model of soft tissues, used herein for the arterial wall, was first proposed by Humphrey and Rajagopal (2002). All the details about an implementation into a finite element framework and related model parameters can be found in Horvat et al. (2019). In short, G&R models continuously account for changes in arterial mass due to the turnover (i.e. degradation and deposition) of collagen and smooth muscle cell (SMC).

The current mass of each constituent k evolves as

$$M^k(s) = M^k(0)Q^k(0) + \int_0^s \dot{m}^k(\tau)q^k(s-\tau)\mathrm{d}\tau, \tag{1}$$

where k represents the specific wall constituent (i.e. elastin, specific direction of collagen and SMC), $q^k(s-\tau)$ is a survival function that defines a percentage of constituent produced at the past time τ that remains at the current time s, with a special case that $Q^k(s) = q^k(s-0)$. Unlike collagen and smooth muscle, elastin is not produced in maturity (Virag et al. 2015). Given that $M^e(0)$ is the initial mass of constituent k, note that $M^e(s) = M^e(0)Q^e(s)$ for elastin. Collagen deposition $\dot{m}^k(\tau)$ is a constant at a basal production rate m^k_B during homeostasis (health) but it may be altered in response to deviations in intramural or wall shear stress (WSS) from respective homeostatic values due to injury, disease, or inflammation. Thus, it can be claimed that

$$\dot{m}^k(\tau) = m_{\mathrm{B}}^k \left(1 + K_\sigma^k \Delta\sigma + K_{\mathrm{C}}^k \Delta\tau_{\mathrm{w}}\right), \tag{2}$$

where $\Delta\sigma$ and $\Delta\tau_{\mathrm{w}}$ are deviations of the mural and WSS from the homeostatic value, and K_σ^k and K_{C}^k are parameters governing constituent level mural and wall shear stress-driven mass production, respectively.

The calculation of intramural stresses is based on the constrained mixture theory and the theory of evolving configurations. The constrained theory assumes that each constituent (elastin, collagen, smooth muscle) experiences the same deformation as the mixture (arterial wall), while simultaneously allowing them individual natural, stress-free configurations. This means that any constituent k deposited at a generic G&R time τ is incorporated within the extant extracellular matrix with its own pre-stretch $\mathbf{G}^k(\tau)$. The theory of the evolving configurations states that the total deformation a constituent produced at time τ feels in a current loaded (*in vivo*) mixture configuration at time s, defined by a deformation gradient $\mathbf{F}^k_{n(\tau)}(s)$. Thus, it depends on the initial deposition pre-stretch $\mathbf{G}^k(\tau)$ and the overall deformation gradient of the mixture ${}^s_\tau\mathbf{F}$ from the deposition time τ to the current time s (see Fig. 1 in Karšaj and Humphrey 2012 for clarification). The associated right Cauchy-Green tensor $\mathbf{C}^k_{n(\tau)}(s)$ is thus $\mathbf{C}^k_{n(\tau)}(s) = \mathbf{F}^k_{n(\tau)}(s)^{\mathrm{T}}\mathbf{F}^k_{n(\tau)}(s)$, and can be used to calculate the inherent strain energy for each constituent $\hat{W}^k$. The isotropic elastin is modeled as a neo-Hookean material, while collagen and smooth muscle are modeled using the standard exponential model in tension and no compressive stiffness, as in Virag et al. (2015). Noting that fibers of the same constituent deposited into the extracellular matrix at different times have different associated deformation gradients, the total strain energy of a constituent is a sum of the unique strain energies of each fiber and their mass fraction, namely

$$W^k(s) = \frac{M^k(0)}{\sum\limits_k M^k(s)} \hat{W}^k Q^k(s) + \int_0^s \frac{\dot{m}^k(\tau)}{\sum\limits_k M^k(s)} \hat{W}^k q^k(s-\tau)\mathrm{d}\tau. \tag{3}$$

The Cauchy stress tensor within the aortic wall can then be calculated as a constrained mixture, i.e.

$$\boldsymbol{\sigma} = \frac{2}{\det\mathbf{F}} \mathbf{F} \frac{\partial W}{\partial \mathbf{C}} \mathbf{F}^{\mathrm{T}} + \sigma^{\mathrm{active}} \boldsymbol{\sigma}_\theta \otimes \boldsymbol{\sigma}_\theta, \tag{4}$$

where $W = \sum_k W^k$ is the overall strain-energy function, and σ^{active} is the active stress contribution from circumferentially oriented smooth muscle contractility.

2.2 *Illustrative Example*

When hemodynamic factors, such as blood pressure or flow, are altered, the arterial wall adapts to the new conditions. In order to verify the G&R model implementation into a finite element environment, but also to illustrate vascular adaptation to perturbations from homeostasis, remodeling of the wall after an abrupt change in flow is here presented. An increase in blood flow causes the widening of arteries, as has been observed clinically both in animal models (e.g., using anastomosis to increase blood flow of rabbits (Brownlee et al. 1991)) and in humans (Fridez et al. 2003). The animal model reported that the time needed for vascular adaptation after an abrupt blood flow increase by 60% was approximately 2 months for rabbits. Interestingly, the increase in diameter of the artery with altered blood flow compared to the diameters of the control arteries was 19%, which reasonably matches the membrane theory which predicts a 17% increase in the arterial diameter. An increase in blood flow in humans is typically caused by exercise, rather than a surgical procedure, making the change less abrupt and measurements more difficult to control. However, the difference in diameter was observed in all arteries between athletes and sedentary subjects, although quantitatively the change varies among arteries: the difference in diameter of femoral arteries was reported to be 11–17.3%, carotid arteries were only slightly dilated in Fridez et al. 2003.

Numerically obtained results are presented in Fig. 1. For the first 400 days of the G&R process, hemostasis is preserved, so there is no change in the luminal diameter (Fig. 1a) or the wall thickness (Fig. 1b). At the G&R time of 400 days the blood flow was abruptly increased by 20%. The increase in flow led to wall shear stress-induced changes in the production of fibers (see Eq. (2)) and active response of SMCs. This triggered growth and a remodeling process. Note that if the WSS driven production rates (K_C) were zero, the vessel would stay in homeostasis, as an increase in blood flow did not affect intramural stress in the wall, and thus the equilibrium would not be disturbed. Change in the flow initially decreased mass production, which was, in turn, reflected by the thinning of the wall and its expansion. Both thinning and dilatation of the wall induced an increase in wall stresses, and thus also the production rate. With the growth of the mass, stresses began to decrease back to the homeostatic

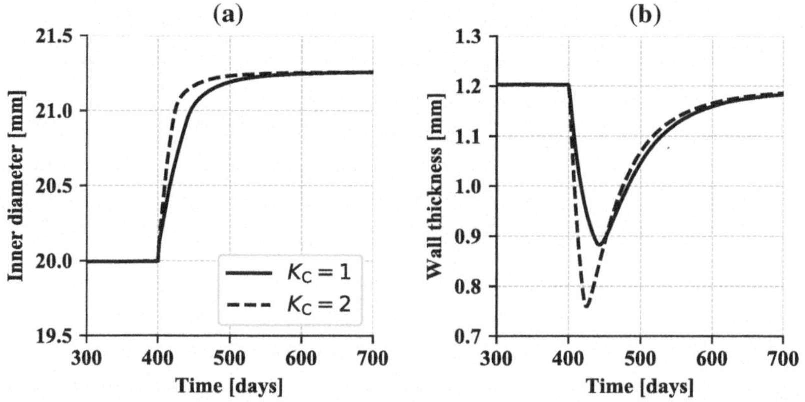

Fig. 1 Evolution of the inner diameter (a), and the wall thickness (b) for different WSS-driven production rates K_C after a 20% increase in the blood flow

value and the new equilibrium was found. Figure 1 also shows that the parameter K_C influences the time needed for the vessel to find a new equilibrium, but it does not change the new equilibrium state, i.e. the final inner radius and thickness. For the larger value of K_C initially lost mass was greater, and, consequently, so were the dilatation and change in wall stresses. This led to a quicker deposition of the new mass, and thus the equilibrium state was reached faster. Depending on the model parameter, the time needed for the vessel to reach a new equilibrium diameter was between 60 and 150 days, which is roughly in accordance with the animal model. For other model parameters and all clarifications, see Horvat et al. (2019).

The new equilibrium state after changes in the blood flow or pressure, but not the G&R process, can be estimated using a membrane theory approximation. According to the membrane theory, when blood flow is 20% above its homeostatic (normal) value, both the inner radius and the thickness should change $1.2^{1/3} = 1.06266$ times compared to the initially, healthy value. Hence, for an aorta with an inner diameter of 20 mm, the terminal inner diameter would give 21.25 mm. As can be seen in Fig. 1, the terminal diameter reached the exact same value, whereas the thickness was slightly overestimated by the membrane theory approximation. In conclusion, the results obtained by the G&R model show great agreement both with clinical observations and the membrane theory.

2.3 G&R of Diseased Arteries

When modeling arteries in disease, the above described G&R model must be slightly adjusted. For example, elastin has a very long half-life in health, estimated to be 40 years. In the previous illustrative example, the mass of elastin was assumed to be constant, since the observed time of several hundred days was relatively short. On the other hand, aneurysms and dissections are characterized by lengthy progression (measured in decades) and local irreversible loss of elastin, which must be taken into account. Along with the loss of elastin, programmed death of SMCs caused by a loss of attachment to the surrounding matrix, including elastin fibers, was also observed in aneurysms. Therefore, the degradation of smooth muscle should be linked to the degradation of elastin. Additionally, phenotypic transitioning of SMCs from a contractile to a synthetic state causing a decrease in maximal active stress should be considered, similar to Valentín et al. (2011).

3 Intraluminal Thrombus Model

The growth process (i.e. change of mass) in the aortic wall happens internally—fibroblasts and smooth muscle in the wall are responsible for the production of collagen. During the growth of an aneurysm, the clinically observed aortic wall thickness does not change significantly, meaning the wall mass typically increases several times compared to a healthy artery because of the expansion. Conversely, a thrombus is continuously being deposited by conversion of the constantly renewed fibrinogen from the blood, with its mass changing significantly—from zero in a healthy artery or small aneurysm to a potentially huge thrombus mass in large AAAs. Although a complete understanding of the spatially evolving deposition of thrombus is still lacking, many medical images of large AAAs suggest that the luminal diameter tends to be maintained. Thus, while it is convenient to model an aortic wall with a constant number of finite elements and keeping track of changes in its mass, for ILT it is more suitable to increase the number of finite elements at each deposition time, but preserve the mass of each element constant.

Within the constant thrombus element mass, it is still possible to monitor the turnover of fibrin and cell apoptosis, and thus change associated mechanical properties. Because thrombus deposition is likely linked to the hemodynamics, in the G&R model the luminal area was assumed to remain constant, meaning that additional ILT was deposited in every time step in which the lesion enlarged sufficiently, burying the prior deposited portions of the luminal layer which then become medial and abluminal layers. Moreover, based on experimental findings (Tong et al. 2011), fibrin, fibrin degradation products (FDPs), erythrocytes (RBCs), and voids were considered to be the most significant space-filling constituents of a layered ILT. Leukocytes, platelets, and non-collagenous extracellular proteins were neglected as load-bearing constituents but were nevertheless crucial for the biological activity of the thrombus. The model was based on diverse experimental and clinical results reviewed in Wilson et al. (2013).

3.1 Fibrin

During fibrinogenesis, platelets release thrombin, which helps to convert soluble fibrinogen into fibrin. Fibrin polymerizes to form a cross-linked mesh. Changes in the mass of fibrin depend on its deposition and degradation, similar to changes in the wall constituents (2). Based on experimental findings, the production rate of fibrin depends on the mass of platelets M_i^{plt} and decreases with increased fibrin density ϕ_i^{f}. Thus, we can define a correlation parameter $K_{\mathrm{plt},\phi}^{\mathrm{f}}$ and describe fibrin production rate as

$$\dot{m}_i^{\mathrm{f}}(\tau) = K_{\mathrm{plt},\phi}^{\mathrm{f}} M_i^{\mathrm{plt}}(\tau)\left(1 - \phi_i^{\mathrm{f}}(\tau)\right). \tag{5}$$

On the other hand, fibrinolysis is governed by plasmin. Moreover, since fibrinolysis has been reported to be inversely proportional to the density of the fibrin mesh and its current stretch (Scott et al. 2011), a mass removal rate-type parameter $K_q^{\mathrm{f}}(\tau)$ used for the survival function

$$q^k(s - \tau) = \exp\left(-\int_\tau^s K_q^f(\tilde{\tau})\mathrm{d}\tilde{\tau}\right) \tag{6}$$

was considered to be

$$K_q^{\mathrm{f}}(\tau) = k_q^{\mathrm{f}} + w_q^{\mathrm{f}} M_i^{\mathrm{pls}}(\tau) \frac{1 - \phi_i^{\mathrm{f}}(\tau)}{\lambda_i(\tau)}, \tag{7}$$

where k_q^{f} is a homeostatic value depending on the natural half-life of fibrin, M_i^{pls} is the mass of plasmin, λ_i is fibrin stretch, and w_q^{f} is a weighting function.

Mechanically, ILT can bear stress and possibly reduce the peak wall stress. Cauchy stress in the ILT was modeled using a strain–energy function for an isotropic fibrin mesh based on Tong et al. (2011) and a neo-Hookean strain–energy function for the FDPs.

3.2 *Cells and Platelets*

Several studies reported that blood-derived cells were found mostly in the luminal layer (Houard et al. 2009; Adolph et al. 1997), presumably because flowing blood can replenish and sustain cells only to a certain depth. Thus, the defined boundary of the luminal layer was a depth beyond which cells could not be replenished, either because a critical fibrin mesh density $\phi_{\mathrm{crit}}^{\mathrm{f,RBC}}$ was reached or because a critical radius $r_{\mathrm{crit}}(\tau)$ was attained. This critical radius was defined in each instant by a critical mass of fibrin $M_{\mathrm{crit}}^{\mathrm{f}}$, such that

$$\sum_{r_l}^{r_{\mathrm{crit}}} M_{\mathrm{i}}^{\mathrm{f}}(r, \tau) = M_{\mathrm{crit}}^{\mathrm{f}}, \tag{8}$$

where M_i^{f} is the mass of fibrin in layer i and r_{l} is the luminal radius. The quantity of blood-borne cells was assumed to be constant in the luminal layer (e.g., $M_{\mathrm{l}}^{\mathrm{plt}}$ for platelets) and decreased with a first-order decay process (with a half-life $\tau_{1/2}^{\mathrm{plt}}$ of 7 to 11 days for platelets and $\tau_{1/2}^{\mathrm{RBC}}$ days for erythrocytes) in deeper layers. In summary, the mass of platelets (and similarly for leukocytes, primarily neutrophils, M_i^{N}) was calculated as

$$M_{\mathrm{i}}^{\mathrm{plt}}(\tau) = \begin{cases} M_l^{\mathrm{plt}}(\tau) & \text{except if} \\ M_l^{\mathrm{plt}}(\tau_{\mathrm{crit}}^{\mathrm{i,1}}) \cdot e^{-K_q^{\mathrm{plt}}(\tau - \tau_{\mathrm{crit}}^{i,1})} & r_{\mathrm{i}}(\tau) \geq r_{\mathrm{crit}}(\tau) \\ M_l^{\mathrm{plt}}(\tau_{\mathrm{crit}}^{\mathrm{i,2}}) \cdot e^{-K_q^{\mathrm{plt}}(\tau - \tau_{\mathrm{crit}}^{i,2})} & \phi_j^{\mathrm{f}}(\tau) \geq \phi_{\mathrm{crit}}^{\mathrm{f}}, \mathrm{plt}, \quad \forall j \geq i \end{cases} \tag{9}$$

where $\tau_{\mathrm{crit}}^{i,1}$ is the time at which a critical mass of fibrin (i.e., r_{crit}) was achieved, thus transitioning the layer from luminal to medial. Similarly, $\tau_{\mathrm{crit}}^{i,2}$ represents the time at which the limit of the luminal layer was achieved by reaching the critical mass fraction of fibrin (i.e., mesh density). Due to their smaller size compared to the erythrocytes, it was assumed that leukocytes did not contribute significantly to the volume of the ILT;

nonetheless, their presence is important for biochemical activity (e.g., production of plasmin, MMPs, or other elastases). The distribution of erythrocytes within the ILT was defined similarly to platelets, with the exception of using a mass fraction instead of the mass, because it is more convenient for the calculation of the overall strain energy of ILT.

3.3 Plasmin, EDPs, and Neovascularization

Conversion of plasminogen to plasmin is a complex process catalyzed by multiple enzymes, including a key step that includes activation either by tissue plasminogen activator (tPA, found on endothelial cells) or urokinase plasminogen activator (uPA, coming mainly from the blood stream via mesenchymal and inflammatory cells). Conversely, activation may be inhibited by plasmin activation inhibitor (PAI-1) and alpha-2 antiplasmin. Noting that not much is known about their distributions, a simplified model was used, capturing the overall increase in activated plasmin phenomenologically. It considered plasmin diffusion from two primary sources: the leukocyte-rich luminal layer of the ILT and the inflamed aneurysmal wall (to which leukocytes invade via the *vasa vasorum*). This increased production of plasmin within the wall is consistent with immunohistological reports (Fontaine et al. 2002). The mass of activated plasmin M_i^{pls} that degrades fibrin is given as

$$M_i^{\text{pls}}(\tau) = \underbrace{K_{\text{N}}^{\text{pls}} M_i^{\text{N}}(\tau) + M_{\text{wall}}^{\text{pls}}(\tau) - \sum_{j=1}^{i} K_{\text{f}}^{\text{pls}} M_j^{\text{f}}(\tau)}_{\geq 0}, \tag{10}$$

where $K_{\text{N}}^{\text{pls}}$ is a correlation factor between plasmin and neutrophils in the thrombus, $K_{\text{f}}^{\text{pls}}$ is the amount of plasmin consumed per unit of fibrin, and $M_{\text{wall}}^{\text{pls}}(\tau)$ is the mass of plasmin in the wall, calculated as $M_{\text{wall}}^{\text{pls}}(\tau) = K_{\text{EDP}}^{\text{pls}} M_{\text{tot}}^{\text{EDP}}(\tau) + K_{\text{VV}}^{\text{pls}} A_{\text{tot}}^{\text{VV}}(\tau)$. The plasmin generated within the wall depended largely on the total mass of elastin degradation products $M_{\text{tot}}^{\text{EDP}}$, which are chemo-attractants for inflammatory cells and stimulate neovascularization of the wall (i.e., increase the area of *vasa vasorum* $A_{\text{tot}}^{\text{VV}}$, (Nackman et al. 1997). Correlations between plasmin and EDPs/*vasa vasorum* were accounted for by $K_{\text{EDP}}^{\text{pls}}$ and $K_{\text{VV}}^{\text{pls}}$, respectively. The production rate of EDPs directly depended on the amount of elastin degraded per time step, while their degradation was modeled as a first-order decay with constant half-life. The total mass of EDPs can then be calculated by (2).

Due to the controversial influence of hypoxia on neovascularization, it was neglected. Instead, based on studies that showed EDPs alone promote neovascularization (Nackman et al. 1997), the development of *vasa vasorum* was modeled solely as a function of the mass of EDPs, such that

$$A_{\mathrm{tot}}^{\mathrm{VV}}(s) = A_0^{\mathrm{VV}} + \int_0^s K_{\mathrm{EDP}}^{\mathrm{VV}} M_{\mathrm{tot}}^{\mathrm{EDP}} \mathrm{d}\tau, \tag{11}$$

where A_0^{VV} is the area of *vasa vasorum* in the healthy aorta, and $K_{\mathrm{EDP}}^{\mathrm{VV}}$ is a correlation factor that relates *vasa vasorum* growth per unit of EDPs per unit of time.

3.4 Fibrin Degradation Products and Voids

The luminal layer contains entrapped erythrocytes, platelets, and other blood-borne molecules within a thick fibrin mesh. Once the luminal layer is buried deeper within the thrombus, erythrocytes cannot be replenished. Likewise, the amount of platelets necessary for fibrin deposition is greatly reduced, which means that the degradation can exceed the deposition. Dissolution of fibrin and apoptosis of red blood cells can leave small interconnected channels ('voids' or 'canaliculi') throughout the ILT (Adolph et al. 1997). Degrading fibrin was assumed to convert partly into fibrin degradation products (FDPs) and partly into canaliculi (as fibrin was removed by macrophages) in a 7:3 ratio.

3.5 Biochemical Interaction of ILT and the Aneurysmal Wall

An increase in proteolytical activity due to the presence of the ILT was integrated via additional terms in the mass removal rate-type parameter K_q^k in the survival function. For example, the elastin in the normally aging aorta depends solely on its natural half-life $\tau_{1/2}^{\mathrm{e}}$ on the order of forty to fifty years: $K_q^{\mathrm{e}}(\tau) = k_q^{\mathrm{e}} = \ln 2/\tau_{1/2}^{\mathrm{e}}$. Increased degradation of elastin in an aneurysm due to inflammation was incorporated via an additional term, i.e.

$$K_q^{\mathrm{e}}(\tau) = k_q^{\mathrm{e}} + w_{q,\mathrm{elas}}^{\mathrm{e}} M^{\mathrm{elas}}(r, \tau), \tag{12}$$

where the elastases M^{elas} were modulated by a weighting factor $w_{q,\mathrm{elas}}^{\mathrm{e}}$. Correspondingly, for collagen, we write (cf. Karšaj and Humphrey 2012)

$$K_q^{\mathrm{c}}(\tau) = \frac{\partial W^{\mathrm{c}} / \partial \mathbf{F}_{n(\tau)}^{\mathrm{c}}}{\partial W^{\mathrm{c}} / \partial \mathbf{F}_{n(0)}^{\mathrm{c}}} k_q^{\mathrm{c}} + w_{q,\mathrm{MMP}}^{\mathrm{c}} M^{\mathrm{MMP}}(r, \tau), \tag{13}$$

where the collagen mass removal parameter K_q^{c} depends on the ratio of the current $\partial W^{\mathrm{c}} / \partial \mathbf{F}_{n(\tau)}^{\mathrm{c}}$ to the homeostatic $\partial W^{\mathrm{c}} / \partial \mathbf{F}_{n(0)}^{\mathrm{c}}$ tension, and k_q^{c} is the rate-type removal parameter associated with the homeostatic half-life; the second terms relate to the distribution of active MMPs (i.e. collagenases).

Similar to plasmin, MMPs may arise from two sources: the luminal layer in the ILT or the *vasa vasorum* in the aortic wall. A first approximation, quasi-static transport, was determined easily from the diffusion equation, i.e. $\partial M^{\text{elas}}/\partial t = \nabla \cdot \left[D \nabla M^{\text{elas}}\right]$, where D is the diffusion coefficient that can be estimated based on available data of the radial distribution of proteases (Fontaine et al. 2004 and Houard et al. 2009). The available elastase/MMPs from the luminal layer was defined as $K_{\text{N}}^{\text{elas}} M_{\text{tot}}^{\text{N}}$ (or $K_{\text{N}}^{\text{MMP}} M_{\text{tot}}^{\text{N}}$ for MMPs), where $K_{N}^{\text{elas/MMP}}$ describes how much elastase/MMP was produced per unit of leukocytes per time step, at a point at radius r_{L} which was defined as

$$r_{\text{L}}(s) = \frac{1}{M_{\text{tot}}^{\text{N}}(s)} \sum_{i} M_{i}^{\text{N}}(s) r_{i}(s), \tag{14}$$

similar to a center of mass. The concentration of elastase/MMPs available at the outer radius depended on the area of the *vasa vasorum* and the number of inflammatory cells: $K_{\text{WBC}}^{\text{elas}} M_{\text{tot}}^{\text{WBC}}(s) + K_{\text{VV}}^{\text{elas}} A_{\text{tot}}^{\text{VV}}(s)$. Thus, by solving the diffusion equation, the overall distribution of proteases was

$$M^{k}(r, s) = \frac{K_{\text{N}}^{k} M_{\text{tot}}^{\text{N}}(s) - K_{\text{WBC}}^{k} M_{\text{tot}}^{\text{WBC}}(s) - K_{\text{VV}}^{k} A_{\text{tot}}^{\text{VV}}(s)}{\ln(r_{\text{L}}(s)/r_{\text{o}})} \ln \frac{r(s)}{r_{\text{o}}} + K_{\text{N}}^{k} M_{\text{tot}}^{\text{N}}(s), \tag{15}$$

where k indicates elastases or collagenases. The mass of inflammatory cells in the wall $M_{\text{tot}}^{\text{WBC}}$, was modeled similar to (2) with a degradation depending on the cellular half-life, and a production $\dot{m}_{\text{tot}}^{\text{WBC}}$ increasing with the growth of the EPD mass and the *vasa vasorum*, such that

$$\dot{m}_{\text{tot}}^{\text{WBC}}(\tau) = K_{\text{EDP,VV}}^{\text{WBC}} \left(\frac{A_{\text{tot}}^{\text{VV}}(\tau)}{A_{0}^{\text{VV}}} \right)^{\alpha} M_{\text{tot}}^{\text{EDP}}(\tau), \tag{16}$$

where $K_{\text{EDP,VV}}^{\text{WBC}}$ is a correlation factor that relates the proliferation of white blood cells with EDPs and neovascularization, and α is a power parameter. A schematic presentation of the whole ILT/wall interaction model is presented in Fig. 2 in Virag et al. 2015.

3.6 Model Predictions—Biochemical Influence of the ILT

The full potential of the described ILT model is presented in Virag et al. (2015, 2017). Very briefly, Fig. 2a shows the predicted distributions of the mass fractions of the constituents within the ILT. The boundary between the luminal and the medial layer is evident by the steep decrease in fibrin and RBCs, as cells could not penetrate further into the thrombus. Similarly, platelets were unable to reach deeper parts of the ILT, disenabling fibrin production. Consequently, with fibrin degradation exceeding its production, the mass fraction of this fibrin decreased. The ILT that is deposited the latest was located near the luminal interface (i.e., near a radius of 10 mm) and is characterized by low amounts of fibrin and large numbers of the RBCs. In the

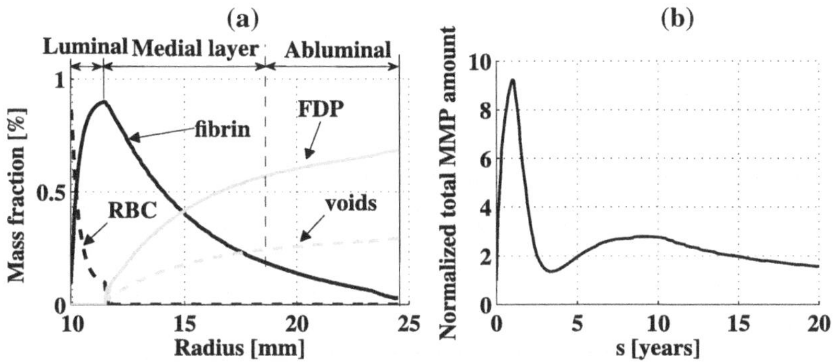

Fig. 2 Predicted radial distribution of mass fractions of structurally important constituents 20 years after aneurysm initialization. The solid black curve represents fibrin, the dashed black curve red blood cells, the gray solid curve voids (canaliculi), and the gray dashed curve fibrin degradation products (FDPs) (a); evolution of the normalized amount of matrix metalloproteinases (MMPs) in the aortic wall (b)

older, buried parts of the luminal layer, the fibrin fraction increased and the RBCs diminished. Unlike the distinctive luminal layer, the border between the intermediate and abluminal layers was less evident; both were characterized by disruption of the fibrin mesh and devoid of blood cells and platelets. These results correspond qualitatively to the reported distribution of constituents from histological analysis of Tong et al. (2011), where four phases of thrombus ages were defined as: phase I (fresh thrombus, new luminal layer in the model) characterized by 90% RBCs and 10% thin fibrin bundles; phase II (matches older luminal layer) with less than 10% RBCs, some condensed proteins (FDPs), and increasing thin and thick fibrin fibers; phases III and IV (i.e., older thrombus) with few RBCs and condensed proteins domination as fibrin is degraded.

It can be seen from Fig. 2a that the thickness of the luminal layer is approximately 2 mm, which is in accordance with clinical observations (Vande Geest et al. 2006; Tong et al. 2011). This is not a predefined value, as luminal layer thickness, or depth to which platelets and blood cells penetrate through the fibrin mesh, is defined by Eq. (9). For quickly expanding aneurysms, the luminal layer thickness increases because the fibrin production is slow compared to the aneurysm expansion. Likewise, for slow expanding or stabilized aneurysm, the luminal layer becomes thinner. During a nonlinearly expanding aneurysm, the luminal layer thickness varies in time.

The evolution of the total normalized metalloproteinases describing proteolytical activity in the underlying wall is highlighted in Fig. 2b. In the early years, when ILT is thin, the amount of MMPs within the adjacent wall changed very steeply. At the time when the thrombus layer that was deposited first was not replenished with the blood cells anymore, it became a much less proteolytically active medial layer. Consequently, the MMP activity sharply decreased. The final MMP count 20 years after an aneurysm initialization was reduced ~3 fold relative to the peak value, but was still higher than the initial value. This happened due to an increased inflammation in the wall itself, due to the *vasa vasorum* and chemotactic EDPs.

3.7 Model Predictions—Biomechanical Influence of the ILT

As noted above, many medical images of large AAAs suggest that the luminal area tends to be maintained constant (see, e.g., Fig. 9 in Zambrano et al. 2016), either at equal or greater than the healthy value. The choice of the new luminal diameter is also an important factor for aneurysm progression and stabilization. Figure 3 shows the evolutions of thrombus-laden aneurysms with different luminal areas compared to AAA without a thrombus. Full black dots in Fig. 3a denote the moment of the first ILT deposition. While AAA continued to grow (and ruptured) exponentially without a thrombus, the radius to which the ILT was continually filled determined whether AAA will stabilize, how long it will take to stabilize, and how much further AAA will expand. If the luminal area was kept constant at a healthy value (i.e. 10 mm), aneurysmal value. The increase in luminal area increased AAA growth and decreased the likelihood of stabilization.

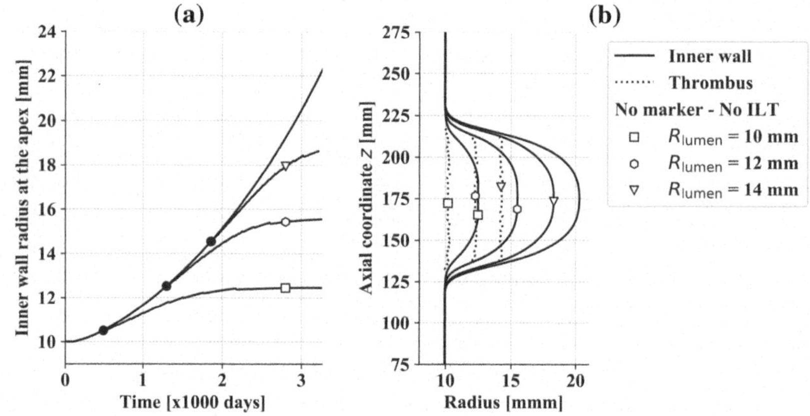

Fig. 3 Evolution of the inner AAA radius (a), contours of the ILT/wall interface and the luminal area (b) for different luminal areas of thrombus deposition

3.8 Model Predictions—Influence of Rupture Risk Factors

The model was further used to investigate the influence of rupture risk factors (smoking, age, sex, hypertension) and stabilization factors on AAA growth (Virag et al.

2017). The focus here is on smoking and MMP inhibition. Smoking is known to elevate oxidative stress, and thus injure the arterial endothelium and promote atherogenesis (Burke and FitzGerald 2003). Furthermore, it also increases inflammation in the vascular wall (Powell 1998) as well as the concentration of fibrinogen in plasma and alters the activity of platelets. Increased platelet activity and increased fibrinogen potentiate thrombosis along the dysfunctional endothelium (Powell 1998). Finally, it correlates strongly with increased aortic blood pressure, heart rate, and arterial stiffness (Mahmud and Feely 2003). Smoking-induced arterial stiffening is suggested to be isotropic, as reflected by a consistent increase in both elastin- and collagen-associated parameters (Enevoldsen et al. 2011).

All of these cardiovascular health disadvantages that smoking brings with it negatively affect the disease progression. Thus, smoking is associated with both larger AAAs (Behr-Rasmussen et al. 2014 and Bhak et al. 2015) and higher expansion rates (Al-Barjas et al. 2006). Quitting smoking does not immediately help lower the AAA expansion rate: it has been shown that arterial stiffness did not decrease two years after quitting smoking (van den Berkmortel et al. 2004) and that it can take up to a decade for stiffness to return to nonsmokers' levels (Jatoi et al. 2007). Lowering inflammation rates also likely takes years, as it was estimated five years of abstinence are necessary to reduce fibrinogen concentrations to the range of never-smokers (Meade et al. 1987).

Based on these studies, smoking was modeled by increased elastin and collagen stiffness, inflammation, and platelet activity. Some other aspects that smoking brings to vascular health, such as endothelial and SMC dysfunction, are also features of non-smoking aneurysms and have been implicitly included in the model for smokers and non-smokers alike. All other G&R parameters were identical between the groups.

Figure 4a shows a numerically predicted 20-year-growth of non-smoker and smoker. The predicted maximum AAA diameter was greater for smokers (5.39 cm) than for non-smokers (4.40 cm). The average increase in the dilatation rate due to smoking can easily be calculated as 0.0495 cm/year. This shows great agreement with clinical the observations, as the increase in expansion rates associated with smoking was 0.05 ± 0.01 cm/year (Bhak et al. 2015). In order to determine which of the smoking-induced alterations introduces rupture risk the most, changes were simulated gradually: first only the influence of an increase in stiffness was simulated, then the increase in both stiffness and inflammation and finally also platelet activity (i.e. smoker). The results suggest that inflammation played the greatest role in early aneurysmal progression, although an increased stiffness also significantly increases the growth rate (Fig. 4a). Fibrinogen concentration and platelet activity (present in the smoker group) did not have a major influence on AAA evolution (dotted vs. dashed curve). Interestingly, the high activity of platelets resulted in a slightly decreased AAA size, because an increased production of fibrin resulted in a slightly stiffer, thinner luminal layer. This influences AAA growth positively because a stiffer luminal layer reduces the peak wall stress more, whereas a thinner luminal layer means it will be distanced from the arterial wall earlier. Note that this does not mean that an increased platelet activity could potentially stabilize AAAs since lower wall stress means that collagen production is decreased with a lower deviation from

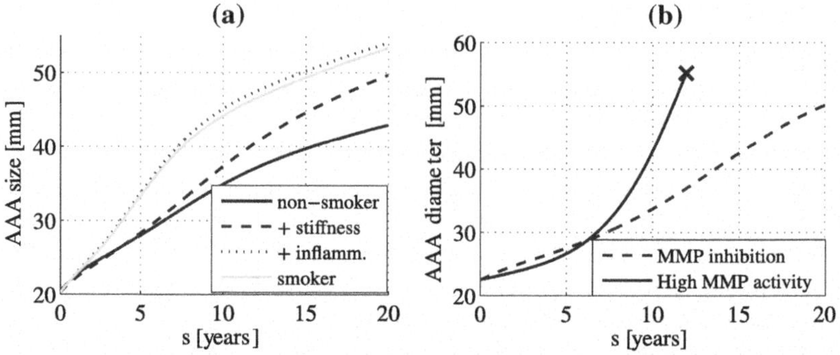

Fig. 4 Simulated evolution of an AAA in a non-smoker (solid curve), in cases of increased stiffness (dashed curve), increased stiffness plus elevated inflammation (dotted curve), and increased stiffness, higher inflammation, and increased platelet activity (gray curve)—that is, full effects of smoking (a), simulated AAA evolution in cases of rapid versus slow collagen degradation (b). Rupture is denoted by '×'

the homeostatic stress value (refer to (2)). In addition, platelets are chemotactic for neutrophils and monocytes (Deuel et al. 1981) and can cause higher inflammation and expansion rates.

3.9 Model Predictions—Influence of Stabilization Factors

Although there is no cure for aneurysm, some studies have proposed the use of anti-inflammatory drugs for the reduction or arrest of the AAA enlargement. Some of the proposed drugs act directly on MMPs (Parodi et al. 2005; Steinmetz et al. 2005) and others reduce extracellular matrix degradation indirectly, e.g., via platelet inhibitors (Owens et al. 2015) or immunosuppressive agents (Lawrence et al. 2004). The hypothesis was investigated on animal models: a study on mice showed that a preservation of the medial elastin can reduce aneurysmal dilatation (Parodi et al. 2005), while another study showed that a non-specific inhibition of collagenases leads to a lesion stabilization (Steinmetz et al. 2005). Unfortunately, non-specific MMP inhibition has not reduced aneurysmal progression in human patients (Arnoud et al. 2013). There is clearly a need for an increased understanding of the biochemical mechanisms of aneurysm formation and evolution.

Figure 4b shows that the thrombus-laden aneurysm model supports results from animal models, with attenuation of collagenase activity (i.e., reducing the parameter that describes collagen degradation by MMPs $w^{c}_{q,\mathrm{MMP}}$) helped to decrease the likelihood of rupture. However, the results also agree with the study showing the failure of drug effects in humans, as after 20 years of disease progression the maximum diameter was 5 cm, at which point surgical treatment is recommended. Thus, the current results suggest that rates of degradation of both collagen and elastin (results not shown here) may strongly influence aneurysmal evolution. The model suggests that high rates of collagen degradation leads to a progressive enlargement, and a reduced inflammation could perhaps prevent rupture, although the reduction of inflammation alone might not suffice.

4 Fluid-Solid-Growth Modeling

Hemodynamic changes are undoubtedly an important factor in the progression of arterial lesions. As discussed in previous sections, they can be sensed by vascular cells and cause the arterial wall to adapt. For example, it is believed that hemodynamics plays a crucial role in IA evolution, and therefore hemodynamic analyses such as computational fluid dynamics (CFD) and fluid-structure-interaction (FSI) simulations, for IAs are far more common than G&R studies. Note that CFD assumes the wall to be rigid and it does not consider the influence of the deformation and movement of the aortic wall on the hemodynamic output, whereas FSI considers the compliance of the wall and, as a result, provides a more accurate description of the hemodynamics (Caballero et al. 2013; Alimohammadi et al. 2014), but it is computationally more expensive. Since ILT formation and deposition is likely to depend on hemodynamics, the logical new phase of numerical modeling is to combine hemodynamic analysis with modeling of G&R of the wall. This loose coupling of CFD or FSI models over a cardiac cycle (time scale in milliseconds) with long-term G&R models of the evolving wall (time scale in days) is referred as fluid-solid-growth (FSG) model. The concept of the FSG framework was first proposed by Humphrey & Taylor in 2008 (Humphrey and Taylor 2008), and implemented by Figueroa et al. (2009) on a basilar artery the following year. There were very few attempts applying a FSG model to IAs (Watton et al. 2011; Selimovic et al. 2013) and AAAs (Grytsan et al. 2015), but all of them neglect the thrombus. Currently, there are no studies that applied the FSG framework to model AD.

In addition to using hemodynamic analysis as an input for calculating the collagen production rate, it can also be used as an indicator of when, where and how much thrombus is being deposited. The general framework is shown in Fig. 5: the long-term G&R model runs until the maximum luminal diameter increases sufficiently (e.g., 1 mm). At that time G&R analysis is interrupted, a luminal surface for CFD simulation is generated, and a hemodynamic analysis over several cardiac cycles is performed. If predefined conditions for thrombus deposition are not met (e.g., time averaged WSS (TAWSS) did not decrease below the threshold value), G&R continues without adjustments. However, when requirements are fulfilled, new thrombus finite elements are added to the finite element aneurysm model at the location defined by a hemodynamic analysis. Interestingly, at lower defined threshold values of TAWSS, which defines the thrombus deposition, not only the luminal area expand more, but the deposition becomes more asymmetrical (Fig. 6a). This asymmetrical deposition consequently leads to a reduced AAA expansion of the shielded distal part of the aneurysmal sac, and the apex is no longer at the middle surface but shifts upward.

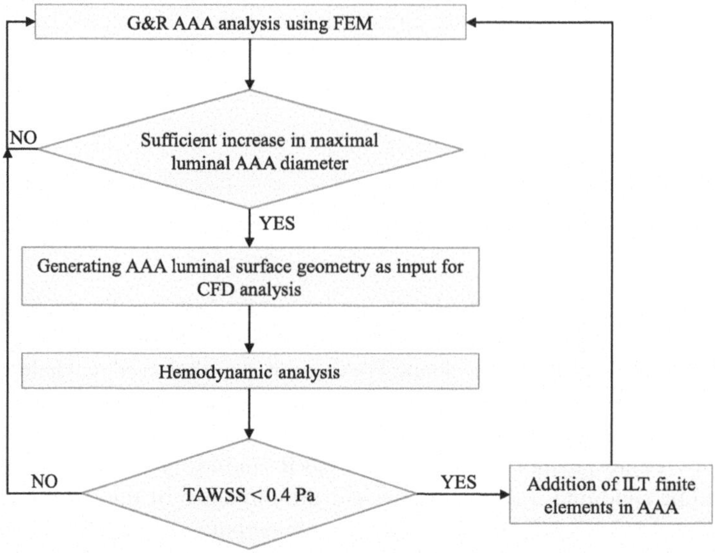

Fig. 5 Schematic representation of an FSG model used in this work

This effect can be seen in Fig. 6b: the first thrombus was deposited at the AAA diameter of 30 mm, up to which point a clear symmetry of the inner wall radius over the axial coordinate $z = 175$ mm can be seen. With ILT deposition, the maximum diameter is at $z = 185$ mm. Moreover, with delayed ILT deposition at the proximal part of the aneurysm, the area covered by thin, proteolytically active ILT is increased. In the results shown here, the biochemical influence of ILT was neglected; however additional expansion of the proximal side and further deformation of the aneurysm sac can be expected due to an increased degradation of the underlying wall.

Figure 6a also shows that the thrombus not only reduces radial but also axial growth. A possible explanation for this is that in AAA without a thrombus, the blood pressure stretches the aneurysm apex a lot, which results in high axial stresses. The formation of a thrombus changes the pressure distribution and significantly reduces axial tensile force at the apex. Interestingly, according to a machine learning model of rupture prediction presented by Shum et al. (2011), the length of the aneurysmal sac is the most important morphologycal characteristic.

5 Conclusions

The ILT has a very complex, controversial, and still unclear role in the progression of cardiovascular diseases. Because many clinical and experimental studies resulted in conflicting conclusions, the ILT presence is usually perceived negatively; however, it might be a natural protection against the rapid expansion of the vascular lesion. The

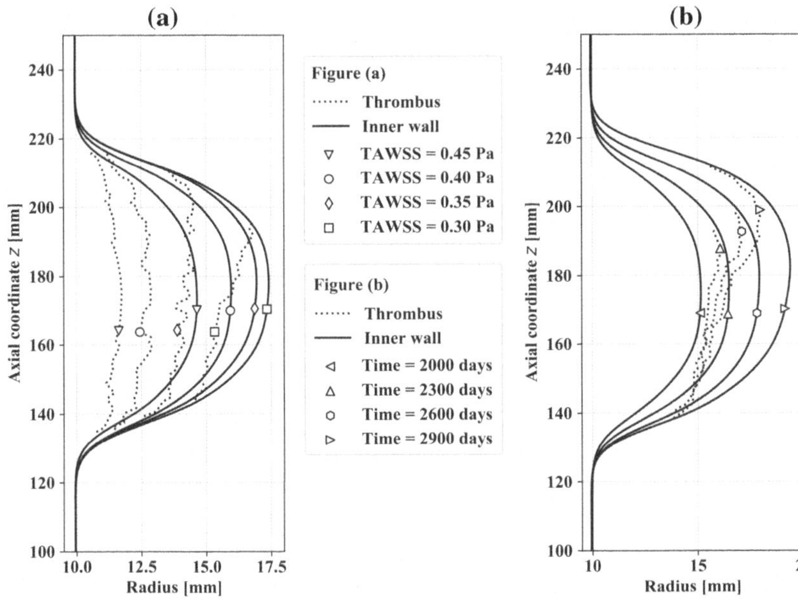

Fig. 6 Contours of the ILT/wall interface radius (solid curves) and the luminal radius (dotted curves) at the same G&R time (2 500 days) for different threshold values of TAWSS (a), and contours at different times during evolutions of AAA when the threshold value is 0.3 Pa (b)

numerical G&R and FSG modeling could be very useful tools for the investigation of the bio-chemo-mechanical processes introduced by thrombus deposition, the guidance of experiments, and the recommendation of more accurate rupture risk factors or better treatments. However, the ILT should not be neglected in the models, as it has a fundamental influence on the disease evolution. For example, during follow-up imaging of AAAs, it was observed that a radial enlargement of aneurysm often occurs via stepwise dilatations, with periods of stability alternating with periods of enlargement (Kurvers et al. 2004). Prior G&R models of aortic aneurysms that were considering only the aortic wall (e.g., Watton and Hill 2009 and Wilson et al. 2012) did not manage to capture this effect. Taking into account the influence of the ILT, the first stagnation could be modeled using a simplified geometry (see Virag et al. 2015 for more details).

The G&R model of the ILT presented in this chapter is partially based on *in vitro* experiments that are true for any blood clot (e.g., information about fibrinogenesis and fibrinolysis), and partially on ILT within AAAs, as histological and mechanical data are available in the literature were needed for the development of the model. Before applying the model to some other vascular lesion, minor adjustments and assumptions would probably be necessary. For example, MMPs from the thrombus in the false lumen of AD would likely diffuse in two directions, both to the intima and the media-adventitia.

On the positive side, the model shows great agreement with clinical observations and animal models. An increase in expansion rates caused by smoking was calculated to be ~0.050 cm/year, and hypertension increased the expansion to ~0.022 cm/year for every 10 mmHg, both are observed clinically (Bhak et al. 2015). The model confirms and offers additional support for the female sex, smoking, and hypertension as factors that increase the risk of rupture. It also supports animal models claiming that inhibiting elastin- and collagen-based inflammation decreases the growth rate, but at the same time agrees with the human experiments that the decrease may not be enough to prevent AAA growth enough with surgery is avoided.

The model not only confirmed already accepted hypotheses, it also brought some new insights into bio-chemo-mechanical processes within AAA. It can explain why a thin thrombus contributes to a rapid expansion, whereas a thick ILT decreases the dilatation rate (Domonkos et al. 2019), despite ILT thickness is proposed to be a rupture risk factor (Vorp et al. 2001). Several hypotheses can be formulated from the numerical results, e.g., that ILT is likely to cause stepwise growth of AAAs or that thrombus indirectly slows down radial growth by prevention of axial expansion.

In conclusion, careful consideration of the mechanical, chemical, and biological influences of proteolytically active, multilayer ILT on vascular lesions is recommended and required in order to maximize the potential of numerical models to improve our understanding of cardiovascular diseases and the risk of rupture and the stabilization factors.

References

Adolph, R., Vorp, D.A., Steed, D.L., Webster, M.W., Kameneva, M.V., Watkins, S.C.: Cellular content and permeability of intraluminal thrombus in abdominal aortic aneurysm. J. Vasc. Surg. **25**, 916–926 (1997)

Al-Barjas, H.S., Ariëns, R., Grant, P., Scott, J.A.: Raised plasma fibrinogen concentration in patients with abdominal aortic aneurysm. Angiology **57**, 607–614 (2006)

Alimohammadi, M., Agu, O., Balabani, S., Díaz-Zuccarini, V.: Development of a patient-specific simulation tool to analyse aortic dissections: assessment of mixed patient-specific flow and pressure boundary conditions. Med. Eng. Phys. **36**, 275–284 (2014)

Arnoud Meijer, C., Stijnen, T., Wasser, M.N.J.M., Hamming, J.F., Van Bockel, J.H., Lindeman, J.H.N.: Doxycycline for stabilization of abdominal aortic aneurysms: a randomized trial. Ann. Intern. Med. **159**, 815–823 (2013)

Behr-Rasmussen, C., Grondal, N., Bramsen, M.B., Thomsen, M.D., Lindholt, J.S.: Mural thrombus and the progression of abdominal aortic aneurysms: a large population-based prospective cohort study. Eur. J. Vasc. Endovasc. Surg. **48**, 301–307 (2014)

Bhak, R.H., Wininger, M., Johnson, G.R., Lederle, F.A., Messina, L.M., Ballard, D.J., Wilson, S.E.: Factors associated with small abdominal aortic aneurysm expansion rate. JAMA Surg. **150**, 44–50 (2015)

Brownlee, R.D., Langille, B.L.: Arterial adaptations to altered blood flow. Can. J. Physiol. Pharmacol. **69**, 978–983 (1991)

Burke, A., FitzGerald, G.A.: Oxidative stress and smoking-induced vascular injury. Prog. Cardiovasc. Dis. **46**, 79–90 (2003)

Caballero, A.D., Laín, S.A.: Review on computational fluid dynamics modelling in human thoracic aorta. Cardiovasc. Eng. Technol. **4**, 103–130 (2013)

Deuel, T.F., Senior, R.M., Chang, D., Griffint, G.L., Heinrikson, R.L., Kaiser, E.T.: Platelet factor 4 is chemotactic for neutrophils and monocytes. Proc. Natl. Acad. Sci. **78**, 4584–4687 (1981)

Di Achille, P., Tellides, G., Figueroa, C.A., Humphrey, J.D.: A haemodynamic predictor of intraluminal thrombus formation in abdominal aortic aneurysms. Proc. R. Soc. A Math. Phys. Eng. Sci. **470**, 20140163–20140163 (2014)

Di Achille, P., Tellides, G., Humphrey, J.D.: Hemodynamics-driven deposition of intraluminal thrombus in abdominal aortic aneurysms. Int. J. Numer. Method Biomed. Eng. **33**, 1–17 (2017)

Domonkos, A., Staffa, R., Kubíček, L.: Effect of intraluminal thrombus on growth rate of abdominal aortic aneurysms. Int. Angiol. **38**, 39–45 (2019)

Enevoldsen, M.S., Henneberg, K.A., Jensen, J.A., Lönn, L., Humphrey, J.D.: New interpretation of arterial stiffening due to cigarette smoking using a structurally motivated constitutive model. J. Biomech. **44**, 1209–1211 (2011)

Figueroa, C.A., Baek, S., Taylor, C.A., Humphrey, J.D.: A computational framework for fluid-solid-growth modeling in cardiovascular simulations. Comput. Methods Appl. Mech. Eng. **198**, 3583–3602 (2009)

Fontaine, V., Jacob, M.P., Houard, X., Rossignol, P., Plissonnier, D., Angles-Cano, E., Michel, J.B.: Involvement of the mural thrombus as a site of protease release and activation in human aortic aneurysms. Am. J. Pathol. **161**, 1701–1710 (2002)

Fontaine, V., Touat, Z., Mtairag, E.M., et al.: Role of leukocyte elastase in preventing cellular re-colonization of the mural thrombus. Am. J. Pathol. **164**, 2077–2087 (2004)

Fridez, P., Zulliger, M., Bobard, F., Montorzi, G., Miyazaki, H., Hayashi, K., Stergiopulos, N.: Geometrical, functional, and histomorphometric adaptation of rat carotid artery in induced hypertension. J. Biomech. **36**, 671–680 (2003)

Frösen, J., Piippo, A., Paetau, A., Kangasniemi, M., Niemelä, M., Hernesniemi, J., Jääskeläinen, J.: Remodeling of saccular cerebral artery aneurysm wall is associated with rupture: histological analysis of 24 unruptured and 42 ruptured cases. Stroke **35**, 2287–2293 (2004)

Grytsan, A., Watton, P.N., Holzapfel, G.A.: A thick-walled fluid-solid-growth model of abdominal aortic aneurysm evolution: application to a patient-specific geometry. J. Biomech. Eng. **137**, 031008 (2015)

Horvat, N., Virag, L., Holzapfel, G.A., Sorić, J., Karšaj, I.: A finite element implementation of a growth and remodeling model for soft biological tissues: verification and application to abdominal aortic aneurysms. Comput. Methods Appl. Mech. Eng. **352**, 586–605 (2019)

Houard, X., Rouzet, F., Touat, Z., et al.: Topology of the fibrinolytic system within the mural thrombus of human abdominal aortic aneurysms. J. Pathol. **212**, 20–28 (2007)

Houard, X., Touat, Z., Ollivier, V., Louedec, L., Philippe, M., Sebbag, U., Meilhac, O., Rossignol, P., Michel, J.B.: Mediators of neutrophil recruitment in human abdominal aortic aneurysms. Cardiovasc. Res. **82**, 532–541 (2009)

Humphrey, J.D, Rajagopal, K.R.: A constrained mixture model for growth and remodelling of soft tissues. Math. Model Methods Appl. Sci. **12**, 407–430 (2002). https://doi.org/10.1142/S0218202502001714

Humphrey, J.D., Taylor, C.A.: Intracranial and abdominal aortic aneurysms: similarities, differences, and need for a new class of computational models. Annu. Rev. Biomed. Eng. **10**, 221–246 (2008)

Inzoli, F., Boschettp, F., Zappa, M., Longo, T., Fumero, R.: Biomechanical factors in abdominal aortic aneurysm rupture. Eur. J. Vasc. Surg. **7**, 667–674 (1993)

Jatoi, N.A., Jerrard-Dunne, P., Feely, J., Mahmud, A.: Impact of smoking and smoking cessation on arterial stiffness and aortic wave reflection in hypertension. Hypertension **49**, 981–985 (2007)

Karšaj, I., Humphrey, J.D.: A mathematical model of evolving mechanical properties of intraluminal thrombus. Biorheology **46**, 509–527 (2009)

Karšaj, I., Humphrey, J.D.: A multilayered wall model of arterial growth and remodeling. Mech. Mater. **44**, 110–119 (2012)

Kurvers, H., Veith, F.J., Lipsitz, E.C., et al.: Discontinuous, staccato growth of abdominal aortic aneurysms. J. Am. Coll. Surg. **199**, 709–715 (2004)

Lawrence, D.M., Singh, R.S., Franklin, D.P., Carey, D.J., Elmore, J.R.: Rapamycin suppresses experimental aortic aneurysm growth. J. Vasc. Surg. **40**, 334–338 (2004)

Mahmud, A., Feely, J.: Effect of smoking on arterial stiffness and pulse pressure amplification. Hypertension **41**, 183–187 (2003)

Meade, T.W., Imeson, J., Stirling, Y.: Effects of changes in smoking and other characteristics on clotting factors and the risk of ischaemic heart disease. Lancet **330**, 986–988 (1987)

Nackman, G.B., Karkowski, F.J., Halpern, V.J., Gaetz, H.P., Tilson, M.D.: Elastin degradation products induce adventitial angiogenesis in the Anidjar/Dobrin rat aneurysm model. Surgery **122**, 39–44 (1997)

Naim, W.N.W.A., Ganesan, P.B., Sun, Z., Liew, Y.M., Qian, Y., Lee, C.J., Jansen, S., Hashim, S.A., Lim, E.: Prediction of thrombus formation using vortical structures presentation in Stanford type B aortic dissection: a preliminary study using CFD approach. Appl. Math. Model. **40**, 3115–3127 (2016)

Owens, A.P., 3rd., Edwards, T.L., Antoniak, S., et al.: Platelet inhibitors reduce rupture in a mouse model of established abdominal aortic aneurysm. Arterioscler. Thromb. Vasc. Biol. **35**, 2032–2041 (2015)

Paliwal, N., Damiano, R.J., Davies, J.M., Siddiqui, A.H., Meng, H.: Association between hemodynamic modifications and clinical outcome of intracranial aneurysms treated using flow diverters. Proc. SPIE Int. Soc. Opt. Eng. **10135**, 101352F (2017)

Parodi, F.E., Mao, D., Ennis, T.L., Bartoli, M.A., Thompson, R.W.: Suppression of experimental abdominal aortic aneurysms in mice by treatment with pyrrolidine dithiocarbamate, an antioxidant inhibitor of nuclear factor-kappaB. J. Vasc. Surg. **41**, 479–489 (2005)

Powell, J.T.: Vascular damage from smoking: disease mechanisms at the arterial wall. Vasc. Med. **3**, 21–28 (1998)

Schriefl, A.J., Collins, M.J., Pierce, D.M., Holzapfel, G.A., Niklason, L.E., Humphrey, J.D.: Remodeling of intramural thrombus and collagen in an Ang-II infusion ApoE-/-model of dissecting aortic aneurysms. Thromb. Res. **130**, e139–e146 (2012)

Scott, D.J.A., Prasad, P., Philippou, H., et al.: Clot architecture is altered in abdominal aortic aneurysms and correlates with aneurysm size. Arterioscler. Thromb. Vasc. Biol. **31**, 3004–3010 (2011)

Selimovic, A., Ventikos, Y., Watton, P.N.: Modelling the evolution of cerebral aneurysms: biomechanics, mechanobiology and multiscale modelling. Procedia IUTAM **10**, 396–409 (2013)

Shadden, S.C., Hendabadi, S.: Potential fluid mechanic pathways of platelet activation. Biomech. Model. Mechanobiol. **12**, 467–474 (2013)

Shum, J., Martufi, G., Di Martino, E., Washington, C.B., Grisafi, J., Muluk, S.C., Finol, E.A.: Quantitative assessment of abdominal aortic aneurysm geometry. Ann. Biomed. Eng. **39**, 277–286 (2011)

Simão Da Silva, E., Rodrigues, A.J., Magalhães Castro De Tolosa, E., Rodrigues, C.J., Villas Boas Do Prado, G., Nakamoto, J.C.: Morphology and diameter of infrarenal aortic aneurysms: a prospective autopsy study. Cardiovasc. Surg. **8**, 526–532 (2000)

Steinmetz, E.F., Buckley, C., Shames, M.L., Ennis, T.L., Vanvickle-Chavez, S.J., Mao, D., Goeddel, L.A., Hawkins, C.J., Thompson, R.W.: Treatment with simvastatin suppresses the development of experimental abdominal aortic aneurysms in normal and hypercholesterolemic mice. Ann. Surg. **241**, 92–101 (2005)

Tong, J., Cohnert, T., Regitnig, P., Holzapfel, G.A.: Effects of age on the elastic properties of the intraluminal thrombus and the thrombus-covered wall in abdominal aortic aneurysms: Biaxial extension behaviour and material modelling. Eur. J. Vasc. Endovasc. Surg. **42**, 207–219 (2011)

Tsai, T.T., Evangelista, A., Nienaber, C.A., et al.: Partial thrombosis of the false lumen in patients with acute type B aortic dissection. N. Engl. J. Med. **357**, 349–359 (2007)

Valentín, A., Humphrey, J.D., Holzapfel, G.A.: A multi-layered computational model of coupled elastin degradation, vasoactive dysfunction, and collagenous stiffening in aortic aging. Ann. Biomed. Eng. **39**, 2027–2045 (2011)

van den Berkmortel, F.W.P.J., Wollersheim, H., van Langen, H., Smilde, T.J., den Arend, J., Thien, T.: Two years of smoking cessation does not reduce arterial wall thickness and stiffness. Neth. J. Med. **62**, 235–241 (2004)

Vande Geest, J.P., Sacks, M.S., Vorp, D.A.: A planar biaxial constitutive relation for the luminal layer of intra-luminal thrombus in abdominal aortic aneurysms. J. Biomech. **39**, 2347–2354 (2006)

Virag, L., Wilson, J.S., Humphrey, J.D., Karšaj, I.: A computational model of biochemomechanical effects of intraluminal thrombus on the enlargement of abdominal aortic aneurysms. Ann. Biomed. Eng. **43**, 2852–2867 (2015)

Virag, L., Wilson, J.S., Humphrey, J.D., Karšaj, I.: Potential biomechanical roles of risk factors in the evolution of thrombus-laden abdominal aortic aneurysms. Int. J. Numer. Method Biomed. Eng. **33**, e2893 (2017)

Vorp, D.A., Lee, P.C., Wang, D.H.J., Makaroun, M.S., Nemoto, E.M., Ogawa, S., Webster, M.W.: Association of intraluminal thrombus in abdominal aortic aneurysm with local hypoxia and wall weakening. J. Vasc. Surg. **34**, 291–299 (2001)

Wang, D.H.J., Makaroun, M.S., Webster, M.W., Vorp, D.A.: Effect of intraluminal thrombus on wall stress in patient-specific models of abdominal aortic aneurysm. J. Vasc. Surg. **36**, 598–604 (2002)

Watton, P.N., Hill, N.A.: Evolving mechanical properties of a model of abdominal aortic aneurysm. Biomech. Model. Mechanobiol. **8**, 25–42 (2009)

Watton, P.N., Selimovic, A., Raberger, N.B., Huang, P., Holzapfel, G.A., Ventikos, Y.: Modelling evolution and the evolving mechanical environment of saccular cerebral aneurysms. Biomech. Model. Mechanobiol. **10**, 109–132 (2011)

Wilson, J.S., Baek, S., Humphrey, J.D.: Importance of initial aortic properties on the evolving regional anisotropy, stiffness and wall thickness of human abdominal aortic aneurysms. J. R. Soc. Interface **9**, 2047–2058 (2012)

Wilson, J.S., Virag, L., Di Achille, P., Karšaj, I., Humphrey, J.D.: Biochemomechanics of intraluminal thrombus in abdominal aortic aneurysms. J. Biomech. Eng. **135**, 021011 (2013)

Zambrano, B.A., Gharahi, H., Lim, C.Y., Jaberi, F.A., Choi, J., Lee, W., Baek, S.: Association of intraluminal thrombus, hemodynamic forces, and abdominal aortic aneurysm expansion using longitudinal CT images. Ann. Biomed. Eng. **44**, 1502–1514 (2016)

Mechanical Characterization and Modeling of Diabetic Aortas

Jianhua Tong

The first time I met Professor Gerhard A. Holzapfel was in a biomechanics course in spring 2006, at a time when he was a Professor of Biomechanics at KTH in Sweden. Thereby I gained a comprehensive understanding of biomechanics as an interdisciplinary field. In 2007 I started my doctorate with a focus on the experimental characterization of aortic aneurysms under the direction of Professor Holzapfel at the Institute for Biomechanics at TU Graz in Austria. As a PhD advisor, Gerhard has an extensive knowledge of the current research, and he also has motivational skills for a PhD student. The experiences I got from him always inspired me to think scientifically in order to solve research problems. During my postdoctoral training and my academic career, we continued to work together and cultivated personal contacts. On the occasion of his 60th birthday, I would like to thank you very much for introducing me to the challenging field of cardiovascular biomechanics. It is a great honor for me to have met a mentor like you in my life.

Tong

Abstract Diabetes mellitus has been recognized as a major risk factor leading to macrovascular disease. Although the precise mechanisms of the diabetic macrovascular disease remain poorly understood, diabetes plays a vital role in the localized remodeling of the descending aorta. The aim of this work is to investigate the mechanical properties of rabbit descending aortas in health and diabetes. A rabbit of type 1 diabetes mellitus was raised by injection of alloxan. Using experimental tools in biomechanics and a material model, we characterized biaxial mechanical behaviors of both aorta types and obtained constitutive parameters. Histological analysis was employed to investigate the underlying microstructure. Results suggest that diabetes impairs elastic properties of the aortas and has significant impact on the local wall mechanics and the microstructure.

J. Tong (✉)
Tongji University School of Medicine, Shanghai, China
e-mail: tongjh@tongji.edu.cn

G. Sommer et al. (eds.), *Solid (Bio)mechanics: Challenges of the Next Decade*, Studies in Mechanobiology, Tissue Engineering and Biomaterials 24,
https://doi.org/10.1007/978-3-030-92339-6_6

1 Introduction

Diabetes mellitus is a major risk factor that may cause severe cardiovascular consequences. The onset of hyperglycemia in patients may enhance formation of glycosylated proteins and advance glycation products, leading to endothelial dysfunction (Brownlee et al. 1988). These pathological changes will contribute to promoting macro- and micro-vascular diseases, in particular, to accelerating atherogenesis (Grundy et al. 1999). The precise mechanisms of the diabetic macrovascular disease remain poorly understood. Atherosclerosis is one of the main vascular complications of diabetes and has significant impact on the health of the affected population. The clinical observations indicate the correlation between the risk of vascular complications in diabetes and poor glycemic control. In hyperglycemic conditions, increased platelet adhesion and hypercoagulability are thought to be a main trigger to initiate formation of atherosclerotic lesions (Beckman et al. 2002).

Both type 1 (insulin-deficient) and type 2 (insulin-resistant with hyperinsulinemia) diabetes impair biomechanical functions of aortas due in large part to changes in extracellular matrix (ECM), particularly in collagen content (Grundy et al. 1999; Akhtar et al. 2014). Enhanced extracellular protease activity in the diabetic aorta induces morphological, compositional and localized mechanical tissue remodeling (Akhtar et al. 2014). Thus, vascular dysfunction often occurs in the early stage of the disease process. Over the last decade researchers have devoted their effort to investigating the effects of diabetes on aortic wall primarily in the areas of biochemistry, morphology, and imaging; relevant biomechanical studies are rare in the literature. For example, Golledge et al. (2008) found that diabetes could strengthen the cross-linking of collagen fibers within aortic media and inhibited secretion of matrix metalloproteinases (MMPs). Thereby, the diabetes may limit overstretch of the aortic wall during expansion in response to intraluminal pressure. Using immunohistochemistry technique, Dua et al. (2010) confirmed that hyperglycemia inhibited plasmin and MMP and thus, would reduce fiber degradation of the aortic wall. Astrand et al. (2007) reported that the measured intima-media layers in the diabetic aortas were significantly thicker than healthy aortas. However, there was no significant difference in aortic diameter between diabetic and healthy groups. In addition, these authors computed the wall stress distribution using Laplace's law, showing that diabetic aortas had a significantly lower stress magnitude than the healthy aortas. Although above studies advance our understanding of changes in structure-function relationship of the aorta when affected by diabetes, quantitative data are lacking in the literature to interpret complex biomechanical alterations of diabetic aortas.

In this study the aim was to systematically investigate biomechanical properties of rabbit aortas in health and diabetes. Using an alloxan-induced type 1 diabetic rabbit model, uniaxial and biaxial tensile tests together with material modeling are performed to characterize changes in the elastic properties of the aortic tissue in response to hyperglycemia. In addition, we seek to identify microstructural alterations of the diabetic aortas in histology and to determine mass fractions of elastin and collagen within aortic walls.

2 Materials and Methods

2.1 Diabetic Animal Model

Thirty female New Zealand white rabbits, 7 weeks old and weighting 1.6–2.2 kg, were purchased from the Experimental Animal Center of Tongji University, Shanghai, China. Animals were housed individually at a constant temperature of 24 °C and were kept in a constant photoperiod (i.e. 7a.m. light/7p.m. dark diurnal light cycle) in plastic cages with free access to food and water for 1 week before experiments. The fasting blood glucose (FBG) of all rabbits was measured after fasting for at least 8 h.

The rabbits were randomly assigned into two groups, i.e. control and diabetic groups. Each group contains 15 rabbits. The diabetic group was then injected a 5% solution of alloxan monohydrate (100 mg/kg of body weight) to raise a diabetic rabbit model, while the control group was given equal amount of normal saline (0.9% w/v of NaCl). The FBG was measured 72 h later and the rabbits with FBG level more than 15 mmol/L were considered to be diabetic. If FBG level failed to exceed 15 mmol/L, the rabbits would be further injected by alloxan (130 mg/kg of body weight) until the FBG level is reached. Note that alloxan could cause massive insulin release from pancreas on the first day. Hence, it was necessary to give all rabbits 5% glucose containing in the drinking water to prevent fatal hypoglycemia in the following 24 h. The rabbits were exposed to diabetic condition (i.e. the FBG level over 15 mmol/L) for four weeks before sacrifice for mechanical testing. All experimental procedures complied with international guidelines for care and use of laboratory animals and were approved by the Animal Ethics Committee of Tongji University, Shanghai, China.

2.2 Specimen Preparation

Rabbits were anesthetized with 2% pentobarbital sodium (2.5 ml/kg) injected into the marginal ear vein. After achieving full anesthesia, midline laparotomy was performed. Abdomens of all rabbits were cut open first and precise location of the aorta in each animal model was identified. To separate the abdominal aorta, three threads were deployed to fix both ends and to elevate the aorta segment. The abdominal aorta was carefully dissected and was cut into two portions with sizes of 20 mm (region 1) for mechanical tests and 5 mm (region 2) for histological investigation.

Given a cylindrical shape we further cut the aorta (harvested from region 1) open axially to obtain plain aortic tissue piece, see Fig. 1. The thicknesses of aortic walls were measured by means of a video-extensometer (Holzapfel et al. 2005). Using a surgical scalpel, two rectangular strips (C and A) with a dimension of approximately 14.0×4.0 mm (length $\times$ width) were extracted in the circumferential and axial directions, respectively (see Fig. 1). These two specimens were used for uniaxial

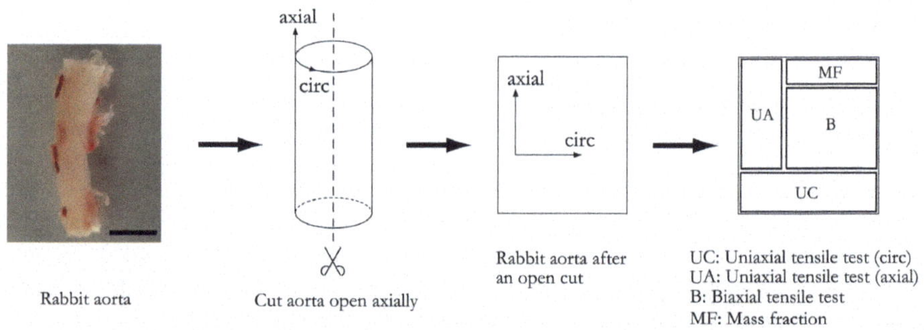

Fig. 1 Schematic illustration of specimen cutting based on a rabbit aorta obtained from region 1. The aorta is cut open axially and four pieces of aortic tissues are extracted to perform uniaxial tensile tests (UC and UA), biaxial tensile test (B), and mass fraction (MF) analysis, respectively. The black bar in the first image indicates a length of 5 mm. Reproduced with permission from Tong et al. (2018)

tensile tests. A square piece (B) with a dimension of approximately 10.0 × 10.0 mm (length × width) was further cut from the remaining tissue, serving for the biaxial tensile test (Fig. 1). The small (rectangular) piece of the aortic tissue (noted by MF) was stored in liquid nitrogen for mass fraction analysis.

For uniaxial tensile tests, pieces of sandpapers were mounted at both ends of the rectangular strips (C and A) with super-adhesive gel to facilitate a defined clamping in the tensile testing machine and to prevent slippage during testing. In addition, two black chips were glued transversely in parallel onto the middle part of the specimens to act as gage markers for deformation measurements. For biaxial tensile tests, four nylon sutures were fixed to each side of the square specimen (B) using fish hooks, and four markers were placed at a distance of approximately 3 mm from each other at the center of the specimen. All specimens were moistened in phosphate-buffered physiological saline (PBS) at 37 °C and tested within 6 h after excision.

2.3 Mechanical Testing and Data Analysis

Uniaxial tensile tests were performed on a computer-controlled, screw-driven high-precision tensile testing machine adapted for biological materials (XJ810-10N, Xiangjie Inc., Shanghai, China). Preconditioning was achieved by executing five loading and unloading cycles at a constant extension rate of 1.0 mm/min to reach 70 kPa (i.e. first Piola-Kirchhoff stress) for each test. Starting from the sixth tensile cycle, the strain was increased with the same extension rate until tissue rupture occurred. The rupture forces and the corresponding stretches were documented.

Details of biaxial testing protocols have been described in a prior study (Tong et al. 2011). Square specimens were mounted in a custom-built biaxial testing device via connecting four carriages by hooked nylon sutures and were then submerged into a bath filled with Dulbecco's Modified Eagle's Medium (DMEM) maintaining at 37.0 ± 1.0 °C. The initial lengths between the two markers in each direction were measured by a CCD camera before the start of the biaxial tensile tests. A stress-driven protocol was employed during the testing procedure. Each specimen was tested using the following protocol for each stress level: $P_{\theta\theta}{:}P_{zz} = 1{:}1$, 1:0.75, 0.75:1, 1:0.5, and 0.5:1, where $P_{\theta\theta}$ and P_{zz} denote the applied engineering stresses in the circumferential and axial directions, respectively. We started at 30 kPa to execute the biaxial protocol for all specimens and continued with a stepwise increase of 10 kPa stress level until tissues' rupture occurred in the equibiaxial protocol. After each increase in stress, five preconditioning cycles were conducted and the sixth was then used for data analysis.

From the uniaxial tensile tests, the ultimate tensile stress σ_{ult} was computed according to

$$\sigma_{\text{ult}} = \frac{f_{\text{rup}}\lambda_{\text{ult}}}{WT}, \tag{1}$$

where f_{rup} is the rupture force, and W and T are the width and the thickness of the specimen in the unloaded configuration, respectively. The ultimate stretch λ_{ult} is defined as l/L, where l and L are the gage lengths in the loaded (state at rupture) and unloaded configurations, respectively.

The Cauchy stress and stretch were computed to quantify the biaxial mechanical responses of the aortic tissues. The shear deformation was treated as negligible. Thus, the Cauchy stresses were determined as

$$\sigma_{\theta\theta} = \frac{f_\theta \lambda_\theta}{TX_z}, \qquad \sigma_{zz} = \frac{f_z \lambda_z}{TX_\theta}, \tag{2}$$

where $\lambda_\theta = x_\theta / X_\theta$ and $\lambda_z = x_z / X_z$ indicate tissue stretches in each direction that are measured by markers. For the equibiaxial stress-controlled protocol ($P_{\theta\theta}{:}P_{zz} = 1{:}1$), the mean peak stretch (MPS) values in the circumferential and axial directions were recorded. Moreover, the maximum tangential modulus (MTM), defined as the maximum slope of the nonlinear stress-stretch curve, was calculated and further averaged to assess aortic stiffness under the equibiaxial stresses of $P_{\theta\theta}{:}P_{zz} = 60$ kPa.

All experimental data were summarized as mean±standard deviation (SD). Prior to statistical analysis, the Shapiro-Wilk and the Kolmogorov-Smirnov tests were used to determine whether each dataset was normally distributed. Comparisons of the quantitative data between control and diabetic groups were performed using the analysis of variance (ANOVA) test followed by the Holm-Sidak test and the Dunn's Method test. For comparison between circumferential and axial mechanical responses, the paired Student's t-test was employed. Non-parametric tests, including the Wilcoxon signed-rank test and Mann-Whitney rank sum test, were used for non-normally distributed sample groups. A p-value of <0.05 determined significance. The statistical analyses were performed using the SPSS statistical package version 21.0 (SPSS Inc./IBM, Chicago, IL).

2.4 Material Modeling

To model the biaxial mechanical behaviors of the control and the diabetic aortas, we used a strain-energy function (Holzapfel et al. 2005) that was developed primarily for collagen-reinforced biological soft tissues such as arterial walls. The strain-energy function has the form

$$\Psi = \mu(I_1 - 3) + \frac{k_1}{k_2}(\exp\{k_2[(1-\rho)(I_1-3)^2 + \rho(I_4-1)^2]\} - 1), \qquad (3)$$

where $\mu > 0$ and $k_1 > 0$ are stress-like parameters (in kPa), and $k_2 > 0$ and $\rho \in [0, 1]$ are dimensionless parameters. The invariants I_1 and $I_4 > 1$ can be expressed as

$$I_1 = \lambda_\theta^2 + \lambda_z^2 + \lambda_r^2, \qquad I_4 = \lambda_\theta^2 \cos^2\varphi + \lambda_z^2 \sin^2\varphi, \qquad (4)$$

where λ_θ, λ_z, λ_r are the principal stretches in the circumferential, axial and radial directions, respectively. The geometrical parameter φ in Eq. (4) denotes the angle between the principal collagen fiber orientation and the circumferential direction. In the present study, however, φ is treated as a phenomenological variable for modeling. Due to the assumption of incompressibility and the vanishing radial stresses, the Cauchy stresses in the circumferential and axial directions are determined.

The experimental data from the five biaxial protocols for the control and the diabetic aortas associated with the circumferential and axial directions were simultaneously fit to the material model (3) using the optimization toolbox in Matlab 8.0. The fitting was carried out following the nonlinear least-squares algorithm that minimizes the sum of squares of the deviation between the predicted stresses and the experimental data. Consequently, the five parameters (μ, k_1, k_2, φ, ρ) were obtained. To measure the 'goodness of fit' the square of the Pearson's correlation coefficient was computed for both the circumferential and the axial Cauchy stresses.

2.5 Histological and Mass Fraction Analysis

Both the control and the diabetic aortas obtained from region 2 were fixed in 4% paraformaldehyde phosphate buffer solution (PFA) overnight at 4 °C, and then embedded in paraffin using standard techniques. The paraffin blocks were cut at 3 μm section on a Leica RM2126 microtome (Leica, Shanghai, China), then stained with Hematoxylin Eosin (HE) and Heidenhain's Azan (Azan) after deparaffinization and hydration through a series of graded alcohol steps. Subsequently, the sections were examined under a Keyence BZ9000 microscope (Keyence Co. Ltd., Japan) to characterize morphological changes of endothelium, collagen fibers and other cellular components in response to diabetes.

The mass fraction analysis quantitatively determined dry weight of elastin and collagen within the intact aortic wall. The MF pieces of the tissue in Fig. 1 were frozen in liquid nitrogen and homogenized by grinding in a mortar with a pestle. After reaching constant weight by drying for 3 to 4 days at 105 °C, the wet and dry weight of the homogenate were determined, respectively; for more details regarding elastin quantitation and collagen analysis see the study by Tong et al. (2013).

3 Results

In total, 15 control and 15 diabetic rabbit aortas were tested and analyzed in the present study. The mean thicknesses of aortic walls for the control and the diabetic groups were measured to be 0.86 ± 0.05 mm and 0.92 ± 0.11 mm ($p = 0.09$). There is no statistically significant increase in wall thickness for the diabetic aortas when compared to the control aortas. The diabetic aortic thicknesses, however, were more variable than the control ones. This is probably due to regional variations of wall thicknesses caused by hyperglycemia.

The Shapiro-Wilk and the Kolmogorov-Smirnov tests showed that the quantified ultimate tensile stresses were normally distributed. The quantified ultimate tensile stresses in the circumferential direction for the control and the diabetic aortas were 384 ± 35 kPa and 417 ± 48 kPa, respectively ($p = 0.07$). The ultimate tensile stresses in the axial directions for the control and the diabetic aortas were computed as 338 ± 39 kPa and 362 ± 41 kPa, respectively ($p = 0.09$). In general, the mean ultimate tensile stresses in both circumferential and axial directions are higher for the diabetic aortas than the control aortas. However, there is no statistically significant difference in ultimate tensile strength between the control and the diabetic aortas.

Representative biaxial mechanical behaviors and corresponding model fit results of the control and the diabetic aortas are shown in Fig. 2. Both aortic tissues exhibited exponential nonlinearity, and anisotropy. The differences in the MPS and MTM values between the control and the diabetic groups are shown by column plots in Fig. 3. Under the equibiaxial stress-controlled protocol (i.e. $P_{\theta\theta}:P_{zz} = 60$ kPa), the circumferential MPS values of the control and the diabetic aortas were measured as 1.13 ± 0.03 and 1.09 ± 0.02, respectively ($p = 0.02$). The MPS values in the axial directions of the control and the diabetic aortas were 1.16 ± 0.02 and 1.13 ± 0.03, respectively ($p = 0.04$); see Fig. 3a. Moreover, the circumferential MTM values of the control and the diabetic aortas were 2.6 ± 0.7 MPa and 3.7 ± 0.9 MPa, respectively ($p = 0.005$). The axial MTM values of the control and the diabetic aortas were computed as 2.2 ± 0.5 MPa and 3.4 ± 0.7 MPa, respectively ($p = 0.003$); see Fig. 3b. It indicates that tissue stiffness of the diabetic aorta is statistically significantly higher than that of the control aorta. Pronounced tissue stiffening occurs in both circumferential and axial directions of the diabetic aortas.

The related constitutive parameters were summarized in Table 1. The material model (Eq. (3)) was suitable to characterize the biaxial mechanical behaviors of both aorta types. The 'goodness of fit' R^2 is 0.95 ± 0.03 and 0.92 ± 0.05 for the control and the diabetic groups, respectively; both values are within the common range for biomaterials. The constitutive parameters μ, which described the average

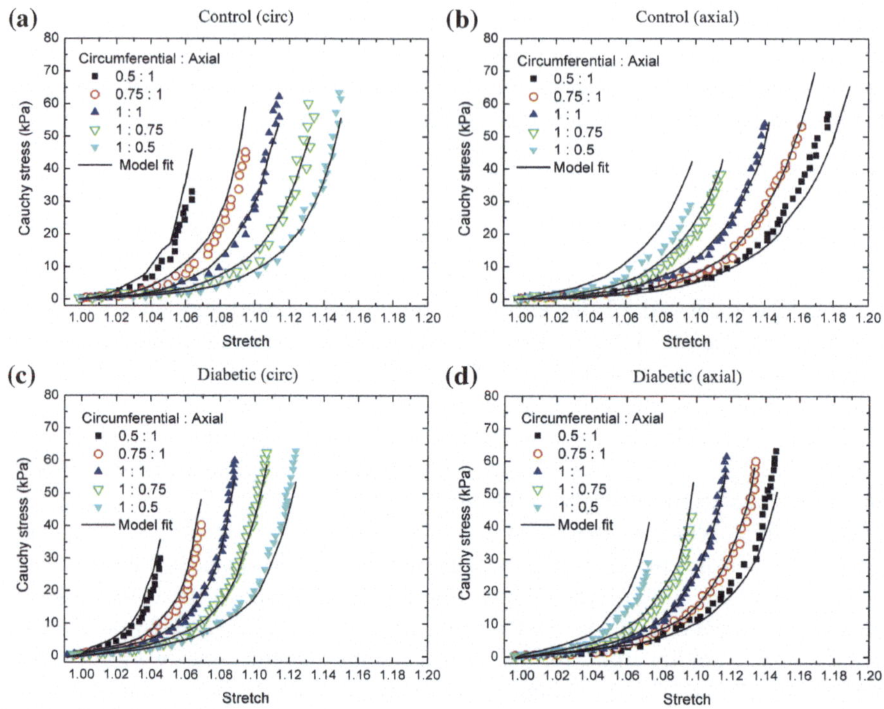

Fig. 2 Representative plots of biaxial experimental data (symbols) with corresponding material model fit (solid curves) for (a), (b) the control and (c), (d) the diabetic aortas in the circumferential and axial directions. Reproduced with permission from Tong et al. (2018)

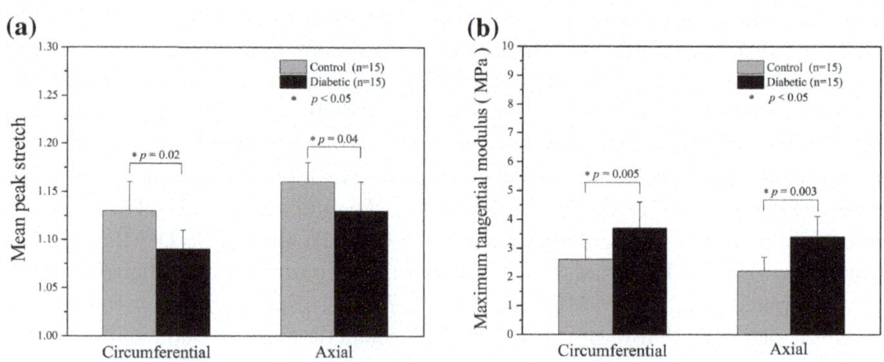

Fig. 3 Column plots (mean values and standard deviations) of (a) the quantified mean peak stretch (MPS) and (b) the calculated maximum tangential modulus (MTM) for the control ($n = 15$) and the diabetic ($n = 15$) rabbit aortas in the equibiaxial stress-controlled protocol ($P_{\theta\theta}:P_{zz} = 60$ kPa). Statistically significant differences are noted in both MPS and MTM values between the control and the diabetic aortas. Reproduced with permission from Tong et al. (2018)

Table 1 Constitutive parameters obtained by material modeling for the control ($n = 15$) and the diabetic ($n = 15$) aortas

	μ (kPa)	k_1 (kPa)	k_2 (–)	φ (°)	ρ (–)	R^2
Control	0.8 ± 0.4	48.6 ± 6.3	31.6 ± 13.2	33.6 ± 5.9	0.26 ± 0.08	0.95 ± 0.03
Diabetic	2.7 ± 0.6	60.5 ± 8.4	73.5 ± 21.6	41.2 ± 8.5	0.18 ± 0.06	0.92 ± 0.05

stiffness of aortic tissues in the low loading domain (i.e. the isotropic matrix material is mainly active), were significantly larger for the diabetic group than the control group ($p = 0.007$). The constitutive parameters k_1 and k_2 governing the average stiffness of the aortic tissues in the high loading domain (i.e. the recruitment of collagenous fibers) were also significantly larger for the diabetic group than the control group (all p-values are less than 0.01). It suggests that the diabetic aortas have significantly higher tissue stiffness in both low and high loading domains when compared to the control aortas. The mean geometrical parameter φ of the diabetic group was larger than that of the control group, however, no significant difference was found ($p = 0.08$). There is no statistically significant difference in the constitutive parameter ρ between the control and the diabetic aortas ($p = 0.11$), suggesting that mechanical anisotropy does not change remarkably for the diabetic aortas when compared with the control aortas.

Representative images of histological sections of the control and the diabetic aortas were shown in Fig. 4. Compared to the control aortas, the HE staining showed that there were diffusive eminences on the arterial dissepiment for the diabetic aortas (Figs. 4a, b). Notably, a number of isolated foam cells were identified within the diabetic aortic layers, see, for example, Fig. 4b. In the Azan staining, collagen hyperplasia was a main microstructural characteristic evidently shown in the most diabetic aorta samples (Figs. 4c, d). This feature may lead to varying distributions and realignment of collagen fibers for the diabetic aortas.

There was no statistically significantly difference in dry weight percentage of elastin between the control and the diabetic groups (control: 9.3% ± 1.8% vs. diabetic: 8.5% ± 2.9%; $p = 0.21$). However, the dry weight percentage of collagen within the diabetic aortas was significantly higher than that within the control aortas (control: 25.4% ± 5.9% vs. diabetic: 33.7% ± 6.8%; $p = 0.02$).

4 Discussion

The animal model used in this work is alloxan-induced type 1 diabetes, which has been well documented in the literature (Sun et al. 2009; Akhtar et al. 2014). Induction of alloxan can essentially damage the pancreatic β-cell so that blood glucose is increased effectively (Mingueneau et al. 2015; Novoselova et al. 2016). Such a method can be applied to different animals to experimentally induce diabetes. The thickness measurements show that there is a trend towards an aortic wall thickening

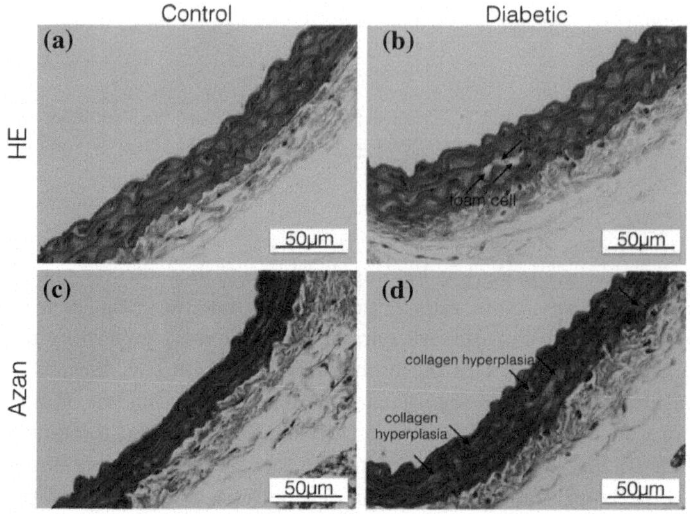

Fig. 4 Representative histological images with HE and Azan staining characterizing the underlying microstructure of (a), (c) the control and (b), Compared to the control aortas in panel (a), the isolated foam cells (denoted by arrows) are identified for the diabetic aortas in panel (b). Panel (c) displays the collagen fibers aligned in the control aortas. By contrast, blue staining of collagen hyperplasia (denoted by arrows) is evidently shown in the diabetic aortas in panel (d). Original magnification 60×. Reproduced with permission from Tong et al. (2018)

for the diabetic aortas, although this phenomenon does not reach statistical significance. This finding is basically consistent with Akhtar et al. (2014) and Kozakova et al. (2014). Note that our thickness data represent the intact diabetic aortas rather than the single medial layers. Therefore, the diabetic aortas may suffer from wall thickening due in large part to tissue fibrosis caused by hyperglycemia.

No significant differences were found in both circumferential and axial ultimate tensile strength between the control and the diabetic groups, suggesting that diabetes has very limited effects on altering the tensile strength of aortas. To the authors' knowledge, no quantitative mechanical data have been measured prior to the present study to investigate changes in tensile strength of the diabetic aortas (Schram et al. 2004; Sun et al. 2009; Akhtar et al. 2014; Kozakova et al. 2014). The previous studies either focused on aortic remodeling (Akhtar et al. 2014) in response to diabetes or investigated changes in the arterial stiffness (Schram et al. 2004; Sun et al. 2009) with a higher glucose exposure.

As can be seen from Fig. 2, biaxial mechanical behaviors of the control and the diabetic aortas are highly nonlinear and anisotropic. It is worth noting that the used biaxial testing protocol covers a large range of deformations including the *in vivo* situation, and hence provides a unique set of mechanical data when compared to uniaxial tensile test. Regarding biaxial mechanical responses, our results suggest that the diabetic aortas have significantly lower extensibility and significantly higher tissue stiffness than the control aortas (Figs. 3a, b). By checking the quantified MTM values in the equibiaxial protocol, we further confirm that tissue stiffening occurs in both circumferential and axial directions of the diabetic aortas. Our finding is basically in agreement with conclusions obtained in the previous studies (Astrand et al. 2007; Sun et al. 2009; Due et al. 2010; Kozakova et al. 2014), in which, however, data are frequently measured by means of observing aortic morphology. For example, Sun et al. (2009) showed an increased aortic stiffness for the diabetic rats by calculating cross-sectional distensibility of aortas. Note that the reduced cross-sectional distensibility of aortas might imply a stiffening trend but does not provide any quantitative data on aortic stiffness. In particular, little is known about the aortic stiffness in the axial direction. Moreover, observing cross-sectional morphological alterations cannot characterize anisotropic mechanical behaviors of aortas.

As can be seen from Fig. 2 and Table 1, the material model is able to characterize the biaxial mechanical behaviors of both aorta types well. Significantly larger values of constitutive parameters μ, k_1 and k_2 indicate that the diabetic aortas have significantly higher tissue stiffness in both low and high loading domains when compared to the control aortas. There is no statistically significant difference in constitutive parameter φ between the control and the diabetic groups. Although φ is phenomenologically determined, it probably indicates that collagen fiber alignment does not alter significantly within the diabetic aortas as compared to the control aortas. However, to verify this statement, a subtle investigation of the microstructure by aid of imaging technique is needed.

Histological analysis, as indicated in Fig. 4, suggests that the diabetic aortas have several changes in the microstructural characteristics when compared to the control aortas. The foam cells embedded in the arterial layers can be a trigger to cause increased wall thicknesses for the diabetic aortas and may promote formation of atherosclerotic lesions (Bjornheden et al. 1999; Katsuda et al. 2004). Note that hyperplasia of collagen may directly or indirectly influence adhesion, alignment and stretches of fiber bundles, and, therefore, plays a key role in tissue's remodeling during pathological process. In particular, varying distributions and realignment of collagen due to hyperplasia will lead to a more disorganized microstructure within the aortic wall, and, thus, will induce changes in mechanical anisotropy of the aortas (Jhun et al. 2009). These factors potentially alter extensibility and tensile strength of the aortic tissue.

Mass fraction analyses suggest that the dry weight percentage of collagen within the diabetic aortas increases significantly as compared to the control aortas, whereas no significant change is found for that of elastin. Such an alteration of collagen content can be much attributed to hyperglycemia-induced vascular remodeling, in which the hyperplasia of collagen is evidently shown for most diabetic aortas by histology.

5 Conclusion

It has been demonstrated that in diabetes there are notable changes in both mechanical properties and microstructure of the rabbit aortas. Regarding biaxial mechanical responses, the diabetic aortas exhibited significantly lower extensibility and significantly higher tissue stiffness than the control aortas. With the constitutive parameters determined in the present study, finite element simulations can be further conducted to analyze the altered stress distributions of the diabetic aortas with respect to the control aortas. From the histological point of view, collagen hyperplasia is identified as the main microstructural characteristics that may potentially lead to aortic wall stiffening. As for future studies, there is a pressing need to quantitatively determine collagen fiber orientations and to investigate hemodynamics within the diabetic aortas. Thus, advanced imaging techniques coupled with computational tools in biomechanics will be required.

Acknowledgements The author gratefully acknowledges funding from the National Natural Science Foundation of China (Grant No. 11702197) and the Shanghai Pujiang Program (16*PJ*1409 200).

References

Akhtar, R., Cruichshank, J.K., Zhao, X., Walton, L.A., Gardiner, N.J., Barrett, S.D., Graham, H.K., Derby, B., Sherratt, M.J.: A localized micro- and nano-scale remodeling in the diabetic aorta. Acta Biomater. **10**, 4843–4851 (2014)

Astrand, H., Rydén-Ahlgren, A., Sundkvist, G., Sandgren, T., Länne, T.: Reduced aortic wall stress in diabetes mellitus. Eur. J. Vasc. Endovasc. Surg. **33**, 592–598 (2007)

Beckman, J.A., Creager, M.A., Libby, P.: Diabetes and atherosclerosis: epidemiology, pathophysiology, and management. J. Am. Med. Assoc. **287**, 2570–2581 (2002)

Bjornheden, T., Levin, M., Evaldsson, M., Wiklund, O.: Evidence of hypoxia areas within the arterial wall in vivo. Arter. Thromb. Vasc. Biol. **19**, 870–876 (1999)

Brownlee, M., Cerami, A., Vlassara, H.: Advanced glycosylation end products in tissue and the biochemical basis of diabetic complications. N. Engl. J. Med. **318**, 1315–1321 (1988)

Dua, M.M., Miyama, N., Azuma, J., Schultz, G.M., Sho, M., Morser, J., Dalman, R.L.: Hyperglycemia modulates plasminogen activator inhibitor-1 expression and aortic diameter in experimental aortic aneurysm disease. Surgery **148**, 429–435 (2010)

Golledge, J., Karan, M., Moran, C.S., Muller, J., Clancy, P., Dear, A.E., Norman, P.E.: Reduced expansion rate of abdominal aortic aneurysms in patients with diabetes may be related to aberrant monocyte-matrix interactions. Eur. Heart J. **29**, 665–672 (2008)

Grundy, S.M., Benjamin, I.J., Burke, G.L., Chait, A., Eckel, R.H., Howard, B.V., Mitch, W., Smith, S.C.J., Sowers, J.R.: Diabetes and cardiovascular disease: a statement for healthcare professionals from the American Heart Association. Circulation **100**, 1134–1146 (1999)

Holzapfel, G.A., Sommer, G., Gasser, C.T., Regitnig, P.: Determination of layer-specific mechanical properties of human coronary arteries with non-atherosclerotic intimal thickening, and related constitutive modeling. Am. J. Physiol. Heart Circ. Physiol. **289**, H2048–H2058 (2005)

Jhun, C.S., Evans, M.C., Barocas, V.H., Tranquillo, R.T.: Planar biaxial mechanical behavior of bioartificial tissues possessing prescribed fiber alignment. J. Biomech. Eng. **131**, 081006 (2009)

Katsuda, S., Miyashita, H., Hasegawa, M., Machida, N., Kusanagi, M., Yamasaki, M., Waki, H., Hazama, A.: Characteristic change in local pulse wave velocity in different segments of the atherosclerotic aorta in KHC rabbits. Am. J. Hypertens. **17**, 181–187 (2004)

Kozakova, M., Morizzo, C., Bianchi, C., Filippi, M.D., Miccoli, R., Paterni, M., Bello, V.D., Palombo, C.: Glucose-related arterial stiffness and carotid artery remodeling: a study in normal subjects and type 2 diabetes patients. J. Clin. Endocrinol. Metab. **99**, 2362–2366 (2014)

Mingueneau, M., Chaix, A., Scotti, N., Chaix, J., Reynders, A., Hammond, C., Thimonier, J.: Hands-on experiments on glycemia regulation and type 1 diabetes. Adv. Physiol. Educ. **39**, 232–239 (2015)

Novoselova, E.G., Glushkova, O.V., Lunin, S.M., Khrenov, M.O., Novoselova, T.V., Parfenyuk, S.B., Fesenko, E.E.: Signaling, stress response and apoptosis in pre-diabetes and diabetes: restoring immune balance in mice with alloxan-induced type 1 diabetes mellitus. Int. Immunopharmacol. **31**, 24–31 (2016)

Schram, M.T., Henry, R.M., van Dijk, R.A., Kostense, P.J., Dekker, J.M., Nijpels, G., Heine, R.J., Bouter, L.M., Westerhof, N., Stehouwer, C.D.: Increased central artery stiffness in impaired glucose metabolism and type 2 diabetes: the Hoorn study. Hypertension **43**, 176–181 (2004)

Sun, H., Zhong, M., Miao, Y., Ma, X., Gong, H.P., Tan, H.W., Zhang, Y., Zhang, W.: Impaired elastic properties of the aorta in fat-fed, streptozotocin-treated rats. vascular remodeling in diabetic arteries. Cardiology **114**, 107–113 (2009)

Tong, J., Cohnert, T., Regitnig, P., Holzapfel, G.A.: Effects of age on the elastic properties of the intraluminal thrombus and the thrombus-covered wall in abdominal aortic aneurysms: biaxial extension behavior and material modeling. Eur. J. Vasc. Endovasc. Surg. **42**, 207–219 (2011)

Tong, J., Schriefl, A.J., Cohnert, T., Holzapfel, G.A.: Gender differences in biomechanical properties, thrombus age, mass fraction and clinical factors of abdominal aortic aneurysms. Eur. J. Vasc. Endovasc. Surg. **45**, 364–372 (2013)

Tong, J., Yang, F., Li, X., Xu, X., Wang, G.X.: Mechanical characterization and material modeling of diabetic aortas in a rabbit model. Annu. Biomed. Eng. **46**, 429–442 (2018)

Biomechanics of the Main Artery in the Lower Limb

Alexey Kamenskiy, Majid Jadidi, Anastasia Desyatova, and Jason MacTaggart

Professor Gerhard A. Holzapfel's work in cardiovascular biomechanics and mechanobiology played a pivotal role in our research careers. His seminal manuscripts and books on constitutive modeling of arteries guided our research for many years, and we had the pleasure to interact with him both as reviewers for his Journal of Biomechanics and Modeling in Mechanobiology and as co-authors on papers. We are very grateful for Dr. Holzapfel's support, and it is an honor to contribute a book chapter to the exciting topic of Mechanobiology, Tissue Engineering, and Biomaterials.

Alexey

Abstract Despite years of technological and pharmacological improvements, failure rates remain high for the lower extremity peripheral arterial disease repairs, particularly when the repair devices cross the knee joint. Though much work has been done investigating the pathological processes associated with this failure, the underlying mechanisms remain insufficiently understood. The main arterial segment within the leg, the femoropopliteal artery, appears to be significantly different from other peripheral arteries, possibly due to lower blood flow, but more importantly, because it experiences large deformations during flexion of the limbs. Understanding the magnitude of these deformations in different postures and arterial segments may help improve the design of repair devices through computational studies of device-artery interactions. These studies rely on comprehensive assessments of arterial mechanics and structure and call for innovative ways of accounting for patient demographics and risk factors to deliver realistic results. In this chapter, we will summarize our recent findings related to the quantification of the biomechanical environment of the lower limb arteries, describe their structure and mechanical properties in the context of age,

A. Kamenskiy (✉) · M. Jadidi · A. Desyatova
Department of Biomechanics, University of Nebraska Omaha, Omaha, NE 68182, USA
e-mail: akamenskiy@unomaha.edu

J. MacTaggart
Department of Surgery, University of Nebraska Medical Center, Omaha, NE 68198, USA

G. Sommer et al. (eds.), *Solid (Bio)mechanics: Challenges of the Next Decade*, Studies in Mechanobiology, Tissue Engineering and Biomaterials 24,
https://doi.org/10.1007/978-3-030-92339-6_7

and propose a computational framework to evaluate device-artery interactions while accounting for the surrounding tissue effects. We will finish by identifying several future directions for this research.

1 Peripheral Arterial Disease of the Femoropopliteal Artery

Peripheral Arterial Disease (PAD) often refers to the atherosclerotic obstruction of the femoropopliteal artery (FPA) that reduces blood flow to the lower limbs. It is a major contributor to public health burden and is associated with high morbidity, mortality, and impairment in quality of life (Mahoney et al. 2010). The FPA begins as the common femoral artery (Fig. 1) and continues as the superficial femoral artery (SFA) into the upper thigh. It becomes the popliteal artery (PA) as it traverses the adductor hiatus (AH), a gap between the adductor magnus muscle and the femur, passing from the anterior thigh posteriorly into the popliteal fossa behind the knee. Clinically, the two principal sites of FPA occlusions are the distal SFA at the AH and the PA behind the knee (Watt 1965) (Fig. 1).

The total annual costs of hospitalizations for patients with PAD are in excess of $21 billion per year, and per-patient costs of PAD are higher than those for both coronary artery disease and cerebrovascular disease (Mahoney et al. 2008). The high cost of PAD is mostly attributed to high numbers of peripheral vascular operations and interventions that fail, resulting in poor clinical outcomes and a frequent need for repetitive interventions (Adam et al. 2005; Conte et al. 2006; Schillinger et al. 2006, 2007; Laird and Yeo 2012; Stavroulakis et al. 2016; Qato et al. 2018). Specifically, primary patency is lost in 30–40% of autogenous vein bypass grafts (Conte et al. 2006) and in 45–58% of synthetic grafts (Lundgren and Study 2013) at 1 year. The results of angioplasty and stenting in the lower extremity are equally bad (Siracuse et al. 2012), with almost half of the patients developing hemodynamically significant restenosis or requiring re-intervention within 2–3 years after treatment (Schillinger et al. 2007; Laird and Yeo 2012; Stavroulakis et al. 2016; Qato et al. 2018).

Though the systemic risk factors for restenosis are the same for the carotid, iliac, and FPA reconstructions, the FPA interventions fail significantly more often than in other locations. This suggests that local factors unique to the FPA strongly influence reconstruction failure. The two major differences between the FPA and other arterial segments are lower blood flow (Parker et al. 2006; Newcomer et al. 2008; Young et al. 2010; Padilla and Fadel 2017; Desyatova et al. 2018a) and the large deformations experienced during flexion of the limbs (Ansari et al. 2013; MacTaggart et al. 2014; Desyatova et al. 2017c, 2018a, 2018b; Maleckis et al. 2018; Poulson et al. 2018). These deformations and the associated intramural stresses and hemodynamic disturbances occurring repetitively with each limb flexion, can injure the arterial wall, alter cell signaling, cause deleterious cellular and biochemical responses, and culminate in disease initiation, progression, or reconstruction failure (Dunlop and Santos 1957; Palma 1959; Watt 1965; Clowes et al. 1983; Wensing et al. 1998; MacTaggart et al. 2014; Maleckis et al. 2018).

2 Mechanical Deformations of the FPA During Limb Flexion

Along its length, the FPA is constrained and tethered by a large number of small branches, bone, fascia, muscle, and nerve that produce non-uniform and locally large deformations (MacTaggart et al. 2014). These deformations include axial compression (Poulson et al. 2018), bending (Poulson et al. 2018), twisting (Desyatova et al. 2017c), and pinching (Desyatova et al. 2018b) that occur in three dimensions and require multiaxial imaging for accurate quantification. Recent studies have used intra-arterial markers and a perfused human cadaver model to measure the magnitudes of these deformations in different FPA segments and common daily postures—standing (straight, 180°), walking (110°), sitting (90°), and gardening (acutely bent, 60°) (Fig. 1). Table 1 summarizes these values representing the maximum deformations experienced by the proximal SFA, artery at the AH, and the PA behind the knee. Deformations were more severe in more acutely bent limb configurations and increased with more distal arterial location, reaching a maximum in the PA behind the knee. Compared with the previously reported values (Ansari et al. 2013) that

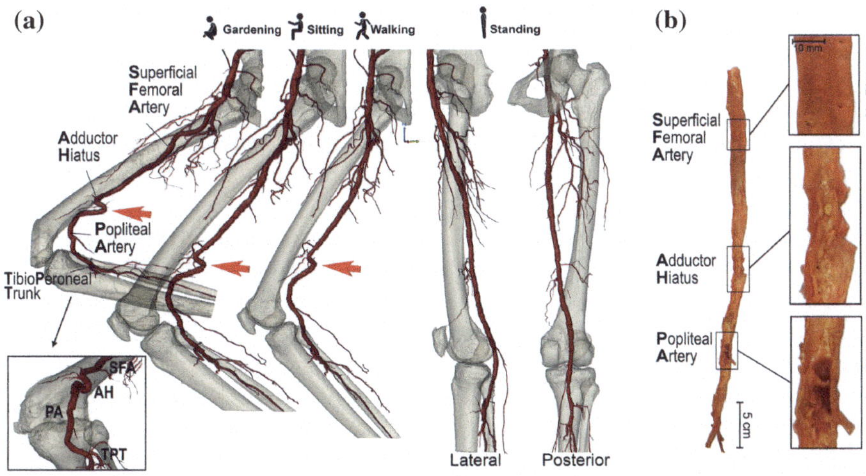

Fig. 1 (a) Deformations experienced by the femoropopliteal artery (FPA) during the limb flexion. Note the severe bending and kinking of the FPA at the adductor hiatus (AH) and in the popliteal artery (PA) behind the knee. (b) Excised and longitudinally-opened FPA with severe PAD at the AH and in the PA

Table 1 Maximum limb flexion-induced deformations experienced by different FPA segments (Desyatova et al. 2017c, 2018b; Poulson et al. 2018; MacTaggart et al. 2019): superficial femoral artery (SFA), artery at the adductor hiatus (AH), and the below-knee popliteal artery (PA)

	Walking			Sitting			Gardening		
	SFA	AH	PA	SFA	AH	PA	SFA	AH	PA
Axial compression (%)	9	11	13	13	14	17	15	19	25
Bending (radius of curvature) (mm)	27	19	17	23	12	11	21	9	8
Twisting (°/cm)	10	8	14	12	10	20	13	16	26
Pinching (aspect ratio)	1.16	1.19	1.24	1.17	1.21	1.26	1.19	1.28	1.35

represented the average deformations experienced by each arterial segment, these newly-determined deformations are 3 to 12-fold more severe (MacTaggart et al. 2014), which raises concerns about the ability of currently used PAD stents to withstand them.

3 Bench-Top and In Situ Evaluation of Commercial PAD Stents

Bench-top evaluation of 12 commercial PAD stents subjected to limb flexion-induced deformations demonstrated significant differences in their ability to accommodate axial compression, bending, and torsion while resisting pinching (Maleckis et al. 2017). High radial stiffness and low resistance to deformations allow the stent to maintain lumen diameter while following deformation that occurs naturally with limb flexion. Overall, some stents were better in accommodating certain deformation modes, but no device was able to demonstrate superior characteristics under all deformations, and many stents experienced instabilities such as global bucking and pinching.

Since stent performance on the bench may differ from that in situ, seven stent designs were evaluated using a perfused human cadaver model (MacTaggart et al. 2019), and the deformations of the stented arteries were compared with their own baselines. The results demonstrated significant effects of stents on limb flexion-induced FPA deformations, and the differences between devices were appreciable. However, similarly to the bench-top results, some designs were able to accommodate several deformation modes, but no device was able to match all FPA deformations. Furthermore, the effects of stenting were not confined to the stented segment and often translated proximal and distal to the target area. For example, stents with high torsional or bending stiffness that significantly restricted arterial deformations in the stented area, often translated these deformations distal to the device, resulting in exacerbated twisting or bending distal to the repaired segment.

Since none of the tested devices were able to accommodate limb flexion-induced deformations without either exacerbating or restricting them, the development of an optimized stent that would have improved compatibility with the artery appears to be of paramount importance. Stents that follow arterial deformations during limb flexion while maintaining lumen diameter to provide adequate blood flow, could minimize arterial injury and improve clinical outcomes (Maleckis et al. 2018), but such development requires a comprehensive understanding of the arterial structure, mechanics, and pathophysiology in the context of patient demographic and risk factors.

4 The FPA Intramural Structure and Its Changes with Age

As a muscular artery, the FPA distributes blood to the tissues of the lower limb by relaxing and contracting the concentric layers of smooth muscle cells (SMCs) that populate most of its tunica media (Humphrey 2002) (Fig. 2).

The density of SMCs in the media decreases with age from ~35 to ~25% (Jadidi et al. 2021b), and the cells undergo morphological changes consistent with a more synthetic phenotype (Bacakova et al. 2018; Jaminon et al. 2019). Older FPAs also contain a significant amount of α-SMA-positive cells in the adventitia, forming a robust network of vasa vasorum (Jadidi et al. 2021b) that is almost absent in younger arteries.

Medial SMCs are surrounded by straight circumferentially-oriented collagen fibers, while longitudinally-oriented undulated collagen is present in the immediate proximity of the external elastic lamina (EEL) that separates media from adventitia (Jadidi et al. 2021b). Collagen density in the media of young arteries is 10–13%, but it increases to 20–30% in older specimens (Jadidi et al. 2021b). Unlike the medial collagen, adventitial collagen appears to be oriented diagonally, thicker, and more undulated. The pitch of adventitial collagen fibers decreases with age, which leads to a shift in the arterial load-carrying capacity (Jadidi et al. 2021b). Unlike the medial collagen content that increases with age, the overall collagen density remains relatively constant at ~23%, likely due to the increase in arterial wall thickness.

Since the FPA is a muscular artery, elastin is not abundant but plays an important physiological role. It facilitates the in situ longitudinal pre-stretch (Kamenskiy et al. 2016), ensures energy-efficient function (Humphrey et al. 2009), and promotes better hemodynamics (Palatini et al. 2011; Desyatova et al. 2018a) by reducing arterial bending and kinking during limb flexion (MacTaggart et al. 2014). Elastin in the FPA is organized primarily in the form of longitudinal fibers in the EEL and a three-dimensional internal elastic lamina (IEL) sheet that separates tunica intima from media. Elastin is a very mechanically and chemically stable protein, formed primarily during the perinatal period (Mithieux et al. 2005; Humphrey et al. 2009). It matures early in life, resulting in considerable tension in maturity. This tension is responsible for >1.5 in situ longitudinal pre-stretch in young arteries, and as the elastin degrades with aging, cyclic mechanical stress, proteolytic destruction, and

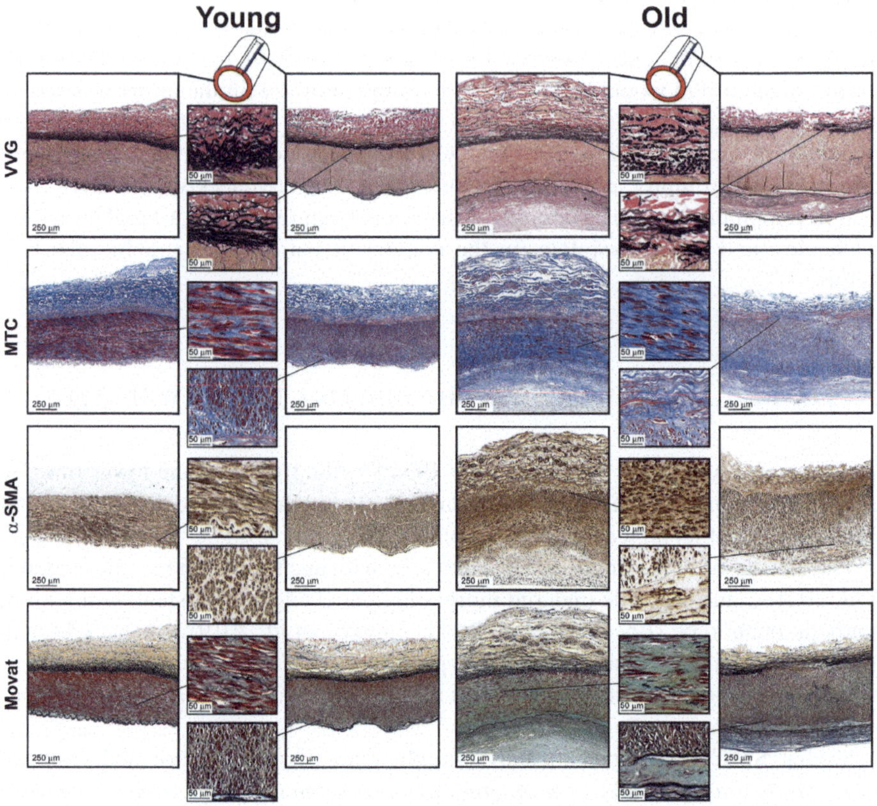

Fig. 2 Representative transverse and longitudinal sections of young (left, 14-year-old) and old (right, 69-year-old) FPA specimens stained with Verhoeff-Van Gieson (VVG, elastin: black), Masson's Trichrome (MTC, collagen: blue), α-smooth muscle cell actin (α-SMA, actin: brown), and Movat's Pentachrome (glycosaminoglycans: greenish-gray)

other disease processes, the pre-stretch reduces to <1.1 in older subjects (Kamenskiy et al. 2016; Jadidi et al. 2019). The overall density of elastin also decreases with age from 7 to 3%, and the density of elastin in the elastin-rich EEL changes from 39 to 20% (Jadidi et al. 2021b).

Apart from the SMCs, collagen, and elastin, the FPA wall also contains glycosaminoglycans (GAGs). GAGs sequester water and are almost absent in young FPAs (Jadidi et al. 2021b). In older arteries, their density is ~5–6% (Jadidi et al. 2021b), and they can typically be found around SMCs in the tunica media, along the collagen fibers in the adventitia, in the thickened intima, and in the immediate vicinity of the IEL, likely contributing to dissection during balloon angioplasty (Anttila et al. 2019).

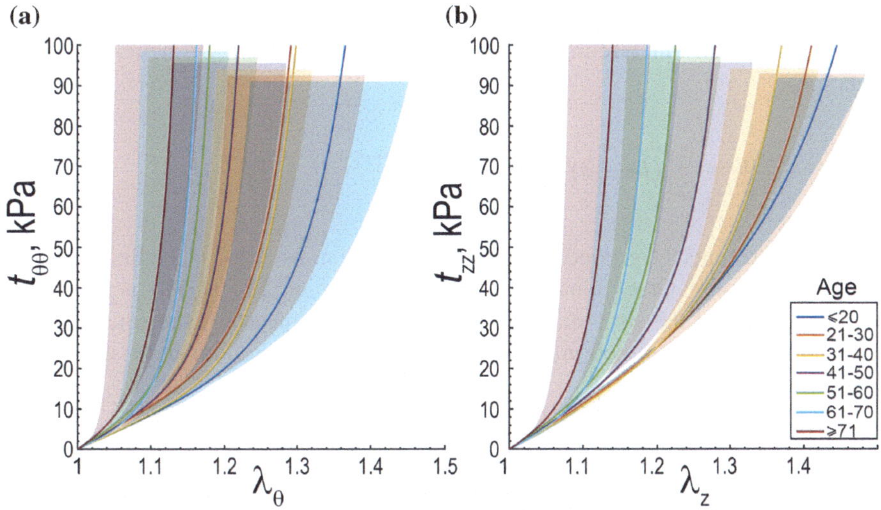

Fig. 3 Average equibiaxial Cauchy stress-stretch response of human FPAs in 7 decadal age groups in (a) circumferential and (b) longitudinal directions. Legend summarizes the range of ages for each group in years. Variability within each age group is demonstrated by the semi-transparent regions that bound the equibiaxial curves corresponding to 25th and 75th percentile ranges. They have different heights for better visualization. Curves are adopted from Desyatova et al. (2017a)

5 The FPA Elastic Properties and Their Evolution with Age

The mechanical properties of human FPAs in relation to age (Kamenskiy et al. 2017; Jadidi et al. 2019) and diabetes (Desyatova et al. 2017a; Kamenskiy et al. 2018) have recently been characterized using planar biaxial extension in >500 arteries 12–99 years old. Across all ages, FPAs demonstrate anisotropy and nonlinearity in their stress-stretch responses, with a predominantly more compliant longitudinal direction (Fig. 3). On average, young FPAs experience >1.4 longitudinal and >1.3 circumferential stretches under 100 kPa equibiaxial stress, while for older specimens, these values are <1.15. Decadal age group separation demonstrates that the mechanical properties in the longitudinal direction have little overlap between young, middle-aged, and older specimens, while more overlap is observed circumferentially. Aging stiffens the artery in both directions, but longitudinal stiffening occurs faster than circumferential, resulting in the moving of the stress-stretch curves closer to each other for older specimens and consequently leading to a more isotropic behavior. Older FPAs also demonstrate higher nonlinearity in their stress-stretch responses, suggesting a shift in the load-carrying capacity from degrading elastin to the accumulating collagen.

A strain-energy function that adequately portrays this passive FPA mechanical behavior in all ages (Jadidi et al. 2019; Kamenskiy et al. 2017; Jadidi et al. 2020) can be written as a summation of contributions from the isotropic ground substance (i.e., GAGs) and the four families of fibers, each phenomenologically accounting for the longitudinal elastin, circumferential SMCs, and two families of helical collagen fibers, i.e.

$$W = \frac{c_{\text{gr}}}{2}(I_{\text{C}} - 3) + \sum_{i=1}^{4} \frac{c_{1i}}{4c_{2i}} \left(e^{c_{2i}\langle IV_{\text{C}i} - 1\rangle^2} - 1 \right) + U(J), \tag{1}$$

where c_{gr} represents the ground substance contribution, $I_{\text{C}} = \text{tr}\mathbf{C}$ is the first invariant of the right Cauchy-Green tensor $\mathbf{C} = \mathbf{F}^{\text{T}}\mathbf{F}$, and $IV_{\text{C}i} = \mathbf{M}_i \cdot (\mathbf{C}\mathbf{M}_i)$ is the fourth invariant of $\mathbf{C}$ and the structural tensor $\mathbf{M}_i \otimes \mathbf{M}_i$ that measures the square of the stretch along the fiber direction $\mathbf{M}_i$ in the reference configuration, and is equal to

$$IV_{\text{C}i} = \lambda_z^2 \cos^2 \varphi_i + \lambda_\theta^2 \sin^2 \varphi_i. \tag{2}$$

Further, λ_z, λ_θ are longitudinal and circumferential stretches, and φ_i is the angle of fiber family i with the longitudinal direction, and specifically, $\varphi_1 = 0°$ phenomenologically accounts for the longitudinal elastic fibers with material parameters c_{11}, c_{21}, and $\varphi_2 = 90°$ corresponds to the passive contribution of circumferential SMCs with material parameters c_{12}, c_{22}, and the two families of helical collagen fibers that are assumed to have the same mechanical properties $c_{13} = c_{14}$, $c_{23} = c_{24}$ are oriented at $\varphi_{3,4} = \varphi$ with the longitudinal direction. The Macaulay brackets $\langle(\bullet)\rangle = \frac{1}{2}[(\bullet) + |(\bullet)|]$ are used to filter positive values, so fibers only contribute to the strain energy during tension. $U(J)$ is a purely volumetric contribution of strain energy function, which in the case of an incompressible material, defines the Lagrange contribution.

The total Cauchy stress tensor $\mathbf{t}$ can then be calculated as

$$\mathbf{t} = -p\mathbf{I} + \bar{\mathbf{t}} = -p\mathbf{I} + \overbrace{2\frac{\partial W}{\partial I_{\text{C}}}\mathbf{B} + 2\sum_{i=1}^{4} \frac{\partial W}{\partial IV_{\text{C}i}}\mathbf{m}_i \otimes \mathbf{m}_i}^{\bar{\mathbf{t}}}, \tag{3}$$

where $\bar{\mathbf{t}}$ contains the deformation dependence, $\mathbf{m}_i = \mathbf{F}\mathbf{M}_i$ is the push-forward of the unit vector $\mathbf{M}_i$, p is the Lagrange multiplier, and $\mathbf{B} = \mathbf{F}\mathbf{F}^{\text{T}}$ is the left Cauchy-Green tensor.

Assuming tissue incompressibility and the absence of shear, the non-zero components of the isochoric Cauchy stress tensor can be written as

$$\begin{aligned}
\bar{t}_{rr} &= c_{\text{gr}}\lambda_r^2 = c_{\text{gr}}\frac{1}{\lambda_z^2\lambda_\theta^2}, \\
\bar{t}_{\theta\theta} &= c_{\text{gr}}\lambda_\theta^2 + c_{12}\left\langle\lambda_\theta^2 - 1\right\rangle e^{c_{22}\langle\lambda_\theta^2-1\rangle^2}\lambda_\theta^2 + \textstyle\sum_{i=3}^4 c_{1i}\left\langle IV_{\text{C}i} - 1\right\rangle e^{c_{2i}\langle IV_{\text{C}i}-1\rangle^2}\lambda_\theta^2\sin^2\varphi, \\
\bar{t}_{zz} &= c_{\text{gr}}\lambda_z^2 + c_{11}\left\langle\lambda_z^2 - 1\right\rangle e^{c_{21}\langle\lambda_z^2-1\rangle^2}\lambda_z^2 + \textstyle\sum_{i=3}^4 c_{1i}\left\langle IV_{\text{C}i} - 1\right\rangle e^{c_{2i}\langle IV_{\text{C}i}-1\rangle^2}\lambda_z^2\cos^2\varphi,
\end{aligned} \tag{4}$$

Table 2 Parameters for the four-fiber family constitutive model describing FPA mechanical properties in 7 decadal age groups. Data are adopted from Jadidi et al. (2019), and the parameters account for the effects of specimen flattening during the planar biaxial test

Age group (yrs)	c_{gr} (kPa)	c_{11} (kPa)	c_{21} (-)	c_{12} (kPa)	c_{22} (-)	$c_{13} = c_{14}$ (kPa)	$c_{23} = c_{24}$ (-)	φ (°)
≤20	3.41	40.78	0.04	0.13	4.18	10.46	0.97	61.66
21–30	12.55	29.80	0.32	6.62	3.37	3.66	3.68	56.55
31–40	9.38	34.43	0.75	6.88	4.03	4.00	4.60	55.11
41–50	9.63	39.41	1.82	10.93	6.31	5.40	7.21	50.58
51–60	5.05	41.09	2.91	14.75	8.24	7.72	10.16	50.23
61–70	3.96	45.24	6.30	15.91	10.51	8.19	18.84	48.35
71≥	3.05	38.53	10.34	26.55	10.42	7.56	24.82	50.96

and considering $t_{rr} = 0$ during the planar biaxial test, the total non-zero stress components can be found as

$$\begin{aligned} t_{\theta\theta} &= c_{\text{gr}}\left(\lambda_\theta^2 - \frac{1}{\lambda_z^2\lambda_\theta^2}\right) + c_{12}\left\langle\lambda_\theta^2 - 1\right\rangle e^{c_{22}\left(\lambda_\theta^2-1\right)^2}\lambda_\theta^2 \\ &+ \sum_{i=3}^{4} c_{1i}\left\langle IV_{\text{C}i} - 1\right\rangle e^{c_{2i}\left(IV_{\text{C}i}-1\right)^2}\lambda_\theta^2\sin^2\varphi, \\ t_{zz} &= c_{\text{gr}}\left(\lambda_z^2 - \frac{1}{\lambda_z^2\lambda_\theta^2}\right) + c_{11}\left\langle\lambda_z^2 - 1\right\rangle e^{c_{21}\left(\lambda_z^2-1\right)^2}\lambda_z^2 \\ &+ \sum_{i=3}^{4} c_{1i}\left\langle IV_{\text{C}i} - 1\right\rangle e^{c_{2i}\left(IV_{\text{C}i}-1\right)^2}\lambda_z^2\cos^2\varphi. \end{aligned} \tag{5}$$

The constitutive parameters are determined by minimizing the difference between the Cauchy stress measured experimentally and its theoretical counterpart given by Eq. (5). To ensure uniqueness, one can perform a non-parametric bootstrap (Ferruzzi et al. 2011; Kamenskiy et al. 2017) with 2000 iterations of random sampling and fitting, followed by the analysis of the probability distribution of each parameter to determine the global minimum. The mechanical properties of human FPAs in 7 decadal age groups are summarized in Table 2 (data are adopted from Jadidi et al. 2019 and account for the effects of specimen flattening during the planar biaxial test).

An alternative way of describing the FPA mechanical properties for a continuum of ages is to incorporate age explicitly into the equations for the Cauchy stresses, i.e. Eq. (5) (Desyatova et al. 2017a):

$$\begin{aligned} t_{\theta\theta}(\tau) &= S_{\text{gr}}(\tau)\,c_{\text{gr}}\left(\lambda_\theta^2 - \frac{1}{\lambda_z^2\lambda_\theta^2}\right) + c_{12}\left\langle(IV_{\text{C}2})^{S_2(\tau)} - 1\right\rangle e^{c_{22}\left((IV_{\text{C}2})^{S_2(\tau)}-1\right)^2}\lambda_\theta^2 + \\ &\sum_{i=3}^{4} c_{1i}\left\langle(IV_{\text{C}i})^{S_i(\tau)} - 1\right\rangle e^{c_{2i}\left((IV_{\text{C}i})^{S_i(\tau)}-1\right)^2}\lambda_\theta^2\sin^2\varphi, \\ t_{zz}(\tau) &= S_{\text{gr}}(\tau)c_{\text{gr}}\left(\lambda_z^2 - \frac{1}{\lambda_z^2\lambda_\theta^2}\right) + c_{11}\left\langle(IV_{\text{C}1})^{S_1(\tau)} - 1\right\rangle e^{c_{21}\left((IV_{\text{C}1})^{S_1(\tau)}-1\right)^2}\lambda_z^2 + \\ &\sum_{i=3}^{4} c_{1i}\left\langle(IV_{\text{C}i})^{S_i(\tau)} - 1\right\rangle e^{c_{2i}\left((IV_{\text{C}i})^{S_i(\tau)}-1\right)^2}\lambda_z^2\cos^2\varphi. \end{aligned} \tag{6}$$

Here $S_{\{gr,i\}}(\tau)$ are stiffening functions of age τ (years) that modify IV_{Ci}. Note that stiffening functions for elastin, SMC, and collagen are incorporated into the power of IV_{Ci}, while the ground substance is assumed to stiffen linearly. Since IV_{Ci} has the physical meaning of the square of stretch in the direction of fiber, S amplifies that stretch for each constituent as a function of age. Consequently, $S > 1$ simulates material stiffening, $S = 1$ recovers the initial elastic response, while $S < 1$ simulates softening. It is important to note that formulating the strain energy density function directly in terms of the modified invariants $(IV_{Ci})^{S_i(\tau)}$ will not produce the Cauchy stresses in Eq. (6) because Cauchy stresses are derived from the strain-energy function by taking derivatives with respect to **C** or its invariants using the chain rule. However, Eq. (6) can be used directly in computational implementations. Using this approach, the constitutive parameters need only be determined for the youngest age group, and the stress-stretch curves will shift according to the aging functions.

The same approach can be used to describe FPA stiffening due to the effects of risk factors, such as diabetes mellitus (DM) (Desyatova et al. 2017a). Table 3 summarizes constitutive parameters for Eq. (6) that describe the aging of non-diabetic and diabetic FPAs in the 41–80 years-old range (Desyatova et al. 2017a). Their comparison demonstrates that FPAs from diabetic subjects in their 40s have already acquired the circumferential stiffness of a 70-year-old non-diabetic artery, while the longitudinal direction appears to be affected to a lesser degree. Recent studies (Kamenskiy et al. 2018) suggest that this effect may be related to the higher prevalence of medial calcification in diabetic FPAs that often manifests as calcium 'rings'. Though the exact mechanism of medial calcification in the FPA is not yet completely understood, it

Table 3 Constitutive parameters describing aging of the FPA with and without diabetes mellitus (DM) (Desyatova et al. 2017a)

	Aging without DM	Aging with DM
τ (yrs)	41–80	41–80
Elastic constitutive parameters		
c_{gr} (kPa)	13.27	12.95
c_{11} (kPa)	17.36	17.46
c_{21} (-)	2.18	1.81
c_{12} (kPa)	6.64	24.33
c_{22} (-)	7.17	14.79
c_{13}, c_{14} (kPa)	2.02	5.15
c_{23}, c_{24} (-)	9.48	15.94
φ (°)	48.34	52.81
Stiffening functions		
S_{gr}	$0.0004\,e^{0.1042\,\tau} + 0.9073$	1.56
S_1	$0.0191\,e^{0.050\,\tau} + 0.8213$	$0.0541\,e^{0.0458\,\tau} + 0.5222$
S_2	1.0	1.0
$S_3 = S_4$	$0.1412\,e^{0.0322\,\tau} + 0.3221$	$0.0056\,e^{0.0615\,\tau} + 0.8855$

could result from the transformation of primarily-circumferentially-oriented SMCs into osteoblast-like cells when exposed to the hyperglycemic environment (Desyatova et al. 2017a; Kamenskiy et al. 2018).

While incorporating age into the constitutive relation allows its description as a continuum, it is important to remember several limitations of this approach. First, it does not account for the growth and remodeling of the tissue and the accumulation of residual stresses that result from it. The development of the appropriate growth and remodeling framework for aging is the future direction of our studies. Second, the constitutive relation and the aging functions are both phenomenological, despite the former being motivated by the FPA intramural structure. The inclusion of structural data on fiber direction and dispersion obtained using multiphoton microscopy may help partially resolve this issue and make the model more structurally-based, which will allow gaining additional insight into the FPA pathophysiology and mechanophysiology (Jadidi et al. 2021c).

6 Assessment of FPA Physiological Characteristics

Historically, physiological arterial characteristics have been assessed using in vivo methods such as pulse wave velocity or diameter changes over the cardiac cycle. While generally straightforward to perform in the clinic, these techniques are limited to the narrow physiological range of pressures, and cannot be used to reliably assess the intrinsic arterial properties, residual stresses, or true physiological intramural stresses that are a function of pre-stretch and the opening angles. Data on the *ex vivo* arterial characteristics combined with the appropriate kinematic framework allows circumventing these issues, and it will be briefly summarized below, along with the demonstration of its use in studying changes to the FPA physiological characteristics with age.

Figure 4 describes deformations that the FPA segment experiences as it progresses from its stress-free, to the load-free, and then to the in vivo loaded states. These deformations can be written as

$$\begin{aligned} \mathbf{F}_{\text{res}} &= \text{diag}\left[\tfrac{\partial \rho}{\partial R}, k\tfrac{\rho}{R}, \lambda_\zeta\right], \\ \mathbf{F}_{\text{load}} &= \text{diag}\left[\tfrac{\partial r}{\partial \rho}, \tfrac{r}{\rho}, \lambda_z^{\text{in situ}}\right], \\ \mathbf{F}_{\text{phys}} = \mathbf{F}_{\text{load}}\mathbf{F}_{\text{res}} &= \text{diag}\left[\tfrac{\partial r}{\partial R}, \tfrac{r}{R}k, \lambda_\zeta \lambda_z^{\text{in situ}}\right], \end{aligned} \tag{7}$$

where R, ρ, and r represent the stress-free radially cut ring, the load-free and the loaded ring radii, respectively, $k = 2\pi/(2\pi - \alpha_\theta)$ is a measure of the opening angle, and λ_ζ and $\lambda_z^{\text{in situ}}$ are the residual and the in situ longitudinal stretches (i.e., pre-stretch), respectively. The longitudinal residual stretch λ_ζ is given as

$$\lambda_\zeta = \frac{\xi}{L}\lambda_z^{\text{flat}}(R_z), \tag{8}$$

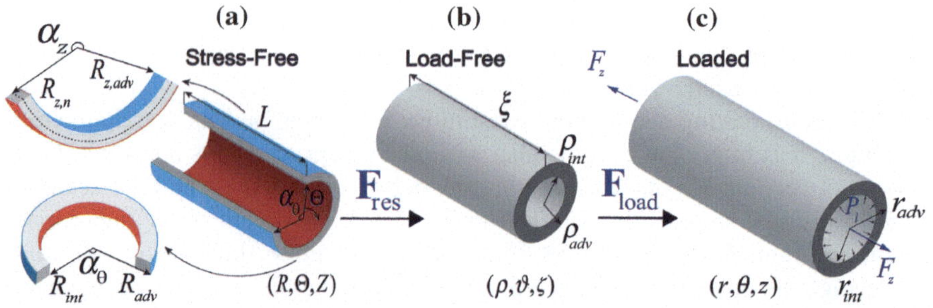

Fig. 4 Kinematics of the artery demonstrating (a) the stress-free configuration obtained by radially opening the artery into a sector, and cutting out a longitudinal strip, (b) the load-free state with no luminal pressure and longitudinal pre-stretch, and (c) the loaded physiological state. Note that the stress-free longitudinal strip curves intima outward, while the radially-cut ring opens intima inward (Jadidi et al. 2019)

where $\lambda_z^{\text{flat}}(R_z)$ is the flattening stretch (Sommer and Holzapfel 2012; Jadidi et al. 2019) that the stress-free longitudinal strip experiences, and assuming that the neutral axis is located in the middle of the section with the radius $R_{z,\text{n}}$, it is equal to

$$\lambda_z^{\text{flat}}(R_z) = \frac{R_{z,\text{n}}}{R_z} = \frac{R_{z,\text{int}}+R_{z,\text{adv}}}{2R_z}. \tag{9}$$

Here, R_z is the radius of the curved longitudinal strip that changes from the intimal (int) to the adventitial (adv) surfaces. Note that in the FPA longitudinal strip curves intima outward and therefore, $R_z^{\text{int}} > R_z^{\text{adv}}$.

Assuming tissue incompressibility, the physiological circumferential stretch can be written as

$$\lambda_\theta^{\text{phys}}(r) = \frac{r}{R}k = \frac{rk}{\sqrt{R_{\text{adv}}^2 - k\lambda_\xi\lambda_z^{\text{in situ}}\left(r_{\text{adv}}^2 - r^2\right)}}, \tag{10}$$

where $R_{\text{int}} \leq R \leq R_{\text{adv}}$ and $r_{\text{int}} \leq r \leq r_{\text{adv}}$ are the radii in the stress-free and loaded physiological state, respectively. Assuming quasi-static motions, in the absence of body forces, the balance of linear momentum results in

$$\text{div}\,\mathbf{t} = \mathbf{0}, \tag{11}$$

where div represents the spatial divergence, and $\mathbf{t}$ is the physiological Cauchy stress tensor. By solving the above equation and assuming no perivascular tethering, i.e., zero pressure on the outer surface of the artery and zero axial force in the load-free state, we can determine ξ/L, $\lambda_\theta^{\text{phys}}(r)$, and the Lagrange multiplier in Eq. (3) (Holzapfel and Ogden 2010; Humphrey 2002).

Table 4 Morphometric characteristics of human FPAs in the stress-free state separated by 7 decadal age groups. Data are adopted from Jadidi et al. (2019)

Age group (yrs)	R_{int} (mm)	R_{adv} (mm)	$R_{\text{z,adv}}$ (mm)	H (mm)	α_θ (°)	α_z (°)
≤20	2.45	3.82	1.83	1.36	126.24	97.04
21–30	3.82	5.23	2.15	1.42	151.24	111.10
31–40	4.12	5.56	2.81	1.45	157.25	154.21
41–50	4.72	6.19	3.71	1.48	166.07	187.05
51–60	4.89	6.55	5.73	1.63	165.73	241.93
61–70	6.62	8.20	7.97	1.79	181.22	271.17
71≥	5.62	7.49	11.80	1.80	154.36	305.83

The averaged through-thickness physiological stress-stretch state can then be determined for any luminal pressure, and we will use 100 mmHg as an example below. Physiological measures can also include the change in the average circumferential stretch during the cardiac cycle $\lambda_\theta^{\text{cardiac}} = r^{\text{sys}}/r^{\text{dias}}$, change in the elastic energy W from diastole (80 mmHg) to systole (120 mmHg), and the physiological circumferential stiffness E_θ defined as the change in the average circumferential physiological stress $t_{\theta\theta}^{\text{phys}}$ between systole and diastole divided by the change in the average circumferential stretch $\lambda_\theta^{\text{phys}}$, i.e.

$$E_\theta = \frac{t_{\theta\theta}^{\text{phys}}(\text{sys}) - t_{\theta\theta}^{\text{phys}}(\text{dias})}{\lambda_\theta^{\text{phys}}(\text{sys}) - \lambda_\theta^{\text{phys}}(\text{dias})}. \tag{12}$$

It is important to distinguish this stiffness from the one measured experimentally using the planar biaxial extension or inflation-extension tests because it incorporates longitudinal pre-stretch and residual deformations.

Tables 4 and 5 summarize radii, opening angles, thickness, and the in situ longitudinal pre-stretch of human FPAs in 7 decadal age groups (Jadidi et al. 2019). Circumferential physiological stretch and stress decrease with age from 1.26 to 1.12, and from 48 to 32 kPa, respectively. In the longitudinal direction, the reduction in the physiological stretch and stress is much more pronounced: 1.57 to 1.1 and 117 kPa to 24 kPa, respectively. Furthermore, the axial force F_z decreases from 1.18 N in the youngest age group to 0.41 N in the oldest age group. The cardiac cycle stretch decreases from 1.04 in the young to 1.02 in the old FPAs and the stored elastic energy reduces 6-fold from 2.3 to 0.4 kPa. The physiological circumferential stiffness increases slightly from 550 to 880 kPa, but stays relatively constant at 750±20 kPa between 25 to 65 years of age (Jadidi et al. 2019), suggesting that the latter could be the FPA's homeostatic target (Humphrey et al. 2014).

Table 5 Morphometric characteristics of human FPAs in the load-free and loaded physiological states separated by 7 decadal age groups. Data are adopted from Jadidi et al. (2019)

Age group (yrs)	ρ_{int} (mm)	ρ_{adv} (mm)	r_{int} (mm)	r_{adv} (mm)	$\lambda_z^{\text{in situ}}$ (-)
≤20	1.37	2.68	2.21	3.00	1.53
21–30	1.66	3.10	2.46	3.29	1.44
31–40	1.83	3.25	2.61	3.52	1.36
41–50	1.81	3.32	2.48	3.51	1.26
51–60	1.82	3.33	2.40	3.44	1.20
61–70	1.90	3.78	2.48	3.74	1.15
71≥	2.22	3.92	2.79	4.07	1.11

7 Inelastic FPA Characteristics

In addition to studying the elastic characteristics of FPAs, it is also important to understand their inelastic behavior (Anttila et al. 2019) under supraphysiological loads, such as those applied during the balloon angioplasty. When taken beyond their elastic limit, arteries demonstrate two distinct features (Balzani et al. 2012; Anttila et al. 2019). First, for each advancing stretch, the stress-stretch curves follow a new trajectory, repeating the unloading part of the previous cycle, but generating a softer response after reaching the stretch maximum. This stress softening response observed for increasing stretches is termed 'damage evolution' since it is assumed to result from microscopic tissue damage. The second characteristic feature is observed when instead of increasing the maximum stretch, the specimen is loaded to the same supraphysiological stretch limit, but the loading is done repeatedly over several cycles. In this case, the artery also demonstrates stress-softening, but the hysteresis eventually reduces, and the stress-stretch curves saturate and become repeatable. This stress softening response observed for the fixed stretch limit during repeated loading is termed 'damage saturation'. To capture both of these effects, the non-isotropic part of the strain energy (Eq. (1)) can be modified as (Anttila et al. 2019)

$$\mathrm{W}^{ti} = \sum_{i=1}^{4} \frac{c_{1i}}{4c_{2i}} \left\{ e^{\left[c_{2i}\langle(1-D_i)IV_{Ci}-1\rangle^2\right]} - 1 \right\}, \tag{13}$$

where D_i are scalar damage functions for each fiber family used to capture remnant strains and the stress-softening effects. This specific form of incorporating the $(1 - D_i)$ term and multiplying it directly with IV_{Ci} inside the strain-energy function allows capturing remnant strains in the fibers due to shifting of the stress-free reference configuration via the damage functions D_i, and permits significant damage hysteresis even for curves that are highly nonlinear in nature. The damage functions D can then be taken in the form (Balzani et al. 2012)

$$D(\beta) = D_s \left[1 - \exp\left(\frac{\ln(1 - r_s)}{\beta_s} \beta \right) \right], \tag{14}$$

where $D_s \in [0,1)$, $r_s \in [0,1)$, $\beta_s > 0$ is the value of the internal variable β that describes damage saturation behavior for a fixed stretch limit reached at $r_s = 0.99$ fraction of the maximum damage value D_s. Higher values of β_s indicate that the tissue can tolerate more cycles of repetitive loading to the same stretch maximum before it reaches a saturated response. Damage activation is controlled by the internal variable β that evolves depending on the loading history

$$\beta = \left\langle \int_0^t \langle I\dot{V}_C(s) \rangle \, ds \right\rangle - \widetilde{\beta}^{ini}, \tag{15}$$

where $\widetilde{\beta}^{ini}$ defines the damage initiation threshold set equal to the value of the integral when the supraphysiological loading level is reached, t defines the current loading situation, and $I\dot{V}_C$ is the first time derivative of IV_C for a fiber family. Lastly, the maximum damage value D_s is defined as

$$D_s(\gamma) = D_\infty \left[1 - \exp\left(\frac{\ln(1 - r_\infty)}{\gamma_\infty} \gamma \right) \right], \tag{16}$$

where $D_\infty \in [0,1)$, $r_\infty \in [0,1)$, and $\gamma_\infty > 0$ is the value of the internal variable γ that describes damage evolution behavior at increasing stretches reached at the fraction $r_\infty = 0.99$ of the convergence limit $D_\infty = 0.99$. Similar to β, variable γ also evolves depending on loading and damage history. Higher values of γ_∞ indicate that tissue can tolerate more damage with increasing stretch limits. In order to ensure that D_s remains unchanged during cyclic loading under a fixed stretch maximum, the internal variable γ is defined as

$$\gamma = \max_{s \in [0,t]} \langle IV_C(s) - IV_{C,ini} \rangle, \tag{17}$$

which can be interpreted as the maximum value over time of the effective fictitiously undamaged function in the current state, and $IV_{C,ini}$ is the value of IV_C at which damage is initiated. The damage progression can then be described as

$$\phi := \langle IV_C(s) - IV_{C,ini} \rangle - \gamma \le 0. \tag{18}$$

The constitutive parameters describing damage saturation and evolution in human FPAs (Anttila et al. 2019) are summarized in Table 6. These data assume damage accumulation in all four fiber families. Damage initiation parameters, i.e., stretches and Cauchy stresses in both longitudinal and circumferential directions, can be calculated using Table 7. They both decrease significantly with age, indicating that older FPAs become much more susceptible to damage.

Table 6 Elastic and damage constitutive parameters representing FPA specimens in 6 age groups (Anttila et al. 2019)

Age group (yrs)	≤30	31–40	41–50	51–60	61–70	71–80
Mean age (yrs)	18.8	35.0	46.0	55.4	65.0	76.3
Elastic parameters						
c_{gr} (kPa)	5.25	9.13	7.78	7.05	3.68	10.99
c_{11} (kPa)	3.21	4.93	4.62	13.72	24.81	29.71
c_{21} (-)	2.40	2.30	4.82	5.19	9.29	11.43
c_{12} (kPa)	4.86	9.01	6.33	14.39	27.96	41.72
c_{22} (-)	4.73	10.42	5.13	6.76	10.90	13.37
c_{13}, c_{14} (kPa)	11.93	3.93	5.33	3.52	1.66	10.74
c_{23}, c_{24} (-)	0.21	4.41	4.28	7.33	9.19	10.83
φ°	35.40	49.53	49.05	33.44	43.74	34.20
Damage parameters						
$\gamma_{\infty}^{\text{col}}$ (kPa)	99.61	19.84	23.63	22.09	21.35	8.77
$\gamma_{\infty}^{\text{el}}$ (kPa)	25.43	40.92	24.13	23.09	15.85	16.97
$\gamma_{\infty}^{\text{smc}}$ (kPa)	19.35	11.40	20.10	18.76	13.09	13.25
β_s^{col} (-)	4.27	48.06	34.53	27.23	43.41	48.45
β_s^{el} (-)	39.25	0.88	42.53	22.64	14.98	16.45
β_s^{smc} (-)	14.89	1.69	25.61	15.75	18.04	15.41

Table 7 Damage initiation stretch $\lambda^{\text{D,ini}}$ and Cauchy stress $t^{\text{D,ini}}$ in two directions: longitudinal (z) and circumferential (θ) estimated by the multiple linear regression model. Here BMI = body mass index; DM = diabetes mellitus; age is measured in years. Data are adopted from Anttila et al. (2019)

Damage initiation	Multivariate model	R^2
$\lambda_z^{\text{D,ini}}$ (-)	$1.847 - 0.003\,\text{BMI} - 0.008\,\text{age}$	0.75
$\lambda_\theta^{\text{D,ini}}$ (-)	$1.795 - 0.002\,\text{BMI} - 0.007\,\text{age}$	0.73
$t_{zz}^{\text{D,ini}}$ (kPa)	$287.6 - 2.992\,\text{age}$	0.32
$t_{\theta\theta}^{\text{D,ini}}$ (kPa)	$239.3 + 2.363\,\text{age} + 51.956\,\text{DM}$	0.29

8 Computational Models of FPA Deformations During Limb Flexion

Descriptions of the FPA mechanical properties and loading conditions imposed by limb flexion allow computational studies of device-artery interactions. These interactions can significantly influence clinical outcomes (Rogers and Edelman 1995; Garasic et al. 2000; Kastrati et al. 2000; Berry et al. 2002; Lally et al. 2005; Timmins et al. 2011; Maleckis et al. 2018; Wang et al. 2018; MacTaggart et al. 2019), and can help provide a rational justification for the choice of a particular repair method or device. But to be useful for surgical planning, computational models must be built with patient and artery characteristics measured at baseline (i.e., prior to the repair), and describe the repair results when using different devices and materials. This is not trivial because the same artery cannot be repaired multiple times to validate modeling predictions. In the FPA, different stents are known to differentially influence arterial limb flexion-induced deformations (MacTaggart et al. 2019), resulting in significantly different arterial shapes, stress-stretch states, and hemodynamics. This means that computationally driving the artery into its deformed state by applying displacements measured at baseline may not necessarily be an accurate description of how this same artery behaves when it is repaired with a stent.

An alternative way of simulating limb flexion-induced FPA deformations is to model the response of the surrounding tissues rather than applying the displacements directly. To accomplish this, the artery is suspended with sets of elastic springs and viscous dashpots (Desyatova et al. 2017b) that represent the surrounding tissues. The reactions in these springs are determined using baseline bent-limb CTA, and since they remain the same after stenting, their re-application to the stented artery produces a new deformed configuration that takes device characteristics into account. The same methodology can also be used to study the effects of aging and disease on limb flexion-induced FPA deformations through substitution of arterial mechanical properties and pre-stretch (Desyatova et al. 2018a). But since such studies are quite complex and computationally intensive, it is important to first discuss the choice of a constitutive formulation that allows accurate description of the FPA behavior while also reducing the number of input parameters and computational time.

In a recent comparison (Desyatova et al. 2017b) of the isotropic neo-Hookean material, the Demiray model (Demiray 1972), the anisotropic two-fiber family Holzapfel-Gasser-Ogden (HGO) model (Holzapfel et al. 2000), and the four-fiber family model (Baek et al. 2007), the four-fiber model most accurately portrayed arterial behavior in all ages, but in subjects younger than 40 years, the performance of all constitutive formulations was similar. In older subjects, the Demiray model and the classic HGO formulation were better than the neo-Hookean model in predicting deformations due to limb flexion, but both significantly overestimated principal stresses compared with the four-fiber family or neo-Hookean models. This information is significant because simpler models have less parameters that need to be determined experimentally, and usually have a more efficient computational implementation, allowing faster calculations.

Another important aspect of FPA computational modeling is the question of whether to account for the variation in the mechanical properties along the artery length. Our recent data comparing proximal SFA to distal PA in the same donors demonstrate that these two segments have different structure, mechanical, and physiological characteristics, but the differences stem primarily from the higher prevalence of vascular disease in the PA (Jadidi et al. 2021a). When the pathology of the two FPA segments is the same, so are the mechanical properties, but not the physiological characteristics that remain different due to the larger diameter of the more proximal SFA compared with the PA. These data suggest that assuming homogeneous mechanical characteristics along the length of the FPA is reasonable when the

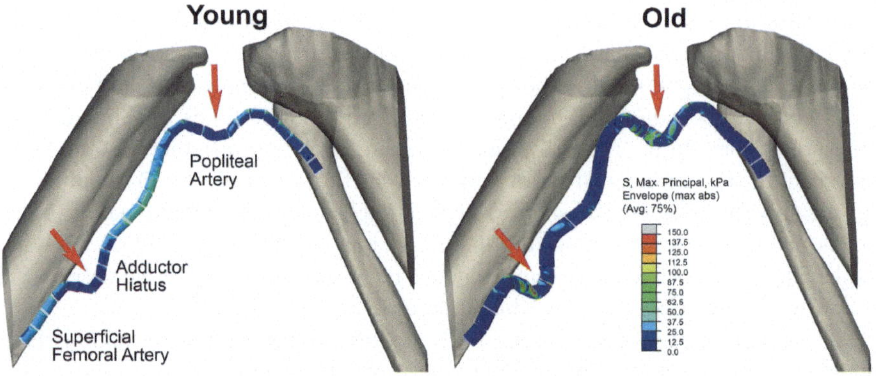

Fig. 5 Computational simulation of limb flexion-induced FPA deformations in a young (left) and old (right) artery demonstrating differences in arterial shape and stress concentrations (Desyatova et al. 2018a)

model accounts for the variation in arterial diameter and investigates similar degrees of vascular disease in the SFA and the PA.

Though much work remains to be done to improve the FPA model of limb flexion, it has recently been used to describe the effects of aging on the arterial shape, the stress-stretch characteristics, and the hemodynamics (Desyatova et al. 2018a). These results have demonstrated that younger arteries assume straighter positions in the flexed limb with less pronounced bends and more uniform stress distribution along the artery length (Fig. 5). Aging resulted in localization of high principal mechanical stresses to the AH and PA below the knee—areas that are prone to the development of vascular pathology (Watt 1965), and has produced a 3–5 fold increase in intramural stresses, with the largest increase observed at the AH (Desyatova et al. 2018a). Assessment of FPA hemodynamics in the flexed limb demonstrated a reduction in atheroprotective wall shear stress after 35 years of age, and an increase in atheroprone and oscillatory shear stresses after the age of 50 years, which may contribute to primary FPA pathology (Desyatova et al. 2018a). This computational model can be expanded to include stent-artery interactions, and can be used to develop an optimized PAD stent design that minimizes cross-sectional pinching during limb flexion while maintaining low intramural stresses and optimal hemodynamics (Desyatova et al. 2018b).

9 Future Directions

Though significant research efforts are focused on the pharmacological treatments and biological mechanisms of PAD, understanding the biomechanics and mechanophysiology of the lower limb arteries is equally important to improve clinical outcomes. The severe biomechanical environment surrounding human FPAs subjects the artery to adverse mechanical deformations, alters flow, and results in complex device-artery interactions that likely contribute to treatment failure. Perfused human cadaver models have been instrumental in understanding the magnitude of these deformations in different postures, but large variation in donor age, risk factors, and disease characteristics complicate their use for device design and initial prototype testing. An alternative strategy that has an advantage of producing consistent results under tightly controlled experimental conditions, is the development of a high-fidelity bench-top physical limb model with surrounding anatomical structures, branch tethering, and realistic arterial properties. If the parameters of this model can be tuned to represent limb flexion data available from human cadavers (MacTaggart et al. 2014, 2019; Desyatova et al. 2017c, 2018b; Poulson et al. 2018) and patients (Gökgöl et al. 2017, 2019; Spinella et al. 2019), it can become a useful and repeatable tool for the initial evaluation of new devices and materials.

Accurate representation of the arterial mechanical properties and longitudinal pre-stretch in these models is a critical component of making them realistic. Advancements in manufacturing and polymer chemistry show promise in achieving this goal, and new developments in machine learning and large data analysis can help better understand the target arterial properties in relation to age, gender, and risk factors. Non-parametric supervised machine learning methods such as Tree-Based algorithms designed for classification and regression tasks, are particularly suitable for this because they can handle a large number of numerical and categorical variables with complex nonlinear interaction, produce highly predictive, stable, and interpretable results, and are capable of dealing with noisy or incomplete data that are common with biological specimens. While often used in business decision making and market analysis, Tree-Based algorithms can also be utilized to describe human arterial characteristics for a particular combination of demographic and risk factor variables, but they require a sizeable dataset to train and test model performance and produce meaningful results. Fortunately, such database of human FPAs has recently been created (Desyatova et al. 2017a; Kamenskiy et al. 2017, 2018; Jadidi et al. 2019), opening new opportunities for studying the effects of demographics and risk factors on the nonlinear anisotropic mechanical properties of human lower extremity arteries.

In addition to using the machine learning approach to guide the development of synthetic arteries with realistic characteristics, it can also be utilized to determine patient-specific FPA properties in live patients. Assessment of these properties in vivo using surrogate measures like pulse wave velocity or distensibility is complicated by the narrow physiological pressure range, and the unknown residual stresses and pre-stretch (Caulk et al. 2018; Jadidi et al. 2020), but combining these metrics with the demographic and risk factor profiles of the patient, and a large database of previously analyzed *ex vivo* specimens, can make models of device-artery interactions more patient-specific and realistic. These high fidelity models can then be used for both the development of new devices with optimized characteristics and for helping the physician make a rational choice of repair from the already existing device options. Design development is a particularly attractive use of computational modeling because it provides a cost-effective way to iterate between design characteristics without manufacturing physical prototypes and before carrying out expensive animal experiments and clinical trials. Recent advances in computational modeling allow performing such design optimizations (Desyatova et al. 2018b), which could lead to the development of an optimal stent capable of minimizing arterial injury while maximizing flow.

In parallel with this development, it is important to address the Achilles' heel of most computational models (Anderson et al. 2007; Henninger et al. 2010; Erdemir et al. 2012), i.e., their clinical validation. Even the most sophisticated models are not substitutes for animal experiments and clinical trials because they cannot recapitulate the full complexity of vascular biology, pathology, physiology, and the external environmental effects. The long time scales and low frequency of recurrent disease in many major vascular beds make this clinical validation challenging, but the FPA provides a unique opportunity in this area because of the high rate of treatment failures that often occur within a few months to a couple of years after treatment (Schillinger et al. 2007; Laird and Yeo 2012; Siracuse et al. 2012; Lundgren and Study 2013; Stavroulakis et al. 2016; Qato et al. 2018).

References

Adam, D.J., Beard, J.D., Cleveland, T., Bell, J., Bradbury, A.W., Forbes, J.F., Fowkes, F.G.R., Gillepsie, I., Ruckley, C.V., Raab, G., Storkey, H.: BASIL trial participants: bypass versus angioplasty in severe ischaemia of the leg (basil): multicentre, randomised controlled trial. Lancet **366**, 1925–1934 (2005)

Anderson, A.E., Ellis, B.J., Weiss, J.A.: Verification, validation and sensitivity studies in computational biomechanics. Comput. Methods Biomech. Biomed. Engin. **10**, 171–184 (2007)

Ansari, F., Pack, L.K., Brooks, S.S., Morrison, T.M.: Design considerations for studies of the biomechanical environment of the femoropopliteal arteries. J. Vasc. Surg. **58**, 804–813 (2013)

Anttila, E., Balzani, D., Desyatova, A., Deegan, P., MacTaggart, J., Kamenskiy, A.: Mechanical damage characterization in human femoropopliteal arteries of different ages. Acta Biomater. **90**, 225–240 (2019)

Bacakova, L., Travnickova, M., Filova, E., Matějka, R., Stepanovska, J., Musilkova, J., Zarubova, J., Molitor, M.: The role of vascular smooth muscle cells in the physiology and pathophysiology of blood vessels. In: K. Sakuma (ed.) Muscle Cell and Tissue – Current Status of Research Field. IntechOpen (2018)

Baek, S., Gleason, R.L., Rajagopal, K.R., Humphrey, J.D.: Theory of small on large: potential utility in computations of fluid-solid interactions in arteries. Comput. Meth. Appl. Mech. Eng. **196**, 3070–3078 (2007)

Balzani, D., Brinkhues, S., Holzapfel, G.A.: Constitutive framework for the modeling of damage in collagenous soft tissues with application to arterial walls. Comput. Meth. Appl. Mech. Eng. **213**, 139–151 (2012)

Berry, J.L., Manoach, E., Mekkaoui, C., Rolland, P.H., Moore, J., Jr., Rachev, A.: Hemodynamics and wall mechanics of a compliance matching stent: in vitro and in vivo analysis. J. Vasc. Interv. Radiol. **13**, 97–105 (2002)

Caulk, A.W., Tellides, G., Humphrey, J.D.: Vascular mechanobiology, immunobiology, and arterial growth and remodeling. In: Mechanobiology in Health and Disease, pp. 215–248. Elsevier (2018)

Clowes, A.W., Reidy, M.A., Clowes, M.M.: Mechanisms of stenosis after arterial injury. Lab. Invest. **49**, 208–215 (1983)

Conte, M.S., Bandyk, D.F., Clowes, A.W., Moneta, G.L., Seely, L., Lorenz, T.J., Namini, H., Hamdan, A.D., Roddy, S.P., Belkin, M., Berceli, S.A., DeMasi, R.J., Samson, R.H., Berman,

S.S.: PREVENT III Investigators: Results of PREVENT III: a multicenter, randomized trial of edifoligide for the prevention of vein graft failure in lower extremity bypass surgery. J. Vasc. Surg. **43**, 742–751 (2006)

Demiray, H.: A note on the elasticity of soft biological tissues. J. Biomech. **5**, 309–311 (1972)

Desyatova, A., MacTaggart, J., Kamenskiy, A.: Constitutive modeling of human femoropopliteal artery biaxial stiffening due to aging and diabetes. Acta Biomater. **64**, 50–58 (2017a)

Desyatova, A., MacTaggart, J., Poulson, W., Deegan, P., Lomneth, C., Sandip, A., Kamenskiy, A.: The choice of a constitutive formulation for modeling limb flexion-induced deformations and stresses in the human femoropopliteal arteries of different ages. Biomech. Model. Mechanobiol. **16**, 775–785 (2017b)

Desyatova, A., Poulson, W., Deegan, P., Lomneth, C., Seas, A., Maleckis, K., MacTaggart, J., Kamenskiy, A.: Limb flexion-induced twist and associated intramural stresses in the human femoropopliteal artery. J. R. Soc. Interface **14**, 20170025 (2017c)

Desyatova, A., MacTaggart, J., Romarowski, R., Poulson, W., Conti, M., Kamenskiy, A.: Effect of aging on mechanical stresses, deformations, and hemodynamics in human femoropopliteal artery due to limb flexion. Biomech. Model. Mechanobiol. **17**, 181–189 (2018a)

Desyatova, A., Poulson, W., MacTaggart, J., Maleckis, K., Kamenskiy, A.: Cross-sectional pinching in human femoropopliteal arteries due to limb flexion, and stent design optimization for maximum cross-sectional opening and minimum intramural stresses. J. R. Soc. Interface **15**, 20180475 (2018b)

Dunlop, G.R., Santos, R.: Adductor-canal thrombosis. N. Engl. J. Med. **256**, 577–580 (1957)

Erdemir, A., Guess, T.M., Halloran, J., Tadepalli, S.C., Morrison, T.M.: Considerations for reporting finite element analysis studies in biomechanics. J. Biomech. **45**, 625–633 (2012)

Ferruzzi, J., Vorp, D.A., Humphrey, J.D.: On constitutive descriptors of the biaxial mechanical behaviour of human abdominal aorta and aneurysms. J. R. Soc. Interface **8**, 435–450 (2011)

Garasic, J.M., Edelman, E.R., Squire, J.C., Seifert, P., Williams, M.S., Rogers, C.: Stent and artery geometry determine intimal thickening independent of arterial injury. Circulation **101**, 812–818 (2000)

Gökgöl, C., Diehm, N., Räber, L., Büchler, P.: Prediction of restenosis based on hemodynamical markers in revascularized femoro-popliteal arteries during leg flexion. Biomech. Model. Mechanobiol. **18**, 1883–1893 (2019)

Gökgöl, C., Schumann, S., Diehm, N., Zheng, G., Buechler, P.: In vivo quantification of the deformations of the femoropopliteal segment: percutaneous transluminal angioplasty vs nitinol stent placement. J. Endovasc. Ther. **24**, 27–34 (2017)

Henninger, H.B., Reese, S.P., Anderson, A.E., Weiss, J.A.: Validation of computational models in biomechanics. Proc. Inst. Mech. Eng. H. **224**, 801–812 (2010)

Holzapfel, G.A., Gasser, T.C., Ogden, R.W.: A new constitutive framework for arterial wall mechanics and a comparative study of material models. J. Elasticity **61**, 1–48 (2000)

Holzapfel, G.A., Ogden, R.W.: Constitutive modelling of arteries. Proc. R. Soc. Lond. A **466**, 1551–1597 (2010)

Humphrey, J.D.: Cardiovascular Solid Mechanics: Cells, Tissues, and Organs. Springer, New York (2002)

Humphrey, J.D., Dufresne, E.R., Schwartz, M.A.: Mechanotransduction and extracellular matrix homeostasis. Nat. Rev. Mol. Cell Bio. **15**, 802–812 (2014)

Humphrey, J.D., Eberth, J.F., Dye, W.W., Gleason, R.L.: Fundamental role of axial stress in compensatory adaptations by arteries. J. Biomech. **42**, 1–8 (2009)

Jadidi, M., Desyatova, A., MacTaggart, J., Kamenskiy, A.: Mechanical stresses associated with flattening of human femoropopliteal artery specimens during planar biaxial testing and their effects on the calculated physiologic stress-stretch state. Biomech. Model. Mechanobiol. **18**, 1591–1605 (2019)

Jadidi, M., Habibnezhad, M., Anttila, E., Maleckis, K., Desyatova, A., MacTaggart, J., Kamenskiy, A.: Mechanical and structural changes in human thoracic aortas with age. Acta Biomater. **103**, 172–188 (2020)

Jadidi, M., Razian, S.A., Anttila, E., Doan, T., Adamson, J., Pipinos, M., Kamenskiy, A.: Comparison of morphometric, structural, mechanical, and physiologic characteristics of human superficial femoral and popliteal arteries. Acta Biomater. **121**, 431–443 (2021a)

Jadidi, M., Razian, S.A., Habibnezhad, M., Anttila, E., Kamenskiy, A.: Mechanical, structural, and physiologic differences in human elastic and muscular arteries of different ages: comparison of the thoracic aorta to the superficial femoral artery. Acta Biomater. **119**, 268–283 (2021b)

Jadidi, M., Sherifova, S., Sommer, G., Kamenskiy, A., Holzapfel, G.A.: Constitutive modeling using structural information on collagen fiber direction and dispersion in human superficial femoral artery specimens of different ages. Acta Biomater. **121**, 461–474 (2021c)

Jaminon, A., Reesink, K., Kroon, A., Schurgers, L.: The role of vascular smooth muscle cells in arterial remodeling: focus on calcification-related processes. Int. J. Mol. Sci. **20**, 5694 (2019)

Kamenskiy, A., Poulson, W., Sim, S., Reilly, A., Luo, J., MacTaggart, J.: Prevalence of calcification in human femoropopliteal arteries and its association with demographics, risk factors, and arterial stiffness. Arterioscler. Thromb. Vasc. Biol. **38**, e48–e57 (2018)

Kamenskiy, A., Seas, A., Bowen, G., Deegan, P., Desyatova, A., Bohlim, N., Poulson, W., MacTaggart, J.: In situ longitudinal pre-stretch in the human femoropopliteal artery. Acta Biomater. **32**, 231–237 (2016)

Kamenskiy, A., Seas, A., Deegan, P., Poulson, W., Anttila, E., Sim, S., Desyatova, A., MacTaggart, J.: Constitutive description of human femoropopliteal artery aging. Biomech. Model. Mechanobiol. **16**, 681–692 (2017)

Kastrati, A., Dirschinger, J., Boekstegers, P., Elezi, S., Schühlen, H., Pache, J., Steinbeck, G., Schmitt, C., Ulm, K., Neumann, F.J., Schömig, A.: Influence of stent design on 1-year outcome after coronary stent placement: a randomized comparison of five stent types in 1,147 unselected patients. Catheter Cardiovasc. Interv. **50**, 290–297 (2000)

Laird, J.R., Yeo, K.K.: The treatment of femoropopliteal in-stent restenosis: back to the future. J. Am. Coll. Cardiol. **59**, 24–25 (2012)

Lally, C., Dolan, F.L., Prendergast, P.J.: Cardiovascular stent design and vessel stresses: a finite element analysis. J. Biomech. **38**, 1574–1581 (2005)

Lundgren, F., Study, S.E.S.: External support of a polytetrafluoroethylene graft improves patency for bypass to below-knee arteries. Ann. Vasc. Surg. **27**, 1124–1133 (2013)

MacTaggart, J., Poulson, W., Seas, A., Deegan, P., Lomneth, C., Desyatova, A., Maleckis, K., Kamenskiy, A.: Stent design affects femoropopliteal artery deformation. Ann. Surg. **270**, 180 (2019)

MacTaggart, J.N., Phillips, N.Y., Lomneth, C.S., Pipinos, I.I., Bowen, R., Baxter, B.T., Johanning, J., Longo, G.M., Desyatova, A.S., Moulton, M.J., Dzenis, Y.A., Kamenskiy, A.V.: Three-dimensional bending, torsion and axial compression of the femoropopliteal artery during limb flexion. J. Biomech. **47**, 2249–2256 (2014)

Mahoney, E.M., Wang, K., Cohen, D.J., Hirsch, A.T., Alberts, M.J., Eagle, K., Mosse, F., Jackson, J.D., Steg, P.G., Bhatt, D.L.: One-year costs in patients with a history of or at risk for atherothrombosis in the United States. Circ. Cardiovasc. Qual. Outcomes **1**, 38–45 (2008)

Mahoney, E.M., Wang, K., Keo, H.H., Duval, S., Smolderen, K.G., Cohen, D.J., Steg, G., Bhatt, D.L., Hirsch, A.T.: Reduction of atherothrombosis for continued health (REACH) registry investigators: vascular hospitalization rates and costs in patients with peripheral artery disease in the United States. Circ. Cardiovasc. Qual. Outcomes **3**, 642–651 (2010)

Maleckis, K., Anttila, E., Aylward, P., Poulson, W., Desyatova, A., MacTaggart, J., Kamenskiy, A.: Nitinol stents in the femoropopliteal artery: a mechanical perspective on material, design, and performance. Ann. Biomed. Eng. **46**, 684–704 (2018)

Maleckis, K., Deegan, P., Poulson, W., Sievers, C., Desyatova, A., MacTaggart, J., Kamenskiy, A.: Comparison of femoropopliteal artery stents under axial and radial compression, axial tension, bending, and torsion deformations. J. Mech. Behav. Biomed. Mater. **75**, 160–168 (2017)

Mithieux, S.M., Weiss, A.S.: Elastin. Adv. Protein Chem. **70**, 437–461 (2005)

Newcomer, S.C., Sauder, C.L., Kuipers, N.T., Laughlin, M.H., Ray, C.A.: Effects of posture on shear rates in human brachial and superficial femoral arteries. Am. J. Physiol. Heart Circ. Physiol. **294**, H1833–H1839 (2008)

Padilla, J., Fadel, P.J.: Prolonged sitting leg vasculopathy: contributing factors and clinical implications. Am. J. Physiol. Heart Circ. Physiol. **313**, H722–H728 (2017)

Palatini, P., Casiglia, E., Gąsowski, J., Głuszek, J., Jankowski, P., Narkiewicz, K., Saladini, F., Stolarz-Skrzypek, K., Tikhonoff, V., Van Bortel, L., Wojciechowska, W., Kawecka-Jaszcz, K.: Arterial stiffness, central hemodynamics, and cardiovascular risk in hypertension. Vasc. Health Risk Manag. **7**, 725–739 (2011)

Palma, E.C.: Hemodynamic arteriopathy. Angiology **10**, 134–143 (1959)

Parker, B.A., Ridout, S.J., Proctor, D.N.: Age and flow-mediated dilation: a comparison of dilatory responsiveness in the brachial and popliteal arteries. Am. J. Physiol. Heart Circ. Physiol. **291**, H3043–H3049 (2006)

Poulson, W., Kamenskiy, A., Seas, A., Deegan, P., Lomneth, C., MacTaggart, J.: Limb flexion-induced axial compression and bending in human femoropopliteal artery segments. J. Vasc. Surg. **67**, 607–613 (2018)

Qato, K., Conway, A.M., Mondry, L., Giangola, G., Carroccio, A.: Management of isolated femoropopliteal in-stent restenosis. J. Vasc. Surg. **68**, 807–810 (2018)

Rogers, C., Edelman, E.R.: Endovascular stent design dictates experimental restenosis and thrombosis. Circulation **91**, 2995–3001 (1995)

Schillinger, M., Sabeti, S., Dick, P., Amighi, J., Mlekusch, W., Schlager, O., Loewe, C., Cejna, M., Lammer, J., Minar, E.: Sustained benefit at 2 years of primary femoropopliteal stenting compared with balloon angioplasty with optional stenting. Circulation **115**, 2745–2749 (2007)

Schillinger, M., Sabeti, S., Loewe, C., Dick, P., Amighi, J., Mlekusch, W., Schlager, O., Cejna, M., Lammer, J., Minar, E.: Balloon angioplasty versus implantation of nitinol stents in the superficial femoral artery. N. Engl. J. Med. **354**, 1879–1888 (2006)

Siracuse, J.J., Giles, K.A., Pomposelli, F.B., Hamdan, A.D., Wyers, M.C., Chaikof, E.L., Nedeau, A.E., Schermerhorn, M.L.: Results for primary bypass versus primary angioplasty/stent for intermittent claudication due to superficial femoral artery occlusive disease. J. Vasc. Surg. **55**, 1001–1007 (2012)

Sommer, G., Holzapfel, G.A.: 3D constitutive modeling of the biaxial mechanical response of intact and layer-dissected human carotid arteries. J. Mech. Behav. Biomed. Mater. **5**, 116–128 (2012)

Spinella, G., Finotello, A., Pane, B., Salsano, G., Mambrini, S., Kamenskiy, A., Gazzola, V., Cittadini, G., Auricchio, F., Palombo, D., Conti, M.: In vivo morphological changes of the femoropopliteal arteries due to knee flexion after endovascular treatment of popliteal aneurysm. J. Endovasc. Ther. **26**, 496–504 (2019)

Stavroulakis, K., Torsello, G., Manal, A., Schwindt, A., Hericks, C., Stachmann, A., Schönefeld, E., Bisdas, T.: Results of primary stent therapy for femoropopliteal peripheral arterial disease at 7 years. J. Vasc. Surg. **64**, 1696–1702 (2016)

Timmins, L.H., Miller, M.W., Clubb, F.J., Moore, J.E.: Increased artery wall stress post-stenting leads to greater intimal thickening. Lab. Invest. **91**, 955–967 (2011)

Wang, L., Tang, D., Maehara, A., Wu, Z., Yang, C., Muccigrosso, D., Zheng, J., Bach, R., Billiar, K.L., Mintz, G.S.: Fluid-structure interaction models based on patient-specific IVUS at baseline and follow-up for prediction of coronary plaque progression by morphological and biomechanical factors: a preliminary study. J. Biomech. **68**, 43–50 (2018)

Watt, J.K.: Origin of femoro-popliteal occlusions. Br. Med. J. **2**, 1455 (1965)

Wensing, P.J.W., Meiss, L., Mali, W.P.T.M., Hillen, B.: Early atherosclerotic lesions spiraling through the femoral artery. Arterioscler. Thromb. Vasc. Biol. **18**, 1554–1558 (1998)

Young, C.N., Deo, S.H., Padilla, J., Laughlin, M.H., Fadel, P.J.: Pro-atherogenic shear rate patterns in the femoral artery of healthy older adults. Atherosclerosis **211**, 390–392 (2010)

Simulation of Arterial Walls: Growth, Fiber Reorientation, and Active Response

Klemens Uhlmann, Anna Zahn, and Daniel Balzani

Gerhard and I started our collaboration on the modeling of the passive response of soft biological tissues in 2005. Based thereon, we worked on a joint project funded by DFG (German Research Foundation) and FWF (Austrian Science Fund) on stress-softening hysteresis as a result of microscopic damage. Our fruitful scientific cooperation led to a series of journal articles and joint presentations at international conferences over the years. Especially in the early stages of my academic career, I gained intellectual momentum by learning from his experience and his way of thinking about biomechanical problems. I very much appreciated his subtle and kind but questioning comments he has provided to me throughout the years. Dear Gerhard, I wish you a happy 60th birthday, good health, and all the best for the future. I am very thankful for all your support and hope for many more years of joint research to come.

Daniel

Abstract Computational simulations of arterial walls in health and disease are considered as a promising toolbox for the improvement of fundamental understanding of how arteries function, for the optimization of medical methods of treatment of several cardiovascular diseases, and for the development of new diagnostic approaches. The complexity of arterial tissue, however, necessitates the incorporation of various aspects of material response, not only including the strongly nonlinear and anisotropic passive behavior. Also, the active response, realistic fiber orientations, as well as suitable residual stress distributions resulting from biomechanical adaptation processes need to be accounted for in numerical simulations to allow for the prediction of realistic distributions of mechanical fields. In the first part of this contribution, a generalized approach for the calculation of residual stresses and fiber orientations in patient-specific arteries is proposed based on phenomenologically imitating bio-

K. Uhlmann · A. Zahn · D. Balzani (✉)
Chair of Continuum Mechanics, Ruhr -Universität Bochum, Universitätsstraße 150, 44801 Bochum, Germany
e-mail: daniel.balzani@ruhr-uni-bochum.de

G. Sommer et al. (eds.), *Solid (Bio)mechanics: Challenges of the Next Decade*, Studies in Mechanobiology, Tissue Engineering and Biomaterials 24,
https://doi.org/10.1007/978-3-030-92339-6_8

logical growth and remodeling. Secondly, a new model for smooth muscle cells is proposed, which accounts for the stretch-dependent calcium influx influencing the chemo-mechanical processes in smooth muscle. The individual model components are separately analyzed and finally combined in a numerical simulation of a simplified artery to enable the analysis of the impact of each model component.

1 Introduction

For the analysis of atherosclerotic arteries with respect to, e.g., potential plaque rupture, patient-specific numerical simulations are considered to be a promising toolbox for the improvement of clinical practice. However, this requires an accurate calculation of the stress and strain distributions throughout the arterial wall. In order to capture the mechanical behavior of arteries with one single model, several aspects have to be taken into account. One of these aspects is the existence of residual stresses, which considerably control the mechanical behavior of arterial walls. Arteries shorten when they are excised from the body, and they spring open when cut in the longitudinal direction. These observations imply that they are not free of stresses in the unloaded state. Without taking these stresses into account, a realistic prediction of load-induced stress distributions in arterial walls is unfeasible. With reference to early experimental investigations in the research groups of Fung and Vaishnav (Vaishnav and Vossoughi 1987; Fung and Liu 1989), the opening angle method was introduced to include circumferential residual stresses in numerical simulations of arterial walls (Choung and Fung 1986). It relies on the idea of using an opened arterial segment as stress-free reference configuration and closing it to a ring prior to applying the loads, which is frequently used (Holzapfel et al. 2000; Alastrué et al. 2007; Balzani et al. 2007). Although this is an elegant approximative approach, there is general consent by now that a single opening angle is not sufficient to characterize the full residual stress state of an artery. Instead, the opening angle is known to depend on various factors like the longitudinal position within the artery, species, age, and gender, on the smooth muscle activity, and the presence of diseases like hypertension, diabetes, or atherosclerosis (Fung and Liu 1989; Matsumoto and Hayashi 1996; Humphrey 2002). Moreover, it has been revealed that the opening angle is layer-specific, which means that a radial cut is not enough to release the entity of the residual stresses (Matsumoto and Hayashi 1996; Humphrey 2002). According to the experiments conducted by Holzapfel et al. (2007), showing that both circumferential and axial strips extracted from the three arterial layers variously deform when releasing the residual stresses, it is rather necessary to consider the residual stress state to be a fully three-dimensional property of arterial walls. Among many other interesting approaches to include residual stresses, e.g., methods which are somehow based on the hypothesis of uniform strain or stress distributions over the wall thickness (Takamizawa and Hayashi 1987; Polzer et al. 2013; Schröder and von Hoegen 2016; Joldes et al. 2018), the idea of simulating the adaptation processes supposed to be responsible for the residual stresses in real arteries is pursued here. The advantage of such approaches is that residual stresses arise automatically without additional technical steps like closing the opened arterial segment to a ring or smoothing of stress distributions. Instead of that, however, a hypothesis on the generation of residual stresses and its implementation are required.

It seems likely that the existence of residual stresses is causally related to the adaptation of biological tissues to changes in their environmental conditions. Whenever deviations from the physiological state occur, e.g., altered blood pressure for more than a few days, growth and remodeling processes with the supposed aim of restoring homeostatic strain or stress levels are initialized (Langille 1993; Rodbard 1975; Matsumoto and Hayashi 1996; Humphrey 2002). The term 'growth' usually refers to changes in mass by adding or removing material, and 'remodeling' describes adaptations of the internal tissue structure. Both processes are thought to be fundamentally controlled by mechanical stimuli, but the exact mechanisms continue to be unknown to a large extent. The large number of experimental observations, e.g., recapitulated by Humphrey (2002), indicate that growth in arteries has to be considered a multiaxial phenomenon involving the radial, circumferential, and axial directions. There seem to be connections between certain functional changes, mechanical quantities, and adaptive processes: (i) altered blood flow mainly affects the wall shear stress and is regulated by an adaptation of the arterial lumen (Langille 1993), (ii) changes in pressure leading to changes in circumferential stress and strain provoke an adaptation of the wall thickness (Matsumoto and Hayashi 1996), and (iii) changes in the axial in vivo stretch are related to adaptations of the vessel length (Jackson et al. 2002). However, normalization of single strain components or stress components by precise unidirectional growth seems not to be possible. For instance, an increase of the wall thickness in hypertensive arteries is indeed accompanied by a reduction of the axial stress, which points out that growth in axial direction might be involved (Matsumoto and Hayashi 1996). The restoration or maintenance of all homeostatic strain or stress values can be assumed to be prevented by interdependencies of different regulative processes. In such cases, the normalization of strains or stresses exceeding their favored level might have priority.

The number of publications on growth and remodeling in arterial tissues in the context of biomechanics is notable, such that an extensive review would go beyond the scope of this contribution. With reference to a review article by Ambrosi et al. (2011) and a more recent overview by Cyron and Humphrey (2017), mainly two different concepts have been established to model growth in arterial walls. The first concept relies on the multiplicative decomposition of the deformation gradient into a growth part and an elastic part, which restores the compatibility of the overall deformation and thereby produces residual stresses (Rodriguez et al. 1994). By choosing different forms of the growth tensor, isotropic, transversely anisotropic, or orthotropic growth in three dimensions can be described, as applied to soft tissues, e.g., in Lubarda and Hoger (2002), Himpel et al. (2005), Ren (2013), Sáez et al. (2014), and Liu et al. (2019). The second concept goes back to the constrained mixture model by Humphrey and Rajagopal (2002), where the tissue is considered as a mixture of different constituents with different deposition stresses and turnover rates but a shared deformation gradient. In contrast to multiplicative growth models, this framework allows analyzing the effects of the characteristic behavior of single constituents like elastin, collagen, or smooth muscle cells on the overall tissue behavior. It also provides more opportunities to investigate different hypotheses on growth and remodeling in arterial walls (Valentín and Holzapfel 2012). As reported by Braeu et al. (2017), the simulation of three-dimensional growth within the concept of constrained mixture theory has been enabled within the last decade but has first been limited to the isotropic case. Combining constrained mixture models with a decomposition of the deformation gradient, the authors extended these attempts to three-dimensional anisotropic growth, which, as motivated above, should be considered essential.

Another important aspect aside from the existence of residual stresses is the active contraction of smooth muscle cells (SMCs) in the arterial wall. Especially in small arteries as arterioles and capillaries, the contraction of SMCs is mechanically predominant to protect the tissue from damage by overstretching. The importance of SMCs decreases in larger arteries. Nonetheless, it can be assumed that smooth muscle contraction remains a crucial component when it comes to the mechanical modeling of arteries since antihypertensive drugs are mostly aiming at suppressing the contractile mechanism. One of the most common modeling approaches for muscle shortening was published in 1938 by Hill (1938), which was later extended by Fung (1970) and is known as the three-element Hill muscle model. This model describes the interaction between the passive (elastin and collagen) and the active (SMCs) material parts by exemplifying them as parallel elements, which allows a superposition of passive and active stresses. While the passive material has been comprehensively described in various material models, e.g., see Holzapfel et al. (2000) and Balzani et al. (2006), the active material is not completely understood yet. As in other muscle types, SMCs contain actin and myosin filaments, which build so-called contractile units. Phosphorylated myosin heads can attach to actin and perform power strokes, which result in a contraction of the cell. Several biochemical processes are involved during the contraction, such as the influx of calcium ions, which initiates the myosin light-chain kinase (MLCK) (Cole and Welsh 2011; Schneider 2013), a decrease of the myosin light-chain phosphatase (MLCP) (Cole and Welsh 2011; Schneider 2013), or polymerization of monomeric myosin (Chitano et al. 2017). However, the majority of mechanical models consider the concentration of cytosolic free calcium ions as the predominant factor to predict the level of contraction (Yang et al. 2003a, 2003b; Stålhand 2011; Böl et al. 2012; Murtada et al. 2010). An influx of calcium ions into the cell can be caused by a release of calcium from the sarcoplasmic reticulum (SR; an ion-storage located within the cell membrane) or from extracellular sources through calcium ion channels. Both sources can be triggered by various biological signals, e.g., calcium waves, junctional calcium transients, calcium sparks, calcium puffs, and L-type calcium channel sparklets (Wray 2010; Amberg and Navedo 2013). The experimental data to identify the contribution of every single mechanism is still very limited. Therefore, existing models use different approaches to rather phenomenologically describe the impact of all biological processes. It is well-accepted to combine the mechanics with the model of cross-bridge phosphorylation proposed by Hai and

Murphy (1988) as a chemical model part that describes the impact of the calcium concentration on myosin heads.

One way to model the changes in calcium concentration is to use an electrochemical model such as a Hodgkin-Huxley membrane model to simulate the influx of calcium (Yang et al. 2003a; 2003b). However, while this might be the most reliable way to capture details of the cell mechanisms, it also requires a detailed amount of experimental data, which cannot be provided sufficiently yet. Other models consider the calcium concentration as an input value and describe the contractile mechanism as dependent on the mechanical behavior of the contractile units (Stålhand 2011; Böl et al. 2012; Murtada et al. 2010). On the one hand, a mechanical approach to model smooth muscle contraction seems to be more attractive since it allows a straightforward way to match mechanical experiments. On the other hand, it neglects the calcium concentration as the predominant factor for the level of contractility. Two more recent publications introduced mechano-electrochemical models for the contraction of SMCs in the uterus and the urinary bladder (Sharifimajd et al. 2021; Seydewitz et al. 2017). Both models consider the membrane potential as the most important activation factor for calcium influx via L-type voltage-gated calcium ion channels, which is closer in line with the suggestions of biological studies than previous models. Nonetheless, it has to be mentioned that SMCs differ in size and functionality depending on their location. Several studies suggest that the influx of calcium in vascular SMCs is mainly mechanosensitive (Cole and Welsh 2011; Schneider 2013). To the authors' knowledge, there is, however, no material model for the arterial wall which considers the dependency of the calcium influx based on mechanical or mechano-electrical effects. In order to capture this mechanism, a stretch-dependent function for the intracellular calcium concentration is introduced in the proposed model.

The mechanical modeling of every aspect of the arterial wall raises individual difficulties. However, when it comes to patient-specific numerical simulations, only a model that embraces all aspects of the arterial wall can finally be considered as reliable. Towards this aim, models for the individual aspects of growth-induced residual stresses, remodeling in terms of the reorientation of the embedded collagen fibers, and active response will be introduced, and their mutual combinations will be analyzed. The numerical simulations are performed on rotationally symmetric arterial rings under intravascular pressure, which allow for efficient comparison of various model combinations. In addition, the simulation of a segment of a patient-specific artery is provided to show that the framework is also applicable to more complex examples.

2 Growth and Fiber Reorientation

As summarized in the introduction, experimental investigations on residual stresses and adaptation in arterial walls suggest that growth and remodeling should be considered anisotropic multiaxial processes that are substantially regulated by mechanical stimuli. The observation that adaptation processes often seem to be governed by the aim of restoring homeostatic stress, not strain levels (Humphrey 2002; Cyron and Humphrey 2017), leads to the estimation that growth and remodeling in arterial walls can be assumed to be stress-driven. Based on this hypothesis, a framework for combined growth and fiber reorientation, addressed for use with phenomenological, hyperelastic material models with two included fiber families, has been formulated in Zahn and Balzani (2018). It is designed to enable approximating residual stresses in arterial wall simulations based on mechanically founded assumptions on the underlying growth processes. Since the collagen fibers are the main load-bearing constituent in arterial walls, they are supposed to be arranged such that a somewhat optimized mechanical behavior leading to an efficient redistribution of the stresses is obtained. In consequence, it is essential for the simulation of stress-induced growth to provide a good approximation of the real fiber arrangement, which on the one hand, is known to vary over the wall thickness and, on the other hand, is subject to changes. Both effects can be captured if the arrangement of the fibers is chosen to adapt in response to the local stress state.

Although growth and fiber reorientation are interconnected to each other, both adaptation processes can be considered separately. In Sect. 2.1, the generalized model for various forms of stress-driven growth related to a set of three preferred directions will be introduced. Stress-driven reorientation of the fibers will be considered in Sect. 2.2, and numerical examples illustrating the basic effects of both phenomena will be presented in Sect. 2.3.

2.1 Generalized Framework for Anisotropic Growth

If growth is modeled based on a multiplicative decomposition of the deformation gradient $\mathbf{F} = \mathbf{F}_e\mathbf{F}_g$, as proposed by Rodriguez et al. (1994), the growth part $\mathbf{F}_g$ describes the deformation from the reference to a stress-free intermediate configuration. This fictitious intermediate state represents the situation at which each material point has been able to grow without any hindrance induced by mechanical load or strain. The remaining elastic part $\mathbf{F}_e$ ensures compatibility by reassembling the individually grown material points into a continuous body, and it accounts for the deformations induced by external loads. Postulating the form of the growth tensor $\mathbf{F}_g$, different assumptions on the growth mechanism can be realized.

2.1.1 Basic Forms of the Growth Tensor

There is still no evidence on the predominating mechanism, leading to the regulation of stress levels observed in arterial walls. From the mechanical point of view, depending on the type of loading, two basic mechanisms are possible for the reduction of stress acting in a direction with unit vector $\mathbf{A}_g$:

1. Increase of the cross-sectional area in the plane normal to $\mathbf{A}_g$.
2. Elongation of the body in the direction of $\mathbf{A}_g$.

If the stress originates from a constant force applied in the direction of $\mathbf{A}_g$, only the first form of growth will be effective, but if the stress arises from a prescribed displacement in the direction of $\mathbf{A}_g$, mainly the second mechanism will lead to a reduction of the stress by compensating the imposed deformation. A simple way to synergize both effects is to assume isotropic growth. This is, however, inconsistent with experimental observations and causes an unnecessarily high volume increase and thus avoidable energetic effort, as numerically shown in Sect. 2.3.1. The three mentioned basic forms of growth can be expressed mathematically by use of the growth tensors

$$\mathbf{F}_g^{\square} = \vartheta \mathbf{I} \qquad \text{for isotropic growth,} \tag{1a}$$

$$\mathbf{F}_g^{\parallel} = \mathbf{I} + (\vartheta - 1)\, \mathbf{A}_g \otimes \mathbf{A}_g \qquad \text{for growth in the direction of } \mathbf{A}_g \text{, and} \tag{1b}$$

$$\mathbf{F}_g^{\perp} = \vartheta \mathbf{I} + (1 - \vartheta)\, \mathbf{A}_g \otimes \mathbf{A}_g \qquad \text{for growth perpendicular to } \mathbf{A}_g, \tag{1c}$$

where the scalar ϑ, denoted as growth factor, defines the amount of growth in the particular directions.

2.1.2 Generalized Growth Tensor

For the description of three-dimensional anisotropic growth in arterial walls, the basic forms of growth according to Eq. (1) are insufficient because they allow taking at most one direction into account. A general stress state, however, is characterized by three principal stresses and their directions. Assuming that each of these stresses is regulated by a specific growth mechanism, a generalized growth tensor

$$\mathbf{F}_g = \mathbf{F}_g^{(3)} \mathbf{F}_g^{(2)} \mathbf{F}_g^{(1)} \tag{2}$$

is formulated. Herein, the three multiplicative parts are associated with the three principal stress directions $\mathbf{A}_g^{(a)}$. They involve three growth factors $\vartheta^{(a)}$ with $a \in [1, 2, 3]$. Since the real growth mechanism is unknown, each of the multiplicative parts of the growth tensor is allowed to adopt one of the basic forms given in Eq. (1) or to equal the unit tensor $\mathbf{I}$. This enables the analysis of a large number of various approaches with up to three preferred directions. Such analysis might lead to a conclusion regarding a realistic mechanism.

By choosing the vectors $\mathbf{A}_g^{(a)}$ to be the eigenvectors of a stress tensor and thus perpendicular to each other, the commutativity of Eq. (2) is ensured automatically. Furthermore, this choice leads to a local formulation, where the directional character of growth is exclusively determined by local mechanical state variables and not by the position of the considered material point within the r-φ-z-coordinate system or other geometric quantities. With reference to Himpel et al. (2005), the Mandel stress tensor

$$\boldsymbol{\Sigma}_e := \mathbf{C}_e\mathbf{S}_e \tag{3}$$

in the intermediate configuration is chosen to be the stress quantity, which controls growth and fiber reorientation. According to Himpel et al. (2005), this particular stress tensor is well suited since it is energetically conjugated to the growth velocity gradient $\mathbf{L}_g = \dot{\mathbf{F}}_g\mathbf{F}_g^{-1}$. The tensors $\mathbf{C}_e$ and $\mathbf{S}_e$, both defined in the intermediate configuration, are the elastic part of the Cauchy-Green tensor, computed as $\mathbf{C}_e = \mathbf{F}_e^T\mathbf{F}_e = \mathbf{F}_g^{-T}\mathbf{C}\mathbf{F}_g^{-1}$, and of the second Piola-Kirchhoff stress tensor. The stresses are obtained from a suitable strain-energy function $\psi(\mathbf{C}_e)$ by differentiation with respect to $\mathbf{C}_e$, i.e. $\mathbf{S}_e = 2\,\partial\psi/\partial\mathbf{C}_e$. If the vectors $\mathbf{A}_g^{(a)}$ are thus computed as principal directions of $\boldsymbol{\Sigma}_e$, only the specific values of the growth factors remain unknown for completeness of the growth tensor. Following Lubarda and Hoger (2002), the evolution of the growth factors is prescribed by a set of evolution equations of the form

$$\dot{\vartheta}^{(a)} = k_\vartheta^{(a)}(\vartheta^{(a)})\,\phi^{(a)}(\boldsymbol{\Sigma}_e) \qquad \text{with} \qquad a \in [1, 2, 3]. \tag{4}$$

Through these equations, the actual value of the growth factor $\vartheta^{(a)}$ depends on the Mandel stress tensor as defined by a driving force function $\phi^{(a)}(\boldsymbol{\Sigma}_e)$ and is subjected to some limiting restrictions imposed by a growth function $k_\vartheta^{(a)}(\vartheta^{(a)})$. Adopting the growth function

$$k_\vartheta^{(a)}(\vartheta^{(a)}) = \begin{cases} k_{\vartheta,(a)}^{+}\left[\dfrac{\vartheta_{(a)}^{+} - \vartheta^{(a)}}{\vartheta_{(a)}^{+} - 1}\right]^{m_{\vartheta,(a)}^{+}} & \text{for} \quad \phi^{(a)}(\boldsymbol{\Sigma}_e) > 0, \\ k_{\vartheta,(a)}^{-}\left[\dfrac{\vartheta^{(a)} - \vartheta_{(a)}^{-}}{1 - \vartheta_{(a)}^{-}}\right]^{m_{\vartheta,(a)}^{-}} & \text{for} \quad \phi^{(a)}(\boldsymbol{\Sigma}_e) < 0, \\ 0 & \text{for} \quad \phi^{(a)}(\boldsymbol{\Sigma}_e) = 0 \end{cases} \tag{5}$$

from Lubarda and Hoger (2002), the growth factors are restricted to a range

$$\vartheta^{(a)} \in [\vartheta_{(a)}^{-}, \vartheta_{(a)}^{+}] \qquad \text{with} \qquad \vartheta_{(a)}^{-} < 1 \;\; \text{and} \;\; \vartheta_{(a)}^{+} > 1, \tag{6}$$

and the growth velocity is scaled by a factor $k_{\vartheta,(a)}^{+/-}$.

2.1.3 Driving Forces for Growth

The internal variable $\vartheta^{(a)}$ is directly related to the amount of growth, which originates from the multiplicative part (a) of the growth tensor. Depending on the specific form of growth, see Eq. (1), different functional forms of $\phi^{(a)}(\boldsymbol{\Sigma}_e)$ might be reasonable. Three possible suggestions are the functions

$$\phi^{(a)}(\boldsymbol{\Sigma}_\mathrm{e}) = \boldsymbol{\Sigma}_\mathrm{e} : \mathbf{I} = \mathrm{tr}\,\boldsymbol{\Sigma}_\mathrm{e}, \tag{7a}$$

$$\phi^{(a)}(\boldsymbol{\Sigma}_\mathrm{e}) = \boldsymbol{\Sigma}_\mathrm{e} : \mathbf{A}_\mathrm{g}^{(a)} \otimes \mathbf{A}_\mathrm{g}^{(a)} \equiv \Sigma_\mathrm{e}^{(a)}, \text{ and} \tag{7b}$$

$$\phi^{(a)}(\boldsymbol{\Sigma}_\mathrm{e}) = \left\langle \boldsymbol{\Sigma}_\mathrm{e} : \mathbf{A}_\mathrm{g}^{(a)} \otimes \mathbf{A}_\mathrm{g}^{(a)} \right\rangle \equiv \left\langle \Sigma_\mathrm{e}^{(a)} \right\rangle, \tag{7c}$$

which are all defined for $a \in [1, 2, 3]$. An often used isotropic measure for the stress acting on the tissue is given by the trace of the Mandel stress tensor in Eq. (7a). For anisotropic forms of growth, a directional scalar measure might be more appropriate. A projection of the Mandel stress tensor in the direction of $\mathbf{A}_\mathrm{g}^{(a)}$ is obtained by Eq. (7b). Due to the fact that $\mathbf{A}_\mathrm{g}^{(a)}$ is defined as the principal direction of $\boldsymbol{\Sigma}_\mathrm{e}$, this projection extracts the eigenvalue $\Sigma_\mathrm{e}^{(a)}$ from the Mandel stress tensor. In Eq. (7c), only tensile values of the principal Mandel stresses are assumed to provoke growth; negative values are excluded with the aid of the Macaulay bracket $\langle \bullet \rangle = \frac{1}{2}\,(\bullet + |\bullet|)$. The in vivo state of arteries is significantly dominated by tensile stresses in the fiber-reinforced, circumferential-axial plane of the wall. In idealized tubular structures, the only compressive stress occurs in the radial direction. Its boundary values are dictated by the internal and external pressures and cannot be regulated by an adaptation of the geometry. Compressive stresses might therefore be excluded from adaptation processes.

2.1.4 Notes on the Algorithmic Treatment

The Mandel stress tensor $\boldsymbol{\Sigma}_\mathrm{e}$ depends on the growth tensor and is thus a function of all growth factors. In consequence of this, Eq. (4) constitutes a coupled set of nonlinear differential equations, which has to be solved for the actual values of the growth factors. Details on this task can be found in the literature, where the implicit Euler backward scheme is applied to similar equations (Himpel et al. 2005; Sáez et al. 2014) and to a former version of the present model (Zahn and Balzani 2018). In order to simplify the algorithmic treatment, the vectors $\mathbf{A}_\mathrm{g}^{(a)}$ and the two orientation vectors $\mathbf{A}^{(1)}$ and $\mathbf{A}^{(2)}$ for the essentially two collagen fiber families are considered as fixed during the local iteration, even if both actually depend on the Mandel stress tensor $\boldsymbol{\Sigma}_\mathrm{e}$, which changes with the growth factors. Once the actual values of the growth factors and the Mandel stress tensor are known, the vectors $\mathbf{A}_\mathrm{g}^{(a)}$ and the new fiber orientation vectors are computed and saved for use in the next time step.

In the context of a finite element framework using the Newton-Raphson scheme for the solution of the nonlinear equations, the linearized stresses are usually given in terms of the tangent moduli

$$\mathbb{C} := 2\,\frac{\mathrm{d}\mathbf{S}}{\mathrm{d}\mathbf{C}} = 2\,\frac{\partial \mathbf{S}}{\partial \mathbf{C}} + 2\sum_{a=1}^{3} \frac{\partial \mathbf{S}}{\partial \vartheta^{(a)}} \otimes \frac{\partial \vartheta^{(a)}}{\partial \mathbf{C}}. \tag{8}$$

The implementation of this tensor involves a number of derivatives, which are provided in Zahn and Balzani (2018). Alternatively, precise numerical differentiation approaches can be applied to the derivation of **S** with respect to **C** with less implementation effort but increased computing time, e.g., by using complex-step derivative approximation schemes (Tanaka et al. 2014) or hyper-dual numbers (Tanaka et al. 2015).

2.2 Fiber Reorientation

Collagen fibers are the central load-bearing components in arterial walls contributing mostly to the entire passive resistance to circumferential and axial loading. In healthy arteries, basically two distinguishable families of collagen fibers, helically wound around the vessel axis, can be identified in the plane of the wall (Schriefl et al. 2012). This arrangement is well in accordance with the fact that fibers are especially appropriate to carry high tensile loads, as occurring in the circumferential-axial plane of pressurized, ideal tubular geometries. Assuming such an idealized geometry, as illustrated on the right-hand side of Fig. 1, the two fiber families can be represented by the vectors $\mathbf{A}^{(1)}$ and $\mathbf{A}^{(2)}$, whose orientation within the circumferential-axial plane is defined by the angles $\beta_{\mathrm{f}}^{(1)}$ and $\beta_{\mathrm{f}}^{(2)} = -\beta_{\mathrm{f}}^{(1)}$ formed with the circumferential direction. However, even in healthy arteries, the fiber angles are known to vary nonlinearly over the wall thickness, with lower angles in the inner and higher angles in the outer arterial layers (Schriefl et al. 2012). Furthermore, the definition of the fiber vectors is complicated in, for instance, patient-specific arteries with irregular geometries deviating from rotationally symmetric tubes. In addition to that, in diseased arteries, the fiber arrangement is yet more complex since it might significantly disperse out of the axial-circumferential plane (Niestrawska et al. 2016; Akyildiz et al. 2017). For these reasons, a mechanically motivated fiber reorientation algorithm is desirable. On the one hand, it can be used to provide realistic fiber orientation vectors at each material point automatically, and on the other hand, it allows an adaptation of this fiber arrangement to changing mechanical conditions.

2.2.1 Target Fiber Orientation

Presuming that the orientation of the collagen fibers in arterial walls is dictated by the stress state, a hypothesis on the favored fiber arrangement can be formulated based on the fact that the predominant purpose of fibers is to bear tensile loads. The stress state at each material point is characterized by three principal stresses and the associated principal directions. Only if at least one of those stresses is positive, the existence of fibers is justified. In the case of one tensile principal stress, the main orientation of the fibers should point in the same direction as this stress. A second distinguishable fiber family gets relevant if two of the principal stresses are positive. Then, a whole plane is dominated by tensile stresses. For an optimized arrangement

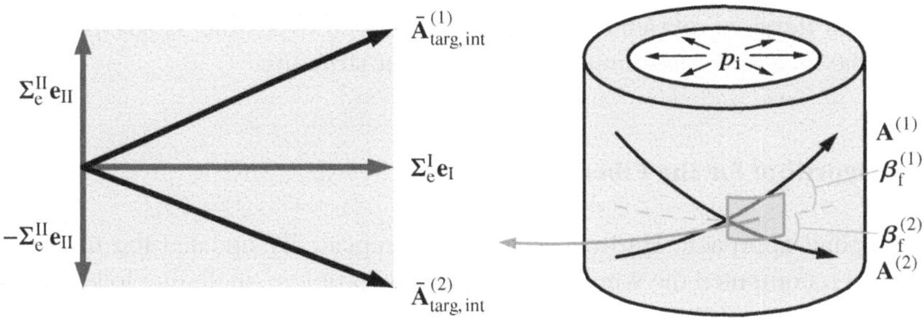

Fig. 1 With reference to Hariton et al. (2007), the target fiber orientation vectors $\bar{\mathbf{A}}^{(1)}_{\mathrm{targ,int}}$ and $\bar{\mathbf{A}}^{(2)}_{\mathrm{targ,int}}$ in the intermediate configuration are defined such that they point into the direction of the resultant of the tensile principal Mandel stresses

of the two fiber families within that plane, the hypothesis postulated by Hariton et al. (2007) is adopted. This hypothesis states that the fibers are arranged symmetrically with respect to the two maximum principal stresses, as shown in Fig. 1. The angle enclosed by the fiber vector and the first principal direction is thereby assumed to be defined by the ratio of the second to the first principal stress, which means that the fiber families are aligned with the direction of the resultant of both principal stresses.

In order to ensure consistency with the growth model described in Sect. 2.1, the fiber arrangement is supposed to be governed by the Mandel stress tensor $\boldsymbol{\Sigma}_{\mathrm{e}}$ in the intermediate configuration instead of the Cauchy stress tensor as proposed in Hariton et al. (2007). After solving the eigenvalue problem $\det[\boldsymbol{\Sigma}_{\mathrm{e}} - \Sigma_{\mathrm{e}}\mathbf{I}] = 0$, which results in the eigenvalues $\Sigma^{\mathrm{I}}_{\mathrm{e}} \geq \Sigma^{\mathrm{II}}_{\mathrm{e}} \geq \Sigma^{\mathrm{III}}_{\mathrm{e}}$ and the corresponding unit eigenvectors $\mathbf{e}_{\mathrm{I}}$, $\mathbf{e}_{\mathrm{II}}$, and $\mathbf{e}_{\mathrm{III}}$, the target fiber orientation vectors

$$\bar{\mathbf{A}}^{(1)}_{\mathrm{targ,int}} = \langle \Sigma^{\mathrm{I}}_{\mathrm{e}} \rangle \mathbf{e}_{\mathrm{I}} + \langle \Sigma^{\mathrm{II}}_{\mathrm{e}} \rangle \mathbf{e}_{\mathrm{II}} \quad \text{and} \quad \bar{\mathbf{A}}^{(2)}_{\mathrm{targ,int}} = \langle \Sigma^{\mathrm{I}}_{\mathrm{e}} \rangle \mathbf{e}_{\mathrm{I}} - \langle \Sigma^{\mathrm{II}}_{\mathrm{e}} \rangle \mathbf{e}_{\mathrm{II}} \tag{9}$$

can be defined if at least $\Sigma^{\mathrm{I}}_{\mathrm{e}}$ is positive. If all principal stresses are compressive, the fibers are ineffective, and an update of the target fiber vectors should be suspended. Performing a pullback operation and adjacent normalization, the target fiber orientation vectors

$$\bar{\mathbf{A}}^{(a)}_{\mathrm{targ}} = \frac{\mathbf{F}_{\mathrm{g}}^{-1}\bar{\mathbf{A}}^{(a)}_{\mathrm{targ,int}}}{\left|\mathbf{F}_{\mathrm{g}}^{-1}\bar{\mathbf{A}}^{(a)}_{\mathrm{targ,int}}\right|} \quad \text{for} \quad a \in [1, 2] \tag{10}$$

in the reference configuration are obtained. Prior to the reorientation, a reasonable assignment of the target vectors to the existing fiber vectors is required. To this end, the target vectors might be exchanged and/or rotated by 180°, such that finally the vectors $\mathbf{A}^{(1)}_{\mathrm{targ}}$ and $\mathbf{A}^{(2)}_{\mathrm{targ}}$ are assigned to the fiber orientation vectors $\mathbf{A}^{(1)}$ and $\mathbf{A}^{(2)}$. With a view to the nonlinear finite element implementation, the reorientation of the fiber vectors has to be performed stepwise because abrupt changes of the stress state almost certainly endanger the convergence of the global Newton iteration. Apart from that, an iterative reorientation, which is delayed over time, is anyway well in line with the idea of a continuously adapting fiber structure.

2.2.2 Algorithm for the Fiber Reorientation

In the original approach, Hariton et al. (2007) repeatedly updated the target fiber vectors and recomputed the whole boundary value problem, including load-stepping for each update until the fiber vectors remained unchanged. This can be avoided if the reorientation is not considered as a purely technical problem but rather as a continuous process over time, as proposed in Zahn and Balzani (2018). Especially in combination with other time-depending processes, for instance, growth or active response, the additional computational effort related to the reorientation of the fibers is then negligible.

The deviation between the existing and the target fiber orientation vectors can be expressed in terms of the angles

$$\eta^{(a)} = \arccos\left(\mathbf{A}^{(a)} \cdot \mathbf{A}^{(a)}_{\text{targ}}\right) \qquad \text{for} \qquad a \in [1, 2], \tag{11}$$

which are expected to evolve towards zero over time. The desired reorientation of the fibers can be achieved by the definition of two scalar evolution equations

$$\dot{\eta}^{(a)} = k_\eta\left(|\eta^{(a)}|\right) \leq 0 \qquad \text{with} \qquad a \in [1, 2] \tag{12}$$

for the angles $\eta^{(1)}$ and $\eta^{(2)}$. Applying the explicit Euler forward time integration to these equations, the actual values of the angles at time step $(n+1)$ can be approximated as

$$\eta^{(a)}_{n+1} \approx \dot{\eta}^{(a)}_n \Delta t + \eta^{(a)}_n \qquad \text{with} \qquad \Delta t = t_{n+1} - t_n \tag{13}$$

and can directly be evaluated since they only depend on known values from the previous time step n. Parallel consideration of growth and fiber reorientation reveals that the time step size Δt needed for convergence of the local Newton iteration for the solution of Eq. (4) is small enough to ensure a stable reorientation of the fibers in spite of explicit, not a priori stable time integration. An implicit integration scheme that would come with a non-zero remodeling part $\mathbb{C}^{\text{r}}$ of the tangent moduli is thus dispensable.

The remodeling function k_η in Eq. (12) is necessarily negative for all non-zero values of $\eta^{(a)}$ and zero for vanishing angles $\eta^{(a)} = 0$. A possible choice is the logarithmic function

$$k_\eta\left(\eta^{(a)}\right) = -k^+_\eta \ln\left(m^+_\eta \left|\eta^{(a)}\right| + 1\right) \;\; \text{with } \eta \text{ in rad}, \;\; k^+_\eta > 0, \;\; m^+_\eta > 0 \tag{14}$$

from Zahn and Balzani (2018), where the parameters k^+_η and m^+_η affect the reorientation velocity. Once Eq. (13) is evaluated, the new fiber orientation vectors can be computed (Zahn and Balzani 2018).

Table 1 Material parameters, adjusted to human tissue by Brands et al. (2008)

	c_1	ϵ_1	ϵ_2	α_1	α_2
Media	17.5 kPa	499.8 kPa	2.4	30 000.9 kPa	5.1
Adventitia	6.6 kPa	23.9 kPa	10.0	1503.0 kPa	6.3

2.3 *Illustrative Numerical Examples*

Prior to applying the framework of combined growth and fiber reorientation to numerical examples, a strain-energy function ψ has to be chosen for the evaluation of the elastic part of the stress tensor $\mathbf{S}_\mathrm{e}$ from the elastic part of the strain tensor $\mathbf{C}_\mathrm{e}$. Here, the passive material response is computed from a strain-energy function within a polyconvex framework for soft biological tissues (Balzani et al. 2006), which includes an isotropic contribution representing the matrix material and ensuring incompressibility, and two transversely isotropic contributions representing two families of embedded collagen fibers. Then, the passive part of the energy reads

$$\psi_\mathrm{p} = c_1 \left(I_1\, I_3^{-\frac{1}{3}} - 3 \right) + \epsilon_1 \left(I_3^{\epsilon_2} + I_3^{-\epsilon_2} - 2 \right) + \sum_{a=1}^{2} \alpha_1 \langle I_1\, J_4^{(a)} - J_5^{(a)} - 2 \rangle^{\alpha_2}, \quad (15)$$

and the energy function is set to $\psi := \psi_\mathrm{p}$ in this section. Note that thereby no active response resulting from smooth muscle cells is taken into account. Based on this energy function, the stresses are computed from the elastic part $\mathbf{C}_\mathrm{e}$ of the strains, which means that the invariants are defined in the intermediate configuration and have to be computed according to

$$I_1 = \mathrm{tr}\,\mathbf{C}_\mathrm{e}, \quad I_3 = \det\mathbf{C}_\mathrm{e}, \quad J_4^{(a)} = \mathrm{tr}[\mathbf{C}_\mathrm{e}\mathbf{M}_\mathrm{int}^{(a)}], \quad J_5^{(a)} = \mathrm{tr}[\mathbf{C}_\mathrm{e}^2\mathbf{M}_\mathrm{int}^{(a)}] \quad (16)$$

with structural tensors $\mathbf{M}_\mathrm{int}^{(a)}$ built from the fiber vectors $\mathbf{A}_\mathrm{int}^{(a)} = \mathbf{F}_\mathrm{g}\mathbf{A}^{(a)}/|\mathbf{F}_\mathrm{g}\mathbf{A}^{(a)}|$ in the intermediate configuration. The material parameters used in the numerical examples are listed in Table 1.

In a first example, some basic aspects of the combined framework for growth and fiber reorientation in arterial walls are illustrated based on an idealized, rotationally symmetric two-layered arterial segment. Afterward, a short section of a patient-specific artery is considered.

2.3.1 Basic Effects of Growth and Fiber Reorientation

In this example, three different forms of growth enclosed in the generalized form of the growth tensor are compared, and the additional effect of reorienting fibers is analyzed. The considered artery is assumed to be an ideal thick-walled tube with inner and outer radii of $r_i = 6.5$ mm and $r_o = 7.6$ mm. It is discretized by 14 quadratic one-dimensional, rotationally symmetric finite elements, which are split into 10 elements in the media of 0.8 mm thickness and 4 elements in the contiguous adventitia. The initial fiber angles are set to $\pm 30°$ in the media and $\pm 50°$ in the adventitia. In a first step, an internal pressure of 120 mmHg and an axial strain of 10% are applied without activation of growth and/or reorientation of the fibers. Then, the arteries are allowed to adapt until a steady state is attained. Thereby, the following three types of growth are considered:

Isotropic: By setting $\mathbf{F}_g^{(3)} = \mathbf{F}_g^{(2)} = \mathbf{I}$ and $\mathbf{F}_g^{(1)} = \mathbf{F}_g^{\square}$, see Eq. (1a), a growth tensor describing isotropic growth is obtained. According to this, the isotropic driving force function from Eq. (7a) is used.

Radial: In order to check if radial growth is a realistic scenario, growth in the direction of the third (compressive) principal stress, which in idealized situations corresponds to the radial direction, is modeled by setting $\mathbf{F}_g = \mathbf{F}_g^{(3)} = \mathbf{F}_g^{\parallel}$, see Eq. (1b). Due to the negative values of the radial stress, only the isotropic driving force from Eq. (7a) is suitable.

Anisotropic: Presuming that growth in arterial walls is anisotropic and multiaxial, one of the many possible examples is considered. Defining each of the multiplicative parts of $\mathbf{F}_g$ according to $\mathbf{F}_g^{(a)} = \mathbf{F}_g^{\perp,(a)}$, see Eq. (1c), and using the driving force function from Eq. (7c) for each of the growth factors, the assumption of growth occurring perpendicular to tensile principal stresses is realized.

In the given situation, parameter $\vartheta^+_{(a)}$, limiting the maximal amount of positive growth related to direction a, is the only parameter among all growth and remodeling parameters, which significantly affects the final result of the adaptation process. To allow for a fair comparison of the three growth mechanisms at the given load level with $\varepsilon_z = 10\%$, the optimal values of this parameter for each direction are determined layerwise using the method described in Zahn and Balzani (2019). The aim of the optimization is an efficient reduction of stress peaks, differences, and gradients. An overview of all parameters related to growth and fiber reorientation is given in Table 2.

In Fig. 2, the distribution of the circumferential and axial stresses over the wall thickness is shown for the three model variants. With reference to a non-adapting artery (dashed lines), it can, in general, be observed that growth leads to a more or less pronounced reduction of the gradient of the circumferential stress and to a reduction of the axial stress in the media. If a reorientation of the fibers is additionally taken into account, mainly the axial stress is affected. In contrast to a considerable extra reduction (Fig. 2a, c), also a strong increase of the axial stress can occur (Fig. 2b). Compared to the events in the media, the effects observed in the adventitia are negligible. This might, however, change at a supra-physiological level of the blood pressure with increasing stiffness of the adventitia.

Since the target fiber orientation is governed by the ratio of the tensile principal stresses, a growth-induced change of the principal stress state has a strong impact on the final fiber arrangement, see Fig. 3. If the axial stress passes the circumferential stress, the fibers reorient towards the axial direction, as obvious from Fig. 3b.

Table 2 Growth and reorientation parameters

Growth model	Layer	a	ϑ^+	ϑ^-	$k_\vartheta^{+/-}$	$m_\vartheta^{+/-}$	k_η^+	m_η^+
Isotropic	Media	1	1.1221	0.98	$1.0\,s^{-1}$	3.0	$0.6\,s^{-1}$	5.0
	Adventitia	1	1.0965	0.98	$1.0\,s^{-1}$	3.0	$0.6\,s^{-1}$	5.0
Radial	Media	3	1.2184	0.98	$1.0\,s^{-1}$	3.0	$0.6\,s^{-1}$	5.0
	Adventitia	3	1.0547	0.98	$1.0\,s^{-1}$	3.0	$0.6\,s^{-1}$	5.0
Anisotropic	Media	1	1.1337	0.98	$1.0\,s^{-1}$	3.0	$0.6\,s^{-1}$	5.0
		2	1.0024	0.98	$1.0\,s^{-1}$	3.0	$0.6\,s^{-1}$	5.0
		3	1.2354†	0.98	$1.0\,s^{-1}$	3.0	$0.6\,s^{-1}$	5.0
	Adventitia	1	1.0563	0.98	$1.0\,s^{-1}$	3.0	$0.6\,s^{-1}$	5.0
		2	1.0067	0.98	$1.0\,s^{-1}$	3.0	$0.6\,s^{-1}$	5.0
		3	1.3042†	0.98	$1.0\,s^{-1}$	3.0	$0.6\,s^{-1}$	5.0

†Exact value not relevant due to zero driving force

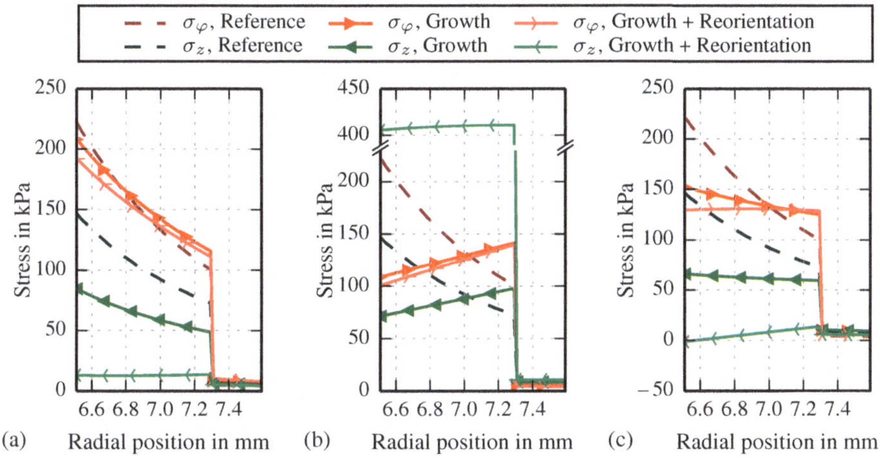

Fig. 2 Final distribution of circumferential and axial stresses over the wall thickness for (a) isotropic, (b) radial, and (c) anisotropic growth with and without reorientation of the fibers. The dashed lines represent the reference stresses prior to adaptation

This contradicts the experimental observation that only collagen fibers in the adventitia are arranged in such a way, whereas those in the media are rather oriented in the circumferential direction. If the axial stress is reduced to such an extent that it approaches zero or becomes compressive, the fibers reorient towards the circumferential direction.

Among the three growth mechanisms analyzed in this example, the anisotropic form of growth seems to be the most realistic scenario: The peaks of circumferential and axial stress and the stress gradients are significantly reduced, and the final arrangement of the fibers qualitatively shows the best accordance with experimental observations (Schriefl et al. 2012), in the sense that almost circumferential fiber orientations are found at the lumen-artery interface with increasing fiber angles when moving further outside of the artery. These results are even obtained with a compar-

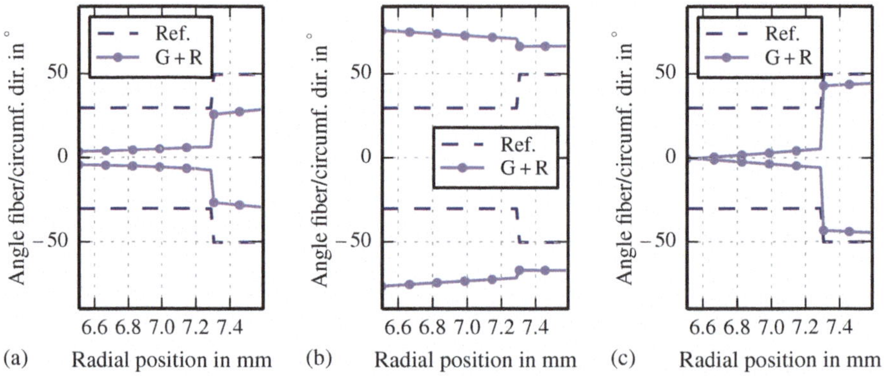

Fig. 3 Final distribution of the fiber angles over the wall thickness for (a) isotropic, (b) radial, and (c) anisotropic growth. The initial distribution of the fiber angles is depicted as dashed lines

Table 3 Growth-induced volume increase

Growth model	Layer	G	G + R
Isotropic	Media	38.17%	37.80%
	Adventitia	24.76%	24.82%
Radial	Media	19.23%	19.97%
	Adventitia	4.18%	4.10%
Anisotropic	Media	25.33%	25.28%
	Adventitia	9.35%	8.59%

atively low growth-induced volume increase, see Table 3, which correlates with low energetic costs. In contrast to the convincing results regarding anisotropic growth, the distributions of circumferential stresses and fiber orientations are qualitatively erroneous for radial growth. In addition to that, axial stresses even increase compared to the reference by applying radial growth and reorientation. Therefore, radial growth should not be considered realistic, although an even lower volume change is observed than for anisotropic growth.

Nevertheless, it should be mentioned that the performance of each growth mechanism is strongly dependent on the loading situation or rather on the ratio of internal pressure and axial strain. Consideration of one single loading situation is thus not sufficient to come to an evaluation regarding the most appropriate growth model. If, e.g., an axial strain of 0% had been applied and the limiting values of the growth factors had been optimized for that situation, radial growth would have produced the most satisfactory results. For that reason, an averaging over different loading situations should be taken into account for the optimization of the growth parameters, as proposed in Zahn and Balzani (2019). Then, the following statement can be formulated:

From the mechanical point of view, purely radial growth in arterial walls can hardly be justified. Other forms of anisotropic growth can lead to improvements compared to isotropic growth: more homogeneous stress distributions at a lower level of growth-induced volume increase.

2.3.2 Patient-Specific Artery

To conclude the section on arterial growth and fiber reorientation, the anisotropic growth model defined in Sect. 2.3.1 is applied to a short segment of a patient-specific coronary artery, which has been obtained from a stack of two-dimensional images using the method of Virtual Histology (Balzani et al. 2012) and has previously been used by Fausten et al. (2016). The artery is discretized by 8401 quadratic, 10-noded tetrahedral finite elements and subjected to boundary conditions that ensure statically determined support. To keep the boundary value problem as simple as possible, all components of the artery—essentially adventitia, media, and plaque—are considered as a homogeneous material with the properties of the medial layer as given in Table 1. Due to this and also because of the pronounced thickness of the wall, the patient-specific artery is expected to be much stiffer and more resistant to the internal pressure than the idealized two-layered arterial segment analyzed before. In order to arrive at a loading situation, which might be comparable to the previous example, an elevated internal pressure of 160 mmHg and a reduced axial strain of 5% are initially applied and held over the entire simulation duration. The limiting values of the growth factors are set to $\vartheta^+_{(1)} = 1.159$, $\vartheta^+_{(2)} = 1.041$, $\vartheta^+_{(3)} = 1.250$, and $\vartheta^-_{(a)} = 0.95$. The initial fiber angles are set to $\pm 30°$ over the entire wall.

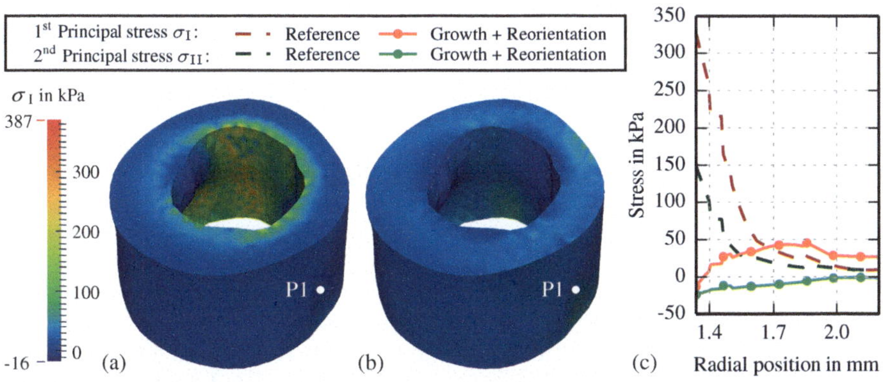

Fig. 4 Contour plots of the first principal Cauchy stress (a) prior to adaptation and (b) after nine seconds of growth and fiber reorientation, and (c) distribution of the first and second principal stresses over the radial position for both situations at an exemplary point P1

As visualized in Fig. 4, the effects of anisotropic growth and fiber reorientation described in the first example can also be observed in the patient-specific artery. The peaks of the first principal stress at the internal surface of the vessel, which are present prior to adaptation (Fig. 4a), are reduced towards a more homogeneous stress distribution (Fig. 4b). The same holds for the second principal stress. For an exemplary point P1, the distribution of both stresses over the wall thickness, prior to and after nine seconds of adaptation, is shown in Fig. 4c. Note that the moderate heterogeneities of stress distributions sometimes described as checkerboard pattern can be reduced if more sophisticated (and thus more expensive) finite element formulations are taken into account, e.g. the Taylor-Hood element.

3 Smooth Muscle Contraction

Biological studies suggest that the predominant factor for the level of contraction of SMCs is the change in intracellular calcium concentration. Furthermore, the influx of calcium into vascular SMCs can be assumed to be mechanosensitive (Wray 2010; Cole and Welsh 2011; Tykocki et al. 2017). The proposed model for the contraction of SMCs consists of a chemical and a mechanical part. For the chemical model part, the model of cross-bridge phosphorylation by Hai and Murphy (1988) is adopted, which describes the influence of calcium ions on myosin filaments. Myosin and actin filaments build so-called contractile units inside of SMCs, where a proportion of myosin heads regulate the contraction of the cell by performing power strokes. A modified version of the model of Murtada et al. (2010) serves as the mechanical model part. It describes the material response of SMCs by taking the chemical state of myosin into account. Additionally, a stretch-dependent calcium function is introduced, which represents the mechanosensitive calcium influx into the cells.

In Sects. 3.1 and 3.2, the chemical and the mechanical model part are described, respectively. Furthermore, an analysis of the influence of active response on growth and remodeling will be presented in Sect. 3.3 by showing results from simulations of arterial rings under intravascular pressure.

3.1 Extension of the Model for Cross-Bridge Phosphorylation

The chemical model describes the influence of calcium influx on the contractile behavior of SMCs. As already mentioned, myosin heads perform power strokes as long as they are phosphorylated and connected to actin. There are in total four functional states, which classify myosin heads in: (A) detached and dephosphorylated, (B) detached and phosphorylated, (C) attached and phosphorylated, and (D) attached and dephosphorylated. The transition of myosin heads from one state into another is modeled by seven reaction rates k_i, describing the rate of phosphorylation (k_1 and k_6), dephosphorylation (k_2 and k_5), attachment (k_3), and detachment (k_4 and k_7). The attachment of myosin heads can only occur when the myosin is phosphorylated. Hence, there is no reaction rate for the transition of myosin heads from state A to state D. The chemical model can be described by four coupled ordinary differential equations

$$\begin{bmatrix} \dot{n}_\mathrm{A} \\ \dot{n}_\mathrm{B} \\ \dot{n}_\mathrm{C} \\ \dot{n}_\mathrm{D} \end{bmatrix} = \begin{bmatrix} -k_1([\mathrm{Ca}^{2+}]) & k_2 & 0 & k_7 \\ k_1([\mathrm{Ca}^{2+}]) & -k_2 - k_3 & k_4 & 0 \\ 0 & k_3 & -k_4 - k_5 & k_6([\mathrm{Ca}^{2+}]) \\ 0 & 0 & k_5 & -k_6([\mathrm{Ca}^{2+}]) - k_7 \end{bmatrix} \begin{bmatrix} n_\mathrm{A} \\ n_\mathrm{B} \\ n_\mathrm{C} \\ n_\mathrm{D} \end{bmatrix}, \tag{17}$$

where $n_\mathrm{A}, n_\mathrm{B}, n_\mathrm{C}$, and n_D constitute the proportions of myosin heads in the respective state. As natural constraints for proportions, they fulfill

$$n_\mathrm{A} + n_\mathrm{B} + n_\mathrm{C} + n_\mathrm{D} = 1, \qquad n_\mathrm{A}, n_\mathrm{B}, n_\mathrm{C}, n_\mathrm{D} \in [0, 1]. \tag{18}$$

While the reaction rates k_2, k_3, k_4, k_6, and k_7 are considered as constant, the reaction rates k_1 and k_6 are calcium-dependent, describing the activity of the activity of the myosin light-chain kinase. Additionally, the calcium concentration $[\mathrm{Ca}^{2+}]$ is modeled as a function of the stretch $\lambda^{(f)} = \sqrt{J_4^{(f)}}$ in the fiber direction of the SMCs, so that

$$k_1 = k_6 = \eta \frac{[\mathrm{Ca}^{2+}]^h}{[\mathrm{Ca}^{2+}]^h + [\mathrm{Ca}_{50}^{2+}]^h} \quad \text{and} \quad [\mathrm{Ca}^{2+}]\left(\lambda^{(f)}\right) = c_1 \left\langle \lambda^{(f)} - \lambda_0 \right\rangle^{c_2}, \tag{19}$$

where η, h, c_1, and c_2 are material parameters, $[\mathrm{Ca}_{50}^{2+}]$ represents the half-activation constant, and λ_0 is the stretch at which the calcium concentration is considered to be zero.

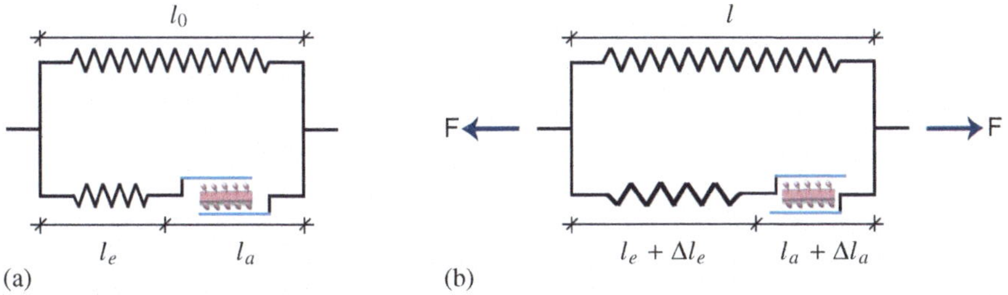

Fig. 5 Three-element Hill muscle model in fiber direction of the SMCs: (a) relaxed state and (b) contracted state. The model consists of a parallel elastic element (elastin and collagen), a serial elastic element (elasticity of SMCs), and a contractile element (myosin filament in red, actin filaments in blue)

3.2 Mechanical Model of Smooth Muscle Contraction

The mechanical part of the model for active material response is based on the approach of Murtada et al. (2010), which takes the three-element Hill muscle model into account (Hill 1938; Fung 1970). Two of those three elements are arranged in series and represent the contractility of SMCs and their elastic material behavior. A third element is set in parallel to the previous two and considers the elastic material response of the passive part, i.e. collagen and elastin. The passive material was already described in Sect. 2.3. An illustration of the three-element Hill muscle model is shown in Fig. 5. There, l_0 defines the length of a tissue segment in fiber direction (f) at rest and l the corresponding length of the same segment in the current state. Hence, l can be expressed as $l = l_0 + \Delta l_e + \Delta l_a$, where Δl_e is the length change of the elastic element of the SMCs and Δl_a is the length change of the active (contractile) element of the SMCs. Subsequently, the stretch in fiber direction (f) is defined as $\lambda^{(f)} = l^{(f)}/l_0^{(f)}$, which can be rewritten as

$$\lambda^{(f)} = \frac{l_0^{(f)} + \Delta l_e^{(f)} + \Delta l_a^{(f)}}{l_0^{(f)}} = 1 + \varepsilon_e^{(f)} + \varepsilon_a^{(f)}, \tag{20}$$

where $\varepsilon_e^{(f)}$ is the elastic strain and $\varepsilon_a^{(f)}$ is the active strain in fiber direction (f). Elastic energy and stress resulting from the active material part are only dependent on the elastic strain $\varepsilon_e^{(f)}$, not the active strain $\varepsilon_a^{(f)}$. In addition, the amount of attached myosin heads determines the level of elastic energy, which can be stored in the smooth muscle tissue. Taking this into account, the active strain-energy function in fiber direction (f) is defined as

$$\psi_a^{(f)} = \frac{\mu_a}{2}\left(n_C + n_D\right)\left(\varepsilon_e^{(f)}\right)^2, \tag{21}$$

where μ_a constitutes a stiffness constant, and n_C and n_D are the proportions of attached myosin heads, i.e. the myosin heads in state C and D. The elastic strain $\varepsilon_e^{(f)}$ in Eq. (21) is obtained from rearranging Eq. (20) to $\varepsilon_e^{(f)} = \lambda^{(f)} - \varepsilon_a^{(f)} - 1$. The active strain $\varepsilon_a^{(f)}$ describes the strain which results from contraction caused by cycling myosin heads in state C. Furthermore, $\varepsilon_a^{(f)}$ can be set into a relation to the contraction velocity $\dot{\varepsilon}_a^{(f)}$ by using Hill's equation for tetanized muscle contraction, which is

$$(F + a)(v + b) = (F_0 + a)\, b. \tag{22}$$

It defines the contraction velocity v of a muscle in a quick-release experiment, where F_0 is the isometric force at which the quick-release is performed, and F is the isotonic afterload. The parameters a and b are used to fit the relation between F and v. By building an analogy to the active material model, v can be replaced by $\dot{\varepsilon}_a^{(f)}$, and F represents the active stress $P_a^{(f)} = \partial\psi_a^{(f)}/\partial\lambda^{(f)}$. Additionally, the driving stress $P_c^{(f)}$

is introduced, which substitutes the force F_0. The driving stress $P_c^{(f)}$ represents the strength at which phosphorylated cross-bridges perform their power strokes. Hence, a time-dependent differential equation for the contraction velocity can be formulated as

$$\dot{\varepsilon}_a^{(f)} = \gamma_1 \frac{P_a^{(f)} - P_c^{(f)}}{P_a^{(f)} + \gamma_2}, \tag{23}$$

where γ_1 and γ_2 are material parameters. The active stress $P_a^{(f)}$ and the driving stress $P_c^{(f)}$ are given by

$$P_a^{(f)} = \mu_a \,(n_C + n_D)\, \varepsilon_e^{(f)} \quad \text{and} \quad P_c^{(f)} = \kappa\, n_C^{(f)}, \tag{24}$$

where κ is the maximally achievable driving stress. In order to include the active as well as the passive response, the strain-energy function is set to $\psi = \psi_p + \sum \psi_a^{(f)}$. In the following section, two fiber directions are considered, i.e., $f \in [1, 2]$.

3.3 *Combined Effects of Growth, Fiber Reorientation, and Active Response*

As mentioned in the introduction, the main goal of this work is to highlight possible influences and interactions of growth, fiber reorientation, and the active material response of the arterial wall. Nonetheless, especially for the modeling of smooth muscle contraction, several details about the functionality are still unclear. One example is the time scale on which SMCs act. Experimental studies mostly investigated the mechanosensitivity of vascular SMCs by applying a constant stretch to a tissue strip or a constant intravascular pressure to an arterial ring (Murtada et al. 2010; Schneider 2013). The data show that SMCs finish the adaptation to a change of the constant load scenario after a time between 30 s and 5 min, where the time frame strongly depends on the applied load. However, the intravascular pressure of an in vivo artery is not constant but fluctuates between the systolic and diastolic blood pressure. To which degree the pressure fluctuation during a single heartbeat triggers a significant reaction of vascular SMCs is not sufficiently investigated yet. Therefore, the simulations analyzed here will focus on a constant load scenario. A similar setup as in Sect. 2.3.1 is used, where rotationally symmetric arterial rings are subjected to a constant intravascular pressure, which represents a medium value between systolic and diastolic blood pressure. As an additional boundary condition, an axial strain of 10% is applied. Since SMCs are only part of the media, and the simulations in Sect. 2.3.1 demonstrated that the adventitia hardly contributed to the resistance of the arterial wall, only the media is considered here with inner and outer radii of $r_i = 6.5$ mm and $r_o = 7.3$ mm. In addition, the same passive material parameters (see Table 1) and the same parameters for growth and fiber reorientation, as in Sect. 2.3.1, are used. To simplify the adaptation process of growth and fiber reorientation, SMCs are here

Table 4 Parameters for active response

Mechanical	μ_a	κ	γ_1	γ_2	c_1	c_2	λ_0
	1000 kPa	700 kPa	0.0001 s^{-1}	26.68 kPa	10000 μM	4.0	0.98
Chemical	$k_{2/5}$	k_3	k_4	k_7	η	h	$[Ca^{2+}_{50}]$
	0.16 s^{-1}	0.07 s^{-1}	0.03 s^{-1}	0.02 s^{-1}	0.8 s^{-1}	4.0	0.35 μM

assumed to be aligned with the collagen fiber directions. For the active response, material parameters are chosen such that a significant but not unrealistic contraction of the arterial wall is obtained (see Table 4).

Growth, fiber reorientation, and active response are modeled as time-dependent processes. However, while the time dependency of the active response represents the contraction of the artery over an actual time, growth and fiber reorientation are only virtually time-dependent as the real time-dependency is not realistically modeled, and only the saturated state is analyzed. Thereby, the two strongly differing time scales, namely the one of the smooth muscle response in the order of seconds or minutes and the one of growth and remodeling in the order of days or weeks, can be taken into account in a single and still feasible simulation. As a first plausibility check, simulations with the active material response are compared to the simulations with the purely passive material model. Here, growth and fiber reorientation are not considered yet. For this purpose, the resulting stress distributions over the entire wall are illustrated in Figs. 6a–c, regarding intravascular pressures of 120, 150, and 180 mmHg, respectively. It can be seen that in all load scenarios, the axial and circumferential stress distributions are qualitatively similar. Additionally, the contraction of SMCs reduces axial stresses recognizably. While also the medium value of circumferential stresses was slightly decreased, a notable increase at the inner side of the artery can be observed. The results indicate that smooth muscle contraction helps to improve the load-bearing capabilities of the artery. However, the increased stress at the inner side of the artery when taking SMCs into account underlines the necessity of residual stresses to reach a more homogeneous stress distribution. It can be expected that a combination of growth and active material response will flatten the stress distribution.

As a second step, it is investigated to what extent the incorporation of active response has an impact on the adaptation of growth and fiber reorientation. Since anisotropic growth showed the most promising results in simulations with the passive material response, isotropic and radial growth are not considered here. Two different types of simulations are regarded. In case A, growth and fiber reorientation are executed while only considering the passive material behavior, and SMCs are activated afterward. In case B, SMCs are activated before the growth and fiber reorientation process. Hence, the simulations are split into the following steps:

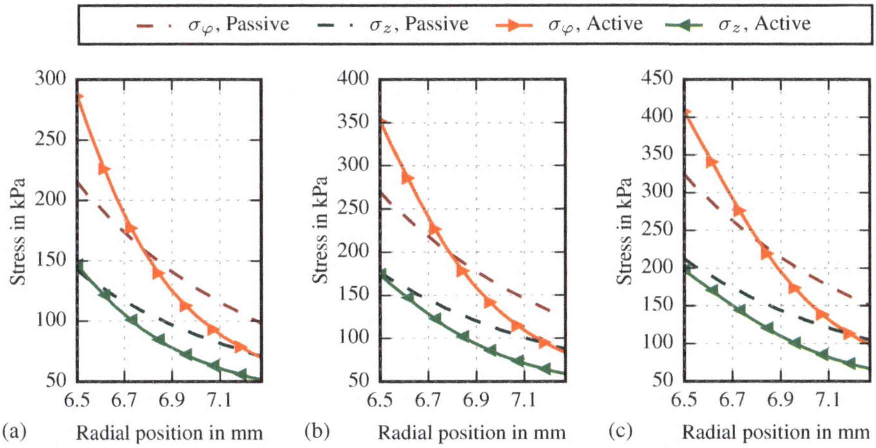

Fig. 6 Active material response without growth and reorientation: Distribution of circumferential and axial stresses over the wall thickness in non-adapting arterial rings with and without active response. Different values of intravascular pressure were used: (a) 120 mmHg, (b) 150 mmHg, and (c) 180 mmHg

Case A:

(i) Intravascular pressure and axial strain are applied without activation of growth, fiber reorientation, and SMCs.
(ii) Activation of growth and fiber reorientation until the saturated state is attained.
(iii) Activation of SMCs and deactivation of further growth and fiber reorientation until a steady state for the active material response is attained.

Case B:

(i) Intravascular pressure and axial strain are applied without activation of growth, fiber reorientation, and SMCs.
(ii) Activation of SMCs until steady state is attained.
(iii) Activation of growth and fiber reorientation and deactivation of further SMC activity until the saturated state is attained.
(iv) Activation of smooth muscle contraction and deactivation of further growth and fiber reorientation.

Both simulation cases are applied for intravascular pressures of 120, 150, and 180 mmHg, as illustrated in Fig. 7. In addition to cases A and B, simulation results, which include growth and fiber reorientation without active response, are presented as a reference. Note that the simulation cases A and B can only be seen as exemplary since more complex interactions between growth and fiber reorientation with smooth muscle contraction were precluded by choice of the specific simulation protocol. Neither the fluctuation of the intravascular pressure during a heartbeat nor an adaptation of all three aspects at the same time has been considered. Nonetheless, the

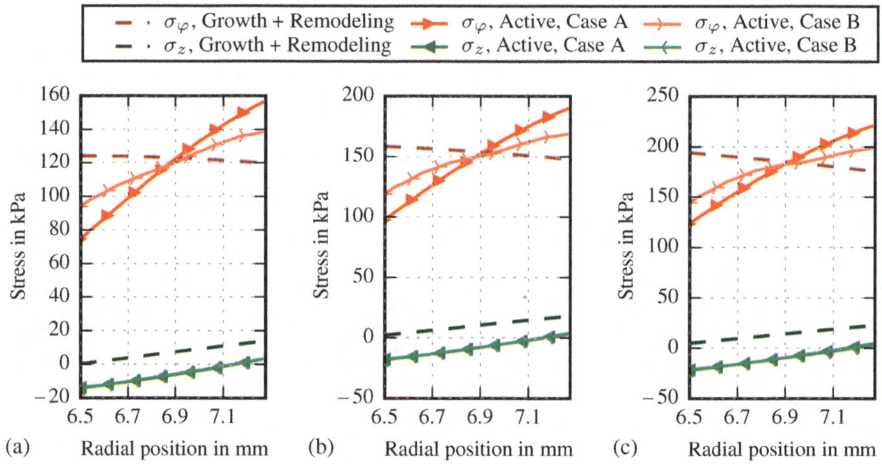

Fig. 7 Active response with anisotropic growth and fiber reorientation: Final distribution of circumferential and axial stresses over the wall thickness for anisotropic growth and an intravascular pressure of (a) 120 mmHg, (b) 150 mmHg, and (c) 180 mmHg. Two different simulation protocols are used: In case A, SMCs are activated after the adaptation by growth and fiber reorientation; In case B, SMCs are activated before the adaptation by growth and fiber reorientation. Growth and reorientation without active material response serve as a reference. *Note that the growth parameters were optimized for simulations, which consider only the passive material response. Hence, stresses in case A and B are not expected to be uniform over the wall thickness*

results show clear tendencies, which will support the construction of a more reliable simulation setup in future investigations. Firstly, it has to be mentioned that neither the active material response nor a different load scenario showed a considerable impact on the volume increase caused by growth. Secondly, in simulations of case A, the final fiber distribution in the media was, as expected, nearly identical with the fiber distribution, as shown in Fig. 3. The fiber directions in simulations of case B were nearly aligned with the circumferential direction. Such a fiber distribution can be explained by taking a look at the axial stresses, which reach even negative values. Since collagen fibers contribute only to the material resistance when they are stretched, the optimal fiber direction for this stress distribution is the circumferential direction. Negative stresses in the axial direction are occurring as a consequence of the smooth muscle contraction in the circumferential direction and the incompressibility of the tissue, which enforces an additional expansion in the axial direction. To avoid negative stresses in the axial direction, higher axial strains could be considered in the boundary conditions. Just as for the volume increase from growth, the impact of the different load scenarios on the fiber reorientation was negligible. Since volume change, fiber reorientation, and axial stresses show no significant difference for the simulation results of case A and case B, the focus is set on the distribution of the circumferential stresses.

For anisotropic growth without active response, a nearly uniform distribution of circumferential stresses can be recognized for all load scenarios. This suggests that the anisotropic growth of an arterial ring is only slightly dependent on the amount of load as long as only the passive material is considered. As can be seen in the graphs for case A and B, the inclusion of the active material response into the simulations disturbs the uniform stress distribution in the circumferential direction.

In this context, note that the growth parameters have not been optimized for a combination of passive and active material yet and were just taken from the optimization for the scenario without active response.

Hence, the residual stresses, which result from anisotropic growth with this parameter set, seem to be disadvantageous when the artery contracts due to SMCs. An adapted parameter set will lead to a smoother stress distribution over the entire wall, which is left for future investigations. A comparison of the stresses in case A and B supports this statement. The graph of the circumferential stresses exhibits a significantly higher slope in case A than in case B. This means that the inclusion of the active material response during the adaptation of growth and fiber reorientation improves the results already notably, even without an adjustment of the growth parameters.

These observations yield the following conclusion.

The active material response has to be considered during the adaptation process of growth and fiber reorientation to obtain reliable results. Optimization of the growth parameters and a matching time scale of all three processes are suggested to enable a realistic, more uniform stress distribution.

4 Conclusion

In the present work, three crucial aspects of the numerical simulation of arterial walls were combined and discussed: growth, fiber reorientation, and active response. By modeling adaptation processes in arterial walls, the approximation of residual stresses and the mechanically founded prediction of fiber orientations were enabled. The inclusion of active contraction of smooth muscle cells is indispensable since it constitutes a relevant part of the mechanics of arterial walls. A combination of growth, fiber reorientation, and active response into a single framework allowed the first study on interactive effects.

For the modeling of growth and fiber reorientation, a stress-driven approach based on the multiplicative decomposition of the deformation gradient was pursued. A generalized formulation of the growth tensor and use of different driving force functions permitted the analysis and comparison of various growth mechanisms. Three exemplary model variants were investigated in a numerical example on rotationally symmetric, two-layered arterial rings subjected to axial strain and intravascular pressure. Due to the physiological level of the blood pressure, the stresses in the adventitia were considered negligible compared to the media, such that only the effects of growth and remodeling on the stress distributions in the media have been analyzed. The results indicated that in the considered arterial segment, neither isotropic nor purely radial growth could be motivated from the mechanical point of view. Although the growth parameters have been optimized with the aim of an effective reduction of stress peaks and gradients, isotropic growth was found to only have a small effect on the distribution of the circumferential stress, and radial growth led to a significant increase of the axial stress. The latter was associated with a reorientation of the fibers towards the axial direction, which conflicts with experimental observations. The results obtained for a more general, anisotropic type of growth turned out to be more promising: even though the growth-induced volume increase was clearly lower than for isotropic growth, nearly uniform distributions of the circumferential and axial stresses in the media could be realized. By applying the considered anisotropic growth to a three-dimensional, patient-specific artery, it could be shown that similar effects can be observed in more complex boundary value problems.

This underlines that anisotropic growth and fiber reorientation led to an improved mechanical state of the artery. However, it should be discussed, if such an improved state can be expected from an in vivo artery. From a biological point of view, it is conceivable that the adaptation of the artery stops as soon as a mechanically stable state of the arterial wall is achieved.

In a second step, an active material model was proposed using a mechanochemical approach. It considered the mechanosensitive influx of calcium ions as the most important trigger for the contraction of SMCs. Simulation results for a rotationally symmetric ring of the media showed that the active material response had a stabilizing effect on the artery, but residual stresses were mandatory to obtain reasonable stress distributions. In addition, smooth muscle contraction was investigated under the aspect of growth and fiber reorientation. For this purpose, only anisotropic growth was considered by applying those growth parameters that had been optimized for the passive material response. Additionally, three different values of intravascular pressure were used to study the dependency of all three processes on the load. The simulations showed qualitatively the same results under different load scenarios. The final fiber distribution and the volume change of the arterial wall were nearly unaffected by the active material response. The most significant changes were observed in the stress distribution. The combination of smooth muscle contraction in the circumferential direction and the incompressibility of the tissue enforced negative axial stresses. To avoid negative stress in axial direction, a higher axial strain in the boundary conditions of the simulations could be considered. Furthermore, the final stress distribution in circumferential direction was less smooth when an active response was added. However, the graph for circumferential stresses showed a lower slope when the active response was already included before the adaptation by growth and fiber reorientation is initiated. This suggests that further adjustments of the simulation setup could lead to a nearly uniform stress distribution, as it was already shown for anisotropic growth under passive material properties.

Then, as a first adjustment, the growth parameters would have to be adopted to obtain an optimal distribution of residual stresses for the combination of active and passive material response. Altogether, it can be outlined that the active response shows a significant impact on the process of growth and fiber reorientation. Therefore, further improvements at a single model part should be analyzed in the context of all aspects of the arterial wall to indeed enable more reliable patient-specific simulations in the future.

Acknowledgements Funding from Deutsche Forschungsgemeinschaft (DFG, German Research Foundation), project ID BA2823/13-2, is greatly appreciated.

References

Akyildiz, A.C., Chai, C.K., Oomens, C.W.J., van der Lugt, A., Baaijens, F.P.T., Strijkers, G.J., Gijsen, F.J.H.: 3D fiber orientation in atherosclerotic carotid plaques. J. Struct. Biol. **200**, 28–35 (2017)

Alastrué, V., Peña, E., Martínez, M.A., Doblaré, M.: Assessing the use of the 'opening angle method' to enforce residual stresses in patient-specific arteries. Ann. Biomed. Eng. **35**, 1821–1837 (2007)

Amberg, G.C., Navedo, M.F.: Calcium dynamics in vascular smooth muscle. Microcirculation **20**, 281–289 (2013)

Ambrosi, D., Ateshian, G.A., Arruda, E.M., Cowin, S.C., Dumais, J., Goriely, A., Holzapfel, G.A., Humphrey, J.D., Kemkemer, R., Kuhl, E., Olberding, J.E., Taber, L.A., Garikipati, K.: Perspectives on biological growth and remodeling. J. Mech. Phys. Solids **59**, 863–883 (2011)

Balzani, D., Neff, P., Schröder, J., Holzapfel, G.A.: A polyconvex framework for soft biological tissues. Adjustment to experimental data. Int. J. Solids Struct.**43**, 6052–6070 (2006)

Balzani, D., Schröder, J., Gross, D.: Numerical simulation of residual stresses in arterial walls. Comput. Mater. Sci. **39**, 117–123 (2007)

Balzani, D., Böse, D., Brands, D., Erbel, R., Klawonn, A., Rheinbach, O., Schröder, J.: Parallel simulation of patient-specific atherosclerotic arteries for the enhancement of intravascular ultrasound diagnostics. Eng. Comput. **29**, 888–906 (2012)

Böl, M., Schmitz, A., Nowak, G., Siebert, T.: A three-dimensionalchemo-mechanical continuum model for smooth muscle contraction. J. Mech. Behav. Biomed. Mater. **13**, 215–229 (2012)

Braeu, F.A., Seitz, A., Aydin, R.C., Cyron, C.J.: Homogenized constrained mixture models for anisotropic volumetric growth and remodeling. Biomech. Model. Mechanobiol. **16**, 889–906 (2017)

Brands, D., Klawonn, A., Rheinbach, O., Schröder, J.: Modelling and convergence in arterial wall simulations using a parallel FETI solution strategy. Comput. Methods Biomech. Biomed. Eng. **11**, 569–583 (2008)

Chitano, P., Wang, L., Tin, G.Y.Y., Ikeba, M., Paré, P.D., Seow, C.Y.: Smooth muscle function and myosin polymerization. J. Cell Sci. **130**, 2468–2480 (2017)

Choung, C.J., Fung, Y.C.: On residual stresses in arteries. J. Biomech. **108**, 189–192 (1986)

Cole, W.C., Welsh, D.G.: Role of myosin light chain kinase and myosin light chain phosphatase in the resistance arterial myogenic response to intravascular pressure. Arch. Biochem. Biophys. **510**, 160–173 (2011)

Cyron, C.J., Humphrey, J.D.: Growth and remodeling of load-bearing biological soft tissues. Meccanica **52**, 645–664 (2017)

Fausten, S., Balzani, D., Schröder, J.: An algorithmic scheme for the automated calculation of fiber orientations in arterial walls. Comput. Mech. **58**, 861–878 (2016)

Fung, Y.C.: Mathematical representation of the mechanical properties of the heart muscle. J. Biomech. **3**, 381–404 (1970)

Fung, Y.C., Liu, S.Q.: Change of residual strains in arteries due to hypertrophy caused by aortic constriction. Circ. Res. **65**, 1340–1349 (1989)

Hai, C.M., Murphy, R.A.: Cross-bridge phosphorylation and regulation of latch state in smooth muscle. J. Appl. Physiol. **254**, C99-106 (1988)
Hariton, I., deBotton, G., Gasser, T.C., Holzapfel, G.A.: Stress-driven collagen fiber remodeling in arterial walls. Biomech. Model. Mechanobiol. **6**, 163–175 (2007)
Hill, A.V.: The heat of shortening and the dynamic constants of muscle. Proc. R. Soc. Lond. B **126**, 136–195 (1938)
Himpel, G., Kuhl, E., Menzel, A., Steinmann, P.: Computational modelling of isotropic multiplicative growth. Comput. Model. Eng. Sci. **8**, 119–134 (2005)
Holzapfel, G.A., Gasser, T.C., Ogden, R.W.: A new constitutive framework for arterial wall mechanics and a comparative study of material models. J. Elast. **61**, 1–48 (2000)
Holzapfel, G.A., Sommer, G., Auer, M., Regitnig, P., Ogden, R.W.: Layer-specific 3D residual deformations of human aortas with non-atherosclerotic intimal thickening. Ann. Biomed. Eng. **35**, 530–545 (2007)
Humphrey, J.D.: Cardiovascular Solid Mechanics. Cells, Tissues, and Organs. Springer, New York (2002)
Humphrey, J.D., Rajagopal, K.R.: A constrained mixture model for growth and remodeling of soft tissues. Math. Models Methods Appl. Sci. **12**, 407–430 (2002)
Jackson, Z.S., Gotlieb, A.I., Langille, B.L.: Wall tissue remodeling regulates longitudinal tension in arteries. Circ. Res. **90**, 918–925 (2002)
Joldes, G.R., Noble, C., Polzer, S., Taylor, Z.A., Wittek, A., Miller, K.: A simple method of incorporating the effect of the uniform stress hypothesis in arterial wall stress computations. Acta Bioeng. Biomech. **20**, 59–67 (2018)
Langille, B.L.: Remodeling of developing and mature arteries: endothelium, smooth muscle, and matrix. J. Cardiovasc. Pharmacol. **21**, S11–S17 (1993)
Liu, H., Zhang, M., Liu, M., Martin, C., Cai, Z., Sun, W.: Finite element simulation of three dimensional residual stress in the aortic wall using an anisotropic tissue growth model. J. Mech. Behav. Biomed. Mater. **92**, 188–196 (2019)
Lubarda, V.A., Hoger, A.: On the mechanics of solids with a growing mass. Int. J. Solids Struct. **39**, 4627–4664 (2002)
Matsumoto, T., Hayashi, K.: Response of arterial wall to hypertension and residual stress. In: Hayashi, K., Kamiya, A., Ono, K. (eds.) Biomechanics, pp. 93–119. Springer (1996)
Murtada, S., Kroon, M., Holzapfel, G.A.: A calcium-driven mechanochemical model for prediction of force generation in smooth muscle. Biomech. Model. Mechanobiol. **9**, 749–762 (2010)
Niestrawska, J.A., Viertler, C., Regitnig, P., Cohnert, T.U., Sommer, G., Holzapfel, G.A.: Microstructure and mechanics of healthy and aneurysmatic abdominal aortas: experimental analysis and modeling. J. R. Soc. Interface **13**, 20160620 (2016)
Polzer, S., Bursa, J., Gasser, T.C., Staffa, R., Vlachovsky, R.: A numerical implementation to predict residual strains from the homogeneous stress hypothesis with application to abdominal aortic aneurysms. Ann. Biomed. Eng. **41**, 1516–1527 (2013)
Ren, J.S.: Growth and residual stresses of arterial walls. J. Theor. Biol. **337**, 80–88 (2013)
Rodbard, S.: Vascular caliber. Cardiology **60**, 4–49 (1975)
Rodriguez, E.K., Hoger, A., McCulloch, A.D.: Stress-dependent finite growth in soft elastic tissues. J. Biomech. **27**, 455–467 (1994)
Sáez, P., Peña, E., Doblaré, M., Martínez, M.A., Kuhl, E.: Computational modeling of hypertensive growth in the human carotid artery. Comput. Mech. **53**, 1183–1196 (2014)
Schneider, H.: Ligandenunabhängige Aktivierung heptahelikaler Transmembranrezeptoren. Ph.D. thesis, Ludwig-Maximilians-Universität München (2013)
Schriefl, A.J., Zeindlinger, G., Pierce, D.M., Regitnig, P., Holzapfel, G.A.: Determination of the layer-specific distributed collagen fiber orientations in human thoracic and abdominal aortas and common iliac arteries. J. R. Soc. Interface **9**, 1275–1286 (2012)
Schröder, J., von Hoegen, M.: An engineering tool to estimate eigenstresses in three-dimensional patient-specific arteries. Comput. Methods Appl. Mech. Eng. **306**, 364–381 (2016)

Seydewitz, R., Menzel, R., Sieber, T., Böl, M.: Three-dimensional mechano-electrochemical model for smooth muscle contraction of the urinary bladder. J. Mech. Behav. Biomed. Mater. **75**, 128–146 (2017)

Sharifimajd, B., Thore, C.J., Stålhand, J.: Simulating uterine contraction by using an electro-chemo-mechanical model. Biomech. Model. Mechanobiol. **15**, 497–510

Stålhand, J., Klarbring, A., Holzapfel, G.A.: A mechanochemical 3D continuum model for smooth muscle contraction under finite strains. J. Theor. Biol. **268**, 120–130 (2011)

Takamizawa, K., Hayashi, K.: Strain energy density function and uniform strain hypothesis for arterial mechanics. J. Biomech. **20**, 7–17 (1987)

Tanaka, M., Fujikawa, M., Balzani, D., Schröder, J.: Robust numerical calculation of tangent moduli at finite strains based on complex-step derivative approximation and its application to localization analysis. Comput. Methods Appl. Mech. Eng. **269**, 454–470 (2014)

Tanaka, M., Sasagawa, T., Omote, R., Fujikawa, M., Balzani, D., Schröder, J.: A highly accurate 1st- and 2nd-order differentiation scheme for hyperelastic material models based on hyper-dual numbers. Comput. Methods Appl. Mech. Eng. **283**, 22–45 (2015)

Tykocki, N.R., Boerman, E.M., Jackson, W.F.: Smooth muscle ion channels and regulation of vascular tone in resistance arteries and arterioles. Compr. Physiol. **7**, 485–581 (2017)

Vaishnav, R.N., Vossoughi, J.: Residual stress and strain in aortic segments. J. Biomech. **20**, 235–239 (1987)

Valentín, A., Holzapfel, G.A.: Constrained mixture models as tools for testing competing hypotheses in arterial biomechanics: a brief survey. Mech. Res. Commun. **42**, 126–133 (2012)

Wray, S.: Calcium signaling in smooth muscle. In: Bradshaw, R.A., Dennis, E.A. (eds.) Handbook of Cell Signaling, 2nd edn, chap. 127, pp. 1009–1025. Elsevier Inc. (2010)

Yang, J., Clark, J.W., Jr., Bryan, R.M., Robertson, C.: The myogenic response in isolated rat cerebrovascular arteries: smooth muscle cell model. Med. Eng. Phys. **25**, 691–709 (2003a)

Yang, J., Clark, J.W., Jr., Bryan, R.M., Robertson, C.S.: The myogenic response in isolated rat cerebrovascular arteries: vessel model. Med. Eng. Phys. **25**, 711–717 (2003b)

Zahn, A., Balzani, D.: A combined growth and remodeling framework for the approximation of residual stresses in arterial walls. Z. Angew. Math. Mech. **98**, 2072–2100 (2018)

Zahn, A., Balzani, D.: Optimization-based comparison of different anisotropic models for growth and remodeling in arterial walls. In: Gleim, T., Lange, S. (eds.) Proceedings of the 8th GACM Colloquium on Computational Mechanics for Young Scientists from Academia and Industry, pp. 405–408. kassel university press GmbH (2019)

Biomechanics of Cardiac Tissues and Various Organs

Advances in Experimental and Computational Biomechanics of the Tricuspid Heart Valve

Chung-Hao Lee, Devin W. Laurence, Colton J. Ross, and Luke T. Hudson

> *We first met at the 8th World Congress of Biomechanics in 2018. During the conference, my postdoctoral fellow, Dr. Anju Babu (who is Dr. Holzapfel's former advisee), presented our pioneering study on the layer-specific tricuspid valve tissue properties that was later well received by our community. Since then, I have been working closely with Dr. Holzapfel in a series of journal publications, including the developments of novel opto-mechanical instrument for quantifying load-dependent collagen fiber microstructures of collagenous tissues and the first tricuspid valve* in silico *model under the isogeometric analysis (IGA) framework for modeling healthy and diseased valve functions. I am fortunate to have such a wonderful mentor like Dr. Holzapfel in my early faculty career, and would like to express my gratitude to him for providing continuous support and advices that help my establishment of a strong and unique research program. Congratulations to Dr. Holzapfel on his 60th birthday!*
>
> *Chung-Hao*

Abstract The tricuspid valve regulates the blood flow between the right atrium and the right ventricle. Although it has long been deemed the *forgotten* valve, the interest in the tricuspid valve has significantly increased over the past five years. This is largely due to the emerging clinical awareness of the importance of tricuspid regurgitation, a valvular disease associated with blood back flow into the right atrium during systole. Many recent advancements have expanded the current understanding of tricuspid valve tissue biomechanics. However, many questions remain to be answered before long-term, durable therapeutics for tricuspid regurgitation can be provided to

C.-H. Lee (✉)
Biomechanics and Biomaterials Design Laboratory (BBDL), School of Aerospace and Mechanical Engineering, University of Oklahoma, 865 Asp Ave., Felgar Hall 219C, Norman, OK 73019, USA
e-mail: ch.lee@ou.edu

D. W. Laurence · C. J. Ross · L. T. Hudson
Biomechanics and Biomaterials Design Laboratory (BBDL), School of Aerospace and Mechanical Engineering, University of Oklahoma, 865 Asp Ave., Norman, OK 73019, USA

G. Sommer et al. (eds.), *Solid (Bio)mechanics: Challenges of the Next Decade*,
Studies in Mechanobiology, Tissue Engineering and Biomaterials 24,
https://doi.org/10.1007/978-3-030-92339-6_9

address several clinical challenges. This book chapter summarizes the recent investigations of tissue mechanics, collagen fiber architecture, and morphology for the subvalvular components of the tricuspid valve, and provides recommendation for future developments in this rising area of cardiovascular biomechanics.

1 Introduction

The tricuspid valve (TV) regulates the blood flow within the right side of the heart. Functionally, the TV is analogous to a check valve, where the TV allows blood flow from the atrium to the ventricle during diastole, and prevents undesired retrograde blood flow into the atrium during systole. This cyclic opening and closing of the TV is facilitated by the four primary subvalvular components—the three TV *leaflets* that seal the atrial-ventricular orifice at systole, the rigid ring-like TV *annulus* that affixes the TV leaflets to the surrounding heart chambers, and the array of string-like *chordae tendineae* that pin the TV leaflets to the *papillary muscles* on the right ventricle wall. Any alterations to the geometry and function of these subvalvular components may cause retrograde blood flow, which is commonly known as tricuspid regurgitation (TR). TR is a complex valvular heart disease that can be classified based on the underlying etiology: (i) *primary* TR occurring when the regurgitation is due to structural alterations of the TV leaflets (e.g., Ebstein's anomaly or chordae rupture-induced tricuspid leaflet flail) (Anwar et al. 2018); (ii) *secondary* TR arising from alterations to the surrounding heart anatomy caused by different etiologies (e.g., pulmonary hypertension, papillary muscle displacement, and annular dilation) (Badano et al. 2013; Nishimura et al. 2014; Sun and O'Gara 2017).

Clinical research suggests that secondary TR accounts for approximately 70–80% of all TR cases (Dreyfus et al. 2005). Likewise, much of the clinical effort has been focused on establishing reliable surgical repair techniques to restore the diseased TV to its normal condition. There are two prominent techniques for TV surgical repair, including ring annuloplasty (Carpentier 1971; Filsoufi et al. 2006) and De Vega suture valvuloplasty (Antunes and Girdwood 1983; De Vega et al. 1972, 1973), both with a common goal of restoring the annulus dimension and geometry to address TR caused by the annular dilation. These techniques are generally successful immediately after the TV repair, but the recurrence of TR has been disappointing, i.e., a recurrence rate approaching 16–18% at 5–8 years following the initial surgery (Huang et al. 2014; Pozzoli et al. 2016). This undesired recurrence of TR suggests that other factors, besides annular dilation, warrant more clinical attention. Although there is an apparent need for improving the long-term outcomes of the current therapeutics, the TV has been historically considered as the '*forgotten valve*'. This could be attributed to the common sentiment among cardiologists and cardiac surgeons that TR would naturally regress after correcting the concomitant left-sided lesion (Braunwald et al. 1967). However, Dreyfus et al. (2005) and Anyanwu and Adams (2010) suggested that the lack of clinical actions addressing TR may be inadequate, and that the TR, *left untreated*, will continue to worsen the prognosis and significantly affect the quality of life of the patient. These new fundamental clinical viewpoints sparked interests in the TV for the clinical community (Badhwar et al. 2017; Chikwe et al. 2015; Desai et al. 2013) and, more recently, the soft tissue biomechanics community (Jett et al. 2018; Khoiy and Amini 2016; Lee et al. 2019; Meador et al. 2020).

This chapter summarizes the recent advances in experimental and computational biomechanics of the TV that have improved our fundamental understanding of the TV and its subvalvular components. We first outline the major questions and challenges for the TV in Sect. 2. Then, the relevant experimental characterizations of the TV leaflets and chordae tendineae are described in Sects. 3 and 4, respectively. The use of the experimental data acquired from these experimental characterizations in constitutive models under the hyperelasticity framework is detailed in Sect. 5, followed by the state-of-the-art *in silico* simulations of the TV in Sect. 6. Finally, we provide our concluding statements as well as our opinion on the future perspectives of TV biomechanics research in Sect. 7.

2 Major Challenges and Questions

From 1990 to 2015, the mitral valve (MV) in the left side of the heart was the primary focus of the two atrioventricular heart valves. Much of our current understanding stems from these seminal studies, such as the mechanical properties (Grashow et al. 2006a, b; May-Newman and Yin 1995; Pham and Sun 2014), microstructural morphology (Kunzelman et al. 1993; Sacks and Yoganathan 2007), collagen fiber architecture (CFA) (Lee et al. 2013, 2015; Liao et al. 2007) or valvular interstitial cell-regulated biosynthetic response (Ayoub et al. 2017; Pierlot et al. 2014; Rego et al. 2016). It is logical to expect the TV to display similar properties as the MV, because they share the same in-vivo roles of regulating the blood flow between the atria and the ventricles. However, there are several inherent differences in their respective in-vivo environments, including the significantly higher systolic transvalvular pressure for the MV (i.e., 120 mmHg) compared to the TV (i.e., 25 mmHg), and the different numbers of valve leaflets (two for the MV versus three for the TV). It is, thus, likely that the properties of the TV's subvalvular components differ from the previous observations for the MV counterparts, leading to the following three motivating questions. Firstly, *what are the mechanical properties of the TV's subvalvular components*? The mechanical properties of the TV may be more compliant due to the less-demanding in vivo transvalvular pressures, yet anatomical differences, such as the different numbers of leaflets, may obscure this. Secondly, *what is the microstructural morphology of the TV's subvalvular components*? Differences (or similarities) in the mechanical properties between the two atrioventricular heart valves may be attributed to the underlying microstructures and tissue morphology. Thirdly, *are the valvular interstitial cell-regulated biosynthetic responses altered due to lower transvalvular pressure for the TV*? A previous study highlighted the lower stiffness of the TV valvular interstitial cells (VICs) that, when combined with the reduced in vivo pressure, may have notable implications for the cellular biosynthetic response to pathological loading. Answering these questions will provide a better fundamental understanding of the function and physiopathology of the TV that will enable the improvements of the existing clinical remedies for TR.

Currently, there are several clinical remedies to address the complex nature of TR. One of the most prominent devices is the annuloplasty ring, which seeks to restrict the annulus size (i.e., circumference), restore the leaflet coaptation, and prevent further regurgitation. Despite the initial success of the annuloplasty ring, there is a 16–18% recurrence of the TR within 5–8 years of the initial repair (Huang et al. 2014; Pozzoli et al. 2016). This leads to several questions to ideally improve the long-term treatment outcomes. Firstly, *how do the individual valvular pathologies alter the TV function and mechanical environment*? Since the current gold standard for TR only addresses the annular dilation, and there is a long-term recurrence of TR, it is possible that other valvular pathology factors should be also addressed. Secondly, *what is the ideal therapeutic option, or combination of therapeutic methods, that best address TR for a given patient's pathological conditions*? Several factors may obscure this decision, such as the patient's unique TV geometry and the severity of the TR. Also, it is uncertain what the target is for a successful TR repair—should it be a certain reduction in the annulus size, the amount of leaflet coaptation, and/or the homeostatic (or healthy) leaflet stresses and strain? *In silico* modeling has emerged as a platform that could be used to address these questions; however, the aforementioned fundamental questions must be carefully answered before the *in silico* simulations can be confidently and reliably used in these clinically-relevant applications.

3 The Tricuspid Valve Leaflets

The three TV leaflets, namely the anterior leaflet (TVAL), the posterior leaflet (TVPL), and the septal leaflet (TVSL), function collectively to seal the orifice between the right atrium and the right ventricle during systole (Fig. 1a). These leaflets are attached to the surrounding heart tissue through the ring-like TV annulus and a series of string-like chordae tendineae that insert into the ventricular surface (Fig. 1b). From the microstructural point of view, these TV leaflet tissues are composed primarily of collagen fibers, elastin, glycosaminoglycans (GAGs) and proteoglycans (PGs), and a population of smooth muscle-like VICs. Morphologically, these constituents are arranged into four distinct layers, described from the atrial to the ventricular surfaces: (i) the *atrialis* (A), composed of circumferentially-oriented elastin; (ii) the *spongiosa* (S), containing non-fibrous GAGs and PGs; (iii) the *fibrosa* (F), a collagen-rich layer; and (iv) the *ventricularis* (V), which is composed primarily of radially-oriented elastin fibers (Kramer et al. 2019; Meador et al. 2020), see Fig. 1c. In addition to these four morphologically-distinct tissue layers, the VICs are distributed throughout the TV tissue and are responsible for the bionsynthetic regulation of the tissue's extracellular matrix. This section summarizes several recent experimental investigations from our laboratory, which aid in advancing our understanding of the TV leaflets' mechanical, collagen fiber architectural, and morphological properties.

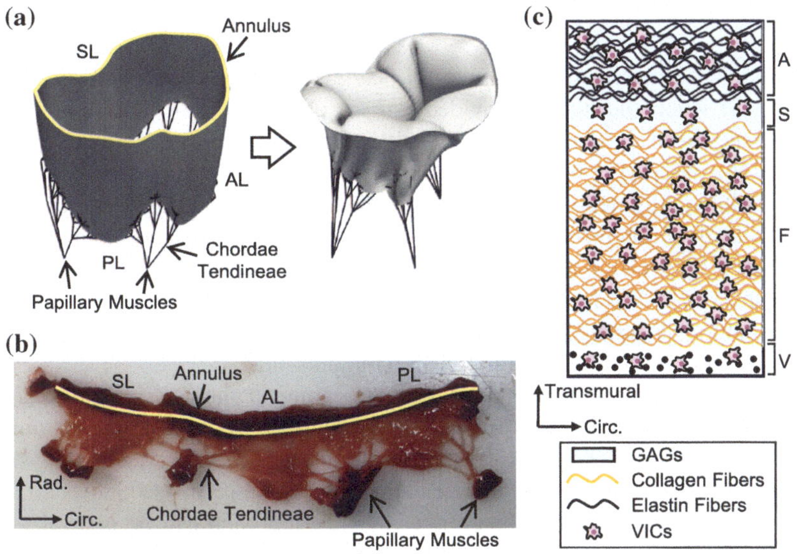

Fig. 1 (a) Schematic of the subvalvular components of the tricuspid heart valve at diastole (*left*) and at systole (*right*) adopted from the finite element simulation of Laurence et al. (2020); (b) a representative excised porcine TV showing the three TV leaflets, the TV annulus, the chordae tendineae, and papillary muscles; (c) illustration of the four distinct tissue layers and VICs of the TV leaflets. (AL = anterior leaflet; PL = posterior leaflet; SL = septal leaflet, rad. = radial; circ. = circumferential; A = atrialis; S = spongiosa; F = fibrosa; V = ventricularis; GAGs = glycosaminoglycans)

3.1 Mechanical Properties of the TV Leaflets

Two approaches have been employed to characterize the mechanical behavior of the TV leaflets. The first is planar biaxial mechanical testing in which orthogonal tensile loads are applied to the tissue specimen using sutures or thin tines (Sacks 2000). Our laboratory has extensively used this experimental method to characterize the planar biaxial mechanical responses of the TV leaflet tissues, which provide useful information for developing TV-specific constitutive models under the hyperelasticity framework (see Sect. 5). One inherent limitation of this testing method is the lack of mimicking the in-vivo loading environment for the tissue. With this in mind, the second characterization approach emulates the in-vivo loading environment via an in-vivo animal model (Sacks et al. 2006) or an in-vitro testing apparatus (Rabbah et al. 2013). This approach provides richer information of the in-vivo functioning environment of the TV. The following focuses primarily on the recent studies from our laboratory on the TV leaflet tissue biomechanics, and briefly reviews the studies that preserved the entire TV apparatus within an in-vitro or in-vivo biomechanical characterization pipeline.

3.1.1 Planar Biaxial Mechanical Characterizations

Mechanical characterizations of the TV leaflets were first provided by Heyden et al. (2015), where the authors provided a survey of the mechanical properties for the leaflet tissues from all four cardiac heart valves. Shortly thereafter, Khoiy et al. (2016) provided the pioneering biaxial mechanical characterization of porcine TV leaflets. Their results highlighted the nonlinear and anisotropic mechanical responses for all three TV leaflets. Interestingly, each leaflet had its distinct mechanical behavior, as later echoed by Meador et al. (2020) and Pham et al. (2017) for ovine and human TVs, respectively. TV leaflet-specific mechanical responses stem from the unique microstructural differences, yet the comprehensive understanding of the intricate mechanics-microstructure interrelationship remains elusive.

Recently, our laboratory expanded these fundamental characterization studies by developing a comprehensive experimental procedure to systematically examine the mechanical behaviors of the TV leaflets in response to varied biaxial loading conditions, as well as the tissue's time-dependent mechanical response (Jett et al. 2018; Ross et al. 2019b). Our investigations using this approach can be generally classified into *bulk-tissue* and *microstructurally-focused* characterizations. The former includes characterizations of the loading-rate effect, the species differences, and spatial variations in the characterized tissue's mechanical properties, while the latter seeks to understand how the microstructural constituents, such as the GAGs or collagen fibers, contribute to the TV leaflet tissue's biaxial mechanical properties. Collectively, these recent results of our laboratory have provided an extensive understanding of the mechanical behavior of the three TV leaflets. The comprehensive biomechanical database we established is useful for informing the development of constitutive models in predictive computational modeling of the TV in vivo function (see Sects. 5 and 6). Our experimental methods and the key findings of these studies are further elaborated as follows.

Bulk-Tissue Characterizations

Our laboratory first aimed to establish a comprehensive biaxial mechanical characterization framework and explore how the strain rate, testing temperature, species, and age influence the observed mechanical behavior (Jett et al. 2018; Ross et al. 2019b). Like the previous studies (Heyden et al. 2015; Khoiy and Amini 2016; Pham et al. 2017), we found that the three TV leaflets had distinct, nonlinear and anisotropic mechanical responses (Fig. 2a, b). Interestingly, the adult ovine specimens were more compliant than their juvenile counterparts, but stiffer than the adult porcine specimens, indicating species and age are important factors for in-vitro biomechanical characterizations. Moreover, the mechanical responses for all three TV leaflets became stiffer with an increased loading rate, with the differences found to be statistically insignificant. This observation agrees with the typical viscoelastic theory for soft tissues (Fung 2013). Finally, the peak radial stretches became larger as the temperature increased; however, no obvious microstructural connection was made.

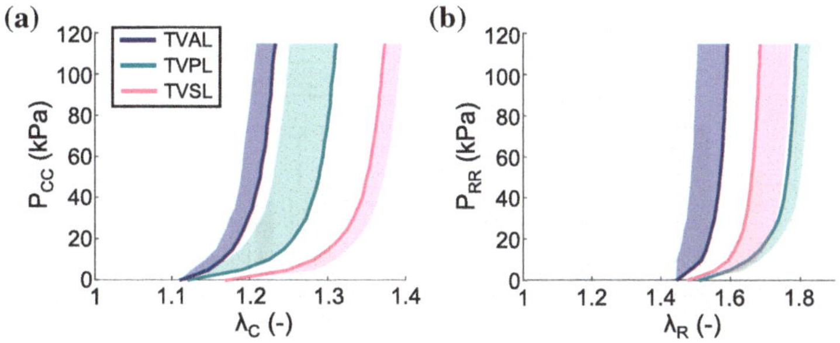

Fig. 2 Characterized biaxial mechanical behaviors of the three TV leaflets ($n = 6$): stress-stretch data in the tissue's (a) circumferential, and (b) radial directions following preconditioning. Results are presented as mean ± SEM

Our study provided a fundamental overview of the TV leaflet mechanical responses and explored how several experimental factors influence the observed mechanical behaviors.

The general consensus within the heart valve biomechanics community is to use the central region of the tissue for biaxial mechanical characterizations (Grashow et al. 2006b; Khoiy and Amini 2016; Pham and Sun 2014). This experimental setup is chosen to avoid chordae insertions that may affect the characterized mechanical properties of the TV leaflet tissues. However, it relies on an assumption that the mechanical properties of the central region accurately represent the entire TV leaflet specimen. Although this assumption may be reasonable for certain applications, no studies have quantified the regionally-varying mechanical properties to evaluate the efficacy of this assumption. In Laurence et al. (2019), our laboratory used a modified experimental technique to characterize the mechanical properties for six smaller regions of the porcine TVAL. From this study, we, *for the first time*, found that the central regions (i.e., Regions B and E) are more anisotropic compared to the edge regions (i.e., Regions A, C, D, and F). This observation was supported by the quantified *anisotropy index* (AI, defined as the ratio of the peak radial stretch to the peak circumferential stretch) (Fig. 3a). Interestingly, we also found that all the six tissue regions appeared to exhibit similar deformations after the specimen was preconditioned (Fig. 3b). Based on Dr. Y. C. Fung's seminal quasi-linear viscoelasticity theory for soft tissues (Fung 2013), this post-preconditioning tissue deformation can be viewed as the tissue's in-vivo functioning deformation, and our novel findings imply the assumption of *homogeneous mechanical properties* commonly adopted in heart valve computational modeling is valid, considering the post-preconditioning state as the reference configuration (i.e., the stress-free configuration). These observations of our work underscore the importance of the chosen stress-free reference configuration (Rausch and Kuhl 2013), and partially verify the previous assumptions that the heart valve leaflet's mechanical properties are *spatially homogeneous* (Amini et al. 2012; Prot and Skallerud 2009; Prot et al. 2010; Sacks et al. 2006).

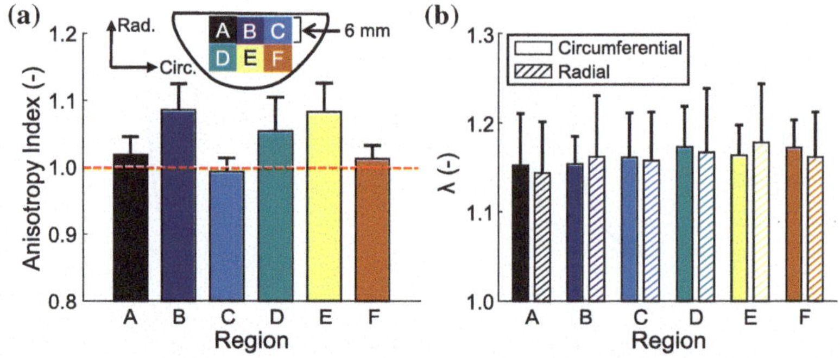

Fig. 3 Quantified regional variations in the TVAL's biaxial mechanical properties: (a) anisotropy index for the six leaflet regions; (b) the circumferential (i.e., solid bars) and radial (i.e., patterned bars) tissue deformations for the six regions. Values are in mean ± SEM. Sample size: Region A ($n = 13$), Region B ($n = 13$), Region C ($n = 13$), Region D ($n = 12$), Region E ($n = 13$), Region F ($n = 10$)

The tissues used in our mechanical characterizations are typically stored in a −14 °C freezer between dissection and testing. In Duginski et al. (2020), our laboratory sought to understand how this temporary freezer storage may influence the TV leaflet's mechanical responses, since the existing literature for various tissues (e.g., cartilage, skin or aortic tissues) had conflicting conclusions. In this study, the fresh TVAL specimens ($n = 10$) were first biaxially characterized using our typical experimental protocols. The tested tissue was then frozen in a −14°C freezer for 48 h, and was subsequently tested using the same experimental characterization. Despite the relatively short duration, the freezer storage appeared to make the leaflet's mechanical response slightly *stiffer* and *more anisotropic*, although we only found a statistically significant difference in the low-tensile mechanical response of the radial tissue direction. This suggests that the short storing period may slightly alter the TV leaflet mechanical response, but in a non-significant manner. Another recent study by Salinas et al. (2020) showed opposite trends (i.e., more compliant tissue behaviors) following a freezing period of 1–3 days at −80 °C. Thus, it is important to emphasize that the freezing-storage method for the TV leaflets should be carefully selected, and it is valuable to examine the differences in the characterized tissue's mechanical behaviors arising from the selected freezer storage technique.

Microstructurally-Focused Characterizations

Our laboratory further elucidated the contributions of microstructural constituents to the TV leaflet mechanical responses. Ross et al. (2019a) used an iterative experimental technique to investigate the role of the GAGs in the TV leaflet tissue biomechanics, where the untreated (i.e., control) specimen was biaxially characterized, then treated with an enzyme solution consisting of hyaluronidase and chrondroitinase to obtain the

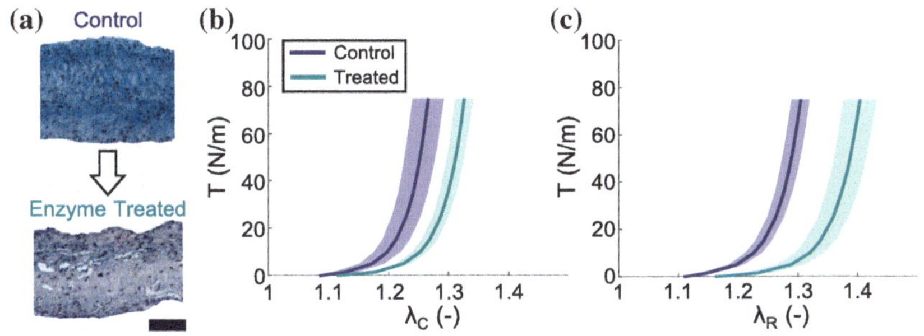

Fig. 4 (a) Comparison of the TV leaflet's microstruture (*top*) before and (*bottom*) after the enzyme degradation. Tension versus stretch responses ($n = 8$), following preconditioning, in the tissue's (b) circumferential, and (c) radial directions before and after the enzyme degradation treatment. Results are presented in mean ± SEM. Scale bar = 100 μm

treated specimen (Fig. 4a), and finally biaxially characterized. This approach enabled us to understand the differences in the mechanical behaviors *by using the same single specimen* to consider the removal of >60% of the GAG content. The tissue's deformations were larger following the enzyme degradation treatment (Fig. 4b, c). Interestingly, we further found that the enzyme degradation-induced discrepancy was primarily attributed to the preconditioning effect. We also conducted stress relaxation tests in the same study, and found a reduced stress relaxation following the removal of the GAGs. This study highlights the intricate contributions of the GAG content to the TV leaflet's biaxial mechanical and viscoelastic properties.

In Kramer et al. (2019), our laboratory used another experimental method to understand the interrelationship between the tissue mechanics and its underlying microstructure of the TV. This approach utilized a micro-dissection procedure to separate the tissue layers of the TVAL and characterize the biaxial mechanical properties of the separated tissue layers. Because the spongiosa layer lacks the structural constituents (such as collagen or elastin fibers) to bear substantial mechanical loading and the ventricularis layer is typically very thin, the TVAL was dissected into two composite layers: namely (i) A/S—the combined atrialis and spongiosa layer; and (ii) F/V—the combined fibrosa and ventricularis layer. These two composite layers were biaxially characterized, and we arrived at two primary observations. Firstly, the tissue deformations with respect to the post-preconditioned configuration were nearly isotropic and much smaller than the anisotropic preconditioning deformations. Secondly, the A/S and F/V layers had similar tissue deformations with respect to the preconditioned configuration, with most of the differences in the observed mechanical response stemming from preconditioning (Fig. 5). The latter observation contradicts an assumption that the fibrosa layer behaves similarly to the intact layer; however, similar investigations for the aortic valve leaflets supported our observation, suggesting that the isotropic behavior of the separated fibrosa is due to the release of

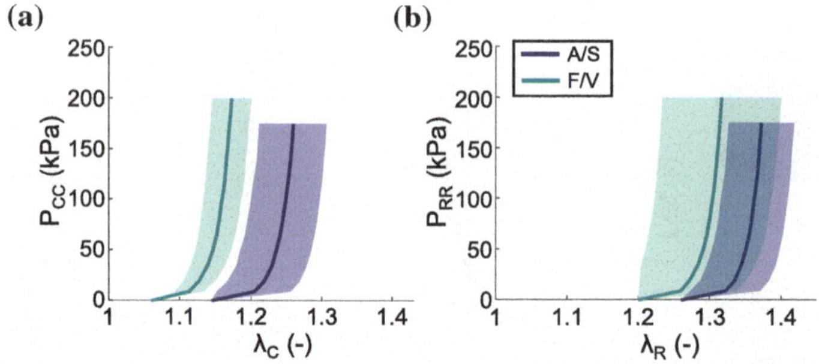

Fig. 5 Biaxial mechanical results for the separated tissue layers of the TVAL ($n = 6$) in (a) the circumferential, and (b) the radial direction, following preconditioning. Values are in mean ± SEM

inter-layer attachments (Stella and Sacks 2007; Vesely and Lozon 1993; Vesely and Noseworthy 1992). Although the observations from Kramer et al. provided an understanding of the layer-specific biomechanical functions of the TV leaflets, this study did not make a direct comparison of the mechanical properties to the intact leaflet (i.e., before micro-dissection). We are currently refining this experimental technique to provide further information for the future developments of microstructure-informed constitutive models for the TV leaflet tissues.

More recently, our laboratory used our iterative enzyme-degradation technique from Ross et al. (2019a) to determine the contributions of the collagen fibers and elastin fibers to the tissue's biomechanical behaviors. The preliminary results from this investigation (Fig. 6) highlight the unique contribution of each microstructural constituent. The differences in the characterized mechanical responses following the collagen removal were most apparent in the circumferential direction. Similar trends were also observed following the removal of elastin fibers, but there was a notable increase in the radial high-tensile modulus. These changes in the mechanical responses may be attributed to the approximate orientation of these structural fibers—the collagen fibers and elastin fibers are aligned near the circumferential and radial directions, respectively—and future *in silico* models with accurate structural-based constitutive models (Rego and Sacks 2017; Zhang et al. 2016) can further corroborate these experimental findings.

3.1.2 In-Vitro and In-Vivo Characterizations

Previous research efforts to characterize the mechanical properties of the TV leaflets in-vitro have followed a distinct methodology: excising the TV apparatus and mounting them to a rigid annulus in a ventricular chamber simulator (Basu and He 2016). Although this technique may give insight into the TV leaflet mechanics and coaptation, utilizing a rigid annulus may not be ideal for mimicking the native valve function, primarily due to the dynamic nature of the cardiac valve annulus and the subsequent

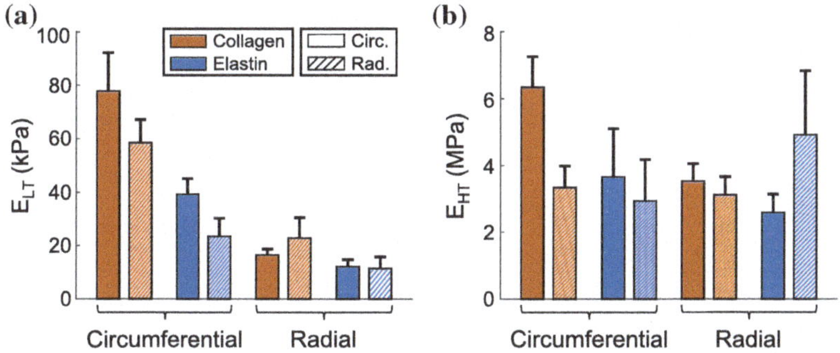

Fig. 6 Comparison of (a) the low-tensile modulus (E_{LT}), and (b) the high-tensile modulus (E_{HT}) before and after the collagenase ($n = 11$) and elastase ($n = 12$) enzyme treatments. Values are in mean ± SEM

alterations in the leaflet strain over several cardiac cycles. In recent years, significant emphasis has been placed on characterizing the in-vitro mechanical properties of the TV leaflets through the utilization of these in-vitro devices.

Khoiy et al. (2016) took the first step towards characterizing the in-vitro mechanical properties of the TV leaflets by quantifying the dynamic deformations of the TV in a beating porcine heart. This was achieved by implementing an *ex vivo* pump apparatus to simulate the right-sided systolic and diastolic pressures exhibited within the cardiac cycle. The results from the use of this system showed that immediately after the coaptation of the TV leaflets, all the leaflet strain values (i.e., circumferential, areal, and radial) increased rapidly and reached their peak values at the maximum right ventricular pressure. Throughout the cardiac cycle, the maximum principal strain remained evenly distributed throughout the leaflet, increasing until reaching the maximum right ventricular pressure, with the strain values as high as 12%. Overall, the strains were shown to be minuscule during diastole and increased during systole. Heterogeneity in the spatial distribution of the circumferential and radial strains was also observed for the TVSL, matching similar trends when the biaxial testing methods are employed. Khoiy et al. (2018a) later expanded this study to investigate the role of chordae tendineae rupture in the disturbance of annular function and leaflet coaptation. When the chordae tendineae ruptured, the flow into the right atrium was found to decrease by 26% and the maximum right ventricular pressure decrease by 17%, highlighting the importance of sub-valvular components in relation to the proper leaflet coaptation and TV function.

Apart from using an intact TV geometry under the in-vitro apparatus, a deeper understanding of TV leaflet function has been underscored by the recent investigations of the in vivo TV leaflet mechanics. Mathur et al. (2019) examined the TV leaflet strains in a beating ovine heart, suturing sonomicrometry crystals to the belly and free-edge portions of each TV leaflet, as well as to the TV annulus. A fiducial marker array was constructed for tracking the deformation of the TV leaflets and for calculating the in vivo leaflet strains. It was found that the strains are qualitatively

Table 1 The mean strain values relative to baseline at end-diastole (ED) and end-systole (ES) for each level of De Vega (DV) cinching (Mathur et al. 2019)

	DV1		DV2		DV3		DV4		DV5	
Strain	ED	ES	ED	ES	ED	ES	ED	ES	ED	ES
Circ. (Mean)	2.45	0.37	1.03	1.12	1.8644	−2.68	2.90	−2.13	2.77	−0.71
Circ. (STD)	5.22	4.764	10.32	10.72	8.88	8.28	9.43	11.72	8.97	18.20
Rad. (Mean)	−3.73	−1.24	−0.94	1.99	−6.61	−8.49	−0.24	−9.80	−2.11	−12.63
Rad. (STD)	13.14	5.51	10.01	14.68	11.12	9.87	21.56	14.21	22.90	16.30
Areal (Mean)	−3.89	−4.86	−2.36	−8.33	−8.15	−19.00	−10.29	−24.53	−11.15	−30.22
Areal (STD)	17.37	16.21	20.99	20.54	20.89	21.98	37.47	32.03	28.36	36.12

different between the belly region and the free edges, with positive strains at systole and little change during diastole in the belly region of each leaflet, whereas the areal and circumferential strains were small at the free edges. Conclusively, no distinct common strain pattern was observed across all the three TV leaflets, although heterogeneity was noted for all strain evaluations (i.e., areal, radial, and circumferential). This study did, however, confirm the anisotropy of the quantified strains within each leaflet, with larger strain in the tissue's radial direction than the circumferential direction. Utilizing the technique, Mathur et al. (2020) later investigated the effects of the De Vega valvuloplasty on the leaflet tissue mechanics when different degrees of cinching were applied. With each degree of cinching, the De Vega technique reduced the range of motion and angular velocity of the TVAL, and, therefore, reduced the areal and radial strains, especially in the belly region of the leaflet. The circumferential strain remained similar throughout each cinching level. These observations concluded that the area strain in the leaflet's belly region was driven by the reduction in radial strains, leading to reduced chordal tethering (Table 1). Furthermore, this mechanical interaction resulted in a reduced range of motion for leaflet coaptation. These investigations are integral for understanding leaflet coaptation and mechanics within the native valve after operation, which will be discussed further in Sect. 7.3.

3.2 Microstructural Quantification for the TV Leaflets

In addition to characterizing the mechanical properties of the TV leaflets, studies on the mesoscopic quantification of the TV leaflet tissues can further unveil the complex underlying microstructure. Specifically, these observations focus primarily on examining the CFA of the TV leaflets, i.e., the reorientation and realignment of the collagen fiber networks in response to the applied mechanical loading (Driessen et al. 2005; Lee et al. 2015; Liao et al. 2007; Meador et al. 2020). Techniques, such as small angle light scattering (SALS) (Pant et al. 2018; Thomas et al. 2019), second-harmonic generation (SHG) (Pokutta-Paskaleva et al. 2019) and two-photon microscopy (Meador et al. 2020), have provided insight to the microstructural components of the TV leaflets. For example, the collagen fibers of the cardiac heart valve leaflets were shown to align predominantly in the circumferential direction in the unloaded state, with varying degrees of fiber dispersion and crimping across different layers of the tissue. In one study, Meador et al. (2020) performed a detailed microstructural observation of the three TV leaflets using both two-photon microscopy and histological quantification. Each microstructural constituent within the TV, including the collagen fibers, elastin, GAGs, and VICs, was found to be heterogeneously distributed with composition differences between the TV leaflets. Although the constituent composition varied from leaflet to leaflet, the CFA was found to be relatively similar, with fibers primarily oriented in the circumferential direction throughout the thickness direction. Depth-discriminated SHG quantification further suggested that the collagen is organized as individual fibers in the atrialis layer with both radial and circumferential orientations, before organizing into higher-order sheets in the fibrosa/ventricularis layers with a distinct circumferential orientation. Although these studies do not quantify the load-dependent changes in the CFA, the microstructural quantification has improved our understanding of collagen organization and composition within the TV leaflets.

In order to bridge this gap in understanding the load-dependent nature of the CFA, our laboratory utilized polarized spatial frequency domain imaging (pSFDI) to quantify the load-dependent changes in the tissue's CFA for the three TV leaflets (Hudson et al. 2020). This technique relies on the polarization-dependent, diattenuating properties of collagen fibers, allowing the quantification of the directional-dependence of the CFA and the degree of optical anisotropy (DOA). Another important aspect of this technique is the pSFDI system's ability to quantify collagen microstructures in a non-destructive manner. In other words, the pSFDI technique does not require the use of optical clearing solutions or fixatives that are known limitations for other collagen imaging modalities, such as SALS (Goth et al. 2016; Plotnikov et al. 2006). The key findings from our study suggest that the TV leaflets display an initial collagen fiber orientation in the circumferential direction at the unloaded state, and that the majority of the crimped fibers rotate towards the radial direction before the applied biaxial tensions reached the asymptotic regime of the stress-strain curve (Fig. 7a, b). Afterwards, the fibers were straightened, exhibiting the high-force regime, which contributed to a better spatial alignment of the collagen fiber networks. More notably, this higher spatial alignment, indicating by the largest changes in the DOA compared to the unloaded state, was found to be the most prominent under radially-dominant loading (i.e., $T_C : T_R = 0.5 : 1$) (Fig. 7b). The non-equibiaxial loading demonstrated a shift of the collagen fibers towards the direction of the dominant applied loading, yielding an increase of DOA following fiber reorientation. Quantifying these collagen microstructural changes for the TV leaflets shed light on the future developments of

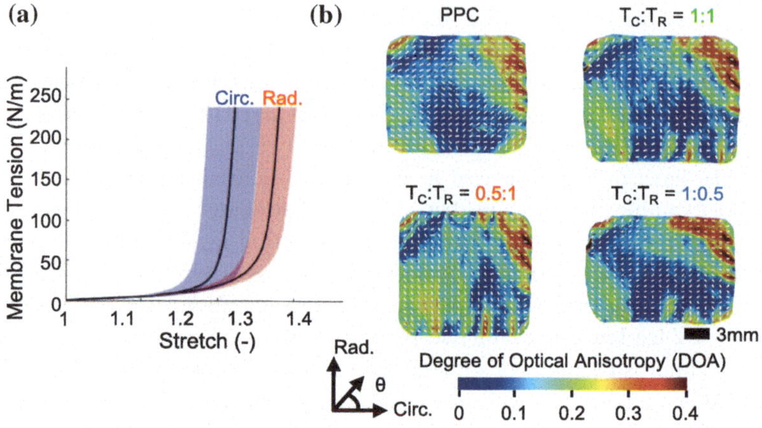

Fig. 7 Quantification of the load-dependent collagen fiber architecture for a representative TVAL specimen using our integrated opto-mechanical instrument: (a) applied force versus tine displacement in the tissue circumferential (blue) and radial (red) directions ($n = 3$); (b) results of collagen fiber orientation and degree of optical anisotropy (DOA). The dash curves represent the predicted fiber orientation angle and the colormaps denote the DOA

microstructure-informed constitutive models, which can be used to better understand complex heart valve diseases.

4 The Tricuspid Valve Chordae Tendineae

The TV chordae tendineae affix the leaflets to the right ventricular walls at the papillary muscle tips, providing supporting forces to the leaflets during systolic closure. The morphology of the TV chordae tendineae have received sparse attention in the previous literature as compared to their MV counterparts, in spite of the noticeable differences between the two atrioventricular heart valves (e.g., anatomy, mechanical loading environment, etc.). Silver et al. (1971) conducted a pioneering study on the TV chordae morphology by analyzing human cadaveric hearts and provided a categorization system that is widely used to date. In their work, they described three primary subsets of the chordae tendineae based on the leaflet insertion locations: (i) the noticeably thicker *strut chordae* that attach to the central, belly region of the TVAL; (ii) the *basal chordae*, affixing at the annulus-leaflet transition (i.e. the base of the leaflet); and (iii) the *marginal chordae*, connecting to the free edge of the leaflets. While the classification system proposed by Silver et al. (1971) provides the basis naming convention for many other studies, researchers have adopted their own categorization schemes due to contentions such as whether or not the strut chordae exist for all the three TV leaflets. For example, some researchers have opted for labeling the chordae based on their cross-sectional area (i.e., small, medium, or large (Smith et al. 2020), or using a more detailed scheme by considering the gross structure, insertion locations, and branching patterns, resulting in over 11 types of TV chordae (Gunnal et al. 2015). While these earlier works provided valuable insight to the anatomical trends in TV chordae morphology, moving forward it is important to adopt a unified and consistent classification/naming convention for the TV chordae tendineae. By using different chordae naming conventions, it becomes difficult to make direct comparisons of the findings between different studies. For example, how do the 'strut chordae in the TVAL' differ from a 'medium chordae in the central region' of the same leaflet?

Additionally, a key aspect of the TV chordae morphology that requires more attention is the comparison between different species. The potential discrepancies in the TV chordae morphology across different animals may affect the translation of research findings from the animal works into the biomechanical understanding of the human chordae. Previous studies have shown more chordae in the porcine TV than the human or ovine TV (Gunnal et al. 2015; Jett et al. 2018; Silver et al. 1971), however other inter-species comparisons are limited or non-existent (e.g., differences in the branching patterns, insertion locations, etc.). The limited interspecies comparisons in the previous works necessitates more comprehensive future works to better understand the differences in TV morphology, including the comparisons of: (i) the tissue donors' ages, (ii) the healthy/diseased state of the valves, and (iii) the chordae branching patterns and insertion locations.

4.1 Mechanical Properties of the TV Chordae Tendineae

Uniaxial Tissue Mechanics of the TV Chordae

The chordae tendineae primarily undergo uniaxial loading in the valve apparatus, and as such the in vitro mechanical characterizations of the chordae tendineae generally involves uniaxial tensile loading. In the chordae characterization experiments, the central segment of the tissue was typically used and mounted between the crossheads of a servo-hydraulic testing device. Then, the chordae were preconditioned via cyclic uniaxial tensions followed by either repeated loading and unloading cycles to a targeted stress, or loading the chordae until rupture. In a recent study from our laboratory, we took a different approach to the aforementioned chordae testing setup that may better emulate the in-vivo function of the TV apparatus (Ross et al. 2020b). In that work, we preserved the insertion to the leaflet and the papillary muscle, and mounted the leaflet-chordae-papillary muscle entity to a tine-based uniaxial testing system for characterizing the uniaxial mechanical properties of the tissue (Fig. 8a). Utilizing the uniaxial mechanical testing of the TV chordae, several key comparisons can be made between: (i) different species, (ii) different leaflet insertion locations, (iii) different cross-sectional areas, and (iv) different experimental boundary conditions.

First, it was observed that the porcine and ovine chordae exhibit similar extensibility in the TVAL (porcine: 1–3% versus ovine: 2–3% at a 1 MPa Cauchy stress),

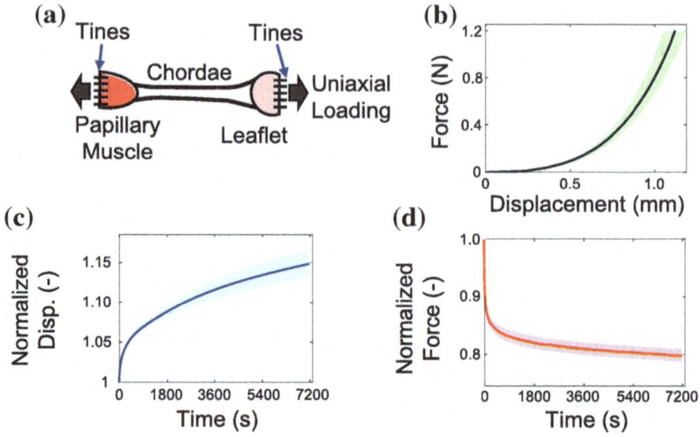

Fig. 8 (a) Schematic of the chordae tendineae tested as a holistic chordae-leaflet-papillary muscle entity using a tine-based mounting system. Mechanical characterization results ($n = 12$): (b) uniaxial tension, (c) 2-h creep, and (d) 2-h stress relaxation. Results are presented as the mean ± SEM

and that for the TVSL and TVPL, the porcine chordae were 80–95% more extensible than the ovine chordae (Pokutta-Paskaleva et al. 2019; Smith et al. 2020). The differences in the tensile properties between the porcine and ovine TVAL and the TVPL/TVSL chordae demonstrate a need for further corroborations of these findings. Second, it has been shown that the strut chordae are generally less extensible than the marginal or basal chordae, and that the marginal and basal chordae appear to have similar tensile characteristics. Third, it has been found that there is an insignificant trend between the TV chordae extensibility and the cross-sectional area; however, chordae with a cross-sectional area $>0.404\,\text{mm}^2$ were shown to have significantly smaller low-tension tangent stiffness than the thicker chordae (Smith et al. 2020). Fourth, comparing our study (Ross et al. 2020b), which considered the porcine TV strut chordae as a leaflet-chordae-papillary muscle entity (Fig. 8b), to an earlier work via the clamp-based testing method (Pokutta-Paskaleva et al. 2019), similar tensile properties were found from both tissue mounting approaches. Although the two testing methods yielded similar uniaxial tissue mechanical behaviors, advantages were demonstrated in the approach using the leaflet-chordae-papillary muscle entity, including the ability to analyze the leaflet and papillary muscle contributions to the chordae stretch, and the ability to image the load-dependent changes in the CFA at the chordae-leaflet insertion (see Sect. 4.2).

While the leaflet-chordae-papillary muscle entity approach is a step closer to emulating the in-vivo deformation of the chordae, even more physiologically representative mechanics could be realized through the use of a right heart simulator, however no studies have been performed in this manner to date. In addition, other useful studies of tissue biomechanics for the TV chordae would involve understanding the effects of disease on the tissue's uniaxial mechanics. One such study was performed by Lim et al. (1983) on the tensile behaviors of a myxomatous human TV chordae and they found that the diseased chordae was more extensible and ruptured at a much lower stress than the healthy chordae. This is a relatively outdated study, and limitations include: (i) only one chordae was tested, (ii) preconditioning was not performed, and (iii) optical strain tracking via fiducial markers was not used, resulting in less accurate measurements of the tissue deformation. Further investigations are warranted to obtain better information about the changes in the mechanical properties and the underlying microstructure of the diseased TV chordae.

4.1.1 Viscoelastic Properties of the TV Chordae Tendineae

Although comprehensive investigations of the quasi-elastic tensile properties of TV chordae are available in the literature, there have been relatively fewer studies on quantification of the viscoelastic properties for the TV chordae. In our laboratory, we are currently investigating the viscoleastic properties of the TVAL strut chordae through creep and stress relaxation tests. For creep testing, the leaflet-chordae-papillary muscle entities were subjected to a 1.2 N force, and the force was maintained through increasing applied displacements over the course of 90 min (Fig. 8c). In stress relaxation testing, the specimens were displaced and held at the displacement associated with a 1.2 N force over a 90-min duration (Fig. 8d). These viscoelastic characterizations are useful for understanding the time-dependent behaviors of the chordae, which will be valuable in the refinement of TV computational models when the time-dependent responses of the TV chordae are imperative.

4.2 Mesoscopic Evaluations of the TV Chordae

In addition to understanding the biomechanical behaviors of the TV chordae, there have been some studies investigating the mesoscale properties and microstructure of the chordae segment or the chordae-leaflet insertion. For visualizing the microstructure of the chordae segments, previous works have used methods such as histology (Pokutta-Paskaleva et al. 2019; Smith et al. 2020), fluorescent dye staining (Smith et al. 2020), SHG imaging (Smith et al. 2020), transmission electron microscopy (Lim 1980) or scanning electron microscopy (Lim 1980). It was found from these studies that the TV chordae are composed of dense collagen fiber bundles that are undulated at their unloaded state, together with inter-dispersed elastin and interstitial cells. On the outer surface of the chordae is the endothelial cell layer where the cell nuclei are fairly circular in shape. Moving from the outer surface to the center, the cell nuclei become more elongated, and undulated collagen fibers and cell nuclei become more directly aligned along the primary axis. Depictions of the depth-discriminated collagen microstructure and the cells of the TV chordae are shown in Fig. 9. Another earlier work on the microstructure also noted the vasculature present in the TV chordae, indicating another potential function of the tissue to provide nutrients to the

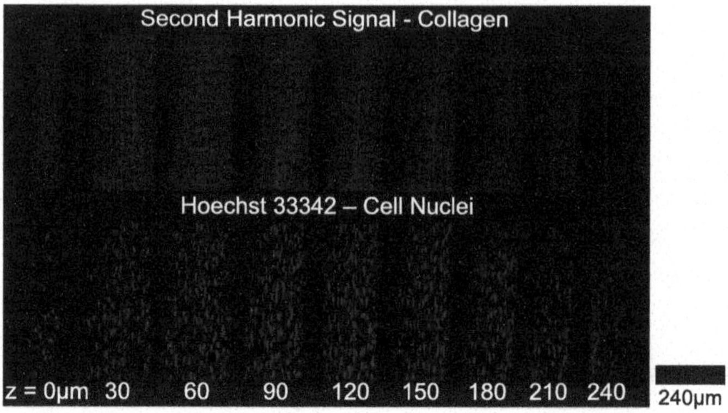

Fig. 9 Second-harmonic generation imaging results reveal the unidirectional, wavy collagen fiber architecture of the chordae at varying tissue depths, while fluorescent dye staining (Hoechst 333432) allows visualization of the cell nuclei through the thickness of the tissue (images adapted from Smith et al. (2020) with copyright permission by Springer Nature)

TV leaflets (Duran and Gunning 1968). Nevertheless, this finding has not been corroborated in recent studies, and additional research efforts are warranted to better understanding this observation.

The TV chordae-leaflet insertion is of particular interest in previous studies due to the complexity of the transition between the leaflet and chordae collagen fiber networks. To better understand and quantify the load-varying CFA changes in the TV chordae-leaflet insertion, in our laboratory, we utilized the pSFDI collagen imaging modality (Ross et al. 2020a) (Fig. 10). We found that with an increased loading (up to 1.2 N), the collagen fibers of the TV chordae-leaflet insertion did not undergo significant reorientation, but underwent significant increases in the DOA, indicating that the uncrimping of the collagen fibers may be the primary mechanism by which the chordae-leaflet insertion bears loading. Furthermore, the changes in the DOA at the leaflet insertion demonstrated the spatial variations in the tissue's collagen microstructure. In contrast, another preliminary study used X-ray diffraction to image the TV chordae-leaflet insertion and noted higher molecular strains at the chordae-leaflet insertion than the chordae segment or the leaflet region (Madhurapantula et al. 2020). From these findings, it is observed that the complexity of the chordae-leaflet insertion microstructure facilitates the distribution of forces from the leaflet to the chordae and papillary muscles. However, the microstructural complexity may also cause some adverse fiber-to-fiber interactions that can lead to tissue rupture.

5 Constitutive Modeling

Constitutive modeling for soft biological tissues, such as the TV leaflets or chordae tendineae, is typically formulated with the hyperelasticity theory (Holzapfel 2000).

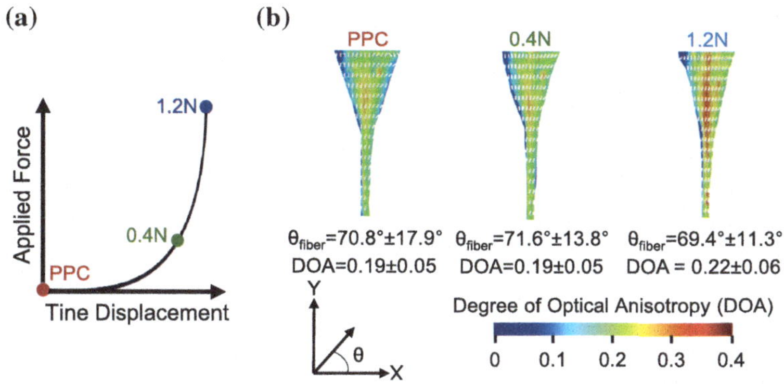

Fig. 10 Quantification of the load-dependent collagen fiber architecture for a representative TV chordae-leaflet insertion using our integrated opto-mechanical instrument: (a) applied force versus tine displacement with 3 configurations used to perform pSFDI-based collagen microstructural imaging (i.e. post-preconditioning (PPC), 0.4 and 1.2 N loading states), and (b) results of collagen fiber orientation and degree of optical anisotropy (DOA). Values are presented as the mean ± SEM, and white dashed lines denote the quantified fiber orientations

First, a strain-energy function (SEF) is chosen that relates the tissue deformations to the stresses of the material. Two commonly-adopted SEFs include the seminal Fung-type model (Tong and Fung 1976) and the Gasser-Ogden-Holzapfel (GOH) model (Gasser et al. 2005). The stress-strain relationship of the material is established by

$$\boldsymbol{\sigma} = 2\mathbf{F}\frac{\partial W}{\partial \mathbf{C}}\mathbf{F}^{\mathrm{T}} - p\mathbf{I}, \tag{1}$$

where $\boldsymbol{\sigma}$ is the Cauchy stress tensor, (F) is the deformation gradient, $\mathbf{C} = \mathbf{F}^{\mathrm{T}}\mathbf{F}$ is the right Cauchy-Green tensor, p is the penalty to enforce incompressibility, and (I) is the identity tensor. The parameters of the selected SEF are typically determined by nonlinear least-squares regression fitting to the mechanical testing data (Aggarwal 2017). Recent advancements in constitutive modeling of the TV leaflets and TV chordae tendineae are discussed in the following subsections.

5.1 *Constitutive Modeling of the TV Leaflets*

Constitutive modeling of the mechanical behavior of the TV leaflets has recently received more and more attention. Khoiy et al. (2018b) considered a Fung-type SEF for the modeling of the biaxial mechanical behavior of the three TV leaflets. The differences in the averaged model parameters and the variances in the estimated model parameters of the three TV leaflets underscored the unique leaflet-specific mechanical properties and inter-specimen variability. Although this work provided the critical first step towards accurately modeling the TV leaflet biaxial mechanical response, the Fung-type SEF is a phenomenological model, which does not consider important microstructural information, and may not be the most representative SEF for the TV leaflet's mechanical behavior.

One promising experimental avenue to determine the appropriate SEF for the TV leaflets could be the use of the constant-invariant testing procedure, such as what was previously adopted for the left ventricle myocardium (Humphrey et al. 1990) and the MV leaflets (May-Newman and Yin 1998). This empirical technique considers carefully designed biaxial mechanical characterization protocols to qualitatively relate the derivatives of the SEF W with respect to the invariants of (C). Following the framework of (Humphrey et al. 1990), the relationship between the SEF derivatives and the experimental data has the form

$$W_{,4} = \frac{1}{2\lambda_1^2}\left(\sigma_{11} - \sigma_{22}\xi_1\xi_2^{-1}\right), \tag{2a}$$

$$W_{,1} = \frac{1}{2}\sigma_{22}\xi_2^{-1}. \tag{2b}$$

Herein, $W_{,1}$ and $W_{,4}$ are the partial derivatives of W with respect to the first and fourth invariants of $\mathbf{C}$, respectively, and ξ_1 and ξ_2 describe various tissue deformations, i.e., $\xi_1 = \lambda_1^2 - (\lambda_1\lambda_2)^{-2}$ and $\xi_2 = \lambda_2^2 - (\lambda_1\lambda_2)^{-2}$. A series of constant I_1 or I_4 characterizations are then used to isolate the contributions of each invariant to the partial derivatives of W and establish an ideal SEF for a specific soft tissue. It is imperative that the CFA is preferentially aligned with one of the biaxial testing axes for I_4 to be maintained at a constant value.

Our laboratory has recently performed pilot constant-invariant tests for the three TV leaflets. The development of our pSFDI system (Hudson et al. 2020; Jett et al. 2020) enables us to ensure the CFA is preferentially aligned with one of the biaxial testing axes. Our preliminary results for a representative TVAL specimen under both the constant I_1 protocols (Fig. 11b) and the constant I_4 protocols (Fig. 11a) highlight the unique contributions of each invariant to the SEF and its partial derivatives. These results will lead to the determination of a TV leaflet-specific SEF.

5.2 *Constitutive Modeling of the TV Chordae Tendineae*

Several SEFs have been proposed for modeling the TV chordae tendineae, including the neo-Hookean, Fung-type, Mooney-Rivlin, Yeoh, and Ogden models (Table 2). Smith et al. (2020) evaluated the effectiveness of each SEF for capturing the uniaxial mechanics of the TV chordae and found that all, except the neo-Hookean and Mooney-Rivlin models, can provide good fits with root mean square errors < 0.05. Their study also suggests that the two-parameter Ogden model would be the most appropriate for describing the TV chordae's mechanical behaviors, as this model avoids over-fitting of the data while preserving model simplicity. This two-

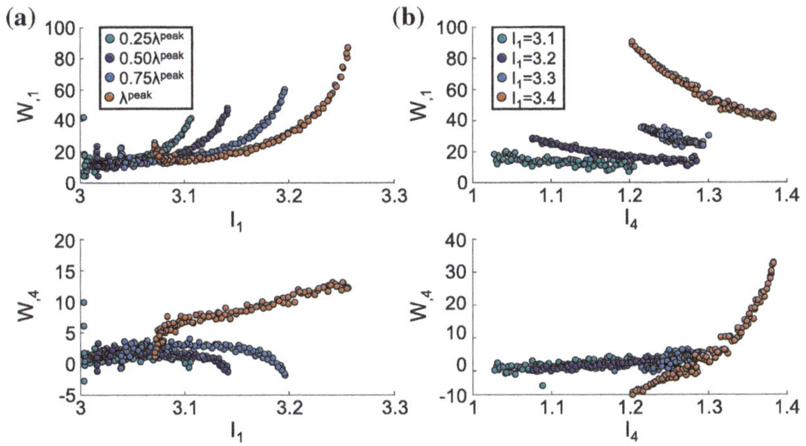

Fig. 11 The relationships between the partial derivatives of W and the invariants of $\mathbf{C}$ in our pilot constant-invariant testing study: (a) under constant I_4 testing protocol; (b) under constant I_1 testing protocol. Note that λ^{peak} is the peak fiber stretch under the equibiaxial loading protocol

Table 2 Existing constitutive models for the uniaxial properties of the chordae tendineae (Smith et al. 2020)

Model	SEF
Neo-Hookean model	$W = C_1[I_1 - 3]$
Fung-type model	$W = C_1[e^{C_2[I_1-3]} - 1]/C_2$
Mooney-Rivlin model	$W = C_1[I_1 - 3] + C_2[I_2 - 3]$
Yeoh model	$W = \sum_{i=1}^{N} C_i[I_1 - 3]^i$
Ogden model	$W = C_1[\lambda_1^{C_2} + \lambda_2^{C_2} + \lambda_3^{C_2} - 3]/C_2 + C_3[\lambda_1^{C_4} + \lambda_2^{C_4} + \lambda_3^{C_4} - 3]/C_4$

parameter Ogden model, with some minor modifications, was used by both Ross et al. (2020b) and Pokutta-Paskaleva et al. (2019) to model the TV chordae's uniaxial mechanics. The evaluation of the SEFs for the TV chordae will be valuable for refining the organ-level simulations of the TV apparatus, resulting in better predicted chordae behaviors and valve closing. While these phenomenological models provide an appropriate description of the TV chordae uniaxial tensile behaviors, they neglect the time-dependent viscoelastic properties or the microstructural information. Future works could focus on the development of a structurally based model of the TV chordae to better understand the time-varying biomechanical tissue behaviors or the microstructure-mechanics relationships.

6 *In Silico* Modeling of the TVs

6.1 Organ-Level Tricuspid Valve Simulations

In silico simulations enable connections between tissue-level mechanical behaviors and in-vivo valvular function of the TV. Prior studies have simulated the TV function to understand patient-specific healthy TV mechanical environment (Kong et al. 2018; Stevanella et al. 2010), explore the impact of papillary muscle displacements on the TV function (Singh-Gryzbon et al. 2019), assess the potential of the MitraClip for restoring healthy TV function (Dabiri et al. 2019), or evaluate the performance of a new contact algorithm (Kamensky et al. 2018). Interested readers are referred to our recent review article of these *in silico* studies for more information (Lee et al. 2019).

Our laboratory has recently provided the first comprehensive *in silico* assessment of how the TV pathologies influence the TV function in Laurence et al. (2020). In brief, the TV geometry from Kamensky et al. (2018) was used in finite element analysis to simulate the TV function, considering four pathological scenarios: pulmonary hypertension, annular dilation, papillary muscle displacement, and rupture of the chordae tendineae. Our simulation results underscored the effects of each valvular pathology on the TV function, for example, how TR could arise from annulus dilation, papillary muscle displacement, and/or the rupture of the chordae tendineae (Fig. 12). Additional quantitative analyses, based on our *in silico* TV simulations, led to three key observations. Firstly, the geometry metrics, such as the tenting area or the coaptation height, depended on which two TV leaflets were in contact. This has important clinical implications, because there is uncertainty regarding which two leaflets are visualized for the clinical quantification of these metrics (Stankovic et al. 2014). Secondly, the isolated scenario of papillary muscle displacement led to TR, suggesting this pathology may be a new target for clinical intervention rather than considering only the annulus dilation as a TR indicator. Thirdly, the chordae tendineae attached to the TVAL or TVSL may be more important to ensure the proper valve closure. Moreover, the leaflet stresses increased for all four pathological scenarios, which may trigger the cellular biosynthetic response and lead to altered leaflet morphology (see Sect. 7.1). This pilot *in silico* study has provided, *for the first time*, the simulation-guided assessment of TV pathologies. However, significant improvements and extensions are warranted to translate our model to the real clinical applications, such as a more accurate description about the in-vivo motions of the TV annulus or papillary muscles.

6.2 Efficacy for the Affine Fiber Kinematics Assumption

Integrating computational and experimental tissue biomechanics enables a cross-validation pipeline for guiding experimental setups and for evaluating the efficacy of a constitutive model to numerically replicate the planar biaxial tests. Thomas et al.

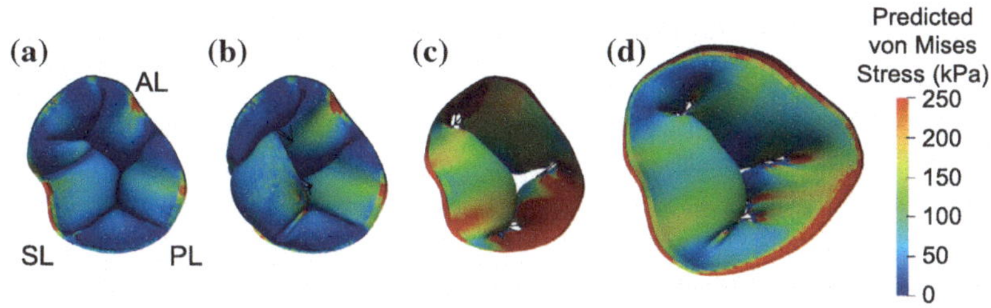

Fig. 12 Finite element simulation results for (a) the healthy, and the diseased TVs: (b) chordae rupture, (c) combined flattened annulus and papillary muscle displacement, and (d) combined flattened annulus and annulus dilation. (AL - Anterior Leaflet; PL - Posterior Leaflet; SL - Septal Leaflet)

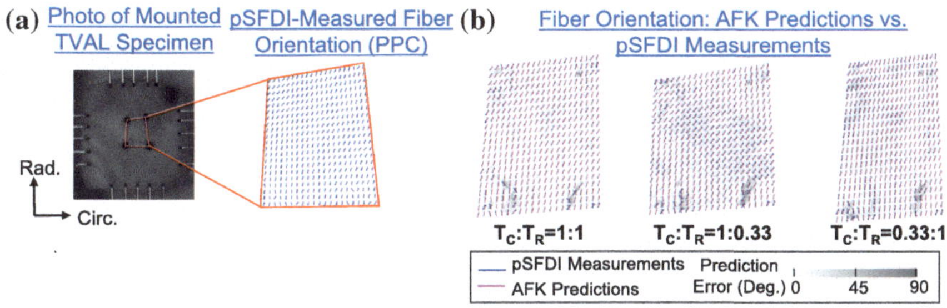

Fig. 13 (a) Combined pSFDI-biaxial testing approach allowed for (b) validation of the affine fiber kinematics (AFK) assumption by comparing the fiber orientation angles (denoted by dashed lines) between the pSFDI measurements and the AFK predictions

(2019) performed such an investigation to validate a microstructurally informed constitutive model for the TV leaflets. In their work, they effectively correlated the tissue biaxial mechanical responses with the collagen fiber kinematics by their combined experimental-computational methods. Their study serves as an example of how an integrated experimental-computational approach can validate the numerical implementation of the computational model in simulating TV function. In our laboratory, we recently performed studies to evaluate the efficacy of the *affine fiber kinematics* assumption—an assumption that has been widely accepted in the heart valve computational biomechanics community—by using our pSFDI modality. Preliminary findings of our work (Fig. 13) suggest that the affine fiber kinematics assumption is a good approximation for TV leaflets under *equi-biaxial* tensions, but larger deviations from this assumption were found for other *non-equibiaxial* loading protocols (i.e., $T_C : T_R = 0.33 : 1$ or $1 : 0.33$). These findings suggest that the affine fiber kinematics assumption may need to be carefully evaluated when it is employed in the organ-level and tissue-level TV simulations to avoid mispredictions of the reorientation and realignment of the underlying collagen fiber networks.

6.3 Simulations of Biaxial Mechanical Testing Experiments

In silico simulations are valuable tools to replicate the biaxial mechanical testing experiments. By simulating the biaxial mechanical testing of the TV leaflets, different experimental conditions can be systematically evaluated to determine the optimal setup to more accurately obtain the TV leaflet biaxial mechanical properties. Examples include the simulations of the MV or other biomimetic materials (Fan and Sacks 2014; Jacobs et al. 2013; Sun et al. 2005), by considering different specimen geometry (e.g., square versus cruciform), different boundary (or tissue mounting) conditions (e.g., rotation-allowed suture, rigid tines, or clamps), and the effective mounting size (e.g., 10×10 mm versus 5×5 mm). Through understanding the effects of various experimental setups on the resultant characterized mechanical properties of the TV leaflet tissues, more representative constitutive model form determination and more accurate model parameter estimation can be ensured, while eliminating potential adverse effects induced by inappropriate experimental setup, such as heterogeneous stress and strain fields.

7 Future Perspectives

7.1 Cell-Mediated Growth and Remodeling (G&R)

The TV leaflets have a population of VICs that regulate the biosynthesis of the tissue's extracellular matrix and constituents. Previous studies for the aortic valve and the MV have highlighted the unique activation from a fibroblast phenotype to a myofibroblast phenotype under pathological conditions (Ayoub et al. 2017, 2020; Balachandran et al. 2006, 2009; Merryman et al. 2007). This phenotypic activation is accompanied by a cellular biosynthetic response associated with an up-regulation of matrix metalloproteinases (MMPs), a down-regulation of the tissue inhibitor of MMP-1, and increases in the contents of collagen fibers, GAGs, and PGs. Analogous studies have not yet been performed for the TV VICs, but we anticipate the TV VICs will have a similar general biosynthetic response to altered loading. Interestingly, the VICs of both the aortic valve and the MV were found to be nearly *twice* as stiff as those of the TV (Merryman et al. 2007), which may alter the corresponding VIC-regulated biosynthetic responses to adapt the TV to the varied mechanical and functional environment. It is obvious that the cellular biosynthetic response should be quantified, separately, for the TV leaflets following previously-established techniques for the mitral VICs. The quantified microstructural changes for the TV leaflets could then be combined with existing G&R theories (Cyron et al. 2016; Humphrey and Rajagopal 2002; Rodriguez et al. 1994) and *in silico* simulations to explore the essential role of VIC-regulated biosynthesis in the recurrence of TR post-surgical repair.

7.2 Constitutive Modeling of the TV Tissues

Constitutive modeling under the hyperelasticity framework has been traditionally used to model the biaxial mechanical response of the heart valve leaflets. Although much of the current focus is on determining the appropriate constitutive model for the TV leaflets, there are many emerging modeling techniques that are promising alternatives to the traditional hyperelasticity-based constitutive modeling approach. The most notable of these techniques is *data-driven computing* that was first introduced by Kirchdoerfer and Ortiz (2016), which directly uses the experimental data with the kinematically admissible states to determine the current stress and strain without a pre-defined SEF. This leads to many distinct advantages for data-driven computational biomechanics, including the capability to represent a wide-range of tissue deformations given sufficient training data or not being limited by the accuracy of a constituent relationship. Our laboratory has recently extended the locally convex data-driven modeling approach (He and Chen 2020) to modeling the MV posterior leaflet biaxial mechanical responses. The results from this study are promising, but extensions are warranted to apply this method to the TV tissues and improve the approach performance, especially when handling noisy data or inter-specimen variability. Other modeling techniques are beginning to draw more attention in the computational biomechanics community, which also aim to forego the *a priori* constitutive model, including Bayesian statistics-motivated (Akintunde et al. 2019; Teferra and Brewick 2019) and stochastic modeling methods (Staber et al. 2019). These approaches must first be extended to multi-axial loading of soft biological tissues with transversely isotropic mechanical behavior before they can be confidently and reliably used for the TV leaflets.

7.3 Other Clinical Challenges

For correcting TR, it is largely recommended that repair should be performed whenever possible, as valvular replacement is associated with high mortality (Alkhouli et al. 2018; Jang et al. 2017). In the case of TV repair, the preferred correction method is to reduce the tricuspid annulus size to facilitate a larger coaptation area of the TV leaflets through either suture valvuloplasty or ring annuloplasty. In the valvuloplasty procedure, a suture is fed through the annulus and pulled to constrict the valve, while in ring annuloplasty a pre-sized rigid or flexible band is sutured to the commissure points of the annulus to change the annulus shape and size. Previous clinical studies have shown that suture annuloplasty is much less effective than ring annuloplasty in preventing recurrent TR; however, some clinicians still perform suture annuloplasty for its procedural simplicity and satisfactory intermediate-term treatment outcome (Lafçi et al. 2019; Tchantchaleishvili et al. 2017). Comparing the rigid and flexible rings used in TV annuloplasty repair, rigid rings provide the benefit of retaining the 3D annulus shape and increasing leaflet coaptation; however these devices have the consequences of restricted annular contraction and an increased chance of ring dehiscence. In contrast, flexible rings have a reduced risk of ring dehiscence and better annular contraction. Nevertheless, they lose the 3D annulus shape and carry an increased potential for sub-optimal leaflet coaptation (Aldea 2019). While TV annuloplasty is generally successful, there is the potential for recurrent TR in the long term, and researchers have recommended the TV annuloplasty ring devices be further improved through the consideration of anisotropic material properties to better retain natural valve contraction, and customization of the ring to each patient's annulus geometries and pathological conditions (Mathur et al. 2020).

Acknowledgements We would like to extend our well wishes to Dr. Gerhard A. Holzapfel for his 60th birthday, and congratulate him on his phenomenal accomplishments and prominent contributions to the soft tissue biomechanics community. Many current research advancements would remain elusive without his groundbreaking experimental, theoretical, and computational investigations. These beautifully-crafted studies from Dr. Holzapfel have provided endless inspirations, which, in addition to his invaluable perspective, have significantly impacted our research community and our laboratory.

We would also like to extend our sincere gratitude to the editors for providing this wonderful opportunity to share our contributions to the heart valve biomechanics community.

We also acknowledge the funding supports from the American Heart Association (AHA) Scientist Development Grant (SDG) Award (16SDG27760143) and the Presbyterian Health Foundation (PHF) Team Science Grant (C5122401) are greatly acknowledged. CHL was in part supported by the institutional start-up fund from the School of Aerospace and Mechanical Engineering, the IBEST-OUHSC Funding for Interdisciplinary Research, and the research funding from the Research Council at the University of Oklahoma. DWL and CJR were supported by the National Science Foundation Graduate Research Fellowship (GRF 2019254233 and 2020307284).

References

Aggarwal, A.: An improved parameter estimation and comparison for soft tissue constitutive models containing an exponential function. Biomech. Model. Mechanobiol. **16**(4), 1309–1327 (2017)

Akintunde, A.R., Miller, K.S., Schiavazzi, D.E.: Bayesian inference of constitutive model parameters from uncertain uniaxial experiments on murine tendons. J. Mech. Behav. Biomed. Mater. **96**, 285–300 (2019)

Aldea, G.S.: Commentary: tricuspid valve ring characteristics: physiologically important, clinically relevant, or too little too late? J. Thorac. Cardiovasc. Surg. **161**, e209–e210 (2021)

Alkhouli, M., Berzingi, C., Kowatli, A., Alqahtani, F., Badhwar, V.: Comparative early outcomes of tricuspid valve repair versus replacement for secondary tricuspid regurgitation. Open Heart **5**(2), e000878 (2018)

Amini, R., Eckert, C.E., Koomalsingh, K., McGarvey, J., Minakawa, M., et al.: On the in vivo deformation of the mitral valve anterior leaflet: effects of annular geometry and referential configuration. Ann. Biomed. Eng. **40**(7), 1455–1467 (2012)

Antunes, M., Girdwood, R.: Tricuspid annuloplasty: a modified technique. Ann. Thorac. Surg. **35**(6), 676–678 (1983)

Anwar, A.M., Folkert, J., Soliman, O.I.: Clinical Recognition of Tricuspid Valve Disease, pp. 25–48. Springer (2018)

Anyanwu, A.C., Adams, D.H.: Functional tricuspid regurgitation in mitral valve disease: epidemiology and prognostic implications. Semin. Thorac. Cardiovasc. Surg. **22**(1), 69–75 (2010)

Ayoub, S., Lee, C.H., Driesbaugh, K.H., Anselmo, W., Hughes, C.T., et al.: Regulation of valve interstitial cell homeostasis by mechanical deformation: implications for heart valve disease and surgical repair. J. R. Soc. Interface. **14**(135), 20170580 (2017)

Ayoub, S., Howsmon, D.P., Lee, C.H., Sacks, M.S.: On the role of predicted in vivo mitral valve interstitial cell deformation on its biosynthetic behavior. Biomech. Model. Mechanobiol. **20**(1), 135–144 (2021)

Badano, L.P., Muraru, D., Enriquez-Sarano, M.: Assessment of functional tricuspid regurgitation. Eur. Heart J. **34**(25), 1875–1885 (2013)

Badhwar, V., Rankin, J.S., He, M., Jacobs, J.P., Furnary, A.P., Fazzalari, F.L., et al.: Performing concomitant tricuspid valve repair at the time of mitral valve operations is not associated with increased operative mortality. Ann. Thorac. Surg. **103**(2), 587–593 (2017)

Balachandran, K., Konduri, S., Sucosky, P., Jo, H., Yoganathan, A.P.: An ex vivo study of the biological properties of porcine aortic valves in response to circumferential cyclic stretch. Ann. Biomed. Eng. **34**(11), 1655–1665 (2006)

Balachandran, K., Sucosky, P., Jo, H., Yoganathan, A.P.: Elevated cyclic stretch alters matrix remodeling in aortic valve cusps: implications for degenerative aortic valve disease. Am. J. Physiol. Heart Circ. Physiol. **296**(3), H756–H764 (2009)

Basu, A., He, Z.: Annulus tension on the tricuspid valve: an in-vitro study. Cardiovasc. Eng. Technol. **7**(3), 270–279 (2016)

Braunwald, N.S., Ross Jr, J., Morrow, A.G.: Conservative management of tricuspid regurgitation in patients undergoing mitral valve replacement. Circulation **35**(4s1), I–63–I–69 (1967)

Carpentier, A.: A new reconstructive operation for correction of mitral and tricuspid insufficiency. J. Thorac. Cardiovasc. Surg. **61**(1), 1–13 (1971)

Chikwe, J., Itagaki, S., Anyanwu, A., Adams, D.H.: Impact of concomitant tricuspid annuloplasty on tricuspid regurgitation, right ventricular function, and pulmonary artery hypertension after repair of mitral valve prolapse. J. Am. Coll. Cardiol. **65**(18), 1931–1938 (2015)

Cyron, C.J., Aydin, R.C., Humphrey, J.D.: A homogenized constrained mixture (and mechanical analog) model for growth and remodeling of soft tissue. Biomech. Model. Mechanobiol. **15**(6), 1389–1403 (2016)

Dabiri, Y., Yao, J., Sack, K.L., Kassab, G.S., Guccione, J.M.: Tricuspid valve regurgitation decreases after mitraclip implantation: fluid structure interaction simulation. MeReC **97**, 96–100 (2019)

De Vega, N.G., Orueta, M., Olivar, E.: La anuloplastia selectiva, regulable y permanente una técnica original para el tratamiento de la insuficiencia tricúspide. Rev. Esp. Cardiol. **25**, 555–556 (1972)

De Vega, N.G., De Rabago, G., Castillon, L., Moreno, T., Azpitarte, J.: A new tricuspid repair short-term clinical results in 23 cases. J. Cardiovasc. Surg. **14**, 384–386 (1973)

Desai, R.R., Abello, L.M., Klein, A.L., Marwick, T.H., Krasuski, R.A., Ye, Y., et al.: Tricuspid regurgitation and right ventricular function after mitral valve surgery with or without concomitant tricuspid valve procedure. J. Thorac. Cardiovasc. Surg. **146**(5), 1126-1132.e10 (2013)

Dreyfus, G.D., Corbi, P.J., Chan, K.M.J., Bahrami, T.: Secondary tricuspid regurgitation or dilatation: which should be the criteria for surgical repair? Ann. Thorac. Surg. **79**(1), 127–132 (2005)

Driessen, N.J.B., Bouten, C.V.C., Baaijens, F.P.T.: Improved prediction of the collagen fiber architecture in the aortic heart valve. J. Biomech. Eng. **127**(2), 329–336 (2005)

Duginski, G.A., Ross, C.J., Laurence, D.W., Johns, C.H., Lee, C.H.: An investigation of the effect of freezing storage on the biaxial mechanical properties of excised porcine tricuspid valve anterior leaflets. J. Mech. Behav. Biomed. Mater. **101**, 103438 (2020)

Duran, C.M.G., Gunning, A.J.: The vascularization of the heart valves: a comparative study. Cardiovasc. Res. **2**(3), 290–296 (1968)

Fan, R., Sacks, M.S.: Simulation of planar soft tissues using a structural constitutive model: finite element implementation and validation. J. Biomech. **47**(9), 2043–2054 (2014)

Filsoufi, F., Salzberg, S.P., Coutu, M., Adams, D.H.: A three-dimensional ring annuloplasty for the treatment of tricuspid regurgitation. Ann. Thorac. Surg. **81**(6), 2273–2277 (2006)

Fung, Y.C.: Biomechanics: Mechanical Properties of Living Tissues. Springer (2013)

Gasser, T.C., Ogden, R.W., Holzapfel, G.A.: Hyperelastic modelling of arterial layers with distributed collagen fibre orientations. J. R. Soc. Interface **3**(6), 15–35 (2005)

Goth, W., Lesicko, J., Sacks, M.S., Tunnell, J.W.: Optical-based analysis of soft tissue structures. Annu. Rev. Biomed. Eng. **18**, 357–385 (2016)

Grashow, J.S., Sacks, M.S., Liao, J., Yoganathan, A.P.: Planar biaxial creep and stress relaxation of the mitral valve anterior leaflet. Ann. Biomed. Eng. **34**(10), 1509–1518 (2006a)

Grashow, J.S., Yoganathan, A.P., Sacks, M.S.: Biaixal stress-stretch behavior of the mitral valve anterior leaflet at physiologic strain rates. Ann. Biomed. Eng. **34**(2), 315–325 (2006b)

Gunnal, S.A., Wabale, R.N., Farooqui, M.S.: Morphological study of chordae tendinae in human cadaveric hearts. Heart Views **16**(1), 1–12 (2015)

He, Q., Chen, J.S.: A physics-constrained data-driven approach based on locally convex reconstruction for noisy database. Comput. Meth. Appl. Mech. Eng. **363**, 112791 (2020)

Heyden, S., Nagler, A., Bertoglio, C., Biehler, J., Gee, M.W., et al.: Material modeling of cardiac valve tissue: experiments, constitutive analysis and numerical investigation. J. Biomech. **48**(16), 4287–4296 (2015)

Holzapfel, G.A.: Nonlinear Solid Mechanics: A Continuum Approach for Engineering Science. Wiley (2000)

Huang, X., Gu, C., Men, X., Zhang, J., You, B., et al.: Repair of functional tricuspid regurgitation: comparison between suture annuloplasty and rings annuloplasty. Ann. Thorac. Surg. **97**(4), 1286–1292 (2014)

Hudson, L.T., Jett, S.V., Kramer, K.E., Laurence, D.W., Ross, C.J., et al.: A pilot study on linking tissue mechanics with load-dependent collagen microstructures in porcine tricuspid valve leaflets. Bioengineering **7**(2), 60 (2020)

Humphrey, J., Strumpf, R., Yin, F.: Determination of a constitutive relation for passive myocardium: I A new functional form. J. Biomech. Eng. **112**(3), 333–339 (1990)

Humphrey, J.D., Rajagopal, K.R.: A constrained mixture model for growth and remodeling of soft tissues. Math. Model. Meth. Appl. Sci. **12**(03), 407–430 (2002)

Jacobs, N.T., Cortes, D.H., Vresilovic, E.J., Elliott, D.M.: Biaxial tension of fibrous tissue: using finite element methods to address experimental challenges arising from boundary conditions and anisotropy. J. Biomech. Eng. **135**(2), 0210041 (2013)

Jang, J.Y., Heo, R., Lee, S., Kim, J.B., Kim, D.H., et al.: Comparison of results of tricuspid valve repair versus replacement for severe functional tricuspid regurgitation. Am. J. Cardiol. **119**(6), 905–910 (2017)

Jett, S.V., Laurence, D.W., Kunkel, R.P., Babu, A.R., Kramer, K.E., et al.: An investigation of the anisotropic mechanical properties and anatomical structure of porcine atrioventricular heart valves. J. Mech. Behav. Biomed. Mater. **87**, 155–171 (2018)

Jett, S.V., Hudson, L.T., Baumwart, R., Bohnstedt, B.N., Mir, A., et al.: Integration of polarized spatial frequency domain imaging (pSFDI) with a biaxial mechanical testing system for quantification of load-dependent collagen architecture in soft collagenous tissues. Acta Biomater. **102**, 149–168 (2020)

Kamensky, D., Xu, F., Lee, C.H., Yan, J., Bazilevs, Y., et al.: A contact formulation based on a volumetric potential: application to isogeometric simulations of atrioventricular valves. Comput. Meth. Appl. Mech. Eng. **330**, 522–546 (2018)

Khoiy, K.A., Amini, R.: On the biaxial mechanical response of porcine tricuspid valve leaflets. J. Biomech. Eng. **138**(10), 104504 (2016)

Khoiy, K.A., Asgarian, K.T., Loth, F., Amini, R.: Dilation of tricuspid valve annulus immediately after rupture of chordae tendineae in ex-vivo porcine hearts. PLOS One **13**(11), e0206744 (2018a)

Khoiy, K.A., Pant, A.D., Amini, R.: Quantification of material constants for a phenomenological constitutive model of porcine tricuspid valve leaflets for simulation applications. J. Biomech. Eng. **140**(9), 094503 (2018b)

Kirchdoerfer, T., Ortiz, M.: Data-driven computational mechanics. Comput. Meth. Appl. Mech. Eng. **304**, 81–101 (2016)

Kong, F., Pham, T., Martin, C., McKay, R., Primiano, C., et al.: Finite element analysis of tricuspid valve deformation from multi-slice computed tomography images. Ann. Biomed. Eng. **46**(8), 1112–1127 (2018)

Kramer, K.E., Ross, C.J., Laurence, D.W., Babu, A.R., Wu, Y., et al.: An investigation of layer-specific tissue biomechanics of porcine atrioventricular heart valve leaflets. Acta Biomater. **96**, 368–384 (2019)

Kunzelman, K.S., Cochran, R.P., Murphree, S.S., Ring, W.S., Verrier, E.D., et al.: Differential collagen distribution in the mitral valve and its influence on biomechanical behaviour. J. Heart. Valve. Dis. **2**(2), 236–244 (1993)

Lafçi, G., Çiçek, Ö.F., Lafçi, A., Esenboğa, K., Günertem, E., et al.: A comparison of three tricuspid annuloplasty techniques: suture, ring, and band. Turk Gogus Kalp Damar Cerrahisi Derg **27**(3), 286–293 (2019)

Laurence, D.W., Ross, C.J., Jett, S.V., Johns, C.H., Echols, A.L., et al.: An investigation of regional variations in the biaxial mechanical properties and stress relaxation behaviors of porcine atrioventricular heart valve leaflets. J. Biomech. **83**, 16–27 (2019)

Laurence, D.W., Johnson, E.L., Hsu, M., Baumwart, R., Mir, A., et al.: A pilot in-silico modeling-based study of the pathological effects on the biomechanical function of tricuspid valves. Int. J. Numer. Method Biomed. Eng. **36**(7), e3346 (2020)

Lee, C.H., Oomen, P.J.A., Rabbah, J.P., Yoganathan, A., Gorman, R.C., et al.: A high-fidelity and micro-anatomically accurate 3D finite element model for simulations of functional mitral valve. In: FIMH Proceedings, pp. 416–424. Springer (2013)

Lee, C.H., Zhang, W., Liao, J., Carruthers, C.A., Sacks, J.I., et al.: On the presence of affine fibril and fiber kinematics in the mitral valve anterior leaflet. Biophys. J. **108**(8), 2074–2087 (2015)

Lee, C.H., Laurence, D.W., Ross, C.J., Kramer, K.E., Babu, A.R., et al.: Mechanics of the tricuspid valve-From clinical diagnosis/treatment, in-vivo and in-vitro investigations, to patient-specific biomechanical modeling. Bioengineering **6**(2), 47 (2019)

Liao, J., Yang, L., Grashow, J., Sacks, M.S.: The relation between collagen fibril kinematics and mechanical properties in the mitral valve anterior leaflet. J. Biomech. Eng. **129**(1), 78–87 (2007)

Lim, K.O.: Mechanical properties and ultrastructure of normal human tricuspid valve chordae tendineae. Japanese J. Physiol. **30**(3), 455–464 (1980)

Lim, K.O., Boughner, D.R., Perkins, D.G.: Ultrastructure and mechanical properties of chordae tendineae from a myxomatous tricuspid valve. Japanese Heart J. **24**(4), 539–548 (1983)

Madhurapantula, R.S., Krell, G., Morfin, B., Roy, R., Lister, K., et al.: Advanced methodology and preliminary measurements of molecular and mechanical properties of heart valves under dynamic strain. Int. J. Mol. Sci. **21**(3), 763 (2020)

Mathur, M., Jazwiec, T., Meador, W.D., Malinowski, M., Goehler, M., et al.: Tricuspid valve leaflet strains in the beating ovine heart. Biomech. Model. Mechanobiol. **18**(5), 1351–1361 (2019)

Mathur, M., Meador, W.D., Jazwiec, T., Malinowski, M., Timek, T.A., et al.: Tricuspid valve annuloplasty alters leaflet mechanics. Ann. Biomed. Eng. **48**(12), 2911–2923 (2020)

May-Newman, K., Yin, F.C.: Biaxial mechanical behavior of excised porcine mitral valve leaflets. Am. J. Physiol. Heart Circ. Physiol. **269**(4), H1319–H1327 (1995)

May-Newman, K., Yin, F.C.: A constitutive law for mitral valve tissue. J. Biomech. Eng. **120**(1), 38–47 (1998)

Meador, W.D., Mathur, M., Sugerman, G.P., Jazwiec, T., Malinowski, M., et al.: A detailed mechanical and microstructural analysis of ovine tricuspid valve leaflets. Acta Biomater. **102**, 100–113 (2020)

Merryman, W.D., Lukoff, H.D., Long, R.A., Engelmayr, G.C., Jr., et al.: Synergistic effects of cyclic tension and transforming growth factor-β1 on the aortic valve myofibroblast. Cardiovasc. Pathol. **16**(5), 268–276 (2007)

Nishimura, R.A., Otto, C.M., Bonow, R.O., Carabello, B.A., Erwin, J.P., et al.: 2014 AHA/ACC guideline for the management of patients with valvular heart disease: a report of the American College of Cardiology/American Heart Association task force on practice guidelines. J. Am. Coll. Cardiol. **63**(22), e57–e185 (2014)

Pant, A.D., Thomas, V.S., Black, A.L., Verba, T., Lesicko, J.G., et al.: Pressure-induced microstructural changes in porcine tricuspid valve leaflets. Acta Biomater. **67**, 248–258 (2018)
Pham, T., Sun, W.: Material properties of aged human mitral valve leaflets. J. Biomed. Mater. Res. A **102**(8), 2692–2703 (2014)
Pham, T., Sulejmani, F., Shin, E., Wang, D., Sun, W.: Quantification and comparison of the mechanical properties of four human cardiac valves. Acta Biomater. **54**, 345–355 (2017)
Pierlot, C.M., Lee, J.M., Amini, R., Sacks, M.S., Wells, S.M.: Pregnancy-induced remodeling of collagen architecture and content in the mitral valve. Ann. Biomed. Eng. **42**(10), 2058–2071 (2014)
Plotnikov, S., Juneja, V., Isaacson, A.B., Mohler, W.A., Campagnola, P.J.: Optical clearing for improved contrast in second harmonic generation imaging of skeletal muscle. Biophys. J. **90**(1), 328–339 (2006)
Pokutta-Paskaleva, A., Sulejmani, F., DelRocini, M., Sun, W.: Comparative mechanical, morphological, and microstructural characterization of porcine mitral and tricuspid leaflets and chordae tendineae. Acta Biomater. **85**, 241–252 (2019)
Pozzoli, A., Lapenna, E., Vicentini, L., Alfieri, O., De Bonis, M.: Surgical indication for functional tricuspid regurgitation at initial operation: judging from long term outcomes. Gen. Thorac. Cardiovasc. Surg. **64**(9), 509–516 (2016)
Prot, V., Skallerud, B.: Nonlinear solid finite element analysis of mitral valves with heterogeneous leaflet layers. Comput. Mech. **43**(3), 353–368 (2009)
Prot, V., Skallerud, B., Sommer, G., Holzapfel, G.A.: On modelling and analysis of healthy and pathological human mitral valves: two case studies. J. Mech. Behav. Biomed. Mater. **3**(2), 167–177 (2010)
Rabbah, J.P., Saikrishnan, N., Yoganathan, A.P.: A novel left heart simulator for the multi-modality characterization of native mitral valve geometry and fluid mechanics. Ann. Biomed. Eng. **41**(2), 305–315 (2013)
Rausch, M.K., Kuhl, E.: On the effect of prestrain and residual stress in thin biological membranes. J. Mech. Phys. Solids **61**(9), 1955–1969 (2013)
Rego, B.V., Sacks, M.S.: A functionally graded material model for the transmural stress distribution of the aortic valve leaflet. J. Biomech. **54**, 88–95 (2017)
Rego, B.V., Wells, S.M., Lee, C.H., Sacks, M.S.: Mitral valve leaflet remodelling during pregnancy: insights into cell-mediated recovery of tissue homeostasis. J. R. Soc. Interface **13**(125), 20160709 (2016)
Rodriguez, E.K., Hoger, A., McCulloch, A.D.: Stress-dependent finite growth in soft elastic tissues. J. Biomech. **27**(4), 455–467 (1994)
Ross, C.J., Laurence, D.W., Richardson, J., Babu, A.R., Evans, L.E., et al.: An investigation of the glycosaminoglycan contribution to biaxial mechanical behaviors of porcine atrioventricular heart valve leaflets. J. R. Soc. Interface. **16**(156), 20190069 (2019a)
Ross, C.J., Laurence, D.W., Wu, Y., Lee, C.H.: Biaxial mechanical characterizations of atrioventricular heart valves. J. Vis. Exp. **146**, e59170 (2019b)
Ross, C.J., Hsu, M.C., Baumwart, R., Mir, A., Burkhart, H.M., et al.: Quantification of load-dependent changes in the collagen fiber architecture for the strut chordae tendineae-leaflet insertion of porcine atrioventricular heart valves. Biomech. Model. Mechanobiol. **20**(1), 223–241 (2020a)
Ross, C.J., Laurence, D.W., Hsu, M.C., Baumwart, R., Zhao, Y.D., et al.: Mechanics of porcine heart valves' strut chordae tendineae investigated as a leaflet-chordae-papillary muscle entity. Ann. Biomed. Eng. **48**(5), 1463–1474 (2020b)
Sacks, M.S.: Biaxial mechanical evaluation of planar biological materials. J. Elast. Phys. Sci. Solids **61**, 199–246 (2000)
Sacks, M.S., Yoganathan, A.P.: Heart valve function: a biomechanical perspective. Philos. Trans. R. Soc. Lond. B. Biol. Sci. **362**(1484), 1369–1391 (2007)
Sacks, M.S., Enomoto, Y., Graybill, J.R., Merryman, W.D., Zeeshan, A., et al.: In-vivo dynamic deformation of the mitral valve anterior leaflet. Ann. Thorac. Surg. **82**(4), 1369–1377 (2006)

Salinas, S.D., Clark, M.M., Amini, R.: The effects of −80°C short-term storage on the mechanical response of tricuspid valve leaflets. J. Biomech. **98**, 109462 (2020)
Silver, M.D., Lam, J.H.C., Ranganathan, N., Wigle, E.D.: Morphology of the human tricuspid valve. Circulation **43**(3), 333–348 (1971)
Singh-Gryzbon, S., Sadri, V., Toma, M., Pierce, E.L., Wei, Z.A., et al.: Development of a computational method for simulating tricuspid valve dynamics. Ann. Biomed. Eng. **47**(6), 1422–1434 (2019)
Smith, K.J., Mathur, M., Meador, W.D., Phillips-Garcia, B., Sugerman, G.P., et al.: Tricuspid chordae tendineae mechanics: insertion site, leaflet, and size-specific analysis and constitutive modelling. Exp. Mech. **61**, 19–29 (2021)
Staber, B., Guilleminot, J., Soize, C., Michopoulos, J., Iliopoulos, A.: Stochastic modeling and identification of a hyperelastic constitutive model for laminated composites. Comput. Meth. Appl. Mech. Eng. **347**, 425–444 (2019)
Stankovic, I., Daraban, A.M., Jasaityte, R., Neskovic, A.N., Claus, P., et al.: Incremental value of the en face view of the tricuspid valve by two-dimensional and three-dimensional echocardiography for accurate identification of tricuspid valve leaflets. J. Am. Soc. Echocardiogr. **27**(4), 376–384 (2014)
Stella, J.A., Sacks, M.S.: On the biaxial mechanical properties of the layers of the aortic valve leaflet. J. Biomech. Eng. **129**(5), 757–766 (2007)
Stevanella, M., Votta, E., Lemma, M., Antona, C., Redaelli, A.: Finite element modelling of the tricuspid valve: a preliminary study. Med. Eng. Phys. **32**(10), 1213–1223 (2010)
Sun, W., Sacks, M.S., Scott, M.J.: Effects of boundary conditions on the estimation of the planar biaxial mechanical properties of soft tissues. J. Biomech. Eng. **127**(4), 709–715 (2005)
Sun, Y.P., O'Gara, P.T.: Epidemiology, anatomy, pathophysiology and clinical evaluation of functional tricuspid regurgitation. Minerva Cardioangiol. **65**(5), 469–479 (2017)
Tchantchaleishvili, V., Rajab, T.K., Cohn, L.H.: Posterior suture annuloplasty for functional tricuspid regurgitation. Ann. Cardiothorac. Surg. **6**(3), 262–265 (2017)
Teferra, K., Brewick, P.T.: A Bayesian model calibration framework to evaluate brain tissue characterization experiments. Comput. Meth. Appl. Mech. Eng. **357**, 112604 (2019)
Thomas, V.S., Lai, V., Amini, R.: A computational multi-scale approach to investigate mechanically-induced changes in tricuspid valve anterior leaflet microstructure. Acta Biomater. **94**, 524–535 (2019)
Tong, P., Fung, Y.C.: The stress-strain relationship for the skin. J. Biomech. **9**(10), 649–657 (1976)
Vesely, I., Lozon, A.: Natural preload of aortic valve leaflet components during glutaraldehyde fixation: effects on tissue mechanics. J. Biomech. **26**(2), 121–131 (1993)
Vesely, I., Noseworthy, R.: Micromechanics of the fibrosa and the ventricularis in aortic valve leaflets. J. Biomech. **25**(1), 101–113 (1992)
Zhang, W., Ayoub, S., Liao, J., Sacks, M.S.: A meso-scale layer-specific structural constitutive model of the mitral heart valve leaflets. Acta Biomater. **32**, 238–255 (2016)

A Bayesian Approach to Parameter Estimation in Cardiac Mechanics

Joakim Sundnes and Rocío Rodríguez-Cantano

I started my research on computational modeling of cardiac mechanics in 2004, and very quickly got familiar with Gerhard's huge contributions in biomechanics. His famous book on nonlinear solid mechanics became my most important reference for learning this fascinating field, and even today it remains an essential reference and handbook in my research. The learning curve of the book is steep, but it is by far the most comprehensive and rigorous introduction to the field that I have found to this date. In 2006 I invited Gerhard to serve on the committee of my first PhD student in cardiac mechanics, and learnt that he is not only a leading expert in the field, but also a very nice person who is eager to share his knowledge with junior researchers. Since then, he has served on several more PhD committees for my students, and we have had a number of inspiring meetings and discussions both in Oslo and in Graz.

Joakim

Abstract Computational models of cardiac mechanics have been shown to capture the general mechanical function of the heart, and have been parameterized based on in vivo patient data to accurately reproduce the heart function of individual patients. Previous attempts for creating such patient-specific problem have typically been based on solving a deterministic inverse problem, which minimizes the misfit between certain model outputs and measured values. While this approach is robust and efficient, and has shown good agreement between model results and patient data, it does not provide any information about the uncertainty in the resulting model parameters. We here present a parameter estimation framework based on a Bayesian approach, which estimates probability density functions (PDFs) of material parameters based on uncertain input values. The method is based on sampling the parameter space and solving the associated forward model for each sample, which may lead to a substantial computational problem if multiple parameters are considered. However,

J. Sundnes (✉) · R. Rodríguez-Cantano
Simula Research Laboratory, P.O. Box 134, 1325 Lysaker, Norway
e-mail: sundnes@simula.no

G. Sommer et al. (eds.), *Solid (Bio)mechanics: Challenges of the Next Decade*,
Studies in Mechanobiology, Tissue Engineering and Biomaterials 24,
https://doi.org/10.1007/978-3-030-92339-6_10

the model also offers additional information in the form of univariate or multivariate PDFs for the estimated parameters. We investigate the potential of the methodology by solving a simple parameter estimation problem in passive left ventricular mechanics, and show that the results are in agreement with previous results obtained with a deterministic parameter estimation method.

1 Introduction

Computational models of cardiac physiology and mechanics have evolved substantially over the last decades, and are able to capture essential features of the heart's mechanical function. Several studies have focused on fitting models to individual patient data (Xi et al. 2013; Balaban et al. 2017; Finsberg et al. 2017), and have demonstrated good agreement between model outputs and recorded measurements. However, the results also indicate that it is difficult to uniquely determine the parameters based on in vivo data alone, and it must be assumed that the model parameter values are associated with considerable uncertainty. This assumption is strengthened by reported parameter values in the literature, which vary by several orders of magnitude (Kallhovd et al. 2019). The potential impact of this uncertainty has been explored in multiple studies (Eck et al. 2016; Campos et al. 2019; Rodríguez-Cantano et al. 2019), but we are not aware of any previous attempts to quantify the uncertainty in the material parameters themselves.

The problem of estimating parameters in a simulation model from observations is fundamentally an inverse problem (Tarantola 2005; Aster et al. 2012). While the forward problem tries to predict the system behavior by solving the governing equations given some parameter values, the inverse problem reverses this relationship by estimating uncertain parameters from measurements or observations. The inverse approach is often formulated as a deterministic nonlinear optimization problem that seeks to minimize the misfit between the observed and predicted outputs in an appropriate norm. This approach has been applied to fitting parameters for active and passive mechanical properties of cardiac tissue, see, e.g., Finsberg et al. (2017) and Balaban et al. (2017).

However, from probability theory, a consistent formulation of the parameter estimation problem can be made using Bayes' theory (Stuart 2010). In the Bayesian inference framework, given observational data and their uncertainty, the governing forward PDEs, and a prior probability distribution describing prior knowledge about the parameters, the posterior PDF over the parameters is obtained as solution of the inverse problem. The main challenge in solving this Bayesian inverse problem stems in computing statistics of this posterior PDF, which may be a surface in high dimensions, and evaluating the probability of each point in the parameter space requires solving the forward problem. Characterizing a multidimensional parameter space will typically require a large number of samples, and in the case of cardiac mechanics the forward problem is a nonlinear finite element model, resulting in a substantial computational problem.

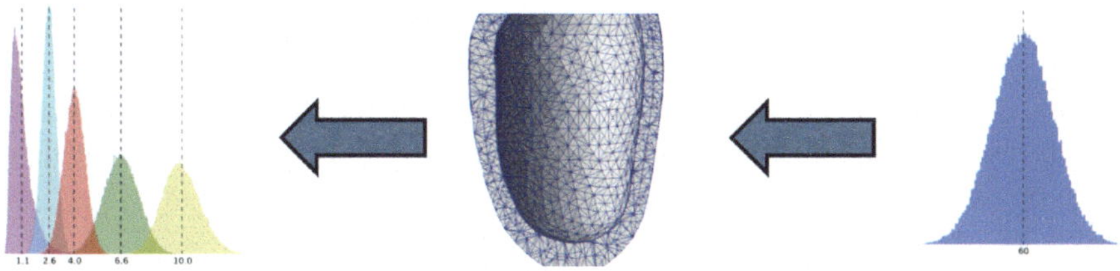

Fig. 1 Illustration of the Bayesian inference problem for passive mechanics of the left ventricle. The forward model is a finite element model linking endocardial pressure to cavity volume. We want to use uncertain observations of cavity volumes to estimate PDFs for the material parameters of the mechanical model

In this study, we present a formulation of the parameter estimation in a Bayesian framework, and apply it to a simple test problem of passive cardiac mechanics. The input to the algorithm will be uncertain measurements of cavity pressures and volumes, and the outputs are probability densities of mechanical material parameters, as illustrated in Fig. 1. The remainder of the paper is organized as follows. In Sect. 2 we describe the Bayesian method for parameter estimation as well as the forward model of cardiac mechanics. Section 3 presents preliminary results, for estimating passive material parameters based on measurements of cavity pressure and volume. Finally, we provide some concluding remarks and outline future research directions in Sect. 4.

2 Models and Methods

As outlined in the introduction, Bayes' theorem provides a framework for estimating uncertain parameters based on prior information and observations. The original formulation of the theorem is in terms of point probabilities given by

$$P(A|B) = \frac{P(B|A)P(A)}{P(B)}. \tag{1}$$

Here, $P(A|B)$ is a conditional probability, i.e., the likelihood of event A given an observation B. Similarly, $P(B|A)$ is the likelihood of observation B given that A is true, $P(A)$ is the prior probability of A, and $P(B)$ is the probability of observing B, often referred to as the marginal likelihood of B. For parameter estimation and uncertainty quantification it is convenient to formulate Bayes' theorem in terms of PDFs (Shao 2003; Keener 2010). For a given vector of model parameters $\mathbf{m}$ and observation y_{obs}, we have

$$\rho_{\text{post}}(\mathbf{m}|y_{\text{obs}}) = \frac{\rho_{\text{pri}}(\mathbf{m}) \times \rho_{\text{lik}}(y_{\text{obs}}|\mathbf{m})}{\rho(y_{\text{obs}})}, \tag{2}$$

where $\rho_{\text{post}}(\mathbf{m}|y_{\text{obs}})$ is the posterior probability density of $\mathbf{m}$ given y_{obs}, $\rho_{\text{lik}}(y_{\text{obs}}|\mathbf{m})$ is the likelihood of observing y_{obs} given $\mathbf{m}$, $\rho_{\text{pri}}(\mathbf{m})$ is the prior probability of $\mathbf{m}$, and $\rho(y_{\text{obs}})$ is the marginal likelihood of the observation. Since the marginal likelihood is simply a scaling factor which turns the posterior probability into a proper PDF, it is often omitted from the theorem formulation, i.e.

$$\rho_{\text{post}}(\mathbf{m}|y_{\text{obs}}) \propto \rho_{\text{pri}}(\mathbf{m}) \times \rho_{\text{lik}}(y_{\text{obs}}|\mathbf{m}). \tag{3}$$

This version simply states that the posterior probability is proportional to the product of the prior probability density, containing prior information and beliefs about $\mathbf{m}$, and the likelihood.

Based on Eq. (3), the inverse problem of identifying parameters $\mathbf{m}$ can be stated as a problem of statistical inference over the space of uncertain parameters, which have to be inferred from the data and a mathematical model. In the most general form, the mathematical model is characterized by a deterministic operator $f : H \to \mathbb{R}^q$, which predicts observables $y_{\text{obs}} \in \mathbb{R}^q$ corresponding to a given parameter $\mathbf{m} \in H$, for a suitable function space H. For the specific case of passive cardiac mechanics, considered below, the mathematical model computes left ventricular cavity volumes resulting from specific choices of material parameters.

The resulting solution to the statistical inverse problem is a posterior distribution, which combines our prior beliefs about $\mathbf{m}$, represented by $\rho_{\text{pri}}(\mathbf{m})$, with the likelihood $\rho_{\text{lik}}(y_{\text{obs}}|\mathbf{m})$ that a given vector $\mathbf{m}$ might have produce the observations y_{obs}.

2.1 The Cardiac Mechanics Model

To explore the properties of the Bayesian approach in cardiac mechanics, we will consider a simple mechanics model describing passive inflation of the left ventricle, which is formulated as a quasi-static and pressure-loaded mechanical equilibrium problem. For a domain H representing the left ventricle, as illustrated in Fig. 1, we want to find the displacement $\mathbf{u} : H \to \mathbb{R}^3$ such that

$$\nabla \cdot (\mathbf{FS}) = \mathbf{0} \quad \text{in } H, \tag{4}$$

where $\mathbf{F}$ is the deformation gradient, i.e., $\mathbf{F} = \nabla\mathbf{u} + \mathbf{I}$, and $\mathbf{S}$ is the second Piola-Kirchhoff stress tensor. The outer surface of the ventricle (epicardium) is assumed to be unloaded, the inner surface (endocardium) is subject to a time varying pressure load, specified below, while the basal plane is subject to a Robin boundary condition (linear spring); see, e.g., Finsberg et al. (2017) for details about the mechanics model.

The passive mechanical response of the tissue was described using a purely incompressible and transversely isotropic version of the Holzapfel-Ogden material model (Holzapfel and Ogden 2009), with an isochoric strain-energy function given by

Table 1 Pressure- and volume data used for the parameter estimation. The unloaded volume ($p = 0\,\text{kPa}$) represents the undeformed reference state of the finite element model. The three other pressure-volume pairs are used in the parameter fitting. Right: Parameter values found by the deterministic parameter estimation. The parameter a was fitted to minimize the mismatch between computed and observed cavity volumes, while keeping the other parameters constant

Pressure (kPa)	Volume (ml)	Parameter	Value
0	78.49	a	0.61 kPa
0.533	123.4	b	5.0
1.06	138.8	a_f	2.581 kPa
1.60	148.7	b_f	5.0

$$\Psi(\mathbf{C}_e) = \frac{a}{2b}\left(\exp\left[b(I_1^E - 3) - 1\right]\right) + \frac{a_f}{2b_f}\left(\exp\left[b_f(I_{4\mathbf{f}_0}^E - 1)^2 - 1\right]\right), \tag{5}$$

where a, b, a_f, b_f are material constants and I_1^E, $I_{4\mathbf{f}_0}^E$ are reduced (pseudo-)invariants that are defined as

$$I_1^E = \text{tr}\overline{\mathbf{C}}_e, \quad I_{4\mathbf{f}_0}^E = \mathbf{f}_0 \cdot (\overline{\mathbf{C}}_e \mathbf{f}_0), \tag{6}$$

where $\overline{\mathbf{C}}_e = J_e^{-2/3}\mathbf{C}_e$ is the volume preserving contribution to the elastic component of the right Cauchy-Green tensor $\mathbf{C}_e = \mathbf{F}_e^T\mathbf{F}_e$, and $J_e = \det(\mathbf{F}_e) > 0$ is the elastic volumetric deformation.

In Finsberg et al. (2017), good agreement with recorded pressure-volume data was obtained by fitting the parameter a in Eq. (5) to minimize the misfit between the observation and the model, keeping all other parameters fixed at values reported in the literature. In the present study we will solve a similar inverse problem within the Bayesian framework, to explore the potential of the methodology and estimate the uncertainty in the material parameters arising from measurement uncertainty. In addition to fitting the single parameter a, we will simultaneously fit a and a_f, to explore the possibility of uniquely determining several parameters based on pressure-volume data.

The left ventricular domain outlined above was represented by a tetrahedral finite element mesh constructed from triangulated segmented surfaces of the endo- and epicardium corresponding to the frame at the beginning of atrial systole, as illustrated in Fig. 1. Myocardial fiber orientations were assigned with the algorithm from Bayer et al. (2012), with endo- and epicardial helix angles set to 60° and −60°, respectively. The equilibrium problem (4) was discretized and solved using a Galerkin finite element method with Taylor-Hood elements, i.e., using piecewise quadratic basis functions for the displacement and piecewise linear for the pressure. The solver was implemented in the finite element framework FEniCS (Logg et al. 2012); see, e.g., Finsberg et al. (2017) for more details on the model formulation. Three separate pressure-volume measurements were recorded and used as observations in the parameter fitting, with values specified in the left panel of Table 1. The approach in

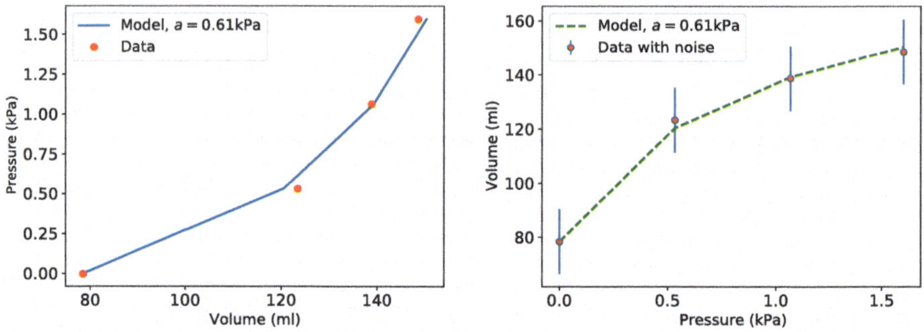

Fig. 2 Data used for parameter estimation. The left panel shows the standard presentation, with pressure as function of volume, while the right panel shows volume as function of input pressure. The latter presentation is used in the parameter estimation, with pressure being a known input and volume assumed to be an measurement with 10% noise

Finsberg et al. (2017) was used to fit the parameter a, holding all other parameters fixed. The resulting parameter values are given in the right panel of Table 1, and a comparison of model results with the data is given in Fig. 2. The left panel shows the pressure as function of volume, which is the usual way to display cardiac pressure-volume data. The right panel shows the cavity volume as a function of pressure, which is how the data will be used in the inverse modeling, since we treat pressure as a given input and volume as the output of the model.

2.2 *Bayesian Inference in Passive Cardiac Mechanics*

To formulate the cardiac mechanics problem in the Bayesian framework, we describe the forward model as an abstract mathematical model linking parameter values $\mathbf{m}$ with observations y_{obs}:

$$f(\mathbf{m}) = y_{\text{obs}}. \tag{7}$$

The observations will be cavity volume measurements for given pressures, while the uncertain parameter vector $\mathbf{m}$ corresponds to the material parameters of the model. As noted above, we will either consider $\mathbf{m}$ to be a single, scalar parameter, a in the model above, or the pair of parameters a, a_{f}. We assume that the uncertainty in the measurements y_{obs} can be modeled by an additive Gaussian noise model, i.e.

$$y_{\text{obs}} = f(\mathbf{m}) + \epsilon \ , \ \text{ with } \epsilon \sim \mathcal{N}(0, \Gamma_{\text{noise}}), \tag{8}$$

where $\epsilon \in \mathbb{R}^q$ is a random variable representing the noise, and $\Gamma_{\text{noise}} \in \mathbb{R}^q$ the noise covariance matrix. Since in our particular case y_{obs} is a scalar, the measurement errors are normally distributed with zero mean and a noise equal to σ^2_{noise}. The noise is assumed to result from a combination of measurement errors in the pressure and volume data.

As we have assumed that observational uncertainty is centered, additive and Gaussian, Eq. (8), we can express the likelihood model as in Petra et al. (2014):

$$\rho_{\text{lik}}(y_{\text{obs}}|\mathbf{m}) \propto \exp\left[-\frac{1}{2}(f(\mathbf{m}) - y_{\text{obs}})^{\mathrm{T}}\,\Gamma_{\text{noise}}^{-1}(f(\mathbf{m}) - y_{\text{obs}})\right]. \tag{9}$$

Again, in the particular case of only one unknown observed variable, the likelihood in Eq. (9) becomes

$$\rho_{\text{lik}}(y_{\text{obs}}|m) \propto \exp\left[-\frac{|f(\mathbf{m}) - y_{\text{obs}}|^2}{2\sigma_{\text{noise}}^2}\right]. \tag{10}$$

In addition to the likelihood, the prior distribution is of utmost practical importance in Bayesian inference, and the most controversial aspect. The prior allows one to incorporate qualitative and quantitative information other than the data. Beyond physical constrains, the information included in the prior may come from heterogeneous sources such as expert knowledge, previous experiments and published literature. Usually, candidate priors are chosen from well known families of distributions such as uniform (bounded from above and below), Gaussian (unbounded parameters), log-normal, or gamma distributions (strictly positive). In the present study we will use uniform priors, assuming no prior knowledge on the parameters except that they are bounded from above and below:

$$\rho_{\text{prior}}^{\text{uniform}}(m) = \begin{cases} \dfrac{1}{b_{\text{high}} - b_{\text{low}}}, & \text{for } b_{\text{low}} < m < b_{\text{high}}, \\ 0, & \text{otherwise}. \end{cases} \tag{11}$$

2.2.1 Repeated Applications of Bayes' Theorem

If we have multiple observations, these can be combined into a single, multi-variate likelihood functions as indicated above. However, if the observations are independent, we may also apply Eq. (3) repeatedly, using the posterior distribution from one observation as the prior for the next. For instance, for the case considered here, we may have measurements of the left ventricular volume for multiple input pressure. With the noise model in Eq. (8), these may be viewed as independent uncertain variables, and we can apply Bayes' theorem repeatedly to improve the accuracy of the parameter estimates. For real-world applications, the volume measurements are typically based on segmentation of medical images, which is usually a semi-manual process that is likely contain systematic and operator-dependent errors. In this case the assumption of independent observations is not strictly correct, and it may be appropriate to include other contributions to the measurement error in addition to the

Gaussian noise in Eq. (8). For simplicity, and to to explore the potential effect of using repeated independent measurements, we will here treat the volume observations as completely independent.

2.2.2 Computing and Characterizing the Posterior PDFs

As noted in the introduction, the main challenge in Bayesian inference is that we need to characterize the posterior PDF by sampling the parameter space. Accurately characterizing a multi-dimensional PDF typically requires a large number of parameters, and the resulting computational load may be substantial. In the present work we limit the number of simultaneously varying parameters to one or two, which significantly reduces the number of samples required, but each sample involves the evaluation of the forward model, which involves solving the finite element model of cardiac mechanics. The computational load may therefore be significant even if we limit the parameter samples to a few hundred.

The standard methods to sample the parameter space and characterize the posterior are the Markov-Chain Monte Carlo (MCMC) methods, see, e.g., Gilks et al. (1995), which come in numerous versions using a variety of different sampling methods. The methods produce a chain of parameter samples concentrated on the highest-probability regions of the PDF, in order to accurately characterize the posterior with a minimal number of samples. Although considered the state of the art in terms of efficiency and robustness, performance of the MCMC methods depends on a careful choice of several method parameters (Gilks et al. 1995). In the present work we are primarily interested in visualizing and characterizing the posterior PDFs, and not necessarily draw large numbers of samples for use in Monte-Carlo simulations or similar methods. We have therefore adopted a simpler approach, since we are only interested in visualizing and describing the posterior PDFs, where we uniformly sample the entire parameter space with no consideration of the expected posterior probability. Since we assume uniform and bounded priors for the parameters, we simply draw uniform (non-random) samples from the non-zero part of the prior distribution. For each parameter sample, say a_i, we compute the resulting cavity volume V_{comp}, and uses the difference $V_{\text{comp}} - V_{\text{obs}}$ and the likelihood to calculate the point probability $P(V_{\text{obs}}|a_i)$. The posterior PDF is then computed by combining and normalizing these point probabilities.

3 Results

The Bayesian framework outlined above was first applied to the problem of inferring a single material parameter from the uncertain measurements. Inspired by the approach in Finsberg et al. (2017), we focused on the parameter a, keeping all other parameters fixed. We applied a uniform prior given by Eq. (11), with $b_{\text{low}} = 0.05$ and $b_{\text{high}} = 2.0$. The likelihood was given by Eq. (10), with y_{obs} equal to the observed cavity

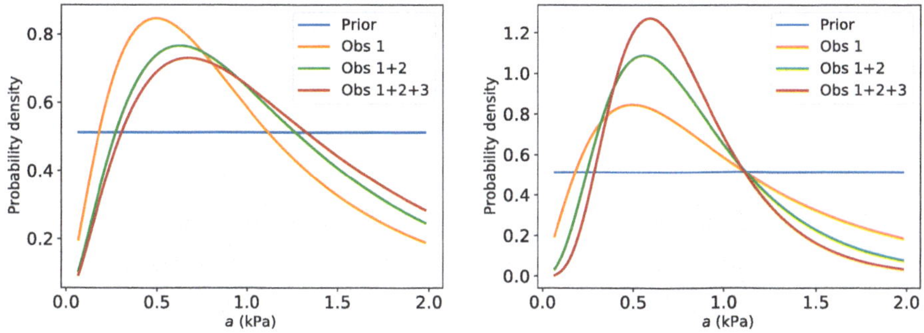

Fig. 3 Initial results of the Bayesian inference. The left panel shows the uniform prior and the three posteriors resulting from treating the pressure-volume observations separately. The right panel shows the result of repeated applications of Bayes' theorem, where the posterior from one observation is used as the prior for the next

volumes specified in Table 1, and 10% noise ($\sigma_{\text{noise}} = 0.1\, y_{\text{obs}}$). For the initial test we treated the three pressure-volume observations separately, all with the same uniform prior, which yielded the three posteriors shown in the left panel of Fig. 3. We see that all three distributions are slightly skewed, and the peaks are close to the value $a = 0.61$ kPa predicted by the deterministic solution method. The right panel of Fig. 3 shows the results of repeated applications of Bayes' theorem, assuming that the three observations are independent. As expected, the uncertainty of the parameter estimate is reduced when more observations are included. The peak value of the final PDF is 0.60, which is close to the result of the deterministic method.

The fact that we could accurately reproduce the pressure-volume curve by fitting a single parameter indicates that it is not possible to uniquely determine multiple material parameters on the basis of pressure-volume data alone. The Bayesian approach outlined here offers the opportunity to explore this non-uniqueness in more detail, and we have performed initial experiments to simultaneously infer the two parameters a and a_{f}. The results are shown in Fig. 4. As for the one-dimensional case we observe that repeated application of all three volume observations, shown in the lower right panel, reduces the variance of the resulting PDF. However, the PDFs also clearly illustrate the problem of non-uniqueness in the estimated parameters, as there appears to be a 'band' of a, a_{f} pairs with very similar posterior probability. Using all three observations may reduce the width of this band, but it is clearly still there. The results in the two-dimensional case are also general less consistent than in the single-parameter experiment, which could be due to under-sampling or not properly spanning the parameter space. For instance, the peak of the PDF varies significantly between the observations, with peak $a = 0.535$, $a_{\text{f}} = 2.375$ for the first observation, $a = 0.39$, $a_{\text{f}} = 4.25$ for the second, $a = 0.825$, $a_{\text{f}} = 1.625$ for the third, and $a = 0.97$, $a_{\text{f}} = 0.5$ when combining all three. The value $a_{\text{f}} = 0.5$ was the chosen lower boundary for our prior distribution, which indicates that a longer interval for the prior would be appropriate.

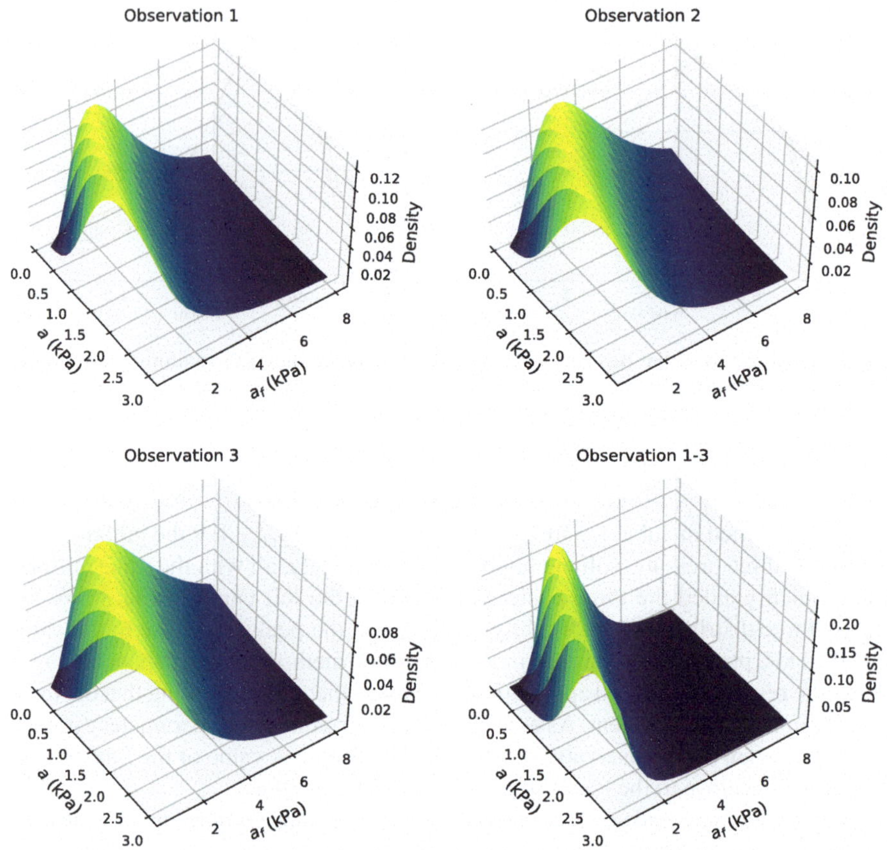

Fig. 4 Preliminary results from the 2D Bayesian inference, showing the posterior PDFs for the parameter pair a, a_f. The lower right panel shows the result of repeatedly applying Bayes' theorem for all three volume observations, while the other three panels result from using a single observation

4 Concluding Remarks

In this chapter we have presented a Bayesian approach for estimating material parameters in a model of passive cardiac mechanics. The results are in agreement with previous results of Finsberg et al. (2017), which solved the parameter estimation problem as a deterministic optimization problem. The method presented here is based on uniform (i.e., 'brute force') sampling of the parameter space, and the computational cost is considerably higher than that of the deterministic approach. However, the method also offers additional into the uncertainty of the obtained parameter values, which in turn can provide valuable input to uncertainty quantification methods for output quantities.

Although the initial results are promising and indicate the added value offered by the Bayesian approach, the present study has several limitations that should be addressed in future works. First, the simple assumption of 10% noise in the measurements should be replaced by more realistic values based on the characteristics of pressure- and volume measurements. The resulting errors are likely to contain both noise and systematic components, and therefore not be completely independent, and may require an extension of the proposed algorithm. However, such a study would likely be the first attempt at quantifying uncertainty in material parameters based on actual measurement uncertainty. Furthermore, the sampling performed here is very coarse, in particular for the two-parameter case, and the analysis of the resulting PDFs is quite limited. A more detailed analysis of the statistical properties of the PDFs, possibly including even more simultaneous parameters, is a natural next step. Finally, combining the results of such studies with forward uncertainty propagation analysis will give additional insight into the uncertainty of material parameters and their impact on output quantities of interest.

References

Aster, R.C., Borchers, B., Thurber, C.H.: Parameter Estimation and Inverse Problems, 2nd edn. Academic Press, Boston (2012)

Balaban, G., Finsberg, H., Odland, H.H., Rognes, M.E., Ross, S., Sundnes, J., Wall, S.: High-resolution data assimilation of cardiac mechanics applied to a dyssynchronous ventricle. Int. J. Numer. Methods Biomed. Eng. **33**, e2863 (2017)

Bayer, J.D., Blake, R.C., Plank, G., Trayanova, N.A.: A novel rule-based algorithm for assigning myocardial fiber orientation to computational heart models. Ann. Biomed. Eng. **40**, 2243–2254 (2012)

Campos, J.O., Sundnes, J., dos Santos, R.W., Rocha, B.M.: Effects of left ventricle wall thickness uncertainties on cardiac mechanics. Biomech. Model. Mechanobiol. **18**, 1415–1427 (2019)

Eck, V.G., Donders, W.P., Sturdy, J., Feinberg, J., Delhaas, T., Hellevik, L.R., Huberts, W.: A guide to uncertainty quantification and sensitivity analysis for cardiovascular applications. J. Num. Methods Biomed. Eng. **32**, e02755 (2016)

Finsberg, H., Balaban, G., Ross, S., Håland, T.F., Odland, H.H., Sundnes, J., Wall, S.: Estimating cardiac contraction through high resolution data assimilation of a personalized mechanical model. J. Comput. Sci. **24**, 85–90 (2017)

Gilks, W., Richardson, S., Spiegelhalter, D.: Markov Chain Monte Carlo in Practice. Chapman & Hall/CRC Interdisciplinary Statistics. Taylor & Francis (1995)

Holzapfel, G.A., Ogden, R.W.: Constitutive modelling of passive myocardium: a structurally based framework for material characterization. Philos. Trans. A Math. Phys. Eng. Sci. **367**, 3445–3475 (2009)

Kallhovd, S., Sundnes, J., Wall, S.: Sensitivity of stress and strain calculations to passive material parameters in cardiac mechanical models using unloaded geometries. Comput. Methods Biomech. Biomed. Eng. **22**, 664–675 (2019)

Keener, R.: Theoretical Statistics: Topics for a Core Course. Springer Texts in Statistics, Springer, New York (2010)

Logg, A., Mardal, K.A., Wells, G. (eds.): Automated Solution of Differential Equations by the Finite Element Method: The FEniCS Book, vol. 84. Springer (2012)

Petra, N., Martin, J., Stadler, G., Ghattas, O.: A computational framework for infinite-dimensional bayesian inverse problems, part II: stochastic newton MCMC with application to ice sheet flow inverse problems. SIAM J. Sci. Comput. **36**, A1525–A1555 (2014)

Rodríguez-Cantano, R., Sundnes, J., Rognes, M.E.: Uncertainty in cardiac myofiber orientation and stiffnesses dominate the variability of left ventricle deformation response. Int. J. Numer. Methods Biomed. Eng. **35**, e3178 (2019)

Shao, J.: Mathematical Statistics. Springer Texts in Statistics. Springer (2003)

Stuart, A.M.: Inverse problems: a Bayesian perspective. Acta Numer. **19**, 451–559 (2010)

Tarantola, A.: Inverse Problem Theory and Methods for Model Parameter Estimation. Society for Industrial and Applied Mathematics SIAM (2005)

Xi, J., Lamata, P., Niederer, S., Land, S., Shi, W., Zhuang, X., Ourselin, S., Duckett, S.G., Shetty, A.K., Rinaldi, C.A., Rueckert, D., Razavi, R., Smith, N.P.: The estimation of patient-specific cardiac diastolic functions from clinical measurements. Med. Image Anal. **17**, 133–146 (2013)

Computational Finite Strain Orthotropic Viscoelasticity of Human Passive Myocardium

Osman Gültekin and Hüsnü Dal

My first face-to-face meeting with Prof. Gerhard A. Holzapfel on a not surprisingly cold February in the old building in Kronesgasse was probably the starting point of a fruitful and inspiring collaboration to which the papers we have pulled off together surely attest. As my supervisor, his unconditional support in every matter, meticulous approach at every stage in scientific work, his enthusiasm for mechanics have always been influential on me. One of the memories that stick to my mind was the one we had when there was a fire drill at the department building, and everybody was outside, waiting for the drill to end, while we saw it not as an excuse and found a decent corner at the nearest Tribeka (a coffee house), where we continued to revise our paper 'On the quasi-incompressible finite element analysis of anisotropic hyperelastic materials', which was accompanied by two cups of 'Verlängerter'. Who knows, one day we may sit around a table again and work on a research article while enjoying several glasses of freshly brewed Turkish tea.

Osman

I first came to know Prof. Gerhard A. Holzapfel through his well-known book 'Nonlinear Solid Mechanics: A Continuum Approach for Engineers' in 2003 as a graduate student attending the interdisciplinary Master of Science program 'Computational Mechanics of Materials and Structures' in Stuttgart. Exactly a decade has passed until we had our first conversation in October 2013 as I was a postdoctoral researcher in Stuttgart. One day, my telephone rang, and I heard a mellifluous voice: 'Hi, I am Gerhard Holzapfel ...'. We had roughly a 40-minute talk about the PhD application of Osman Gültekin. Two years later, I was invited to Graz for a collaboration on the modeling of failure in biological tissues with the phase-field approach. Since then, we have met on various occasions. The most

O. Gültekin · H. Dal (✉)
Department of Mechanical Engineering, Middle East Technical University, Dumlupınar Bulvarı No.1, 06800 Çankaya, Ankara, Turkey
e-mail: dal@metu.edu.tr

G. Sommer et al. (eds.), *Solid (Bio)mechanics: Challenges of the Next Decade*,
Studies in Mechanobiology, Tissue Engineering and Biomaterials 24,
https://doi.org/10.1007/978-3-030-92339-6_11

remarkable impressions I had about Gerhard Holzapfel are his unending enthusiasm for research, his kind and persuasive voice, and the expressions of joy that I grasped on his face during our scientific discussions. As time passed, Gerhard Holzapfel has been among the most inspiring figures in my academic career.

Hüsnü

Abstract This study presents the theoretical and modeling aspects of the two recently published orthotropic viscoelastic constitutive frameworks by Cansız et al. (Comput. Methods Biomech. Biomed. Eng. 18: 1160–1172, 2015) and Gültekin et al. (Comput. Methods Biomech. Biomed. Eng. 19: 1647–1664, 2016). Recent revelations clearly predicate the viscoelastic response of the human passive myocardium in terms of stress hysteresis and relaxation. The first contribution establishes a canonical representation, formulates a rheological model with Maxwell elements considered along fiber, sheet, and normal directions in the logarithmic strain space. The second contribution presents a convolution-type evolution of the viscous stresses, which are incorporated in a rheological model with Maxwell elements featuring matrix, fiber, sheet, and fiber-sheet terms akin to the hyperelastic contributions. In the sequel, we address the performance of both models by fitting them to cyclic and stress relaxation test data obtained from triaxial shear tests on human passive myocardium.

1 Introduction

Being an electro-mechanical pump, the heart plays an indispensable role in human life, as revealed by the share of cardiovascular diseases in the number of deaths globally. According to the *Global Burden of Disease* study (2020) on the causes of deaths, an estimated 18.6 million people died from cardiovascular diseases (CVDs) in 2019, representing 32.8% of all global deaths. Such a figure leaves no room for doubt on the priority of the medical treatment of CVDs, which in turn entails a thorough understanding of the biomechanics and mechanobiology of the human heart. Therefore, electro-mechanical investigations supported by predictive constitutive models are crucial in replacing the empirical approaches common in clinics (Dal et al. 2013). Experiments conducted by Lecarpentier et al. (1985) showed that muscle force reaches its peak point before the sarcomere is fully shortened, and sarcomere lengthening lasts longer after the muscle force returns to the preload level under isometric loading conditions. Furthermore, the isotonic test revealed the relation between afterload magnitude and time to peak velocity of muscle lengthening. Both of these peculiarities, which cannot be explained by elasticity, are attributed to viscosity. Additionally, viscoelasticity manifests itself in terms of hysteresis in the stress-strain curves of in vitro experiments. A canonical formulation for the electro-visco-elasticity of heart tissue has been recently proposed in terms of a modified Hill model (Cansiz et al. 2017). Moreover, finite element analysis based on bidomain models of cardiac electromechanics clearly demonstrates that viscosity significantly alters the fibrillation patterns observed in the myocardium (Cansiz et al. 2018).

1.1 Histological and Structural Features of Human Passive Myocardium

The human heart is regarded as a helical network of cardiac muscle fibers structured in laminar sheets. The pumping of the blood circulating in the body is conducted with the help of four chambers, i.e. the left and right *atria* and *ventricles*. The right ventricle (RV) pumps the venous blood into the *pulmonary circulation* while the left ventricle (LV) pumps the arterial blood to the *systemic circulation*. The high blood pressure in the LV is resisted by its relatively thicker wall compared to the RV.

The ventricular wall comprises three layers: the innermost *endocardium*, the middle *myocardium*, and the outermost *epicardium*. The epicardium and the endocardium are largely made of elastin and epimysial collagen, whereas the myocardium is composed of laminae of parallel myocytes involving endomysial and perimysial collagen and accordingly form a locally orthotropic structure with three distinct material axes (Humphrey 2002; Holzapfel and Ogden 2009).

Myocytes in muscle fibers are aligned across the wall that they course a right-handed helical pattern from the endocardium towards the mid-wall, while they change their course to form a left-handed helix between the mid-wall and the epicardium (Streeter Jr. et al. 1969; Sands et al. 2005, 2008). Besides, the laminae of fiber bundles, also called sheets, show similar transmural variations, see Sands et al. (2008) and LeGrice et al. (1995). In a recent study, Rohmer et al. (2007) visualized human heart samples and found out that the orientations of fibers and sheets differ in regard to the sub-locations in the LV, namely the *anterior*, *posterior*, *lateral*, and *septal* walls. Studies on animal samples are abundant, addressing the dispersion of both fiber and sheet arrangements in the heart, see Karlon et al. (1998), Usyk et al. (2001), and Strijkers et al. (2009). A more recent study by Sommer et al. (2015) shed light on the phenomena which quantitatively measures the in-plane and out-of-plane dispersions from mean fiber and sheet orientations for samples taken from human myocardium.

1.2 Mechanical Features of Human Passive Myocardium

The myocardial tissue is considered to exhibit incompressible material behavior under mechanical loading. The myocytes are particularly responsible for the active contraction of the heart, whereas the endomysial and perimysial collagen appear as the primary structural elements of the extracellular matrix, engendering passive mechanical response. The tissue features a locally orthotropic structure due to the specific arrangement of the endomysial and perimysial collagen in laminar sheets. The corresponding highly nonlinear orthotropic material behavior was observed in Dokos et al. (2002) and Sommer et al. (2015) via simple shear tests, i.e. the tested six simple shear modes yielded six distinct mechanical responses.

While the perimysial collagen network, essentially made of type I collagen (high strength), sits on the cleavage planes and connects the contiguous lamina,

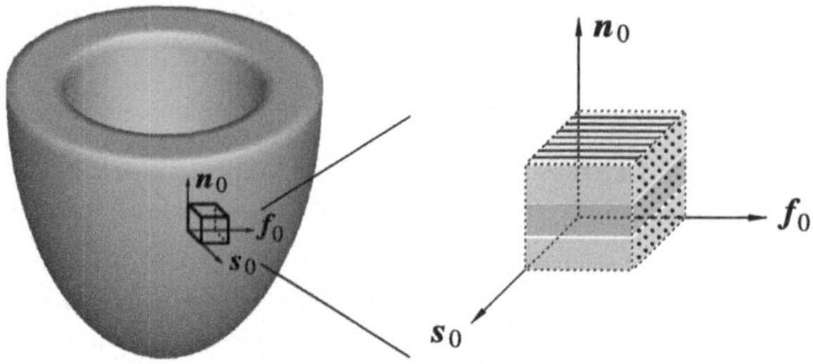

Fig. 1 Schematic diagram of a human left ventricle and the idealized orthotropic structure with its orthonormal basis vectors $\boldsymbol{f}_0$, $\boldsymbol{s}_0$, and $\boldsymbol{n}_0 = \boldsymbol{s}_0 \times \boldsymbol{f}_0$, defined by the endomysial and perimysial collagen, and the normal to cleavage planes, respectively

the endomysial fibers, mainly consisting of Type III collagen (highly deformable), are woven around the long axes of the myocytes, see Rohmer et al. (2007) and references therein. Given the idealized laminar organization of the myocardial tissue, the orthotropic material behavior is characterized by three orthonormal basis vectors: *fiber* $\boldsymbol{f}_0$, *sheet* $\boldsymbol{s}_0$, and *normal* $\boldsymbol{n}_0$, as illustrated in Fig. 1. The myocardial tissue has long been considered as a nonlinear hyperelastic material due to the lack of experimental evidence for its viscoelasticity. The first tangible evidence came from an experimental study on explanted pig hearts in Dokos et al. (2002). Hystereses observed in all shear modes for cyclic triaxial shear tests along with the stress relaxation during step tests indicated the viscoelastic features of the myocardium under shear deformation. Recently, a more affirmative investigation was conducted on the human myocardium in Sommer et al. (2015), where the viscoelastic mechanical response manifests itself in both biaxial and shear experiments with various stretch amplitudes. In addition, stress relaxation for biaxial extension and triaxial shear tests was addressed in that study.

The viscoelastic response of the passive myocardium is attributed to the extracellular fluid filtrating through the elastic body, see Tsaturyan et al. (1984) and Yao et al. (2012). Since the myocardial active behavior has negligible effects on the stress relaxation phenomenon, a purely passive process sounds logical. The extracellular matrix components *proteoglycans*, which regulate the fluid flow in and out of the articular cartilage, are probably responsible for the viscous flow in the myocardium, although ample evidence is yet to be found (Fomovsky et al. 2010).

1.3 Hyperelastic Modeling of Human Passive Myocardium

The directional dependence of the mechanical behavior of the myocardial tissue has been gradually phased in over the years, beginning with the *isotropic* models, see for instance Demiray (1972), continuing with the *transversely isotropic* models based on invariants, e.g., Humphrey et al. (1990), and on the Green-Lagrangian strain tensor, e.g., Guccione et al. (1991). Later, *orthotropic* models were proposed, see, e.g., Nash and Hunter (2000), Costa et al. (2001), and Schmid et al. (2006), in terms of a modified Fung-type model (Fung 1993). Nevertheless, the afore-stated models lack either the micro-mechanical motivation or the property to be polyconvex, a desirable aspect in the sense of the stability during finite element analyses. The model suggested by Holzapfel and Ogden (2009) is, however, an incompressible, thick-walled, orthotropic, polyconvex, and a micro-structurally motivated hyperelastic constitutive model for the passive myocardium based on invariants. The numerical performance of the model was addressed among others by, e.g., Göktepe et al. (2011). Afterwards, the dispersion in the collagen orientation was incorporated into that model in Eriksson et al. (2013). In addition, Bischoff et al. (2002) proposed another microstructurally-based finite hyperelastic model for a general class of polymeric materials.

1.4 Viscoelastic Modeling of Human Passive Myocardium

The earliest endeavor to model the viscoelasticity of myocardial tissue was elucidated in Löffler and Sagawa (1975), where a one-dimensional viscoelastic model accounts for both passive and active responses of the tissue. This model considers Maxwell elements for the passive branch, whereas either a Maxwell or a Voigt element describes the active branch. Later on, biphasic (fluid and solid) viscoelastic material laws were established for the passive myocardium that account for the possible effects of the extracellular fluid on the viscous properties. First of all, Yang and Taber (1991) proposed an isotropic biphasic model which was followed by transversely isotropic laws (Huyghe et al. 1991, 1992) based on the quasi-linear approach of Fung (1993), where the evolution of stresses is represented by convolution integrals. This model regards the elastic response as transversely isotropic, while the relaxation function reflects isotropy.

More descriptive finite strain linear anisotropic viscoelastic models were developed in Kaliske (2000), Holzapfel and Gasser (2001), and Holzapfel et al. (2002), where the stress update algorithm is obtained by means of the convolution integral representation. The evolution equations are ultimately motivated by a rheological analog, see Simo (1987) and Holzapfel (1996). Nonetheless, they provide inadequate validity from the perspective of material characterization.

The most recent orthotropic viscoelastic models on the passive myocardium were documented in Cansız et al. (2015) and Gültekin et al. (2016). The former adopts the hyperelastic model given in Holzapfel and Ogden (2009) and uses one-dimensional Maxwell-type rheological structures in the logarithmic strain space in $\boldsymbol{f}_0$, $\boldsymbol{s}_0$, and $\boldsymbol{n}_0$ directions for the viscous response. The predictive capability of the model is tested against the shear data obtained from the passive ventricular myocardium of pig hearts (Dokos et al. 2002). However, the latter features a more realistic baseline hyperelastic response as proposed in Eriksson et al. (2013) for the hyperelastic response, while the non-equilibrium (viscous) response is governed by stress evolution equations derived from the three-dimensional analogy of the generalized Maxwell model (Holzapfel 1996). Therein, each Maxwell element corresponds to a distinct response reflected by a specific invariant in the free-energy function, namely matrix, fiber, sheet, and relative shear between fiber and sheet. The performance of the model is assessed in accordance with the biaxial and triaxial shear test data in Sommer et al. (2015) obtained from human myocardium reflecting the stress hystereses along with the stress relaxation responses.

1.5 Scope of the Work

In this work, our aim is to assess the modeling capacity of the recently proposed orthotropic viscoelastic formulations, i.e. Cansız et al. (2015) and Gültekin et al. (2016), by comparing their respective fits against the triaxial shear test data documented in Sommer et al. (2015) on human passive myocardium. To this end, the baseline hyperelastic response of the tissue needs to be identical for both models, and therefore described by Eriksson et al. (2013), which takes into account the dispersion of fiber and sheet orientations from their mean directions.

The quasi-incompressible response of the myocardium is accounted for by the multiplicative decomposition of the deformation gradient into volumetric and isochoric parts in the sense of Flory (1961). This leads to the additive split of the Helmholtz free-energy function into *equilibrium* (hyperelastic) and *non-equilibrium* (viscous) contributions. Herein, it is assumed that the viscous contributions are purely volume-preserving.

The present chapter is organized as follows. In Sect. 2, we outline the general finite strain viscoelastic continuum body framework that applies to both models with an emphasis given to the *Lagrangian* configuration. Section 3 deals with the specific hyperelastic contribution due to the free-energy function and the stress and moduli expressions thereof. Section 4 is devoted to the distinct non-equilibrium responses of Cansız et al. (2015) and Gültekin et al. (2016), their corresponding evolution of stresses together with the algorithmic viscous tangent moduli. In the sequel, Sect. 5 is concerned with the elastic and viscous parameter estimations in accordance with the experimental data published in Sommer et al. (2015). Section 6 concludes the chapter.

2 Continuum Mechanics for Viscoelastic Passive Myocardium

In this section, we introduce the necessary kinematics of finite deformation and propose a general framework of viscoelasticity for human passive myocardium formulated in the Lagrangian description.

2.1 Kinematics of Finite Deformation for Myocardium

The mechanical response of a viscoelastic body is assumed to be described by two independent sets of field variables, State$(\boldsymbol{X}, t) := \{\boldsymbol{\varphi}(\boldsymbol{X}, t), \boldsymbol{\mathcal{I}}(\boldsymbol{X}, t)\}$, where the deformation map $\boldsymbol{\varphi}(\boldsymbol{X}, t)$ represents the nonlinear deformation field that transforms the material points $\boldsymbol{X} \in \text{Sym}[B]$ at time $t_0 \in \text{Sym}[T] \subset \mathbb{R}_+$ onto spatial points $\boldsymbol{x} = \boldsymbol{\varphi}_t(\boldsymbol{X}) \in \text{Sym}[S]$ at time $t \in \text{Sym}[T] \subset \mathbb{R}_+$, which is also called the current/Eulerian configuration of the material points. The internal field variables $\boldsymbol{\mathcal{I}}(\boldsymbol{X}, t)$, however, describe the time-dependent dissipative material response. Next, the deformation gradient $\mathbf{F} = \nabla_X \boldsymbol{\varphi}_t(\boldsymbol{X})$ maps the tangents of material curves onto those of spatial ones. We hereinafter follow Flory (1961) and use the multiplicative decomposition of the deformation gradient, i.e. $\mathbf{F} = \mathbf{F}_{\text{vol}}\overline{\mathbf{F}}$, into volumetric (spherical) $\mathbf{F}_{\text{vol}}$ and isochoric (distortional) $\overline{\mathbf{F}}$ parts such that

$$\mathbf{F}_{\text{vol}} = J^{1/3}\mathbf{1} \quad \text{and} \quad \overline{\mathbf{F}} = J^{-1/3}\mathbf{F}, \tag{1}$$

where $J = \det \mathbf{F} > 0$ is the volume ratio. The incompressibility condition is satisfied for $J = \det \mathbf{F}_{\text{vol}} = 1$. Therein, $\mathbf{1}$ indicates the second-order identity tensor. In addition, we introduce the right Cauchy-Green tensor and its unimodular counterpart

$$\mathbf{C} = \mathbf{F}^{\text{T}}\mathbf{F} \quad \text{and} \quad \overline{\mathbf{C}} = \overline{\mathbf{F}}^{\text{T}}\overline{\mathbf{F}}, \tag{2}$$

respectively. The invariant theory of hyperelasticity based on the *material objectivity* and the *principle of material frame invariance* demands the following isotropic invariants

$$\bar{I}_1 = \text{tr}\overline{\mathbf{C}}, \quad \bar{I}_2 = \frac{1}{2}\left[(\text{tr}\overline{\mathbf{C}})^2 - \text{tr}(\overline{\mathbf{C}}^2)\right], \quad \text{and} \quad \bar{I}_3 = \det \overline{\mathbf{C}} = 1, \tag{3}$$

for the description of the ground matrix response (Holzapfel 2000). The orthotropic material behavior is characterized by three orthonormal basis vectors: *fiber* $\boldsymbol{f}_0$, *sheet* $\boldsymbol{s}_0$, and *normal* $\boldsymbol{n}_0$ as indicated in Fig. 1. The associated unimodular invariants can be expressed as

$$\begin{aligned}
\bar{I}_{4\text{f}} &= \boldsymbol{f}_0 \cdot \overline{\mathbf{C}}\boldsymbol{f}_0, & \bar{I}_{4\text{s}} &= \boldsymbol{s}_0 \cdot \overline{\mathbf{C}}\boldsymbol{s}_0, & \bar{I}_{4\text{n}} &= \boldsymbol{n}_0 \cdot \overline{\mathbf{C}}\boldsymbol{n}_0, \\
\bar{I}_{5\text{f}} &= \boldsymbol{f}_0 \cdot \overline{\mathbf{C}}^2\boldsymbol{f}_0, & \bar{I}_{5\text{s}} &= \boldsymbol{s}_0 \cdot \overline{\mathbf{C}}^2\boldsymbol{s}_0, & \bar{I}_{5\text{n}} &= \boldsymbol{n}_0 \cdot \overline{\mathbf{C}}^2\boldsymbol{n}_0, \\
\bar{I}_{8\text{fs}} &= \boldsymbol{f}_0 \cdot \overline{\mathbf{C}}\boldsymbol{s}_0, & \bar{I}_{8\text{fn}} &= \boldsymbol{f}_0 \cdot \overline{\mathbf{C}}\boldsymbol{n}_0, & \bar{I}_{8\text{sn}} &= \boldsymbol{s}_0 \cdot \overline{\mathbf{C}}\boldsymbol{n}_0.
\end{aligned} \tag{4}$$

The unimodular invariants $\bar{I}_{4\text{f}}$ and $\bar{I}_{4\text{s}}$ need to be modified in order to account for the dispersion of collagen orientations along the $\boldsymbol{f}_0$ and $\boldsymbol{s}_0$ direction, a common phenomenon for a diseased myocardium. According to Eriksson et al. (2013), κ_f and κ_s are the two introduced scalar structural parameters representing fiber and sheet distributions, respectively. As a result, the modified unimodular fourth-invariants have the following form

$$\bar{I}_{4\mathrm{f}}^{*} = \kappa_{\mathrm{f}}\bar{I}_1 + (1 - 3\kappa_{\mathrm{f}})\bar{I}_{4\mathrm{f}} \quad \text{and} \quad \bar{I}_{4\mathrm{s}}^{*} = \kappa_{\mathrm{s}}\bar{I}_1 + (1 - 3\kappa_{\mathrm{s}})\bar{I}_{4\mathrm{s}}, \tag{5}$$

where $\kappa_i \in [0, 1/3]$, $i \in \{\mathrm{f}, \mathrm{s}\}$. For $\kappa_i = 0$, the fibers are perfectly aligned, i.e. $\bar{I}_{4\mathrm{f}}$ or $\bar{I}_{4\mathrm{s}}$ is recovered. On the other hand, isotropy is recovered for $\kappa_i = 1/3$. Let the respective structure tensors have the following form,

$$\begin{aligned} &\mathbf{H}_{\mathrm{f}} = \boldsymbol{f}_0 \otimes \boldsymbol{f}_0, &&\mathbf{H}_{\mathrm{s}} = \boldsymbol{s}_0 \otimes \boldsymbol{s}_0, &&\mathbf{H}_{\mathrm{n}} = \boldsymbol{n}_0 \otimes \boldsymbol{n}_0, \\ &\mathbf{H}_{\mathrm{fs}} = \mathrm{sym}(\boldsymbol{f}_0 \otimes \boldsymbol{s}_0), &&\mathbf{H}_{\mathrm{fn}} = \mathrm{sym}(\boldsymbol{f}_0 \otimes \boldsymbol{n}_0), &&\mathbf{H}_{\mathrm{sn}} = \mathrm{sym}(\boldsymbol{s}_0 \otimes \boldsymbol{n}_0), \end{aligned} \tag{6}$$

along with the modified structure tensors, i.e.

$$\mathbf{H}_{\mathrm{f}}^{*} = \kappa_{\mathrm{f}}\mathbf{1} + (1 - 3\kappa_{\mathrm{f}})\boldsymbol{f}_0 \otimes \boldsymbol{f}_0 \quad \text{and} \quad \mathbf{H}_{\mathrm{s}}^{*} = \kappa_{\mathrm{s}}\mathbf{1} + (1 - 3\kappa_{\mathrm{s}})\boldsymbol{s}_0 \otimes \boldsymbol{s}_0. \tag{7}$$

Note that Eqs. (6) and (7) are double-contracted with the unimodular part of the Cauchy-Green tensor $\overline{\mathbf{C}}$ yielding the corresponding invariants given in Eqs. (4) and (5), respectively. For the description of the canonical viscous material response in Sect. 4.2, the following logarithmic strains are introduced

$$\varepsilon_{\mathrm{f}} := \frac{1}{2}\ln(\sqrt{\bar{I}_{4\mathrm{f}}}), \qquad \varepsilon_{\mathrm{s}} := \frac{1}{2}\ln(\sqrt{\bar{I}_{4\mathrm{s}}}), \qquad \varepsilon_{\mathrm{n}} := \frac{1}{2}\ln(\sqrt{\bar{I}_{4\mathrm{n}}}), \tag{8}$$

in the sense of Miehe and Göktepe (2005) and Cansız et al. (2015). The specific choice gives rise to the strain-like internal variables that are $\mathcal{I}(\boldsymbol{X}, t) = \{\alpha_{\mathrm{f}}, \alpha_{\mathrm{s}}, \alpha_{\mathrm{n}}\}$ for the canonical representation, whereas there is no specific form required for the convolutional representation, see Sect. 4.1.

2.2 General Continuum Framework of Orthotropic Viscoelasticity

The local balance of the linear and angular momentum serve as basic sets of equations in continuum mechanics to set up the quasi-static equilibrium

$$\mathrm{DIV}\,(\mathbf{FS}) + \rho_0\boldsymbol{\gamma} = \mathbf{0}, \quad \mathbf{S} = \mathbf{S}^{\mathrm{T}}, \tag{9}$$

where $\mathbf{S}$ is the second Piola-Kirchhoff stress tensor and $\rho_0\boldsymbol{\gamma}$ is the body force per unit volume in the reference configuration. Next, we define the *Clausius-Planck* inequality for an isothermal process, i.e.

$$\mathcal{D}_{\mathrm{loc}} := \frac{1}{2}\mathbf{S} : \dot{\mathbf{C}} - \dot{\Psi} \geq 0, \tag{10}$$

from which the stress response can readily be derived, see Holzapfel (2000). Therein, $\mathcal{D}_{\text{loc}}$ and $\dot{\Psi}$ stand for the local part of the dissipation and the temporal evolution of the free-energy function. In the sequel, the free-energy for a dissipative material is defined in an additively split form characterizing the volumetric and isochoric parts,

$$\Psi := \hat{\Psi}(\mathbf{C}, \mathbf{H}; \mathcal{I}) = \hat{\Psi}_{\text{vol}}(J) + \hat{\bar{\Psi}}(\overline{\mathbf{C}}, \mathbf{H}; \mathcal{I}), \tag{11}$$

for which the structure tensors and internal variables entering the sets $\mathbf{H}$ and $\mathcal{I}$ are determined both in Sect. 3 and Sect. 4, respectively. Substitution of the material time derivative of Eq. (11) into Eq. (10) leads to

$$\mathcal{D}_{\text{loc}} := (\mathbf{S} - 2\partial_{\mathbf{C}}\Psi) : \dot{\mathbf{C}}/2 - \partial_{\mathcal{I}}\Psi : \dot{\mathcal{I}} \geq 0. \tag{12}$$

The Coleman-Noll exploitation, see Coleman and Gurtin (1967), of Eq. (12) implies the particular form for the constitutive equations, i.e.

$$\mathbf{S} = 2\partial_{\mathbf{C}}\Psi = 2\partial_{\mathbf{C}}\hat{\Psi}(\mathbf{C}, \mathbf{H}; \mathcal{I}) \quad \text{and} \quad \boldsymbol{\beta} := -\partial_{\mathcal{I}}\Psi = -\partial_{\mathcal{I}}\hat{\Psi}(\mathbf{C}, \mathbf{H}; \mathcal{I}), \tag{13}$$

where $\boldsymbol{\beta}$ designates per definition the *internal force*, which can also be regarded as the *viscous overstress*. Then, the reduced dissipation inequality reads

$$\mathcal{D}_{\text{loc}}^{\text{red}} := \boldsymbol{\beta} : \dot{\mathcal{I}} \geq 0. \tag{14}$$

Note that this constraint must be satisfied by choosing thermodynamically consistent evolution equations for the internal variables. The second Piola-Kirchhoff stress tensor in Eq. $(13)_1$ consists of volumetric and isochoric contributions, according to

$$\mathbf{S} = \mathbf{S}_{\text{vol}} + \mathbf{S}_{\text{iso}}, \quad \text{where} \quad \mathbf{S}_{\text{vol}} := 2\partial_{\mathbf{C}}\hat{\Psi}_{\text{vol}}(J) \quad \text{and} \quad \mathbf{S}_{\text{iso}} := 2\partial_{\mathbf{C}}\hat{\bar{\Psi}}(\overline{\mathbf{C}}, \mathbf{H}; \mathcal{I}). \tag{15}$$

The volumetric part of the second Piola-Kirchhoff stress tensor can be recast into the following form

$$\mathbf{S}_{\text{vol}} = \hat{\Psi}'_{\text{vol}}(J)J\mathbf{C}^{-1}, \quad \text{where} \quad \hat{\Psi}'_{\text{vol}}(J) = \partial_J\hat{\Psi}_{\text{vol}}(J). \tag{16}$$

The isochoric part of the second Piola-Kirchhoff stress tensor can be rewritten as

$$\mathbf{S}_{\text{iso}} = \bar{\mathbf{S}} : \mathbb{Q} \quad \text{with} \quad \bar{\mathbf{S}} := 2\,\partial_{\overline{\mathbf{C}}}\hat{\bar{\Psi}}(\overline{\mathbf{C}}, \mathbf{H}, \mathcal{I}) \quad \text{and} \quad \mathbb{Q} := \partial_{\mathbf{C}}\overline{\mathbf{C}}, \tag{17}$$

where $\bar{\mathbf{S}}$ is the second Piola-Kirchhoff stress tensor conjugate to the unimodular part of the right Cauchy-Green tensor $\overline{\mathbf{C}}$, whereas $\mathbb{Q}$ stands for the Lagrangian deviatoric projection tensor that reads

$$\mathbb{Q} = J^{-2/3}\left(\mathbb{I} - \frac{1}{3}\mathbf{C} \otimes \mathbf{C}^{-1}\right), \quad \text{where} \quad \mathbb{I} := \frac{1}{2}\left(\mathbf{1}\overline{\otimes}\mathbf{1} + \mathbf{1}\underline{\otimes}\mathbf{1}\right) \tag{18}$$

is the fourth-order symmetric identity tensor.[1] $\bar{\mathbf{S}}$ is further decomposed into the elastic (equilibrium) and viscous (non-equilibrium) parts, i.e.

$$\bar{\mathbf{S}} = \bar{\mathbf{S}}^{\mathrm{e}} + \bar{\mathbf{S}}^{\mathrm{v}}. \tag{19}$$

Accordingly, we can decompose the isochoric free-energy function into the elastic (equilibrium) and viscous (non-equilibrium) parts as well, i.e.

$$\hat{\bar{\Psi}}(\overline{\mathbf{C}}, \mathbf{H}; \mathcal{I}) = \hat{\bar{\Psi}}^{\mathrm{e}}(\overline{\mathbf{C}}, \mathbf{H}) + \hat{\bar{\Psi}}^{\mathrm{v}}(\overline{\mathbf{C}}, \mathbf{H}; \mathcal{I}). \tag{20}$$

Then, the elastic (equilibrium) $\bar{\mathbf{S}}^{\mathrm{e}}$ and the viscous (non-equilibrium) $\bar{\mathbf{S}}^{\mathrm{v}}$ parts can be defined as

$$\bar{\mathbf{S}}^{\mathrm{e}} := 2\partial_{\overline{\mathbf{C}}}\hat{\bar{\Psi}}^{\mathrm{e}}(\overline{\mathbf{C}}, \mathbf{H}) \quad \text{and} \quad \bar{\mathbf{S}}^{\mathrm{v}} := 2\partial_{\overline{\mathbf{C}}}\hat{\bar{\Psi}}^{\mathrm{v}}(\overline{\mathbf{C}}, \mathbf{H}; \mathcal{I}). \tag{21}$$

Next, we turn our focus on the algorithmic tangent moduli, i.e.

$$\Delta\mathbf{S} := \mathbb{C}^{\mathrm{algo}} : \frac{1}{2}\Delta\mathbf{C}, \quad \text{where} \quad \mathbb{C}^{\mathrm{algo}} = \mathbb{C}_{\mathrm{vol}} + \mathbb{C}_{\mathrm{iso}}^{\mathrm{algo}} \tag{22}$$

with volumetric and isochoric parts. The volumetric part of the tangent moduli result from the second derivative of the $\hat{\Psi}_{\mathrm{vol}}(J)$ with respect to $\mathbf{C}$,

$$\mathbb{C}_{\mathrm{vol}} := 4\,\partial^2_{\mathbf{CC}}\hat{\Psi}_{\mathrm{vol}}(J) = \left[J^2\hat{\Psi}''_{\mathrm{vol}}(J) + J\hat{\Psi}'_{\mathrm{vol}}(J)\right]\mathbf{C}^{-1}\otimes\mathbf{C}^{-1} - 2J\mathbb{I}_{\mathbf{C}^{-1}}, \tag{23}$$

where we have defined $\hat{\Psi}''_{\mathrm{vol}}(J) := \partial^2_{JJ}\hat{\Psi}_{\mathrm{vol}}(J)$. In Eq. (23), $\mathbb{I}_{\mathbf{C}^{-1}}$ denotes a geometric transformation tensor and can be expressed as

$$\mathbb{I}_{\mathbf{C}^{-1}} := \frac{1}{2}\left(\mathbf{C}^{-1}\overline{\otimes}\mathbf{C}^{-1} + \mathbf{C}^{-1}\underline{\otimes}\mathbf{C}^{-1}\right). \tag{24}$$

In what follows, we describe the isochoric part of the algorithmic moduli

$$\mathbb{C}_{\mathrm{iso}}^{\mathrm{algo}} = \mathbb{Q}^{\mathrm{T}} : \overline{\mathbb{C}}^{\mathrm{algo}} : \mathbb{Q} + \frac{2}{3}\left(\bar{\mathbf{S}} : \overline{\mathbf{C}}\right)\mathbb{P}_{\mathbf{C}^{-1}} - \frac{2}{3}\left(\mathbf{S}_{\mathrm{iso}}\otimes\mathbf{C}^{-1} + \mathbf{C}^{-1}\otimes\mathbf{S}_{\mathrm{iso}}\right) \tag{25}$$

along with the definition of the geometric transformation tensor

$$\mathbb{P}_{\mathbf{C}^{-1}} := \mathbb{I}_{\mathbf{C}^{-1}} - \frac{1}{3}\mathbf{C}^{-1}\otimes\mathbf{C}^{-1}. \tag{26}$$

Analogous to Eq. (19), the algorithmic tangent conjugate to the unimodular part of the right Cauchy-Green tensor $\overline{\mathbf{C}}$ in Eq. (25) is also split into elastic (equilibrium) and viscous (non-equilibrium) parts, i.e.

---

[1] Different dyadic products used in Eqs. $(18)_2$ and (24) are defined as $(\boldsymbol{A}\overline{\otimes}\boldsymbol{B})_{ijkl} = A_{ik}B_{jl}$ and $(\boldsymbol{A}\underline{\otimes}\boldsymbol{B})_{ijkl} = A_{il}B_{jk}$.

$$\overline{\mathbb{C}}^{\text{algo}} = \overline{\mathbb{C}}_{\text{e}} + \overline{\mathbb{C}}_{\text{v}}^{\text{algo}}, \tag{27}$$

where the elastic and the algorithmic viscous parts can be delineated as

$$\overline{\mathbb{C}}_{\text{e}} = 4\,\partial^2_{\overline{\mathbf{C}}\,\overline{\mathbf{C}}}\hat{\overline{\Psi}}^{\text{e}}(\overline{\mathbf{C}}, \mathbf{H}) \quad \text{and} \quad \Delta\mathbf{S}_{\text{v}} = \overline{\mathbb{C}}_{\text{v}}^{\text{algo}} : \frac{1}{2}\Delta\mathbf{C}. \tag{28}$$

Recall that the closed-form expression for the elastic part of the material tangent can be obtained from the free-energy function, whereas the viscous part of the tangent can be derived numerically in a time-discrete setting.

3 Equilibrium Constitutive Relations for Passive Myocardium

In this section, the volumetric and the isochoric elastic constitutive relations are determined in accordance with the orthotropic hyperelastic response of human passive myocardium. The specific form of the volumetric part of the free-energy function is stated, i.e.

$$\hat{\Psi}_{\text{vol}}(J) = \kappa(J - \ln J - 1), \tag{29}$$

enforcing the incompressibility through the bulk modulus κ. The following derivatives of Eq. (29) with respect to the Jacobian J,

$$\hat{\Psi}'_{\text{vol}}(J) = \kappa(1 - 1/J) \quad \text{and} \quad \hat{\Psi}''_{\text{vol}}(J) = \kappa(1/J^2), \tag{30}$$

are substituted in Eqs. (16) and (23) in order to obtain the final expressions for the volumetric parts of the stress and tangent moduli. The definition of the isochoric elastic free-energy function in Eq. (20) follows from Eriksson et al. (2013), i.e.

$$\begin{aligned}\hat{\overline{\Psi}}^{\text{e}}(\overline{\mathbf{C}}, \mathbf{H}) = {} & \frac{a}{2b}\left\{\exp[b(\bar{I}_1 - 3)] - 1\right\} + \sum_{i=\text{f,s}} \frac{a_i}{2b_i}\left\{\exp[b_i(\bar{I}^*_{4i} - 1)^2] - 1\right\} \\ & + \frac{a_{\text{fs}}}{2b_{\text{fs}}}\left\{\exp[b_{\text{fs}}\bar{I}^2_{8\text{fs}}] - 1\right\},\end{aligned} \tag{31}$$

reflecting ground matrix, fiber, sheet, and fiber-sheet contributions, where a, a_{f}, a_{s}, a_{fs} denote the stress-like material constants, while b, b_{f}, b_{s}, b_{fs} are dimensionless constants. Next, the elastic (equilibrium) stress tensor $\bar{\mathbf{S}}^{\text{e}}$ in Eq. (21)$_1$ can be expressed by

$$\bar{\mathbf{S}}^{\text{e}} = \hat{\overline{\Psi}}^{\text{e}}_1\,\mathbf{1} + \hat{\overline{\Psi}}^{\text{e}}_{4\text{f}}\,\mathbf{H}^*_f + \hat{\overline{\Psi}}^{\text{e}}_{4\text{s}}\,\mathbf{H}^*_{\text{s}} + \hat{\overline{\Psi}}^{\text{e}}_{8\text{fs}}\,\mathbf{H}_{\text{fs}}, \tag{32}$$

where deformation-dependent scalar elastic stress coefficients read

$$
\begin{aligned}
\hat{\bar{\Psi}}^{\mathrm{e}}_{1} &:= 2\partial_{\bar{I}_1}\hat{\bar{\Psi}}^{\mathrm{e}} = a\,\exp[b(\bar{I}_1 - 3)],\\
\hat{\bar{\Psi}}^{\mathrm{e}}_{4\mathrm{f}} &:= 2\partial_{\bar{I}^*_{4\mathrm{f}}}\hat{\bar{\Psi}}^{\mathrm{e}} = 2a_{\mathrm{f}}(\bar{I}^*_{4\mathrm{f}} - 1)\,\exp[b_{\mathrm{f}}(\bar{I}^*_{4\mathrm{f}} - 1)^2],\\
\hat{\bar{\Psi}}^{\mathrm{e}}_{4\mathrm{s}} &:= 2\partial_{\bar{I}^*_{4\mathrm{s}}}\hat{\bar{\Psi}}^{\mathrm{e}} = 2a_{\mathrm{s}}(\bar{I}^*_{4\mathrm{s}} - 1)\,\exp[b_{\mathrm{s}}(\bar{I}^*_{4\mathrm{s}} - 1)^2],\\
\hat{\bar{\Psi}}^{\mathrm{e}}_{8\mathrm{fs}} &:= 2\partial_{\bar{I}_{8\mathrm{fs}}}\hat{\bar{\Psi}}^{\mathrm{e}} = 2a_{\mathrm{fs}}\bar{I}_{8\mathrm{fs}}\,\exp(b_{\mathrm{fs}}\bar{I}^2_{8\mathrm{fs}}).
\end{aligned}
\tag{33}
$$

In the subsequent treatment, we identify the elastic unimodular part of the tangent moduli $\overline{\mathbb{C}}_{\mathrm{e}}$ in Eq. (28)$_1$, i.e.

$$
\overline{\mathbb{C}}_{\mathrm{e}} = \hat{\bar{\Psi}}^{\mathrm{e}\prime}_{1}\,\mathbf{1}\otimes\mathbf{1} + \hat{\bar{\Psi}}^{\mathrm{e}\prime}_{4\mathrm{f}}\,\mathbf{H}^*_{\mathrm{f}}\otimes\mathbf{H}^*_{\mathrm{f}} + \hat{\bar{\Psi}}^{\mathrm{e}\prime}_{4\mathrm{s}}\,\mathbf{H}^*_{\mathrm{s}}\otimes\mathbf{H}^*_{\mathrm{s}} + \hat{\bar{\Psi}}^{\mathrm{e}\prime}_{8\mathrm{fs}}\,\mathbf{H}_{\mathrm{fs}}\otimes\mathbf{H}_{\mathrm{fs}}. \tag{34}
$$

A closer look into the deformation-dependent scalar elastic moduli coefficients yields

$$
\begin{aligned}
\hat{\bar{\Psi}}^{\mathrm{e}\prime}_{1} &:= 4\partial^2_{\bar{I}_1\bar{I}_1}\hat{\bar{\Psi}}^{\mathrm{e}} = 2ab\,\exp[b(\bar{I}_1 - 3)],\\
\hat{\bar{\Psi}}^{\mathrm{e}\prime}_{4\mathrm{f}} &:= 4\partial^2_{\bar{I}^*_{4\mathrm{f}}\bar{I}^*_{4\mathrm{f}}}\hat{\bar{\Psi}}^{\mathrm{e}} = 4a_{\mathrm{f}}[1 + 2b_{\mathrm{f}}(\bar{I}^*_{4\mathrm{f}} - 1)^2]\,\exp[b_{\mathrm{f}}(\bar{I}^*_{4\mathrm{f}} - 1)^2],\\
\hat{\bar{\Psi}}^{\mathrm{e}\prime}_{4\mathrm{s}} &:= 4\partial^2_{\bar{I}^*_{4\mathrm{s}}\bar{I}^*_{4\mathrm{s}}}\hat{\bar{\Psi}}^{\mathrm{e}} = 4a_{\mathrm{s}}[1 + 2b_{\mathrm{s}}(\bar{I}^*_{4\mathrm{s}} - 1)^2]\,\exp[b_{\mathrm{s}}(\bar{I}^*_{4\mathrm{s}} - 1)^2],\\
\hat{\bar{\Psi}}^{\mathrm{e}\prime}_{8\mathrm{fs}} &:= 4\partial^2_{\bar{I}_{8\mathrm{fs}}\bar{I}_{8\mathrm{fs}}}\hat{\bar{\Psi}}^{\mathrm{e}} = 4a_{\mathrm{fs}}(1 + 2b_{\mathrm{fs}}\bar{I}^2_{8\mathrm{fs}})\,\exp(b_{\mathrm{fs}}\bar{I}^2_{8\mathrm{fs}}).
\end{aligned}
\tag{35}
$$

4 Non-Equilibrium Constitutive Relations for Passive Myocardium

In this section, we elaborate on two distinct constitutive formulations for the orthotropic viscous response of myocardial tissue in the sense of Cansız et al. (2015) and Gültekin et al. (2016), respectively. The formalism employed in Gültekin et al. (2016) pursues a convolution representation of the viscous overstresses, while Cansız et al. (2015) delivers a canonical representation in terms of strain-like internal variables conjugate to driving viscous stresses.

4.1 Convolution Representation of Orthotropic Viscoelasticity

The convolution description of the viscous response omits the specific characterization of the internal variables $\mathcal{I}(\boldsymbol{X}, t)$, thereby circumventing the need for the specific free-energy function for the viscous part in Eq. (20)$_2$. Hence, instead of what has been defined in Eq. (21)$_2$, the following first-order linear differential equation (rate equation) together with the initial values govern the evolution of the viscous overstresses in line with the orthotropic response within a semi-closed time interval $t \in (0, T]$, i.e.

$$\dot{\bar{\mathbf{S}}}^{\mathrm{v}}_i + \frac{\bar{\mathbf{S}}^{\mathrm{v}}_i}{\tau_i} = \xi_i \dot{\bar{\mathbf{S}}}^{\mathrm{e}}_i, \quad \bar{\mathbf{S}}^{\mathrm{v}}_i\big|_{t=0} = \mathbf{0}, \quad \text{and} \quad \bar{\mathbf{S}}^{\mathrm{v}} = \sum_{i=\mathrm{m,f,s,fs}} \bar{\mathbf{S}}^{\mathrm{v}}_i, \tag{36}$$

where all the viscous effects a priori vanish at the stress-free reference configuration. The parameter constants specify the non-dimensional *free-energy factors* $\xi_i \in [0, \infty)$ and *relaxation/retardation times* $\tau_i \in (0, \infty)$. A closed-form solution of Eq. (36) is the convolution integral expression

$$\bar{\mathbf{S}}^{\mathrm{v}}_i = \int_0^t \exp\left(-\frac{t-s}{\tau_i}\right) \xi_i \dot{\bar{\mathbf{S}}}^{\mathrm{e}}_i \mathrm{d}s. \tag{37}$$

Let $[t_n, t_{n+1}]$ be the sub-interval of time for which the time increment $\Delta t = t_{n+1} - t_n$ describes the algorithmic formulation. Following the derivation in Simo (1987) and Holzapfel (2000), the integral in Eq. (37) is split up between $[0, t_n]$ and $[t_n, t_{n+1}]$. Upon some elementary mathematical manipulations, we obtain by using the *mid-point rule* the current viscous (non-equilibrium) stress tensor $\bar{\mathbf{S}}^{\mathrm{v}}_i$ in Eq. $(36)_3$, i.e.

$$\bar{\mathbf{S}}^{\mathrm{v}}_{i,\,n+1} = \exp\left(-\frac{\Delta t}{2\tau_i}\right)\left[\exp\left(-\frac{\Delta t}{2\tau_i}\right)\bar{\mathbf{S}}^{\mathrm{v}}_{i,\,n} - \xi_i \bar{\mathbf{S}}^{\mathrm{e}}_{i,\,n}\right] + \exp\left(-\frac{\Delta t}{2\tau_i}\right)\xi_i \bar{\mathbf{S}}^{\mathrm{e}}_{i,\,n+1}, \tag{38}$$

associated with the ground matrix, fiber, sheet, and fiber-sheet contributions that directly arise from the hyperelastic response Eq. (31). The corresponding viscous unimodular part of the tangent moduli $\overline{\mathbb{C}}^{\mathrm{algo}}_{\mathrm{v}}$ in Eq. $(28)_2$ can be obtained as follows

$$\overline{\mathbb{C}}^{n+1}_{\mathrm{v},\,i} = \exp\left(-\frac{\Delta t}{2\tau_i}\right)\xi_i \overline{\mathbb{C}}^{n+1}_{\mathrm{v},\,i} \quad \text{and} \quad \overline{\mathbb{C}}^{\mathrm{algo}}_{\mathrm{v}} = \sum_{i=\mathrm{m,f,s,fs}} \overline{\mathbb{C}}^{n+1}_{\mathrm{v},\,i}. \tag{39}$$

4.2 Canonical Representation of Orthotropic Viscoelasticity

The specific form of the viscous (non-equilibrium) part of the isochoric free-energy function in Eq. (20) can be described in a quadratic form, i.e.

$$\hat{\bar{\Psi}}^{\mathrm{v}}(\overline{\mathbf{C}}, \mathbf{H}; \mathcal{I}) = \hat{\bar{\Psi}}^{\mathrm{v}}(\boldsymbol{\varepsilon}; \mathcal{I}) = \sum_{i=\mathrm{f,s,n}} \hat{\bar{\Psi}}^{\mathrm{v}}_i(\varepsilon_i, \alpha_i), \quad \text{where} \quad \hat{\bar{\Psi}}^{\mathrm{v}}_i(\varepsilon_i, \alpha_i) = \frac{1}{2}\mu^{\mathrm{v}}_i(\varepsilon_i - \alpha_i)^2, \tag{40}$$

in line with the logarithmic strain space, see Eq. (8), corresponding to the fiber $\boldsymbol{f}_0$, sheet $\boldsymbol{s}_0$, and normal $\boldsymbol{n}_0$ directions. Therein, μ^{v}_i and α_i are the non-equilibrium shear moduli and the strain-like internal variables, respectively. The logarithmic form in Eq. (40) maps the multiplicative characteristics of large-strain viscoelasticity to the additive structure of the geometrically linear theory, see Miehe and Göktepe (2005) for further details. The logarithmic stresses and the thermodynamical forces conjugate to the internal variables are obtained according to

$$\sigma_i := \partial_{\varepsilon_i} \hat{\bar{\Psi}}_i^{\mathrm{v}}(\varepsilon_i, \alpha_i) \quad \text{and} \quad \beta_i := -\partial_{\alpha_i} \hat{\bar{\Psi}}_i^{\mathrm{v}}(\varepsilon_i, \alpha_i), \tag{41}$$

where $\sigma_i = \beta_i$ due to the particular choice in Eq. (40). It should be emphasized that the non-equilibrium free-energy function in Eq. (40) embodies the energy stored in the spring of a Maxwell element. Hence, we now elucidate the dissipated energy in the dashpot of the same Maxwell element via a generic power-type dissipation potential,

$$\Phi(\dot{\mathcal{I}}) = \sum_{i=\mathrm{f,s,n}} \Phi_i(\dot{\alpha}_i), \quad \text{where} \quad \Phi_i(\dot{\alpha}_i) = \frac{\hat{\beta}_i^2}{\eta_i} \frac{1+m_i}{2+m_i} \left(\frac{\eta_i}{\hat{\beta}_i} |\dot{\alpha}_i| \right)^{\frac{2+m_i}{1+m_i}} \tag{42}$$

governed by the material constants $\hat{\beta}_i, \eta_i$, and m_i designating the *activation stress*, *viscosity*, and *linear-nonlinear transition* parameters, respectively. The use of Eqs. (40) and (42) in conjunction with Biot's equation (Biot Biot (1965)), which states

$$\partial_{\alpha_i} \hat{\bar{\Psi}}_i^{\mathrm{v}}(\varepsilon_i, \alpha_i) + \partial_{\dot{\alpha}_i} \Phi_i(\dot{\alpha}_i) = 0 \quad \text{and} \quad \alpha_i(0) = \alpha_i^0, \tag{43}$$

give rise to the evolution equation for the internal variables, i.e.

$$\dot{\alpha}_i = \gamma_i^{\mathrm{v}}(\beta_i), \quad \text{where} \quad \gamma_i^{\mathrm{v}}(\beta_i) = \frac{1}{\eta_i} \left| \frac{\beta_i}{\hat{\beta}_i} \right|^{m_i} \beta_i, \tag{44}$$

through some mathematical operations. Note that for $m > 0$, γ_i^{v} becomes a nonlinear function of β_i and the update of the internal variables α_i must be evaluated by local Newton iterations in a time-discrete setting.

In order to compute the viscous overstress $\bar{\mathbf{S}}^{\mathrm{v}}$ in Eq. $(21)_2$, the current value of the internal variables α_i must be calculated from the evolution equation, see Eq. (44). To this end, we first consider discrete time steps in an incremental interval $[t_n, t_{n+1}]$. In a deformation-driven framework, all state variables are assumed to be known at time t_n, and the total deformation at time t_{n+1} is prescribed. For the computation of the internal variables $\alpha_{i,n+1}$ at the current time-step, let us introduce the time-discrete form of Eq. $(44)_1$ by resorting to the unconditionally stable backward-Euler integration scheme yielding

$$\alpha_{i,n+1} = \alpha_{i,n} + \Delta t \gamma_i^{\mathrm{v}}(\beta_{i,n+1}). \tag{45}$$

For $m_i > 0$, the evolution equation, Eq. $(44)_2$ becomes a nonlinear function of α_i with no closed-form solution. Hence, the computation of the internal variables demands a Newton-type iterative solution scheme. To this end, we define a residual expression r_i as a function of the internal variable, i.e.

$$r_i(\alpha_{i,n+1}) := \alpha_{i,n+1} - \alpha_{i,n} - \Delta t\, \gamma_i^{\mathrm{v}}(\beta_{i,n+1}) = 0, \tag{46}$$

which needs to be linearized and solved for the new local Newton iteration step $k+1$,

$$\alpha_{i,n+1}^{k+1} = \alpha_{i,n+1}^{k} - r_i^k \,/\, \mathcal{K}_i^k, \tag{47}$$

in which the local tangent term $\mathcal{K}_i$ is defined as

$$\mathcal{K}_i^k := \partial_{\alpha_{i,n+1}} r_i \big|_{\alpha_{i,n+1}^k} = 1 + \Delta t \frac{\mu_i^{\mathrm{v}}(m_i+1)}{\eta_i} \left| \frac{\beta_{i,n+1}^k}{\hat{\beta}_i} \right|^{m_i}. \tag{48}$$

Further details on the Newton-type iteration scheme can be found in Cansız et al. (2015). For the sake of conciseness, $n+1$ henceforth drops out of the equations. Then, the current viscous (non-equilibrium) stress $\bar{\mathbf{S}}^{\mathrm{v}}$ reads

$$\bar{\mathbf{S}}^{\mathrm{v}} = \hat{\bar{\Psi}}_{4\mathrm{f}}^{\mathrm{v}}\, \mathbf{H}_{\mathrm{f}} + \hat{\bar{\Psi}}_{4\mathrm{s}}^{\mathrm{v}}\, \mathbf{H}_{\mathrm{s}} + \hat{\bar{\Psi}}_{4\mathrm{n}}^{\mathrm{v}}\, \mathbf{H}_{\mathrm{n}}, \tag{49}$$

via the substitution of Eq. (40) into Eq. $(21)_2$. Therein, the deformation-dependent scalar viscous stress coefficients are defined as

$$\hat{\bar{\Psi}}_{4i}^{\mathrm{v}} := 2\partial_{\bar{I}_{4i}} \hat{\bar{\Psi}}^{\mathrm{v}}(\boldsymbol{\varepsilon}; \mathcal{I}) = \frac{\sigma_i}{\bar{I}_{4i}}, \quad \text{where} \quad \sigma_i = \mu_i^{\mathrm{v}}(\boldsymbol{\varepsilon}_i - \alpha_i). \tag{50}$$

The algorithmic viscous tangent moduli, defined in Eq. $(28)_2$, can be further elaborated as

$$\overline{\mathbb{C}}_{\mathrm{v}}^{\mathrm{algo}} = 2 \sum_{i=\mathrm{f,s,n}} \left(\partial_{\bar{I}_{4i}} \bar{\mathbf{S}}^{\mathrm{v}} \otimes \partial_{\overline{\mathbf{C}}} \bar{I}_{4i} + \partial_{\alpha_i} \bar{\mathbf{S}}^{\mathrm{v}} \otimes \partial_{\overline{\mathbf{C}}} \alpha_i \right), \tag{51}$$

by exploiting the definitions in Eqs. (40) and (49). The latter expression in Eq. (51) requires the use of *implicit function theorem* on the residual expression in Eq. (46) leading to

$$\overline{\mathbb{C}}_{\mathrm{v}}^{\mathrm{algo}} = \hat{\bar{\Psi}}_{4\mathrm{f}}^{\mathrm{v}\,\prime}\, \mathbf{H}_{\mathrm{f}} \otimes \mathbf{H}_{\mathrm{f}} + \hat{\bar{\Psi}}_{4\mathrm{s}}^{\mathrm{v}\,\prime}\, \mathbf{H}_{\mathrm{s}} \otimes \mathbf{H}_{\mathrm{s}} + \hat{\bar{\Psi}}_{4\mathrm{n}}^{\mathrm{v}\,\prime}\, \mathbf{H}_{\mathrm{n}} \otimes \mathbf{H}_{\mathrm{n}}. \tag{52}$$

The deformation-dependent scalar viscous moduli coefficients in Eq. (52) are defined as follows

$$\hat{\bar{\Psi}}_{4i}^{\mathrm{v}\,\prime} := \frac{(c_i - 2\sigma_i)}{\bar{I}_{4i}^2}, \quad \text{where} \quad c_i = \frac{\mu_i^{\mathrm{v}}}{\mathcal{K}_i}. \tag{53}$$

5 Representative Example: Triaxial Shear Tests

In the preceding sections, both the convolution and canonical representation of finite orthotropic viscoelasticity have been covered in terms of their continuum mechanical and algorithmic framework. The objective of this section is to assess the modeling

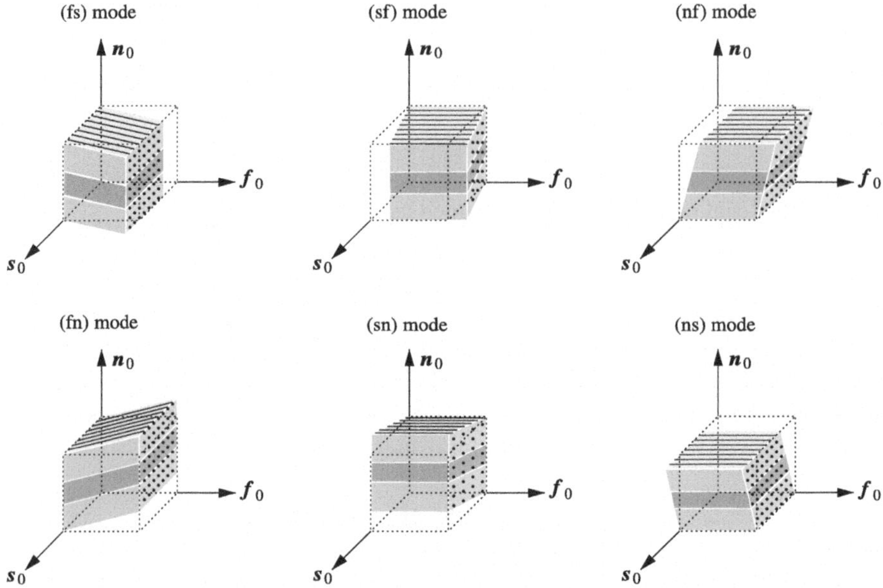

Fig. 2 Specimens subjected to triaxial shear loading with six different shear modes identified, i.e. (fs), (fn), (sf), (sn), (nf), (ns)

capabilities of these models by fitting them to experimental data, in particular triaxial shear test data, and compare the results they produce.

The average of 18 (in case of cyclic) and 5 (in case of relaxation) tests conducted on cubic samples excised from the mid-layer of human myocardium on a triaxial shear testing machine were reported in Sommer et al. (2015). Figure 2 shows the respective configurations of a specimen undergoing triaxial shear associated with six different modes, i.e. (fs), (fn), (sf), (sn), (nf), and (ns) – hereby, the first symbol relates to the plane, while the second symbol indicates the loading direction. A sinusoidal displacement load pattern is considered in order to replicate the cyclic shear experiments. Thus,

$$\gamma(t) = \gamma_{\max} \sin(2\pi t \, / \, \tau), \tag{54}$$

where $\gamma_{\max} = 0.5$ and the period τ is equal to 56 s. On the other hand, we apply constant deformations in the case of relaxation shear test throughout the entire test time, i.e. $\gamma(t) = \gamma(t_0) = 0.5$. As a first step, the elastic parameters corresponding to triaxial shear tests are identified by considering the baseline hyperelastic incompressible material response of the tissue according to Gültekin et al. (2016), through a nonlinear least-squares analysis, see Table 1. Note that the dispersion parameters $\kappa_{\mathrm{f}} = 0.08$ and $\kappa_{\mathrm{s}} = 0.09$ have already been identified via an experimental process in Sommer et al. (2015). Subsequently, the viscoelastic parameters are identified by means of a cyclic and relaxation type of loading. The obtained parameter sets for the

Table 1 Elastic material parameters in Eq. (31) obtained from triaxial shear tests, whereby a, a_f, a_s, a_{fs} are in [kPa], and b, b_f, b_s, b_{fs} are dimensionless

a	b	a_f	b_f	a_s	b_s	a_{fs}	b_{fs}
0.4	6.55	3.05	29.05	1.25	36.65	0.15	6.28

Table 2 Viscous material parameters for the convolution representation in Sect. 4.1 obtained from triaxial shear tests (cyclic and relaxation), whereby τ_m,τ_f, τ_s, τ_{fs} are in [s], and ξ_m, ξ_f, ξ_s, ξ_{fs} are dimensionless

ξ_m	τ_m	ξ_f	τ_f	ξ_s	τ_s	ξ_{fs}	τ_{fs}
2	16	2.2	25	1	14	2	15

Table 3 Viscous material parameters for the canonical representation in Sect. 4.2 obtained from triaxial shear tests (cyclic and relaxation) for the linear and nonlinear cases, whereby μ_f^v, μ_s^v, μ_n^v, $\hat{\beta}_f$, $\hat{\beta}_s$, $\hat{\beta}_n$ are in [kPa], η_f, η_s, η_n are in [kPa s], and m_f, m_s, m_n are dimensionless

	μ_f^v	η_f	μ_s^v	η_s	μ_n^v	η_n	$\hat{\beta}_f$	m_f	$\hat{\beta}_s$	m_s	$\hat{\beta}_n$	m_n
Linear case	200	2500	130	1300	75	750	–	–	–	–	–	–
Nonlinear case	200	2500	130	1300	75	750	10	2	6	2	4	2

convolution representation in Sect. 4.1 and the canonical representation in Sect. 4.2 are listed in Tables 2 and 3. Figure 3 illustrates the cyclic Cauchy stress versus the amount of shear under six distinct shear modes when the convolution type viscoelasticity is used. Salient hystereses distinct for each mode are observed that excellently mimic the experimental data, though simulations display small deviations in (fs) and (fn) shear modes around the moderate values of shear. The hystereses lack, in a sense, a small amount of counter-clockwise rotation around the center, which would otherwise make the results perfect for these modes too. Nevertheless, the proposed model excellently captures the dissipated energy realized by the hystereses, outperforming the linear and nonlinear cases of the canonical representation, compare with Fig. 4a, b. Focusing on the stress relaxation behavior arisen from the convolution representation, see Fig. 5a, some disparities between the model results and the experimental data are highlighted, which are associated with each mode during the descent of the curves before reaching thermodynamical equilibrium. They, nonetheless, fully recover the thermodynamic equilibrium stresses in time. It should be mentioned that the more Maxwell branches (a spring connected serially to a dashpot) delineating the viscous response of the separate contributions, i.e. matrix, fiber, sheet, and fiber-sheet, we use, the more coherent hysteresis and stress relaxation behavior we obtain. However, this makes the analysis less favorable (too many material parameters) from a modeling point of view, and therefore the plots are not included here.

Figure 4a, b depict the cyclic Cauchy stress versus the amount of shear under six distinct shear modes when the canonical viscoelastic formulation is employed for linear and the nonlinear case, respectively. Stress hystereses are evident and distinct for each mode, fairly reflecting the experimental data, though some discrepancies exist in all shear modes. In particular, both the linear and the nonlinear versions fall short of describing central hystereses around low and moderate amounts of shear due to the peculiar crimped responses they exhibit. Nevertheless, the overall

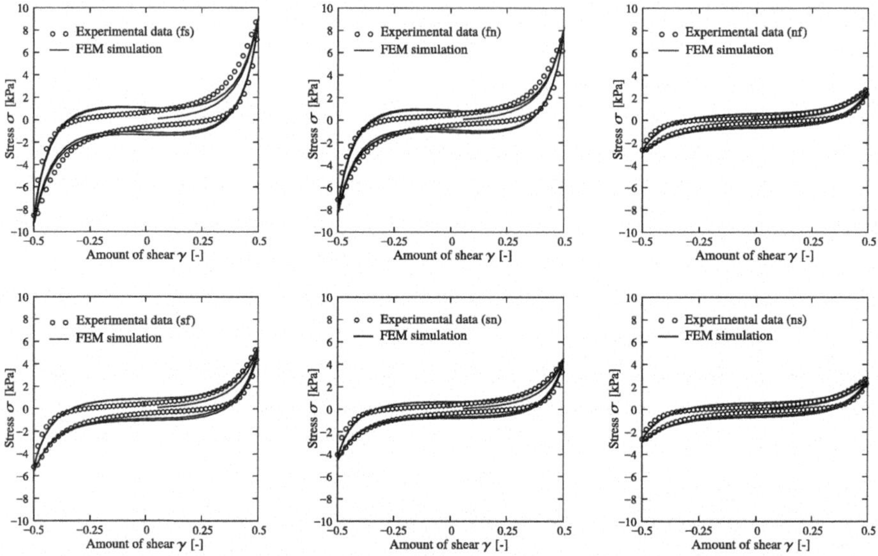

Fig. 3 Cauchy stress σ versus amount of shear γ response of the convolution representation in Sect. 4.1 against the cyclic triaxial shear test data for modes (fs), (fn), (sf), (sn), (nf), (ns)

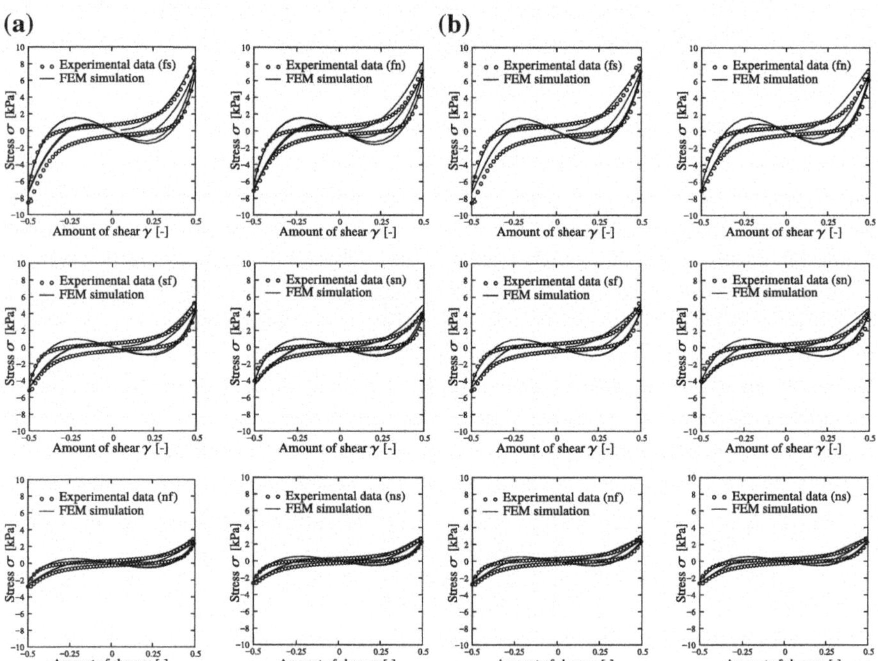

Fig. 4 Cauchy stress σ versus amount of shear γ response of: (a) linear and (b) nonlinear canonical representation in Sect. 4.2 against the cyclic triaxial shear test data for modes (fs), (fn), (sf), (sn), (nf), (ns), respectively

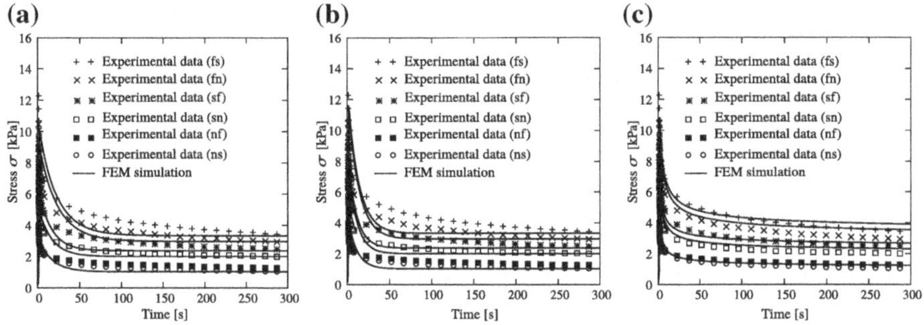

Fig. 5 Cauchy stress σ versus time response for: (a) convolution representation in Sect. 4.1; (b) linear and (c) nonlinear case of the canonical representation in Sect. 4.2 with respect to the relaxation test data obtained for shear modes (fs), (fn), (sf), (sn), (nf), (ns). The solid curves for (nf) and (ns) modes overlap each other

dissipated energy is sufficiently captured by the model. The real difference between the linear and nonlinear cases comes to the surface when their stress relaxation responses are peered deeper, see Fig. 5b, c. The nonlinear case pertaining to the canonical representation outperforms the rest with 12 viscous parameters used in the analysis. In fact, the nonlinear case fits the relaxation data in an excellent manner. On the other hand, the linear case results in a quite similar response to that of the convolution representation, compare with Fig. 5a. In the end, we emphasize that the canonical representation of the orthotropic viscoelasticity can be improved as regards the cyclic triaxial responses, provided that a logarithmic term consisting of the matrix contribution such as $\varepsilon_{\mathrm{m}} = \ln(\sqrt{I_1/3})/2$ is incorporated into the non-equilibrium free-energy response.

6 Conclusion

In this work, we have covered two recently proposed orthotropic viscoelastic material model; one by Gültekin et al. (2016) in the sense of convolution integral of the viscous stresses, the other one by Cansız et al. (2015) dealing with the strain-like internal variables and their conjugate thermodynamical forces in a canonical sense. Later, we have assessed the modeling capacity of both models by comparing their respective cyclic and stress relaxation responses under triaxial shear deformation with respect to the experimental data documented in Sommer et al. (2015) on human passive myocardium. The baseline hyperelastic response of the tissue is identical for both models, as described by Eriksson et al. (2013). The construction of the models is based on the additive split of the free-energy function composed of an equilibrium and a non-equilibrium part. The viscous overstress response is dictated by rate equations for ground matrix, fiber, sheet, and fiber-sheet contributions in the convolution representation. However, the canonical representation involves an explicit description of the energy storage and dissipation occurring in a Maxwell branch of the rheological structure. The viscous overstress response is then dictated by Biot's equations along fiber, sheet, and normal directions. The demonstrated capability of both orthotropic viscoelastic models to capture experimental data provides sufficient validity from quantitative and qualitative aspects.

References

Biot, M.A.: Mechanics of Incremental Deformations. Wiley, New York (1965)

Bischoff, J.E., Arruda, E.M., Grosh, K.: A microstructurally based orthotropic hyperelastic constitutive law. J. Appl. Mech. **69**, 570–579 (2002)

Cansız, B., Dal, H., Kaliske, M.: Computational cardiology: a modified Hill model to describe the electro-visco-elasticity of the myocardium. Comput. Methods Appl. Mech. Eng. **315**, 434–466 (2017)

Cansız, B., Dal, H., Kaliske, M.: Computational cardiology: the bidomain based modified Hill model incorporating viscous effects for cardiac defibrillation. Comput. Mech. **62**, 253–271 (2018)

Cansız, F.B.C., Dal, H., Kaliske, M.: An orthotropic viscoelastic material model for passive myocardium: theory and algorithmic treatment. Comput. Methods Biomech. Biomed. Eng. **18**, 1160–1172 (2015)

Coleman, B.D., Gurtin, M.E.: Thermodynamics with internal state variables. J. Chem. Phys. **47**, 597–613 (1967)

Costa, K.D., Holmes, J.W., McCulloch, A.D.: Modeling cardiac mechanical properties in three dimensions. Philos. Trans. R. Soc. Lond. A **359**, 1233–1250 (2001)

Dal, H., Göktepe, S., Kaliske, M., Kuhl, E.: A fully implicit finite element method for bidomain models of cardiac electromechanics. Comput. Methods Appl. Mech. Eng. **253**, 323–336 (2013)

Demiray, H.: A note on the elasticity of soft biological tissues. J. Biomech. **5**, 309–311 (1972)

Dokos, S., Smaill, B.H., Young, A.A., LeGrice, I.J.: Shear properties of passive ventricular myocardium. Am. J. Physiol. **283**, H2650–H2659 (2002)

Eriksson, T.S.E., Prassl, A.J., Plank, G., Holzapfel, G.A.: Modeling the dispersion in electromechanically coupled myocardium. Int. J. Numer. Method Biomed. Eng. **29**, 1267–1284 (2013)

Flory, P.J.: Thermodynamic relations for highly elastic materials. Trans. Faraday Soc. **57**, 829–838 (1961)

Fomovsky, G., Thomopoulos, S., Holmes, J.W.: Contribution of extracellular matrix to the mechanical properties of the heart. J. Mol. Cell Cardiol. **48**, 490–496 (2010)

Fung, Y.C.: Biomechanics. Mechanical Properties of Living Tissues, 2nd edn. Springer, New York (1993)

Global Burden of Disease Collaborative Network: Global Burden of Disease Study 2019 (GBD 2019). Seattle, United States: Institute for Health Metrics and Evaluation (IHME) (2020). Available via http://ghdx.healthdata.org/gbd-results-tool

Göktepe, S., Acharya, S.N.S., Wong, J., Kuhl, E.: Computational modeling of passive myocardium. Commun. Numer. Methods Eng. **27**, 1–14 (2011)

Guccione, J.M., McCulloch, A.D., Waldman, L.K.: Passive material properties of intact ventricular myocardium determined from a cylindrical model. J. Biomech. Eng. **113**, 42–55 (1991)

Gültekin, O., Sommer, G., Holzapfel, G.A.: An orthotropic viscoelastic model for the passive myocardium: continuum basis and numerical treatment. Comput. Methods Biomech. Biomed. Eng. **19**, 1647–1664 (2016)

Holzapfel, G.A.: On large strain viscoelasticity: continuum formulation and finite element applications to elastomeric structures. Int. J. Numer. Methods Eng. **39**, 3903–3926 (1996)

Holzapfel, G.A.: Nonlinear Solid Mechanics. A Continuum Approach for Engineering, Wiley, Chichester (2000)

Holzapfel, G.A., Gasser, T.C.: A viscoelastic model for fiber-reinforced composites at finite strains: continuum basis, computational aspects and applications. Comput. Methods Appl. Mech. Eng. **190**, 4379–4403 (2001)

Holzapfel, G.A., Ogden, R.W.: Constitutive modelling of passive myocardium: a structurally based framework for material characterization. Philos. Trans. R. Soc. Lond. A **367**, 3445–3475 (2009)

Holzapfel, G.A., Gasser, T.C., Stadler, M.: A structural model for the viscoelastic behavior of arterial walls: continuum formulation and finite element analysis. Eur. J. Mech. A/Solids **21**, 441–463 (2002)

Humphrey, J.D.: Cardiovascular Solid Mechanics. Cells, Tissues, and Organs, Springer, New York (2002)

Humphrey, J.D., Strumpf, R.K., Yin, F.C.P.: Determination of a constitutive relation for passive myocardium: I. A new functional form. J. Biomech. Eng. **112**, 333–339 (1990)

Huyghe, J.M., van Campen, D.H., Arts, T., Heethaar, R.M.: The constitutive behaviour of passive heart muscle tissue. A quasi-linear viscoelastic formulation. J. Biomech. **24**, 841–849 (1991)

Huyghe, J.M., Arts, T., van Campen, D.H., Reneman, R.S.: Porous medium finite element model of the beating left ventricle. Am. J. Physiol. Heart Circ. Physiol. **262**, H1256–H1267 (1992)

Kaliske, M.: A formulation of elasticity and viscoelasticity for fibre reinforced material at small and finite strains. Comput. Methods Appl. Mech. Eng. **185**, 225–243 (2000)

Karlon, W.J., Covell, J.W., McCulloch, A.D., Hunter, J.J., Omens, J.H.: Automated measurement of myofiber disarray in transgenic mice with ventricular expression of ras. Anat. Rec. **252**, 612–625 (1998)

Lecarpentier, Y., Martin, J.L., Claes, V., Chambaret, J.P., Migus, A., Antonetti, A., Hatt, P.Y.: Real-time kinetics of sarcomere relaxation by laser diffraction. Circ. Res. **56**, 331–339 (1985)

LeGrice, I.J., Smaill, B.H., Chai, L.Z., Edgar, S.G., Gavin, J.B., Hunter, P.J.: Laminar structure of the heart: ventricular myocyte arrangement and connective tissue architecture in the dog. Am. J. Physiol. Heart Circ. Physiol. **269**, H571–H582 (1995)

Löffler, L., Sagawa, K.: A one-dimensional viscoelastic model of cat heart muscle studied by small length perturbations during isometric contraction. Circ. Res. **36**, 498–512 (1975)

Streeter, D.D., Jr., Spotnitz, H.M., Patel, D.P., Ross, J., Jr., Sonnenblick, E.H.: Fibre orientation in the canine left ventricle during diastole and systole. Circ. Res. **24**, 339–347 (1969)

Miehe, C., Göktepe, S.: A micro-macro approach to rubber-like materials. Part II: the micro-sphere model of finite rubber viscoelasticity. J. Mech. Phys. Solids **53**, 2231–2258 (2005)

Nash, M.P., Hunter, P.J.: Computational mechanics of the heart. J. Elasticity **61**, 113–141 (2000)

Rohmer, D., Sitek, A., Gullberg, G.T.: Reconstruction and visualization of fiber and laminar structure in the normal human heart from ex vivo diffusion tensor magnetic resonance imaging (DTMRI) data. Investig. Radiol. **42**, 777–789 (2007)

Sands, G.B., Gerneke, D.A., Hooks, D.A., Green, C.R., Smaill, B.H., LeGrice, I.J.: Automated imaging of extended tissue volumes using confocal microscopy. Microsc. Res. Tech. **67**, 227–239 (2005)

Sands, G.B., Smaill, B.H., LeGrice, I.J.: Virtual sectioning of cardiac tissue relative to fiber orientation. In: Proceedings of the 30th Annual International Conference of the IEEE Engineering in Medicine and Biology Society EMBS, pp. 226–229 (2008)

Schmid, H., Nash, M.P., Young, A.A., Hunter, P.J.: Myocardial material parameter estimation-a comparative study for simple shear. J. Biomech. Eng. **128**, 742–750 (2006)

Simo, J.C.: On a fully three-dimensional finite-strain viscoelastic damage model: formulation and computational aspects. Comput. Methods Appl. Mech. Eng. **60**, 153–173 (1987)

Sommer, G., Schriefl, A.J., Andrä, M., Sacherer, M., Viertler, C., Wolinski, H., Holzapfel, G.A.: Biomechanical properties and microstructure of human ventricular myocardium. Acta Biomater. **24**, 172–192 (2015)

Strijkers, G.J., Bouts, A., Blankesteijn, W.M., Peeters, T.H.J.M., Vilanova, A., van Prooijen, M.C., Sanders, H.M.H.F., Heijman, E., Nicolay, K.: Diffusion tensor imaging of left ventricular remodeling in response to myocardial infarction in the mouse. NMR Biomed. **22**, 182–190 (2009)

Tsaturyan, A.K., Izacov, V.J., Zhelamsky, S.V., Bykov, B.L.: Extracellular fluid filtration as the reason for the viscoelastic behavior of the passive myocardium. J. Biomech. **17**, 749–755 (1984)

Usyk, T.P., Omens, J.H., McCulloch, A.D.: Regional septal dysfunction in a three-dimensional computational model of focal myofiber disarray. Am. J. Physiol. Heart Circ. Physiol. **281**, H506–H514 (2001)

Yang, M., Taber, L.A.: The possible role of poroelasticity in the apparent viscoelastic behaviour of passive cardiac muscle. J. Biomech. **24**, 587–597 (1991)

Yao, J., Varner, V.D., Brilli, L.L., Young, J.M., Taber, L.A., Perucchio, R.: Viscoelastic material properties of the myocardium and cardiac jelly in the looping chick heart. J. Biomech. Eng. **134**, 1–7 (2012)

Towards Surgical Training Phantoms Obtained by Additive Manufacturing: Mechanical Characterization of Abdominal and Pelvic Organs. A Literature Review

Stefania Marconi, Erika Negrello, Valeria Mauri, Gianluca Alaimo, and Ferdinando Auricchio

Unfortunately time flies and there are so many on-going events and activities that it is, in general, difficult to have a clear recollection of past events. However, I clearly remember the first time I had a conversation with Gerhard.
It was 1999, we were in Bolder for 5th U.S. National Congress on Computational Mechanics, and we happened to be in the same Mini-Symposium on Cardiovascular Computational Mechanics. It was the start of a long friendship and many visits of me in Graz and of Gerhard in Pavia.
I always enjoyed spending time with Gerhard talking about his great vision and efforts in bringing together continuum mechanics, biomechanics, experimental evidences, computational aspects, etc. etc. and I appreciate his vision in building up a new and active group of researchers in Graz able to really boost some of the mentioned activities. I think that all this work has been extremely inspiring for many of us. So, dear Gerhard, thanks, many happy new celebrations (together), and let us enjoy (again together) a toast with a good glass of wine.
Ferdinando

Abstract Among the many advantages brought by Additive Manufacturing technologies in the clinical context, the possibility of producing training phantoms is of prominent importance. Within such a context, soft tissues, as those of abdominal and pelvic organs, pose major challenges in the reproduction of their biomechanics. The starting point for the choice of the most suitable material is a comprehensive understanding of the target tissue biomechanics. This can be done thanks to mechanical characterization testing, which is not trivial to perform on biological tissues for a

S. Marconi · E. Negrello · V. Mauri · G. Alaimo · F. Auricchio (✉)
Department of Civil Engineering and Architecture, University of Pavia, Pavia, Italy
e-mail: auricchio@unipv.it

S. Marconi
Fondazione IRCCS Policlinico San Matteo, Pavia, Italy

G. Sommer et al. (eds.), *Solid (Bio)mechanics: Challenges of the Next Decade*, Studies in Mechanobiology, Tissue Engineering and Biomaterials 24,
https://doi.org/10.1007/978-3-030-92339-6_12

number of reasons. The present study presents a literature review of the approaches for the mechanical characterization of main abdominal and pelvic organs, focusing on the methodology applied for the testing as well as on the protocol for samples provision, storage, and preparation for the specific mechanical test.

1 Introduction

In recent years, clinical and research activities in the medical context started to progressively focus on the topic of *personalization*. This topic can be declined in many different fields of application; consider, for example, the growing number of personalized solutions for prostheses and orthoses or the attention paid to a detailed patient-specific imaging and 3D reconstructions in the therapeutic path establishment or in the surgical planning. Similarly, an increased attention to the training of healthcare personnel is seen, with the aim of providing the novices with tools having a high degree of realism, thus enabling a safe preparation before direct clinical experience.

In this context, Additive Manufacturing (AM), mostly known as 3D printing, has played an increasingly important role, allowing the production, at a sustainable cost, of highly complex geometries such as patient-specific anatomical models. Geometries reproducing anatomical organs would be in many cases impossible to produce with any other production technology rather than AM and, even if feasible, would have extremely high costs.

Today, AM technologies play a relevant role in the medical field with a number of destinations of use: the production of anatomical models for surgical planning (Pietrabissa et al. 2016), training (Marconi et al. 2019), teaching (Spinelli et al. 2017; Marconi et al. 2017), consulting with the patient and his/her family to retrieve an informed consent (Pietrabissa et al. 2016), development of patient-specific intraoperative tools and instruments (Canzi et al. 2020) as well as personalized prostheses and aids (Auricchio and Marconi 2016). Many clinical specialties have already benefit from AM, with a major use from orthopaedic and maxillo-facial surgical specialties for surgical planning, being the first clinical users of the technology thanks to an easier image processing to retrieve the bony structures of interest. Moreover, a reliable replica of mechanical properties of the bone can be obtained thanks to the wide range of materials available in AM machines.

The biomechanics of soft tissues are more challenging to reproduce: nowadays available AM materials are still not capable of perfectly reproducing the behavior of soft tissues. Rubber-like photopolymers and hydrogels are doing the best job, but further research is still required on this topic. With that said, it becomes clear why the field of abdominal and pelvic surgery is less involved in the use of 3D-printed models for training purposes (Pietrabissa et al. 2020). Beside the choice of the most suitable material, a fundamental issue comes from the comparison of the mock tissue with the real one, for which a mechanical characterization is required. Unfortunately, such testing campaign is usually not trivial to carry out for a number of reasons:

- *in vivo* methods—the most accurate approach—can be logistically and ethically challenging to be applied on humans,
- *ex vivo* methods—the most common solution—bring a lot of uncertainties due to conservation issues and due to the *post-mortem* degradation, lack of blood perfusion and different environmental conditions, often not easy to recreate during the testing,
- *ex vivo* samples retrieved from surgery are often altered by the presence of the pathology for which the intervention is required,
- the use of animal models brings significant financial and logistical challenges as well and results are not guaranteed to be in line with human tissues.

Furthermore, data coming from mechanical characterization of living tissues are fundamental also for the determination of a suitable constitutive model to describe their response to stimuli.

Accordingly, the present study reports a literature review of the approaches for the mechanical characterization of main abdominal and pelvic organs, focusing on the methodology applied for the testing as well as on the protocol for samples provision, storage, and preparation for the specific mechanical test.

2 Materials and Methods

Scopus database was employed for the search combining keywords related to the anatomical region of interest (e.g., abdominal organs, liver, kidney, stomach, pelvis, etc.) and to the mechanical tests (e.g., mechanical characterization, tensile test, compression test, etc.). Queries provided a total of 119 papers, divided into 116 full papers, two conference papers and one review. Afterwards, papers were analyzed according to the exclusion criteria listed in the following:

- mechanical tests are not described,
- mechanical tests are performed only on mock tissues.

The screening was first run relying only on the publication's title and abstract, then on a full text basis. It resulted in a final set of 18 papers. Results are reported divided first according to the anatomical structure and then to the type of mechanical test.

3 Characterization Approaches

3.1 Liver

The liver is the largest gland in the human body, it is connected with the digestive system and is made up of four lobes (The Johns Hopkins University, The Johns Hopkins Hospital, and Johns Hopkins Health System 2021). It has a complex structure, made up of 3 different tissues: (i) the parenchyma, which makes up most of its volume; (ii) the vascular and biliary structures, which form a branched network within the parenchyma; (iii) the Glisson's capsule, which envelops the entire organ and ensures its cohesion. It is located in the upper part of the abdominal cavity below the diaphragm and it is supplied by the hepatic artery, the main arterial vessel of the liver. It is part of the endocrine system and participates in the production of hormones, in addition to the bile, which is very important in the digestion of fats. Among abdominal organs, the liver is the most extensively studied structure, accounting for three studies dealing with tensile tests (Brunon et al. 2010; Umale et al. 2011; Untaroiu and Lu 2013), four with compression tests (Melvin et al. 1973; Miller 1999; Conte et al. 2010; Tejo-Otero et al. 2020), two with indentation (Kim et al. 2003, 2008), and two with dynamic tests (Estermann et al. 2020; Tejo-Otero et al. 2020). While most of the studies are carried out on *post mortem* specimens, three papers (Miller 1999; Kim et al. 2003, 2008) performed in vivo analyzes, which can provide reliable information not affected by the conservation and preparation of the test specimens, which are the major issues when dealing with biological samples. The same authors performed the analyzes also on kidney specimens and the related results are reported in Sect. 3.2.

3.1.1 Tensile Tests

In the study of Untaroiu et al. (2013), uniaxial tensile tests were performed on the parenchyma of 10 fresh bovine livers within 36 h after slaughter. In order to preserve the tissue between the time of procurement and specimen preparation, the livers were immersed in a bath of Dulbecco's Modified Eagle Medium (DMEM) to maintain specimen hydration, and cooled (without freezing) with wet ice. *Dog-bone* shape specimens with length, middle width and thickness close to 55.5 mm, 10 mm, and 5 mm, respectively, were obtained from the parenchyma. Each specimen was tested once until failure at room temperature. The testing system consisted of two motor driven linear stages (Parker Daedal MX80S, Irwin, PA) mounted to a vertically oriented aluminum plate. The 3D geometry of each specimen was obtained using a FARO Laser Scan Arm to acquire a point-cloud representation of the whole tissue surface. A collinear and equidistant pattern of paint (optical) markers was applied to each specimen prior to testing with 4 mm between each marker and tracked throughout the duration of the test using motion analysis software. In the majority of the tests, the tissue tears propagated gradually until complete disruption of the tissue without a sudden drop of the tensile force. Defining the failure time was challenging because some tears occurred inside of the tissue or on its back surface and therefore were not video-recorded. Moreover, in few specimens, tears did not result in a complete tissue disruption. Therefore, the failure point was defined as the point of the force-time curve where the force reached a maximum and then decreased more than 3% of its peak value, and the corresponding time was defined as the time of failure. A total of

30 tensile tests were conducted on parenchyma specimens at three different loading rates.

The data from this study showed that the response of liver parenchyma in tension is nonlinear and strain rate dependent. An increase in stiffness and failure force when loading rate increased was observed, a typical viscoelastic characteristic of biological tissues. The Green-Lagrangian strain at failure was recorded at average values close to 50% and showed no significant dependence on the loading rates. However, the second Piola-Kirchhoff stress showed rate-dependence, and the average failure stress ranged from 33 to 94 kPa. Marker, analytical and specimen-specific models were employed to identify the parameters of the first-order Ogden material model. The Finite Element (FE) simulations adopting marker and analytical models usually showed stiffer responses and lower values of failure stress/strain than the specimen-specific model. For the specimen-specific model, the failure forces resulted approximately equal to 80, 95 and 140 kPa for a loading rate of 0.01, 0.1 and 1 Hz, respectively.

Umale et al. (2011) used 12 livers of female pigs to perform uniaxial quasi-static tensile tests; the experimental procedure started with a total hepatectomy, which was performed on porcine to extract the liver from the body. The organ was then wrapped in a mob soaked with saline solution, packed in a polythene bag inserted and transported in an ice box, maintained at 4–6 °C. The liver was taken to the laboratory for the experiments within a maximum of 30 min, and then cut lobe by lobe to prepare the samples. The capsule tissue was thus first spread over the plastic coated side of a thin paper. Then, to prepare the samples, the capsule was cut along with the paper, sample dimensions being maintained at 25 mm length and 15 mm width. The moisture content in prepared samples was maintained by thawing them in a saline solution until the tests were performed. The paper was removed once the capsule sample is fixed in the clamps of a Bose ElectroForce 3200 biomaterials test instrument for the elongation test. The clamps were also covered with the same plastic coated paper and changed for every sample, to avoid slipping or rupture to the clamped part of the capsule sample. To minimize the effect of strain rate, the experiments were performed at a strain rate of 0.1 mm/s. Moreover, major hepatic vein and intra-parenchymal veins have been tested: moderate size veins (3–6 mm diameter) were detected inside the liver and then isolated with a scalpel, ensuring that the vein is not damaged. The sample was then tightly clamped on the Bose machine and an elongation test was carried out at 0.1 mm/s. All tests were done within 4–6 h of euthanasia of the pig and within 30 min after the samples were prepared. A total of 12 livers were used for the study from which 30 capsule samples and 20 vein samples have been extracted and tested.

Regarding the capsules, about 22 samples out of 30 failed at or just below 30% deformation (i.e. at 27.73%) and the remaining 8 samples failed after 30% deformation (i.e. at 36.72%). The elastic modulus obtained between 0% and 7% strain was 8.22 ± 3.42 MPa and between 20 and 30% strain was 48.15 ± 4.5 MPa. The hepatic veins were tested up to 35% strain and it is observed that the force required for 20% elongation is much less. Afterwards, the samples suddenly stiffen causing an abrupt increase in force for further elongation. Major hepatic veins were observed to have some longitudinal linings on the internal surface making them anisotropic. The smaller veins inside the liver were found to be more homogeneous and anisotropy can be assumed to play a less important role. For both Glisson's capsules and hepatic vein tissues the Ogden constitutive model was proposed as the best fit model.

Brunon et al. (2010) run quasi-static tensile tests on parenchyma and capsule samples until failure. Seven adult pig livers were obtained from the local butchery, within 4 or 5 days after euthanasia. Whole livers or parts of them were either stored in the freezer until the cutting phase (from 24 h to a few days, −18 °C), or immediately cut into samples. Four livers were removed from *post mortem* human subjects. The first was entirely frozen, while the three others were divided in two parts, which were, respectively, kept fresh and frozen. Parts kept fresh were tested within five days after death. Mechanical characterization of the capsule was therefore carried out by using biomaterial samples made of parenchyma and capsule for both human and porcine livers, to ensure the similarity of the experimental procedure between species. From frozen samples, a 5 mm thick slice was cut using a butcher electrical slicer, parallel to the liver surface. From fresh samples, the slice was cut with a butcher knife, as regularly as possible. I-shaped samples were then cut with a scalpel along a pattern, while parenchyma was too compliant to be cut with a stamp. The capsule face of the samples was covered with a random pattern of white dots made of matt paint with a toothbrush. Samples were kept under moist paper to ensure their hydration, and then tested within 5 min at room temperature. The two ends of the sample were clamped; sliding was prevented by using sand-paper stuck on the clamps. Clamps were set up vertically on a Deltalabs EM550 electromechanical testing machine. Normalized load as well as failure properties—ultimate load per width unit and ultimate strain—are determined. Digital Image Correlation (DIC) was used to measure the strain field on the capsule. Samples were loaded with a displacement speed of 0.5 mm/s, which implied a range of longitudinal strain rate of 10^{-3}–$10^{-2}\,s^{-1}$ depending on the sample initial length, remaining in quasi-static domain. Elongation of the sample was recorded with a Visarios 1500 Speed Cam digital video camera. Clamp-to-clamp displacement and tensile load on the upper clamp were measured by using a cable actuated position sensor and a 3 daN single point load cell, with a data acquisition frequency of 100 Hz. The digital high-speed video camera, positioned perpendicularly to the sample, recorded the tensile test, so as to follow the random pattern and to measure the current dimensions of the sample at any time, with an accuracy of ±0.3 mm. The load-displacement curve obtained from the test exhibited several drops representing capsule and parenchyma failures. This failure mechanism allowed to break down the load measurement into parenchyma load and capsule load. A bilinear behavior, observed in previous studies, was also observed for part of the tests, but not for all. The normalized load of the linear part occurring at large strains was 1.5 ± 1.9 N/mm ($n = 29$) and 2.8 ± 2.3 N/mm ($n = 27$) for porcine and human tissues, respectively.

According to the study results, the behavior and failure properties of the liver capsule depend significantly on the biological origin once the capsule has been thawed. Moreover, it has been observed that freezing can affect differently the tensile behavior of capsules from various biological origins, despite their apparently similar constitution, with porcine capsule more damaged by freezing than human capsules, thus remarking the importance of the conservation of biological tissues.

3.1.2 Compression Tests

The study of Conte et al. (2010) deals with the uniaxial quasi-static rear-to-front compression characterization of the liver. The experimental test was performed using a hydraulic press on a whole liver sampled from a *post mortem* human subject female, 81 years old. The sample was placed posteriorly and was compressed between two horizontal plates, coated with petroleum jelly to avoid friction with the organ surface, at a uniform velocity of 80 mm/min. The sample was treated with Winckler solution to ensure proper preservation of soft tissues and then held at 3 °C until liver removal. To prevent organ degradation, the test was performed a few hours after organ removal from the cadaver. Compression was carried out until the 60% of the initial thickness with no initial preconditioning.

The proposed hyperelastic parenchyma model successfully fitted the experimental data, obtained through the previous described compression tests, and it was found to be robust enough to capture the experimental variability.

Miller (1999) developed the mathematical models of liver and kidney tissue deformation, based on in vivo compression experimental results previously proposed by Melvin et al. (1973). Melvin et al. conducted in vivo compression tests on 17 livers and six kidneys of anaesthetized Rhesus monkeys. Results here presented are quite unique for the methodology employed. Each organ was laid onto a load cell while still being perfused by the living animal. Load and impactor displacement were measured. In the calculation of stresses, Melvin et al. approximated the test configuration as that of an uniaxial compression. The adopted loading velocities were 5, 250 and 500 cm/s which corresponded to the nominal strain rates of approximately 0.225, 11.25 and 22.5 s^{-1} for the liver, and 0.385, 19.24, 38.47 s^{-1} for the kidney. Melvin's results should be understood as average across the organs, being performed on the organs in vivo, not on tissue samples extracted from the specific locations. The stress-strain curves were convex for all compression rates containing no linear portion from which a meaningful elastic modulus could be determined. The tissue response stiffened with the increasing loading speed, indicating a strong stress-strain rate dependence. A sufficiently accurate agreement between the theoretical single-phase, hyperviscoelastic model and experiment was found for compression levels reaching 35% and for loading velocities varying over two orders of magnitude.

Otero et al. (2020) performed uniaxial compression on lamb livers procured from a local supplier within 24 h *post mortem*. The tissue was cut using a biopsy punch, getting 6 samples of 16.22 ± 0.61 mm diameter and 7.04 ± 1.81 mm height. The measured Young's Modulus was in the range of 1.4 ± 0.8 kPa.

3.1.3 Indentation Tests

Kim et al. (2003, 2008) analyzed the mechanical behavior of porcine livers and kidneys through in vivo experiments. Pigs were placed in a supine position on the operating table and adjusted to have indentation stimuli normal to the organ surface at the point of contact. An incision was made in the abdominal region and the anatomical structures were dissected to identify the organs. The study employed a mechanical indenter to facilitate the manipulation and programming of the motion required for the various notching modes. Indentation stimuli were delivered using the tactile interface device Phantom Premium-T 1.5 (SensAble Technologies), which was programmed to function as a mechanical stimulator. The forces were measured using the 6-axis force transducer Nano 17 (ATI Industrial Automation), which was connected to the tip of the Phantom. The indenter corresponded to a flat-tipped cylindrical probe, 2 mm in diameter, attached to the tip of the Phantom, with the force transducer mounted in the middle, to accurately detect reaction forces.

The static force responses and the dynamic responses to the ramp and to its maintenance at the entrance were then measured. In order to observe the intrinsic force responses of the organs to the indentation stimulus, the pig was saturated with oxygen during the periods of application of the stimulus. Effective management of respiration was a crucial issue in measuring force responses in vivo. The time required for each indentation stimulus and the time for which the animal's breathing was suspended had to be optimized. An increase in the breath-blocking period increased the CO_2 concentration in the bloodstream and caused the animal to breathe spontaneously. From the steady state force responses of the organs and the induced displacements, it was possible to draw the graphs representing the force-displacement relationships. Among the various nonlinear elastic models (nonlinear constitutive equation), an exponential model of the material was selected, because the exponential curve looked very natural as far as the description of the experimental data is concerned. One of the material models that uses the exponential relationship is the so-called Blatz model. The exponential model with two independent parameters was used for the characterization of the nonlinear behavior of the tissues. Using a nonlinear optimization technique, authors found the parameters that provide the model predictions most in agreement with the experimental data. For the characterization of mechanical behavior, both the Kelvin model and a nonlinear spring model are used, with the nonlinear spring model showing a better prediction than the Kelvin model.

3.1.4 Dynamic Tests

Besides uniaxial compression on lamb livers, Otero et al. (2020) also performed Dynamic Mechanical Analysis (DMA) and Shore Hardness measurements. Regarding the Shore hardness, no sample preparation was needed. DMA applies an oscillating force to a sample and analyzes the response of the material to that force. DMA calculates not only the complex modulus E^*, the storage modulus $E^{'}$ and the loss modulus $E^{''}$, but also the dynamic viscosity η and the axial stiffness of the specimens

K as well as the stress-strain curves. Additionally, the complex modulus is obtained by relating the storage modulus with the loss modulus. Organs were tested using a DMA Q800 equipment of TA Instruments at 37 °C, 1 Hz and a preload force of 0.001 N. Additionally, strain amplitudes covered a range between 10 and 100 μm. The samples were kept in the linear elastic regime, and hence, the stress-strain relationship is linear. Regarding the Shore hardness, the ASTM D2240-Durometer Hardness method was used (ASTM 2005). Shore Durometer Type 00 and 000 were used for measuring Shore hardness of the biological tissues.

Average results for the lamb liver are: storage modulus $E^{'} = 1.7 \pm 0.46\,\text{kPa}$, loss modulus $E^{''} = 0.28 \pm 0.15\,\text{kPa}$, dynamic viscosity $\eta = 0.016 \pm 0.008\,\text{kPa·s}$, stiffness $K = 67 \pm 38\,\text{N/m}$ and average shore hardness = 18 Shore 00 or 47 Shore 000. Being $E^{'}$ significantly higher than $E^{''}$ the material can be regarded as mainly elastic rather than viscous.

Estermann et al. (2020) proposed a method for measuring hyperelasticity and viscoelasticity of bovine and porcine hepatic parenchyma in tension. First, nonlinear stress-stretch curves resulting from ramp loading and unloading, were interpreted based on a hyperelastic framework, using a Veronda-Westmann strain-energy function. Strain-specific elastic moduli, such as initial stiffness, were defined in different segments of the stress-stretch curves. Furthermore, dissipated and stored energy density were calculated. Next, the viscoelastic nature of liver tissue was examined with 2 different methods: stress relaxation and dynamic cyclic testing, finding that in tension, stress relaxation was experimentally more convenient than dynamic cyclic testing. Hepatic samples were taken from 8 porcine, and 2 bovine livers, obtained from a local butchery about 24 h after slaughtering and prepared for testing immediately upon arrival in the laboratory. Between slaughtering and sample preparation, whole livers were stored at ~4 °C in sealed plastic bags. All samples were kept well hydrated by submerging them in 0.9% saline solution. Mechanical testing was conducted immediately after sample preparation at room temperature (~23 °C), thus the tissues were never frozen. Rectangular tensile test specimens were extracted with their long axis orientated parallel to the diaphragmatic and the visceral surface of the organs.

The resulting samples (36 porcine, 36 bovine) consisted of parenchyma tissue without the Glisson's capsule, free of large blood vessels and bile ducts to ensure relative homogeneity. Sample thickness was chosen as ~5 mm. Tensile testing was conducted on a zwickiLine testing machine endowed with a load cell measuring up to 100 N and an accuracy of 0.02% (S2M/100N, Hottinger Baldwin Messtechnik GmbH, Darmstadt, Germany). Custom-built tissue clamps enabled manual fine-tuning of the gripping force with a screw. The tissue clamping surface was covered with sand paper (grit P80) to prevent sample slippage during tension. For the evaluation of the dynamic cyclic tests, extremely small phase shifts between force and displacement of a few milliseconds need to be measured, requiring perfectly synchronous measurements.

Each sample was exposed to one of the following mechanical tests: ramp test, stress relaxation, or dynamic cyclic test. All 36 samples of the porcine and bovine tissue were divided among the three testing methods, resulting in 12 samples per test and tissue type.

Ramp test. The crosshead of the testing machine was displaced by 10 mm (resulting in a maximum engineering tissue strain of 0.10–0.14) at a speed of 5 mm/min (corresponding to a strain rate of $\sim 0.001\,s^{-1}$) and then moved back to its initial position at the same speed. Force and distance between the markers (from DIC) for calculating strains, were both recorded at 10 Hz sampling frequency.

Stress relaxation. A cross-head displacement of 5 mm (~0.06 tissue strain based on DIC measurement) was applied within ~0.4 s and held for 300 s. Force and marker's distance were both recorded at a rate of 10 Hz.

Dynamic Cyclic Test. Sinusoidal displacements of 0.5 mm (0.006 tissue strain) were applied at 0.5, 1.0, 1.5, and 2.0 Hz for 100 cycles at each frequency. Samples were allowed to relax at 6 mm for 250 s between each tested frequency to enable comparison of the frequency-dependent tissue response.

The ramp stress-stretch curves were interpreted based on a hyperelastic modeling approach, utilizing a Veronda-Westmann strain-energy function. In order to quantify the viscous properties of liver tissue that already became visible as hysteresis and, with that, energy dissipation in the ramp tests, stress relaxation experiments were performed. Thus, stress relaxation and dynamic cyclic test were conducted for extracting storage and loss moduli, as well as loss tangent. Regarding the resulting elastic moduli, initial stiffness was similar for porcine and bovine liver (~30 kPa), while porcine liver was stiffer for higher strains. Comparing stress relaxation with dynamic cyclic testing, loss tangent of porcine and bovine liver was the same for both methods (0.05–0.25 at 1 Hz). As for the material model selection, authors found that the utilized Veronda-Westmann model was appropriate for representing the hyperelasticity of liver tissue seen in ramp tests. Concerning viscoelasticity, both chosen testing methods—stress relaxation and dynamic cyclic testing—yielded comparable results for storage, loss moduli and loss tangent as long as elasticity nonlinearities were heeded. Storage and loss moduli matched well for bovine, but not as well for porcine tissue.

3.2 Kidney

The kidneys are bilateral bean-shaped organs whose main function is to filter and excrete waste products from the blood (National Institutes of Health 2021; Hoffman 2021). They are also responsible for water and electrolyte balance in the body. They are positioned in the posterior area of the abdominal cavity, on the sides of the spine. The right kidney is located below the liver, the left kidney behind the spleen. The kidneys are encased in complex layers of fascia and fat. They are arranged—from deep to superficial—as: (i) Renal capsule—tough fibrous capsule; (ii) Perirenal fat—collection of extraperitoneal fat; (iii) Renal fascia (also known as Gerota's fascia or perirenal fascia)—encloses the kidneys and the suprarenal glands; (iv) Pararenal fat—mainly located on the posterolateral aspect of the kidney. Internally, the kidneys have an intricate and unique structure. The renal parenchyma can be divided into two main areas —the outer cortex and inner medulla. The cortex extends into the medulla, dividing it into triangular shapes, known as renal pyramids (TeachMe Series 2020).

Only four papers performed mechanical characterization of the kidney. The most comprehensive study is from Farshad et al. (1999), which focused on an investigation of the material properties of the fresh pig kidney and a parametric characterization of both its elastic and inelastic behavior. The investigation included density measurements, uniaxial as well as triaxial compression tests, tensile tests, and shear tests. Triaxial compression tests were performed on the cortex tissues placed in a compression chamber. Tensile tests were carried out on the outer capsule. Shear tests were performed by punching a cylinder into a slice of the cortex.

Specimens were taken from pig kidneys, which were taken out from the body of the cadaver of the animal. To prevent dehydration of the tissues, the organ and later on the samples were stored in a saline solution until testing occurred. They were prepared and tested within 12 h after anesthesia of the animal; testing was carried out at room temperature (23 °C) and 50% relative humidity. The specimens that were to be tested later than after 6 h were kept at a temperature of 4 °C otherwise, they were kept at room temperature. Due to the compression test up to first rupture or higher, a reduction of the sample weight of less than 4% was measured. For each test and for each testing condition, a minimum of five samples was used.

Tensile tests. To carry out tensile tests on the cortex, dog-bone shaped specimens were cut out from 8 mm thick slices of the kidney. The specimens had a gauge length of about 15 mm, width of 5 mm, and thickness of about 8 mm. They were mounted on the testing machine by saw-faced grips.

Uniaxial compression tests. First series of compression tests were made on small cubes ($\sim$10 × 10 × 10 mm), which were extracted from different locations and different orientations of the cortex. Uniaxial compression tests were performed by pressing the sample between two polished and lubricated parallel surfaces, employing a load cell of 20 N for the full loading range. Compression tests were performed with various loading rates ranging from static loading (creep) to quasi-static loading having a speed of 500 mm/min. To assess the apparent anisotropy of the cortex, compression tests were performed on specimens, which were oriented in radial and tangential directions.

Triaxial compression tests. Triaxial compression tests were preliminary performed with a very simple compression chamber consisting of a transparent polymer tube and a piston, with pig kidney cortex samples immersed in a physiological saline solution. The load was applied uniaxially on the top of the piston that was placed in the testing machine, at a rate of 1 mm/min.

Shear tests. For shear-punching tests, samples were also extracted from the sagittal slices of the cortex. The diameter of the shear sample was 40 mm and its thickness was about 8–10 mm; the diameter of the shear-punching cylinder, which had sharp edges, was 20 mm. Shear-punching tests were carried out on the undisturbed surface of the kidney (with the capsula being removed) and on the cut section of the cortex (undisturbed surface up-side down).

The extensive experimental investigation on the fresh pig kidney and the related theoretical modeling performed by Farshad et al. allowed to derive the following conclusions. The general behavior of samples under compression is nonlinear, as typically happens for soft biological tissues, marked by a very flat part up to about 30% relative elongation followed by stiffening up to rupture. Uniaxially compressed specimens from the cortex fail by radial rupturing. The maximum nominal rupture strain of uniaxially compressed pig kidney samples may reach to about 50%. Uniaxial compression tests on the radial and tangential specimens from the cortex under various loading rates show an increase of the rupture stress with the increase in the loading rate and a decrease of the corresponding strain. They also show that increasing the loading speed the difference of fracture strain between the radial and the tangential direction would become more pronounced. Uniaxial compression tests on the specimens from pig kidney depict a creep response which eventually approaches an a strain asymptote. For the characterization of the nonlinear viscoelastic behavior of the pig kidney a two parameter Blatz model has been assumed. Accordingly, two independent material parameters needed to be identified. It was possible to derive the material parameters embedded in this model through experimental correlation. For characterization of the time-dependent material behavior of the cortex tissue, a four parameter linear viscoelastic model composed of a Kelvin element connected in series with a Maxwell element has been assumed. The four viscoelastic parameters of the cortex were obtained by a curve fitting process.

Uniaxial compression tests have been analyzed also by Miller (1999) whose methodology is described in Sect. 3.1. The stress-strain curves are convex for all compression rates containing nonlinear portion. The tissue response stiffened with the increasing loading speed, indicating a strong stress-strain rate dependence as in (Farshad et al. 1999). The proposed model is valid for compressive nominal strains up to 35% and fast (impact) strain rates between 0.385 and 38.47 Hz.

Moreover, Kim et al. (2008) performed an **indentation test** on kidney, modeling the viscoelastic behavior of these organs; the constitutive model was fitted to the in vivo data using the Levenberg-Marquardt optimization algorithm, combined with a three-dimensional nonlinear FE simulation.

3.3 Stomach

The stomach is located in the upper left part of the abdomen and has the main function of allowing the transit of food to the intestine and thanks to the gastric juices it participates in digestion (Canadian Cancer Society 2021; OER services 2020). From an anatomical point of view, it has an elongated sac shape and is connected both to the esophagus, from which it receives nutrients, and to the small intestine. It is about 25–28 cm long and has elastic properties, given by the folds that make up the wall. The stomach wall is made up of three layers: gastric mucosa, muscular robe and serous robe. The stomach is divided into four main regions: the cardia, fundus, body (or *corpus*), and pyloric antrum. The cardia is the point where the esophagus connects to the stomach and through which food passes into the stomach. Located inferior to the diaphragm, above and to the left of the cardia, is the dome-shaped fundus. Below the fundus is the body, the main part of the stomach. The funnel-shaped pylorus connects the stomach to the duodenum.

Only the study of Aydin et al. (2017) dealt with the mechanical testing of stomach tissue, performing a **biaxial tensile cyclic testing** campaign on porcine stomachs. Samples were collected from 11 female domestic pigs, immediately flushed with water to remove remaining chyme and then stored in isotonic saline solution in a fridge at 4 °C to prevent dehydration and enzymatic degradation processes. The tests were conducted on average 19 ± 10 h after harvest. At the time of the experiment, no tissue deterioration was evident. Square tissue patches were excised from the three major regions of the stomach (fundus, corpus, and antrum) close to the greater curvature. After the above described preparation, the tissue patches were subjected to biaxial mechanical tests at room temperature (20 °C), using a biaxial testing machine (mediX 0.1—Messphysik Materials Testing, Austria). The tissue patches were perforated at a distance of 4 mm from the four edges with 9 equidistant steel wires, respectively. These wires were mounted freely rotating on four custom-made plastic brackets. The distance between each pair of opposite plastic brackets was computer-controlled during the experiments in order to sample a representative set of biaxial strain states.

To measure the tissue deformation, four markers were glued on each tissue patch such that they formed a square with edge length 10 mm in the center of the tissue patch in the unloaded initial configuration (Fig. 1). During the biaxial tests, the tissue patches were not continuously submerged in the isotonic saline solution in which they had been stored before the experiment but were kept hydrated by periodic moistening with isotonic saline solution (0.9%). Since during the experiments, the yield strain was found to vary significantly between the different regions of the stomach, in order to ensure that all biaxial tests were performed in the elastic regime and to avoid damage to the tissue, patches from the fundus, corpus, and antrum were stretched in both directions only up to a maximum value of 1.5, 1.2 and 1.4, respectively, according to preliminary tests.

After the preconditioning, biaxial global stretch ratios between the two directions of 1:1, 3:4, 4:3, 2:1, 1:2, 4:1, and 1:4 were imposed on the tissue patches, following sinusoidal displacement-controlled cycles. The displacement speed was set to ensure an upper bound of around $2.2 \cdot 10^{-2}$ s^{-1} for the strain rate, which had been confirmed by previous experiments to ensure a quasi-static regime with negligible viscoelastic effects. Authors performed biaxial mechanical tests of porcine gastric tissue samples from 11 pigs and examined the differences between the three major anatomical regions in the stomach, fundus, corpus, and antrum. The yield strain was found to vary significantly between the different regions of the stomach. Fitting the experimental data to a biaxial Fung exponential model, authors obtained material parameters characterizing the elasticity of the gastric wall. The experiments performed by Aydin et al. reveal considerable anisotropy and different mechanical properties in the three stomach regions. Furthermore, they show that the mechanical

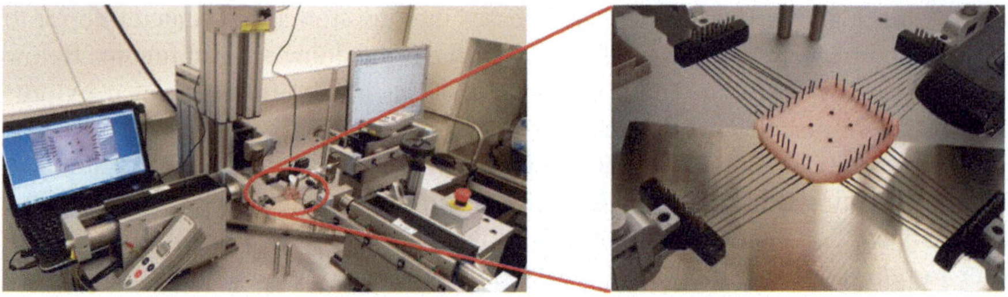

Fig. 1 Biaxial testing machine with a tissue patch subjected to biaxial strain. Adopted from Aydin et al. (2017)

properties of the gastric wall and the physiological function of the different regions of the stomach are closely related.

3.4 Pancreas

The pancreas (Hoffman 2021) is an elongated cone-shaped gland that secretes both endocrine and exocrine substances. This gland is made up of four parts: head, neck, body and tail.

In the literature, few studies deal with the mechanical characterization of the pancreas, mainly due to its great inter- and intra-patients' variability. Among the few articles, we mention the results obtained by Rubiano et al. (2017). They performed a thorough characterization of human resected pancreatic tissue by means of **indentation testing**, comparing normal, pancreatitis affected and cancer affected tissue. After surgery for pancreatic tumor resection, a portion of the resected tissue went to pathology and, if enough tissue was left after clinical testing, the remaining portion was prepared for indentation testing. Resected tissue samples were placed in DMEM/F12-10% FBS culture media on ice in the operating room and transported to the laboratory for testing. The size of the sample varied with each patient, but generally volumes ranged between 125 and 600 mm^3. A custom cantilever-based indenter was used to record force-displacement values during testing (see Fig. 2). A piezoelectric stage (P-628.1CD, Physik Instrumente) controlled the base of a soft titanium cantilever with a 1 mm or 4 mm diameter rigid tip while a custom program in LabVIEW (National Instruments) read cantilever tip deflection with a capacitive sensor (C8S-3.2-2.0 and compact driver CD1-CD6, Lion Precision). Tip deflection was converted into force using cantilever stiffness, which was calibrated to be 79.8 N/m. Submerged patient tissue samples (1–3 mm thick) were indented to 10% of sample thickness (100–300 μm) at an indentation rate of 10 mm/s, after which the probe was kept in place for 90–210 s to allow for the tissue to achieve >90% relaxation. Sample thickness was measured using the indenter to determine appropriate inden-

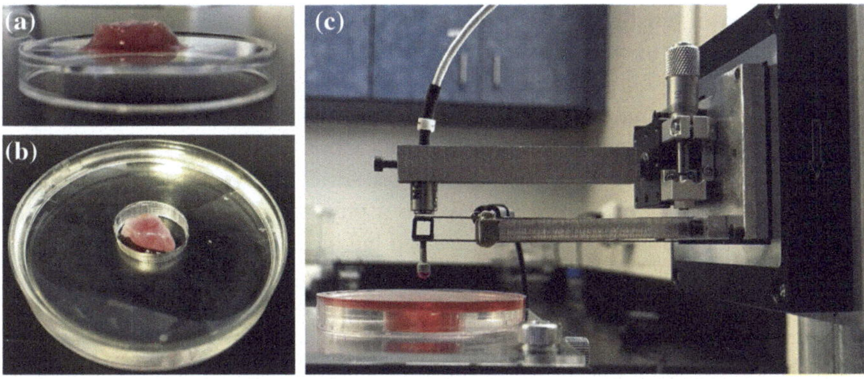

Fig. 2 Pancreatic tissue are sliced to obtain a flat top surface (a), placed into a custom silicone well (b) and flooded with room temperature media to fully submerge samples during indentation (c). Adopted from Rubiano et al. (2017)

tation depth. Samples were indented at 6–13 distinct locations along the length of the tissue, depending on the size of the sample and availability of flat regions.

As results, both pancreatitis (2.15 ± 0.41 kPa) and tumors (5.46 ± 3.18 kPa) have higher significantly moduli than normal tissue (1.06 ± 0.25 kPa). Pancreatitis samples have significantly lower time-dependent properties than both normal and cancer tissue. Both the viscosity and characteristic time follow a trend for pancreatitis values to decrease and cancer values to increase. For viscosity, pancreatitis (63.2 ± 26.7 kPa·s) is significantly lower than both normal tissue (252 ± 134 kPa·s) and tumors (349 ± 222 kPa·s). Similarly, characteristic times for pancreatitis (27.6 ± 14.0 s) are significantly lower than both normal tissue (92.7 ± 46.4 s) and tumors (66.1 ± 20.8 s).

3.5 *Bladder and Rectum*

The bladder is a hollow and distensible muscular organ in humans and other vertebrates that stores urine from the kidneys before disposal by urination. In humans, it sits on the pelvic floor: during the urinary process, the urine enters the bladder via the ureters and exits via the urethra. On the other hand, the human rectum is the final straight portion of the large intestine and it is about 12 cm long. It begins at the rectosigmoid junction, the end of the sigmoid colon, at the level of the third sacral vertebra. Its caliber is similar to that of the sigmoid colon at its commencement, but it is dilated near its termination, forming the rectal ampulla. Few studies are available in the literature about the mechanical characterization of these anatomical structures; Boubaker et al. (2015) elaborated an experimental protocol to characterize the mechanical properties of the bladder and rectum wall porcine tissues, by means of **uniaxial tensile tests**. Urinary bladder and rectum were dissected from a six-month old male pig, washed with phosphate-buffered saline and freed from the para-vesical tissue and mesothelium, respectively. Sample strips (30 × 5 mm) were resected along the circumferential and longitudinal axis, which runs from the apex to neck for the bladder and proximal to distal direction for the rectum. The average thickness was measured to be 2.9 and 2.1 mm for vesical and rectal samples, respectively. Afterwards, samples were moved into a physiological salt solution (Krebs solution), maintained at body temperature and bubbled with a gas mixture of 95% O_2 and 5% CO_2 at pH 7.4. A single-column zwickiLine quasi-static testing machine was employed. Both ends of each tissue strip were glued between sandpaper using Cyanoacrylate for no more than 5 mm at each end, such that they were firmly gripped during testing and no slippage can occur. The sample was then slightly preloaded with a small tension force of 0.005 N and subsequently a uniaxial tensile loading to failure was performed at a displacement rate of 0.5 mm/s. Authors observed that pelvic organ wall tissues reveal a nonlinear (hyperelastic) anisotropic behavior. These tissues also exhibited a viscoelastic behavior through creep and relaxation responses when a time-dependent loading was applied.

3.6 Prostate

The prostate (Institute for Quality and Efficiency in Health Care 2020) (or prostate gland) is a dense fibromuscular gland that is part of the male genital system. Its shape is like an inverted cone having a base above surrounding the neck of the urinary bladder, apex below situated on the external urethral sphincter. The prostate lies directly inferior to the bladder and wraps around the proximal urethra in the lesser pelvis. A fibrous capsule encloses the gland with the nerves and vascular plexus, which is further surrounded by a visceral layer of pelvic fascia.

With respect to other structures, for prostatic tissue more studies are available on human tissue. Ma et al. (2015) performed a **uniaxial cyclic tensile test**, Phipps et al. (2005) and Zhang et al. (2007) studied the mechanical behavior at (compression), while Ahn et al. (2010) performed an (indentation) characterization.

In the study of Ma et al. (2015) prostatic human tissue was resected at autopsy within 24 h after death (one case) and periurethral prostate tissues were retrieved from prostatectomy surgery (28 cases). All tissues were received in 10% Roswell Park Memorial Institute (RPMI) medium and processed immediately for mechanical and histological studies. Samples were processed to take 9 tissue slices parallel and 9 perpendicular to the urethra and trimmed to rectangular shape. All trimmed tissues were subjected to uniaxial load-unload mechanical testing using a constant strain rate of 0.01 s^{-1} to attain a maximum strain of 0.3 with a custom-build tensile tester.

In the 9 sections parallel to the mean tangent of the urethra, compared to the 9 holds perpendicular to it, the modulus was 450 ± 77 kPa versus 560 ± 180 kPa, the mean stromal content was 78 ± 12% versus 88 ± 12% and the mean collagen content was 59 ± 9% versus 66 ± 10%. The tangent modulus for these tissues ranged from 9 to 2390 kPa and the average collagen content of these tissues was 32%. Periurethral prostate tissues demonstrated nonlinear viscoelastic mechanical behavior. Tissue from men with lower urinary tract symptoms was significantly stiffer with significantly higher collagen content and lower glandularity than that from men without lower urinary tract symptoms.

In the study of Phipps et al. (2005), fresh tissue specimens were collected from patients undergoing transurethral resection of the prostate for benign or malignant prostatic enlargement and tested by applying a dynamic compressive strain to the samples. The mechanical test rig consisted of an electromechanical shaker (LDS V201 with amplifier PA-25E) capable of inducing a cyclic load between 1 and 50 Hz at displacements between 0.5 and 1 mm. A load cell (Sensotec Model 31) with signal conditioning unit (Measurements Group 2311) was used to measure the load resulting from the cyclic loading using a 2 mm diameter ball end probe. Results demonstrated that the mechanical characteristics of prostatic tissue can be described by a viscoelastic model and have confirmed that significant correlations exist between prostatic tissue morphology and mechanical characteristics. They have also shown that measurable differences exist between the mechanical characteristics of benign and malignant prostatic tissue.

Zhang et al. (2007) focused on the characterization of human prostates obtained from radical prostatectomy, by means of stress-relaxation and compression tests. The prostate was sectioned and cylindrical cores (approximately 8 mm in diameter and 7 mm in length) were obtained from the posterior zone of the prostates where cancer often occurs. A total of 17 fresh tissue specimens were collected from 8 patients. The core samples were soaked in saline before mechanical testing, which was performed within 2 h. A 1/S mechanical device (MTS Systems Co, Eden Prairie, MN, USA) with 5 N load cell was used to test the core samples. The upper and lower platens were coated with vegetable oil before testing. To minimize the dehydration effect, the side of each sample was coated with a thin layer of petroleum jelly. After 2 min for tissue recovery from precompression, a uniaxial unconfined compression was conducted to measure the time-domain stress relaxation data at room temperature. The compression rate and the strain value were adjusted to 0.5 mm/s and 5%, respectively. Tests lasted about 700 s. The stress-relaxation curve of each sample during the hold period was fit to a fractional viscoelastic model which micro-mechanical scheme is a spring in parallel to a Scott-Blair element (Scott Blair 1947; Di Paola et al. 2013) involving the Caputo's fractional derivative (Podlubny 1999). Nonlinear least-squares fitting was applied to each curve. The averaged three model parameters were then obtained. This study systematically investigated viscoelastic properties of normal and cancerous human prostate tissues. The stress relaxation testing and the proposed fractional viscoelastic modeling approaches provide the frequency-dependent storage and loss moduli, from which both elastic and viscous behavior can be extracted. From in vitro conditions, the elastic contrast between cancer and normal appears to be 2.6:1 at 150 Hz, although the in vivo contrast could be higher as a result of the additional effects of elevated interstitial pressure.

Finally, Ahn et al. (2010) analyzed the mechanical property of prostatic tissues through indentation and evaluated whether measurable differences in mechanical properties exist between cancerous and noncancerous tissues in an *ex vivo* experiment. A total of 552 sites from 46 prostate specimens taken during radical prostatectomy underwent an indentation experiment with a minimally motorized indenter, within 30 min after removal. The minimally motorized indenter was designed to have a maximal outer diameter of 8 mm and a length of 150 mm. It was controlled by a micro-DC motor, and the position was sensed with an integrated magnetoresistive sensor. For each prostate specimen, the indentation experiment was performed at 12 sites across the posterior surface of the prostate, similar to those of double-sextant needle biopsies. The indenter contacted the posterior surface of the prostate and induced a 3 mm deep indentation at 1 mm/s. The mean elastic modulus at a 3 mm indentation of the regions containing cancerous and noncancerous tissue was 24.1 ± 14.5 kPa and 17.0 ± 9.0 kPa, respectively. These results showed increased elastic modulus for the region containing cancerous tissue. Changes in the mechanical properties reflect the changes in the integrity of the biological tissue because pathological changes alter the properties of tissue owing to transformation of the integrity of cells or intercellular matrixes.

4 Conclusion

The present review provides an insight on the methodologies applied for the challenging activity of biological tissues mechanical testing, focusing on the low-covered target of abdominal and pelvic structures. Among 18 papers, 9 studies employed a porcine model, 7 studies a human model, two studies a bovine model and one study employed a monkey and a lamb model, respectively. Interestingly, a porcine-human (Brunon et al. 2010) and a porcine-bovine (Estermann et al. 2020) comparison is provided for the liver. The most extensively studied organ is the liver (11 papers), followed by the kidney and the prostate (four papers each). The majority of the studies performed mechanical tests to identify the most suitable constitutive model to simulate the mechanical behavior of the tissues. Viscoelastic models are mainly employed, in some cases also including hyperelastic effects, to take large strains into account.

References

Ahn, B.M., Kim, J., Ian, L., Rha, K.H., Kim, H.J.: Mechanical property characterization of prostate cancer using a minimally motorized indenter in an ex vivo indentation experiment. Urology **76**, 1007–1011 (2007)

ASTM D2240-05: Standard Test Method for Rubber Property–Durometer Hardness. Book of Standards Volume: 09.01 (2005)

Auricchio, F., Marconi, S.: 3D printing: clinical applications in orthopaedics and traumatology. EFORT Open Rev. **1**, 121–127 (2016)

Aydin, R.C., Brandstaeter, S., Braeu, F.A., Steigenberger, M., Marcus, R.P., Nikolaou, K., Notohamiprodjo, M., Cyron, C.J.: Experimental characterization of the biaxial mechanical properties of porcine gastric tissue. J. Mech. Behav. Biomed. Mater. **74**, 499–506 (2017)

Boubaker, M.B., Haboussi, M., Ganghoffer, J.F., Aletti, P.: Predictive model of the prostate motion in the context of radiotherapy: a biomechanical approach relying on urodynamic data and mechanical testing. J. Mech. Behav. Biomed. Mater. **49**, 30–42 (2015)

Brunon, A., Bruyére-Garnier, K., Coret, M.: Mechanical characterization of liver capsule through uniaxial quasi-static tensile tests until failure. J. Biomech. **43**, 2221–2227 (2010)

Canadian Cancer Society (CCS): Anatomy and physiology of the stomach. https://www.cancer.ca/en/cancer-information/cancer-type/stomach/stomach-cancer/the-stomach/?region=on. Retrieved on Feb 3, 2021

Canzi, P., Avato, I., Marconi, S., Del Maestro, M., Giotta Lucifero, A., Magnetto, M., Carlotto, E., Auricchio, F., Luzzi, S., Benazzo, M.: A 3D printed custom-made mask model for frameless neuronavigation during retrosigmoid craniotomy. A preclinical cadaveric feasibility study. Ann. Ital. Chir. **91**, 526–533 (2020)

Conte, C., Masson, C., Arnoux, P.J.: Inverse analysis and robustness evaluation for biological structure behaviour in FE simulation: application to the liver. Comput. Meth. Biomech. Biomed. Eng. **15**, 993–999 (2012)

Di Paola, M., Pinnola, F.P., Zingales, M.: Fractional differential equations and related exact mechanical models. Comput. Math. Appl. **66**, 608–620 (2013)

Estermann, S.J., Pahr, D.H., Reisinger, A.: Hyperelastic and viscoelastic characterization of hepatic tissue under uniaxial tension in time and frequency domain. J. Mech. Behav. Biomed. Mater. **112**, 104038 (2020)

Farshad, M., Barbezat, M., Füeler, P., Schmidlin, F., Graber, P., Niederer, P.: Material characterization of the pig kidney in relation with the biomechanical analysis of renal trauma. J. Biomech. **32**, 417–425 (1999)

Hoffman, M.: Human Anatomy – Picture of the Kidneys. https://www.webmd.com/kidney-stones/picture-of-the-kidneys. Retrieved on Feb 4, 2021

Hoffman, M.: Human Anatomy – Picture of the Pancreas. https://www.webmd.com/digestive-disorders/picture-of-the-pancreas. Retrieved on Feb 5, 2021

Institute for Quality and Efficiency in Health Care (IQWiG): How does the prostate work? https://www.ncbi.nlm.nih.gov/books/NBK279291/. Retrieved on Nov 21, 2020

Kim, J., Tay, B.K., Stylopoulos, N., Rattner, D.W., Srinivasan, M.A.: Characterization of intra-abdominal tissues from in vivo animal experiments for surgical simulation. In: Ellis, R.E., Peters, T.M. (eds.) Medical Image Computing and Computer-Assisted Intervention - MICCAI 2003, pp. 206–213. Springer, Berlin, Heidelberg (2003)

Kim, J., Ahn, B., De, S., Srinivasan, M.A.: An efficient soft tissue characterization algorithm from in vivo indentation experiments for medical simulation. Int. J. Med. Robot. **4**, 277–285 (2008)

Ma, J., Gharaee-Kermani, M., Kunju, L., Hollingsworth, J.M., Adler, J., Arruda, E.M., Macoska, J.A.: Prostatic fibrosis is associated with lower urinary tract symptoms. J. Urol. **188**, 1375–1381 (2012)

Marconi, S., Pugliese, L., Botti, M., Peri, A., Cavazzi, E., Latteri, S., Auricchio, F., Pietrabissa, A.: From CT scanning to 3-D printing technology for the preoperative planning in laparoscopic splenectomy. Surg. Endosc. **31**, 4102–4110 (2017)

Marconi, S., Negrello, E., Mauri, V., Pugliese, L., Peri, A., Argenti, F., Auricchio, F., Pietrabissa, A.: Toward the improvement of 3D-printed vessels' anatomical models for robotic surgery training. Int. J. Artif. Organs **42**, 558–565 (2019)

Melvin, J.W., Stalnaker, R.L., Roberts, V.L.: Impact injury mechanisms in abdominal organs. SAE Trans. 115–126 (1973)

Miller, K.: Constitutive modelling of abdominal organs. J. Biomech. **33**, 367–373 (2000)

National Institutes of Health (NIH): Your Kidneys & How They Work. https://www.niddk.nih.gov/health-information/kidney-disease/kidneys-how-they-work. Retrieved on Feb 4, 2021

OER services: The Stomach. Module 7: The Digestive System. Anatomy and Physiology II. https://courses.lumenlearning.com/suny-ap2/chapter/the-stomach/. Retrieved on Dec 3, 2020

Phipps, S., Yang, T.H., Habib, F.K., Reuben, R.L., McNeill, S.A.: Measurement of the mechanical characteristics of benign prostatic tissue: a novel method for assessing benign prostatic disease. Urology **65**, 1024–1028 (2005)

Pietrabissa, A., Marconi, S., Peri, A., Pugliese, L., Cavazzi, E., Vinci, A., Botti, M., Auricchio, F.: From CT scanning to 3-D printing technology for the preoperative planning in laparoscopic splenectomy. Surg. Endosc. **30**, 366–371 (2016)

Pietrabissa, A., Marconi, S., Negrello, E., Mauri, V., Peri, A., Pugliese, L., Marone, E.M., Auricchio, F.: An overview on 3D printing for abdominal surgery. Surg. Endosc. **34**, 1–13 (2020)

Podlubny, I.: Caputo's fractional derivative in Fractional Differential Equation, pp. 78–80. Academic Press (1999)

Rubiano, A., Delitto, D., Han, S., Gerber, M., Galitz, C., Trevino, J., Thomas, R.M., Hughes, S.J., Simmons, C.S.: Viscoelastic properties of human pancreatic tumors and in vitro constructs to mimic mechanical properties. Acta Biomater. **67**, 331–340 (2018)

Scott Blair, G.W.: The role of psychophysics in rheology. J. Colloid Sci. **2**, 21–32 (1947)

Spinelli, D., Marconi, S., Caruso, R., Conti, M., Benedetto, F., De Beaufort, H.W., Auricchio, F., Trimarchi, S.: 3D printing of aortic models as a teaching tool for improving understanding of aortic disease. J. Cardiovasc. Surg. **60**, 582–588 (2019)

Tejo-Otero, A., Lustig-Gainza, P., Fenollosa-Artes, F., Valls, A., Krauel, L., Buj-Corral, I.: 3D printed soft surgical planning prototype for a biliary tract rhabdomyosarcoma. J. Mech. Behav. Biomed. Mater. **109**, 103844 (2020)

TeachMe Series: TeachMe Anatomy: The Kidneys. https://teachmeanatomy.info/abdomen/viscera/kidney/. Retrieved on Nov 21, 2020

The Johns Hopkins University, The Johns Hopkins Hospital, and Johns Hopkins Health System: Liver: Anatomy and Functions. https://www.hopkinsmedicine.org/health/conditions-and-diseases/liver-anatomy-and-functions. Retrieved on Feb 5, 2021

Umale, S., Chatelin, S., Bourdet, N., Deck, C., Diana, M., Dhumane, P., Soler, L., Marescaux, J., Willinger, R.: Experimental in vitro mechanical characterization of porcine Glisson's capsule and hepatic veins. J. Biomech. **44**, 1678–1683 (2011)

Untaroiu, C.D., Lu, Y.C.: Material characterization of liver parenchyma using specimen-specific finite element models. J. Mech. Behav. Biomed. Mater. **26**, 11–22 (2013)

Zhang, M., Nigwekar, P., Castaneda, B., Hoyt, K., Joseph, J.V., di Sant'Agnese, A., Messing, E.M., Strang, J.G., Rubens, D.J., Parker, K.J.: Quantitative characterization of viscoelastic properties of human prostate correlated with histology. Ultrasound Med. Biol. **34**, 1033–1042 (2007)

Three-Dimensional Multi-Scale Modeling of Electro-Chemomechanical Gastric Smooth Muscle Contraction

Lisa Klemm, Robert Seydewitz, and Markus Böl

I first met Gerhard in person at one of the first summer schools held in Graz. Since then we have organized various mini-symposia at conferences and are in constant contact. I appreciate Gerhard's professional and personal manner and look forward to working with him.

Markus

Abstract The stomach, to be a J-shaped hollow organ, undergoes large deformations during the digestion of a meal. Hereby, different types of contractions across the stomach wall accomplish different tasks, including the storage, mixing, and transport of ingested food, leading to a complex motility pattern. To coordinate the different gastric contractions, specialized pacemaker cells, so-called interstitial cells of Cajal (ICC), generate electrical slow waves. These slow waves propagate in an electrically-coupled network of ICCs and depolarize neighboring smooth muscle cells. Impaired stomach motility can be caused by a failure of the electrical regulation underlying the activation of gastric smooth muscle contraction, namely gastric dysrhythmias, and can be evoked by any disturbance in the electrical network. Multi-scale models are potential tools to predict gastric motility based on underlying events of ICC network electrophysiology. In the present study, a three-dimensional multi-scale model of the electro-chemomechanical activation of gastric smooth muscle contraction is presented. Additionally, a mechano-electrical feedback mechanism is included. Simulations are performed on a realistic, physiological stomach geometrical model and a postsurgical stomach, studying the effects of a surgical excision on the stomach wall electrophysiology and motility.

L. Klemm · R. Seydewitz · M. Böl (✉)
Institute of Mechanics and Adaptronics, 38106 Braunschweig, Germany
e-mail: m.boel@tu-braunschweig.de

G. Sommer et al. (eds.), *Solid (Bio)mechanics: Challenges of the Next Decade*, Studies in Mechanobiology, Tissue Engineering and Biomaterials 24,
https://doi.org/10.1007/978-3-030-92339-6_13

1 Introduction

The stomach is a J-shaped hollow organ in the gastrointestinal tract, which accounts for the storage of ingested food, the mechanical mixing and crushing down of ingested food to generate chyme, and the transport of this chyme into the duodenum. These tasks are accomplished by the stomach wall's (SW's) microstructure and its ability to contract and relax, leading to large deformations for the stomach to undergo. Hereby, depending on the region of the SW, different functions are performed by different types of gastric smooth muscle contractions, resulting in a complex motility pattern across the whole SW. Generally, the upper part of the stomach features mostly tonic contractions, which regulate the emptying of digested food into the duodenum by changing the stomach's capacity (Azpiroz 1994). In the stomach's lower part, strong peristaltic contraction waves propagate distally while they mix and transport larger particles and crush them down until they are liquified enough for emptying. After emptying, the stomach's shape returns completely back to its reference shape. These deformable characteristics are facilitated by the microstructure of the SW, which consists of several layers (from the inside out): The tunica mucosa, tunica submucosa, tunica muscularis, and serosal layer, see also Fig. 1a. Depending on the stomach's region, these layers vary significantly in thickness, waviness, and fiber orientation, thus adapting to the specific region-dependent contractions. At micro-scale, a robust electro-chemical mechanism coordinates the different gastric contractions. Between and within the muscular layers of the SW lie the interstitial cells of Cajal (ICC), which are specialized pacemaker cells organized in an electrically coupled network and which generate propagating electrical signals, also called slow waves. Gastric smooth muscle cells (SMC) in the muscular layer are electrically coupled to the ICC and respond to the electrical slow waves by altering the intracellular calcium concentration, which can ultimately result in a contraction or relaxation of the muscular layer (Sanders 2008).

Failure of the regulatory mechanisms for gastric contractions can lead to different impairments of the SW's motility, associated with several diseases. Hereby, the coordination of gastric contractions through an electrically-coupled ICC network is of importance. Any loss of connectivity in this network, whether it is caused by a defective ICC pacemaker function or by surgical incisions, can lead to gastric electrical dysrhythmias and therefore to abnormal motility patterns (Kelly and Code 1971; Sarna et al. 1972; Hinder and Kelly 1977; Streutker et al. 2007; Farrugia 2008; Sanders et al. 2014b; Du et al. 2015). Frequently occuring effects of gastric dysrhythmias are delayed gastric emptying, also known as gastroparesis (Chen et al. 1994, 1996; Sanders et al. 2014b), or other defects in gastric emptying, which may be accompanied by, e.g., nausea or vomiting (Sanders et al. 2006; Du et al. 2015).

Multi-scale models of electro-chemomechanical activation of gastric smooth muscle contraction are potential computational tools to test hypotheses on gastric motility in health and disease. However, only a few theoretical models are available that describe the active, electro-chemomechanical contraction behavior of the stomach. So-called cell models (Buist et al. 2006; Edwards and Hirst 2006; Hirst and Edwards

2006; Hirst et al. 2006; Corrias and Buist 2007; Corrias and Buist 2008; Du et al. 2009; Buist et al. 2010; Gajendiran and Buist 2011; Poh et al. 2012; Du et al. 2013a; Sathar et al. 2014; Yeoh et al. 2016) describe micro-scale processes of the mechano-electrochemical contraction. Herein, most approaches consider the electrical field only, while the modeling concepts of Corrias and Buist (2007, 2008) and Yeoh et al. (2016) use the electrical and calcium field. Only one model describes the dependence of the active mechanical response on the calcium field (Gajendiran and Buist 2011). Further, various multi-scale models focus on a detailed description of the stomach's electrophysiology, but do not include any mechanical contribution (Cheng et al. 2003; Pullan et al. 2004; Cheng et al. 2007; Du et al. 2010, 2013a, 2015; Corrias et al. 2012; Sathar et al. 2015). Little has been reported on the electro-chemomechanical modeling of stomach smooth muscle contraction with regard to three-dimensional boundary value problems. An intensive investigation of the literature revealed only one model (Akhmadeev and Miftahof 2010; Miftahof 2017) that is able to describe the complete electro-chemomechanical contraction process inside stomach tissue. On the other hand, the model contains a large number of material parameters and constants that were not identified experimentally. In another approach by Brandstaeter et al. (2018), an electro-mechanical constitutive framework of gastric motility is presented that includes pacemaker electrophysiology and smooth muscle contractility. Among other things, the authors realized three-dimensional simulations on a very simple tube-like geometry to demonstrate the general capability of their modeling approach.

In the present study, a three-dimensional multi-scale model for the description of the electro-chemomechanical activation of gastric smooth muscle contraction is proposed. For the active formulation, we incorporate information about the electrical activation of the SMC by pacemaker activity of the ICC, modeled by using the FitzHugh-Nagumo-type equations (Aliev et al. 2000). The dependency of the intracellular calcium dynamic on the SMC membrane potential is modeled via a Gaussian function, while the chemo-mechanical coupling in the SMC is modeled with the extended Hai-Murphy model (Gajendiran and Buist 2011). A mechano-electrical feedback mechanism is considered, taking into account the mechano-sensitivity of the ICC and SMC. To enhance the predictive ability of the presented modeling concept, appropriate active and passive model parameters are identified based on experiments performed on porcine stomach tissue strips. The predictive behavior of the model is investigated in simulations of a healthy and a postsurgical stomach, demonstrating numerical examples in health and disease.

2 Structural and Functional Aspects of Stomach Soft Tissue at Different Scales

This section gives an overview of the SW's structure and function at the organ-, tissue-, and cell-level. A key aspect is the coordination of active smooth muscle contractions of the SW, whose failure can result in several gastric motility impairments. Thus, the complex mechanisms underlying gastric motility need to be highly regulated, which is accomplished at different scales.

2.1 Structure and Function at Organ-Level

Anatomically, the stomach can be divided into three regions (Fig. 1a, from cranial to caudal): The fundus, corpus, and antrum. The stomach is held in position via ligaments attached to the lesser and greater curvature, see Fig. 1b. Functionally, the stomach is divided into two parts (Masaoka and Tack 2009; Ferrua and Singh 2010). The proximal stomach, consisting of the fundus and upper corpus, functions as a gastric reservoir and is responsible for the storage of food and the emptying of liquids. In the empty stomach a high gastric tone is maintained in the proximal part, resulting in a contracted SW and a reduced capacity of the stomach (Schulze 2006). When food enters the gastrointestinal tract, the proximal SW region starts to relax due to a reduced tone, enabling a larger filling without a significant increase in the internal pressure (Schulze-Delrieu and Shirazi 1987; Wilson and Stevenson 2019). After some time, the SW gradually regains its tonic contraction, accompanied by a reduction of the stomach's capacity and the emptying of liquids (Azpiroz 1994; Notivol et al. 1995; Tack et al. 2003). Solid food is then moved to the distal part of the stomach, which consists of the lower corpus and antrum. The distal stomach serves as a muscular pump and features strong peristaltic contraction waves (Guyton and Hall 2006). At a rate of three cycles per minute (cpm), these phasic contraction waves originate in the corpus at the greater curvature and move to the antrum, while getting stronger and faster. While propagating, they transport and mix the food, which is retained at the pylorus and pushed back until it is liquified enough to be emptied into the duodenum (Kwiatek et al. 2006; Schulze 2006).

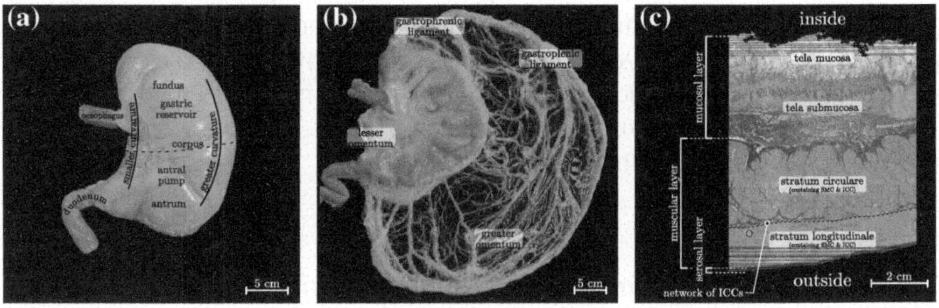

Fig. 1 Structure of the stomach: (a) stomach inflated with physiological solution; (b) deflated stomach with extended ligaments; (c) histological cross-sectional presentations stained with Picro-Sirius red

2.2 Structure and Function at Tissue-Level

The SW consists of three main layers (Fig. 1c, from the inside out): The mucosal layer, muscular layer, and serosal layer. These layers show regional differences in the microstructure, for a detailed description the reader is referred to Bauer et al. (2020). Generally, the mucosal layer, which can be subdivided into tela mucosa and tela submucosa, mainly consists of a network of collagen fibers. Functionally, it is responsible for secretion of acids to digest the meal. The mucosal layer features only passive mechanical characteristics and restricts the SW's inflation at large stretches. Large gastric folds serve to enlarge the surface of the gastric mucosa. The muscular layer, connected to the mucosal layer via connective tissue, consists of the stratum circulare and the stratum longitudinale. Herein, longitudinal and circumferential orientated smooth muscle fibers are responsible for the active contraction behavior of the SW. Between the longitudinal and circular muscle layers lies the plexus myentericus, which is a network of nerves controlling the activation of the muscular layer. In the plane of the myenteric plexus between the longitudinal and circular muscle layers, but also interspersed in these two layers, special pacemaker cells, the ICCs, are embedded (Fig. 1c). They build a pacing network which will be described in detail in Sect. 2.3. A very thin layer, named the serosal layer, covers the outer surface of the stomach. This layer will be neglected within the present modeling approach, as it shows almost no influence on load-bearing processes, e.g., in the case of the urinary bladder (Natali et al. 2015).

2.3 Structure and Function at Cell-Level

In terms of gastric smooth muscle activation, two types of cells play a key role: The ICCs and the SMCs. The ICCs are connected to each other, consequently building an electrically coupled network. ICCs are pacemaker cells which generate an electrical signal, the so-called slow waves, which are expressed through an oscillation of the membrane potential (Sanders et al. 2006). Isolated ICCs generate slow waves with an individual pacing frequency, but in a coupled network the frequencies in the ICCs each adapt to the highest natural frequency occurring in the network, which is 3 cpm in the human stomach (Hinder and Kelly 1977; O'Grady et al. 2010). This process is called entrainment. Thus, the ICCs work as an electrical syncytium through which they coordinate the global propagation of the electrical slow waves along the SW. Because the highest frequency occurs at the greater curvature, this region is called the dominant pacing site of the stomach. From here, the electrical slow waves originate and propagate distally to the antrum, owing to a decreasing frequency gradient in this direction. Because the slow waves propagate faster in the circumferential direction (Moriya and Miyazaki 1985), they distally move in ring waves, terminating at the pylorus.

Via gap junctions, the ICC is electrically coupled to the neighbouring SMC (Sanders 2008), being the contractile unit of the SW. Due to this coupling, the omnipresent slow waves also depolarize the SMC membrane in an oscillatory fashion.

Generally, the peak potential of the SMC slow waves is just below the mechanical threshold of −40 mV, being the voltage where contractions start to occur (Sanders et al. 2014a). In the presence of an additional input, e.g., a neural or mechanical signal, the depolarization of the SMC membrane through the electrical slow waves can result in action potentials superimposed on the peak potential (PP) of the SMC membrane (Huizinga and Lammers 2009; Johnson 2012). The SMC membrane potential then exceeds the mechanical threshold to trigger contractions. Thus, a contraction can only occur at the peak of a slow wave. One key feature is the gradient of SMC resting membrane potential (RMP) along the longitudinal axis of the stomach (Hirst and Edwards 2006). The RMP is the most negative potential, which occurs between the peaks of two slow waves. As the RMP and PP in the fundus are equal and generally above the mechanical threshold, a maintained gastric tone is present in this region (Azpiroz 1994), as mentioned in Sect. 2.1. Distally, the RMP gets more negative from the corpus to the antrum. Thus, the contractions get more phasic and stronger.

3 Modeling

To investigate gastric smooth muscle excitation-contraction, a modeling approach with two coupling-pathways is used in the present work, see Fig. 2. In the electrochemical coupling-pathway, the ICC slow waves, described by the ICC membrane potential Φ_{ICC}, i.e. Fig. 2a, lead to the depolarization of the SMC membrane potential Φ_{SMC}, Fig. 2b. Following the descriptions in Sect. 2.3, the depolarization leads to the opening of voltage-gated ion channels, see Fig. 2c, and to an increase in the intracellular calcium concentration $\left[\text{Ca}^{2+}\right]$ of the SMC, see Fig. 2d. As $\left[\text{Ca}^{2+}\right]$ increases, the concentration of activated myosin light-chain kinase also increases, see Fig. 2e, which is an enzyme responsible for the creation of cross- and latch-bridges between myosin and actin. As a consequence, gastric contractions are activated, leading to a deformation φ of the SW, see Fig. 2f.

The second, mechano-electrochemical coupling-pathway is considered to cover the ability of the ICC and SMC to respond to mechanical stimuli. Both, ICCs and SMCs contain mechanosensitive ion channels, which open during cell stretching (Kirber et al. 2000; Kraichely and Farrugia 2007). A stretch signal φ can cause an increase in the pacing frequency in the ICC (Won et al. 2005), while it modulates the PP and RMP of the SMC, as already mentioned in Sect. 2.3. Summarizing, smooth muscle stretch φ influences the ICC and SMC membrane potential Φ_{ICC} and Φ_{SMC}, see Figs. 2g and h, respectively.

Having introduced the four primary variables, the electro-chemomechanical state

$$\mathcal{S}(\mathbf{X}, t) = \{\Phi_{\text{ICC}}(\mathbf{X}, t), \Phi_{\text{SMC}}(\mathbf{X}, t), \left[\text{Ca}^{2+}\right](\mathbf{X}, t), \varphi(\mathbf{X}, t)\} \tag{1}$$

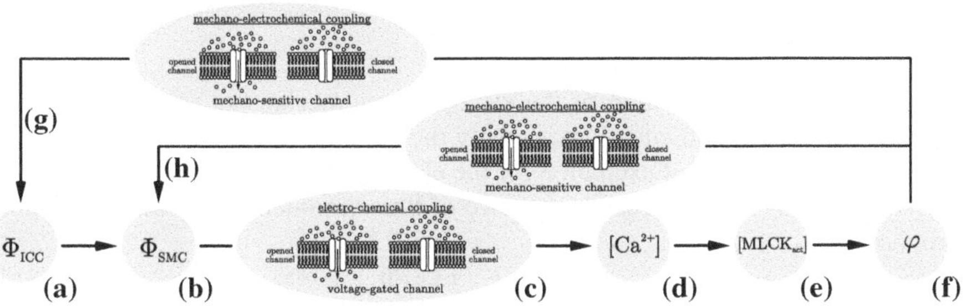

Fig. 2 Electro-chemomechanical (a)–(f) and mechano-electrochemical (g) and (h) coupling pathways of gastric smooth muscle contractions, see also Klemm et al. (2020) (® Copyright 2020 Elsevier B.V.). Four fields are considered, namely, the membrane potentials of the interstitial cells of Cajal Φ_{ICC} and the smooth muscle cells Φ_{SMC}, the intracellular calcium ion concentration $[\text{Ca}^{2+}]$, and the deformation $\boldsymbol{\varphi}$

of a material points' placement $\mathbf{X}$ at time t is given.

3.1 Kinematics

We consider a contractive body, in which the continuous, nonlinear mapping $\boldsymbol{\varphi}(\mathbf{X}, t)$ relates the vector $\mathbf{X}$ of the coordinates in the reference configuration $\mathcal{B}_0$ to the vector $\mathbf{x} = \boldsymbol{\varphi}(\mathbf{X}, t)$ of the coordinates in the current configuration $\mathcal{B}$. In this regard, $\mathbf{F} = \nabla_X \boldsymbol{\varphi}(\mathbf{X}, t)$ is the corresponding deformation gradient, with $\nabla_X(\bullet)$ being the spatial derivative with respect to the coordinates $\mathbf{X}$. Further, $J = \det \mathbf{F} > 0$ denotes the determinant of the deformation gradient, characterizing volumetric changes.

To account for the different smooth muscle (index: s) and collagen (index: c) fiber orientations, we introduce the structural tensors

$$\mathbf{Z}_{\text{c/s}} = \mathbf{M}_{\text{c/s}} \otimes \mathbf{M}_{\text{c/s}}, \tag{2}$$

describing the in-wall dispersions in the reference configuration, with the direction vector $\mathbf{M}_{\text{c/s}}(\mathbf{X})$. Thus, the corresponding stretches

$$\lambda_{\text{c/s}} = \sqrt{I_{4,\text{c/s}}} = \sqrt{\mathbf{C} : \mathbf{Z}_{\text{c/s}}} \tag{3}$$

depend on the scalar-valued fourth invariants $I_{4,\text{c/s}}$, and $\mathbf{C} = \mathbf{F}^{\text{T}}\mathbf{F}$ describes the right Cauchy-Green tensor. Note, unlike the isotropic part in Eq. (13), where the determination of the first invariant makes use of the isochoric part of $\mathbf{C}$, i.e., $\widehat{\mathbf{C}}$, the formulation of the anisotropic part here is based on the right Cauchy-Green tensor of the total deformation. For more details the reader is referred to Seydewitz et al. (2017).

3.2 Governing Equations

The spatio-temporal evolution of the electro-chemomechanical state $\mathcal{S}$ in Eq. (1) is governed by two field equations, namely the balance of linear momentum and a diffusion-type equation of excitation for the electrical ICC membrane potential.

Excluding any body forces, the balance of linear momentum in the local spatial form reads

$$J\,\mathrm{div}\,(J^{-1}\boldsymbol{\tau}) = \mathbf{0}, \tag{4}$$

defining the quasi-static stress equilibrium. Herein, div (•) is the spatial divergence and $\boldsymbol{\tau}$ denotes the Kirchhoff stress tensor. The boundary conditions on the disjoint surfaces $\partial\mathcal{B}_\varphi$ and $\partial\mathcal{B}_\sigma$ are given by

$$\boldsymbol{\varphi} = \breve{\boldsymbol{\varphi}} \text{ on } \partial\mathcal{B}_\varphi \text{ and } \mathbf{t} = \breve{\mathbf{t}} \text{ on } \partial\mathcal{B}_\sigma. \tag{5}$$

The traction vector $\breve{\boldsymbol{t}}$ is defined via the Cauchy stress theorem $\breve{\mathbf{t}} = \boldsymbol{\sigma}\,\mathbf{n}$, where $\mathbf{n}$ denotes the outward surface normal and $\boldsymbol{\sigma}$ is the Cauchy stress tensor.

For the spatio-temporal description of the ICC membrane potential, the following diffusion reaction-type equation

$$\dot{\Phi}_{\mathrm{ICC}} - J\mathrm{div}\left(J^{-1}\mathbf{q}^{\Phi_{\mathrm{ICC}}}\right) - F^{\Phi_{\mathrm{ICC}}} = 0, \tag{6}$$

is considered. Herein, the material time derivative $\dot{(\bullet)} = \partial(\bullet)/\partial t|_X$ of the action potential Φ_{ICC} is related to the reference configuration. The source term $F^{\Phi_{\mathrm{ICC}}}$ controls the activation and inactivation dynamics of the ICC membrane potential and will be described in Sect. 3.3.2. The field equation of excitation depends on the corresponding boundary conditions

$$\Phi_{\mathrm{ICC}} = \breve{\Phi}_{\mathrm{ICC}} \text{ on } \partial\mathcal{B}_{\Phi_{\mathrm{ICC}}} \text{ and } q^{\Phi_{\mathrm{ICC}}} = \breve{q}^{\Phi_{\mathrm{ICC}}} \text{ on } \partial\mathcal{B}_{q^{\Phi_{\mathrm{ICC}}}}. \tag{7}$$

Similar to the Cauchy stress theorem of the mechanical description, the surface flux is defined by the relationship $\breve{q}^{\Phi_{\mathrm{ICC}}} = \mathbf{q}^{\Phi_{\mathrm{ICC}}}\,\mathbf{n}$.

To account for the propagation and entrainment of the slow waves in the ICC-network, the flux term

$$\mathbf{q}^{\Phi_{\mathrm{ICC}}} = \mathbf{d}\nabla_x\Phi_{\mathrm{ICC}} \tag{8}$$

relates the spatial gradient of the action potential $\nabla_x\Phi_{\mathrm{ICC}}$ via the deformation- and location-dependent anisotropic conductivity tensor

$$\mathbf{d} = d_{\mathrm{iso}}\mathbf{I} + \sum_{i=1}^{n_s} d_{\mathrm{aniso},i}\,\mathbf{Z}^{\mathrm{R}}_{\mathrm{s},i}, \tag{9}$$

additively decomposed in an isotropic and anisotropic contribution. Herein, d_{iso} and $d_{\mathrm{aniso},i}$ are conductivity coefficients, n_s denotes the number of considered directions,

Table 1 Mechanical parameters for porcine SW tissue identified from active and passive experiments

Mucosal layer		
Parameter	Value	Unit
c_{c1}	0.5335	kPa
c_{c2}	0.0123	–
$f_{c,1}/f_{c,2}$ [†]	0.3781/0.6219	–
μ_e	0.0903	kPa
Muscular layer		
Parameter	Value	Unit
c_{s1}	0.2894	kPa
c_{s2}	0.0188	–
$f_{s,1}/f_{s,2}$ [†]	0.5788/0.4212	–
μ_{ECM}	0.1217	kPa
λ_{opt}	2.5403	–
ξ_s	0.8571	–
P_{opt}	24.5345	kPa
K_{vol}	121.6594	kPa

[†] Indices 1 and 2 correspond to the longitudinal and circumferential fascicle directions, respectively

$\mathbf{I}$ defines the identity tensor, and

$$\mathbf{Z}^{R}_{s,i} = \frac{\mathbf{F}\,\mathbf{M}_i}{|\mathbf{F}\,\mathbf{M}_i|} \otimes \frac{\mathbf{F}\,\mathbf{M}_i}{|\mathbf{F}\,\mathbf{M}_i|} \tag{10}$$

are the rotated structural tensors. Considering the electric conductivity of $d_{iso}\mathbf{I}$ as isotropic with respect to the spatial configuration, the pull-back of $d_{iso}\mathbf{I}$ leads to the anisotropic conductivity in the reference configuration (Cherubini et al. 2012).

3.3 Constitutive Equations

The two essential load-bearing parts of the SW are the (inner) mucosal and the (middle) muscular layer. Consequently, these two layers were considered in this modeling approach, while the (outer) serosal layer was neglected. To determine the mechanical parameters for the constitutive models of both layers, see Table 1, active and passive experiments on fresh porcine tissue strips were conducted. The protocol and results of the passive and active experiments can be viewed in the Appendix. Based on the collected experimental data, the mechanical parameters in Table 1 for both layers were identified based on an optimization procedure, see Eq. (34) in the Appendix.

3.3.1 Mucosal Layer

Featuring only passive material characteristics, the main constituents of the mucosal layer (index: MCL) are elastin (index: e) and collagen (index: c). Consequently, the strain-energy function

$$\psi_{\mathrm{MCL}} = \widehat{\psi}_{\mathrm{e}} + \psi_{\mathrm{c}} + U \tag{11}$$

is decomposed into a mixed contribution, including the components for elastin ($\widehat{\psi}_{\mathrm{e}}$) and collagen ($\psi_{\mathrm{c}}$), and a purely volumetric part

$$U = \frac{K_{\mathrm{vol}}}{4}(J^2 - 1 - 2\ln(J)), \tag{12}$$

which controls the degree of incompressibility by means of the parameter K_{vol}.

As a load-bearing protein, elastin builds three-dimensional networks of isotropic character. Thus, the isotropic neo-Hookean material model

$$\widehat{\psi}_{\mathrm{e}} = \frac{\mu_{\mathrm{e}}}{2}(\widehat{I}_1 - 3) \tag{13}$$

is used, with μ_{e} to be the shear modulus and the isochoric first invariant $\widehat{I}_1 = \mathrm{tr}\widehat{\mathbf{C}}$, defined as the trace of the isochoric part of the right Cauchy-Green tensor $\widehat{\mathbf{C}}$.

As a second load-bearing protein, the nonlinear mechanical behavior of collagen is described by an exponential stress-stretch relationship along the anisotropically orientated fiber direction

$$\psi_{\mathrm{c}} = \sum_{i=1}^{n_{\mathrm{c}}} f_{\mathrm{c},i}\, \psi_{\mathrm{c},i}, \tag{14}$$

where the strain-energy functions for SM (Holzapfel 2001), i.e.

$$\psi_{\mathrm{c},i} = \begin{cases} \dfrac{c_{\mathrm{c}1}}{2\,c_{\mathrm{c}2}} \left\{\exp\left[c_{\mathrm{c}2}\left(\lambda_{\mathrm{c},i}^2 - 1\right)^2\right] - 1\right\} & \text{if } \lambda_{\mathrm{c},i} > 1, \\ 0 & \text{else} \end{cases} \tag{15}$$

depend on two material constants, $c_{\mathrm{c}1}$ and $c_{\mathrm{c}2}$, weighted in every incorporated collagen fiber direction i with the corresponding fraction $f_{\mathrm{c},i}$. This fraction fulfils the constraint $\sum_{i=1}^{n_{\mathrm{c}}} f_{\mathrm{c},i} = 1$, with the total number of incorporated directions $n_{\mathrm{c}} = 2$ in case of longitudinal and circumferential oriented collagen fibers. The restriction in Eq. (15) ensures that only fibers under tension contribute to the mechanical response.

3.3.2 Muscular Layer

The muscular layer (index: MSL) consists of anisotropic oriented smooth muscle fascicles (index: s) embedded in an extracellular matrix (index: ECM). Thus, based on earlier modeling approaches (Schmitz and Böl 2011; Böl et al. 2012; Böl and Schmitz 2013; Seydewitz et al. 2017; Klemm et al. 2020), the strain-energy function

$$\psi_{\mathrm{MSL}} = \widehat{\psi}_{\mathrm{ECM}} + \psi_{\mathrm{s}} + U \tag{16}$$

is additively decomposed into an isochoric and purely passive contribution ($\widehat{\psi}_{\mathrm{ECM}}$) describing the mechanical behavior of the ECM, a contribution for the anisotropically aligned SM fascicles (ψ_{s}), and a volumetric contribution (U), in analogy to Eq. (12).

The ECM consists of elastin and collagen forming an isotropic load-bearing network. Thus, similar to Eq. (13), the isotropic neo-Hookean material model

$$\widehat{\psi}_{\mathrm{ECM}} = \frac{\mu_{\mathrm{ECM}}}{2}(\widehat{I}_1 - 3), \tag{17}$$

which depends on the shear modulus μ_{ECM}, is used.

The strain-energy function for the active smooth muscle fascicles, oriented in longitudinal and circumferential directions, reads

$$\psi_{\mathrm{s}} = \sum_{j=1}^{n_{\mathrm{s}}} f_{\mathrm{s},j}\,(\psi^{\mathrm{a}}_{\mathrm{s},j} + \psi^{\mathrm{p}}_{\mathrm{s},j}), \tag{18}$$

including active (index: a) and passive (index: p) contributions. Based on the anisotropic distribution of these fascicles, the strain-energy functions are weighted with SMC volume fractions $f_{\mathrm{s},j}$ in every direction $j = 1, ..., n_{\mathrm{s}}$, where $n_{\mathrm{s}} = 2$ in case of longitudinal and circumferential directions.

According to the collagen in the mucosal layer, see Eq. (15), the passive strain-energy functions (Holzapfel 2001)

$$\psi^{\mathrm{p}}_{\mathrm{s},j} = \begin{cases} \dfrac{c_{\mathrm{s}1}}{2\,c_{\mathrm{s}2}} \left\{\exp\left[c_{\mathrm{s}2}\left(\lambda^2_{\mathrm{s},j} - 1\right)^2\right] - 1\right\} & \text{if } \lambda_{\mathrm{s},j} > 1, \\ 0 & \text{else} \end{cases} \tag{19}$$

feature two material parameters $c_{\mathrm{s}1}$ and $c_{\mathrm{s}2}$ and are functions of SM fascicle stretches $\lambda_{\mathrm{s},j}$.

For the active contribution, a Gaussian function

$$\psi^{\mathrm{a}}_{\mathrm{s},j}(\Phi_{\mathrm{SMC}}, \left[\mathrm{Ca}^{2+}\right], \mathbf{C}) = \eta(\mathrm{Ca}^{2+})\; P_{\mathrm{max}} \int \exp\left[\frac{-(\lambda_{\mathrm{s},j} - \lambda_{\mathrm{opt}})^2}{2\xi^2_{\mathrm{s}}}\right] \mathrm{d}\lambda_{\mathrm{s},j} \tag{20}$$

mimics the bell shaped active stress response, which depends on $\lambda_{\mathrm{s},j}$, the constant ξ_{s}, controlling the width of the active stress generation, and the optimum stretch λ_{opt} at which the generated stress P_{max} reaches its maximum value. Furthermore, the strain-energy functions depend on the chemical degree of activation $\eta(\mathrm{Ca}^{2+})$, obtained by the equations presented in the following paragraph.

The electrical source term $F^{\Phi_{\rm ICC}}$ in Eq. (6) represents the electrophysiology of a single ICC. To mimic this behavior, we make use of a phenomenological two-variable FitzHugh-Nagumo type model (Aliev et al. 2000; Pullan et al. 2004). In doing so, it is common to express the model equations and variables in terms of the dimensionless potential

$$\phi = \frac{\Phi_{\rm ICC} - \Phi^r_{\rm ICC}}{\beta_{\rm ICC}}, \tag{21}$$

which is obtained from the ICC membrane potential $\Phi_{\rm ICC}$. Herein, the conversion coefficient

$$\beta_{\rm ICC} = \Phi^p_{\rm ICC} - \Phi^r_{\rm ICC} \tag{22}$$

equals the difference between the plateau potential $\Phi^p_{\rm ICC}$ and the resting membrane potential $\Phi^r_{\rm ICC}$ of the ICC slow waves. The formulation of the electric source

$$F^{\Phi_{\rm ICC}} = \bar{\beta}_{\rm ICC} F^{\phi}(\phi, r) \tag{23}$$

involves a multiplicative decomposition into $\bar{\beta}_{\rm ICC}$, a temporally normalized scaling factor of $\beta_{\rm ICC}$, and the dimensionless electric source

$$F^{\phi}(\phi, r) = \kappa\phi(\phi - \alpha)(1 - \phi) - r, \tag{24}$$

which is based on the two-variable FitzHugh-Nagumo modeling approach. Herein, r is the second, so-called recovery variable, governed by an additional differential equation

$$\dot{r} = \varepsilon(\gamma(\phi - \zeta) - r). \tag{25}$$

In Eqs. (24) and (25), κ, α, γ, and ζ are constant material parameters controlling the oscillation of the electrical slow waves. The parameter

$$\varepsilon = \varepsilon_0 + \sum_{j=1}^{n_s} f_{s,j}\, \theta^{\varepsilon}_{s,j} \tag{26}$$

is a position- and stretch-dependent term controlling the intrinsic frequency of the ICC. Herein, $\varepsilon_0(\mathbf{X})$ sets the proximal-to-distal frequency gradient along the SW in the absence of mechanical stimulation, see Table 2. Consequently, the term

$$\theta^{\varepsilon}_{s,j} = \begin{cases} \dfrac{w_f}{1 + \exp[-u_f(\lambda_{s,j} - v_f)]} & \text{if } \lambda_{s,j} > 1, \\ 0 & \text{else} \end{cases} \tag{27}$$

controls the stretch-dependency of the ICC slow wave frequency, with the parameters w_f, u_f, and v_f. Hereby, in a presented phenomenological approach, a smooth heaviside function relates the frequency to the stretch.

With the dimensionless ICC membrane potential ϕ at hand, we obtain the SMC membrane potential

$$\Phi_{\mathrm{SMC}} = (\Phi^{p}_{\mathrm{SMC}} - \Phi^{r}_{\mathrm{SMC}})\,\phi + \Phi^{r}_{\mathrm{SMC}} \tag{28}$$

by scaling ϕ with the location- and stretch-dependent resting membrane potential Φ^{r}_{SMC} and the stretch-dependent plateau potential Φ^{p}_{SMC} of the SMC slow waves. This simplification enables us to handle the SMC membrane potential Φ_{SMC} as a local variable.

For the RMP of the SMC, the following relation

$$\Phi^{r}_{\mathrm{SMC}} = \Phi^{r}_{\mathrm{SMC}\,0} - \sum_{j=1}^{n_{\mathrm{s}}} f_{\mathrm{s},j}\,\theta^{r}_{\mathrm{s},j} \tag{29}$$

is applied. The propagation of a negative RMP gradient along the longitudinal axis of the stomach within the SMC network implies that the value of the electrical potential in the absence of stretch $\Phi^{r}_{\mathrm{SMC}\,0}(\mathbf{X})$ is a function of the material coordinates $\mathbf{X}$, cp. Table 2. Further, the stretch sensitivity is controlled by

$$\theta^{r}_{\mathrm{s},j} = \begin{cases} \dfrac{w_a(1-\omega)}{1+\exp[-u_a(\lambda_{\mathrm{s},j}-v_a)]} & \text{if } \lambda_{\mathrm{s},j} > 1, \\ 0 & \text{else} \end{cases} \tag{30}$$

with w_a, u_a, and v_a being potential-like and dimensionless parameters. Herein, the dimensionless function $\omega(\mathbf{X}) \in [0, 1]$ ensures a smooth transition from stretch-induced hyperpolarization in the fundus to stretch-induced depolarization in the antrum along the curved longitudinal axis of the stomach.

Unlike the RMP, the SMC plateau potential in absence of any deformation $\Phi^{p}_{\mathrm{SMC}\,0}$ remains constant along the curved longitudinal axis of the stomach, and the overall expression for the electric PP reads

$$\Phi^{p}_{\mathrm{SMC}} = \Phi^{p}_{\mathrm{SMC}\,0} - \sum_{j=1}^{n_{\mathrm{s}}} f_{\mathrm{s},j}\theta^{p}_{\mathrm{s},j} \tag{31}$$

including the stretch-dependent contribution

$$\theta^{p}_{\mathrm{s},j} = \begin{cases} \dfrac{w_a(1-2\omega)}{1+\exp[-u_a(\lambda_{\mathrm{s},j}-v_a)]} & \text{if } \lambda_{\mathrm{s},j} > 1, \\ 0 & \text{else.} \end{cases} \tag{32}$$

In the undeformed state of the fundus, the PP and the RMP are approximately identical. The potential difference remains constant during stretching, while both values decrease, reducing the electric tonic to allow relaxation. The potential difference in

Table 2 Material parameters for the electrical excitation

Parameter	Value	Unit	Reference
κ	7.0	–	Pullan et al. (2004)
α	0.5	–	Pullan et al. (2004)
β	0.7	–	Pullan et al. (2004)
γ	8.0	–	Pullan et al. (2004)
ε_0	0.04-0.0025	–	Pullan et al. (2004)
Φ^p_{ICC}	−20	mV	Pullan et al. (2004)
Φ^r_{ICC}	−70	mV	Pullan et al. (2004)
$\Phi^p_{\text{SMC 0}}$	−38.5	mV	Buist et al. (2010)
$\Phi^r_{\text{SMC 0}}$	−39 to −68	mV	Lin et al. (2006), Buist et al. (2010)
d_{iso}	20-4250	mm^2 s^{-2}	Chosen
$d_{\text{aniso},1}/d_{\text{aniso},2}$	20-250/1500	mm^2 s^{-2}	Chosen
w_f/w_a	0.03/80	mV	Chosen
u_f/u_a	15/35	–	Chosen
v_f/v_a	1.4/1.16	–	Chosen

the stretch-free state of the antrum is more distinct owing to a more negative value of the RMP, and stretching initiates an increase in the PP. The parameters for the electrical excitation are given in Table 2.

Assuming that the dynamics of $\left[\text{Ca}^{2+}\right]$ in the gastric SMC follows a bell-shaped curve (Fukuta et al. 2002), the relationship between the SMC membrane potential and the intracellular calcium concentration is modeled as a Gaussian function (Seydewitz et al. 2017)

$$\left[\dot{\text{Ca}^{2+}}\right] = q_0 \exp\left[\frac{-\Phi^2_{\text{SMC}}}{2\xi^2}\right] - k\left(\left[\text{Ca}^{2+}\right] - \left[\text{Ca}^{2+}\right]_{\text{rest}}\right). \tag{33}$$

The constant ξ serves as a degree of freedom to limit the sensitivity of the calcium release to a certain range of the SMC membrane potential. The material parameters q_0 and k control the release and return of the calcium, and $\left[\text{Ca}^{2+}\right]_{\text{rest}}$ denotes the resting calcium concentration.

The calcium concentration $\left[\text{Ca}^{2+}\right]$ is used as an input signal to calculate the concentration of activated myosin light-chain kinase, and the fraction of cross and latch bridges represented by $\eta(\text{Ca}^{2+})$ between myosin and actin filaments. For this purpose, the extended Hai-Murphy model by Gajendiran and Buist (2011) is used. The fraction $\eta(\text{Ca}^{2+})$ is used in Eq. (20) to determine the active stress response of the muscular layer of the SW. The parameters used to determine the calcium dynamics and the fraction $\eta(\text{Ca}^{2+})$ are presented in Table 3.

Table 3 Material parameters for the chemical processes inside the SMC

Parameter	Value	Unit	Reference
q_0	0.8	$\mu M\ s^{-1}$	Klemm et al. (2020)
k	1.0	s^{-1}	Klemm et al. (2020)
$[Ca^{2+}]_{rest}$	0.15	μM	Gajendiran and Buist (2011)
$f_{1/2/3/4/5/6/7/8/9}$	12/480/5/840/28/ 120/7.5/5/7.6	$\mu M^{-1} s^{-1}$	Gajendiran and Buist (2011)
$r_{1/2/3/4/5/6/7/8/9}$	12/1200/135/45.4/ 0.0308/4/3.75/25/22.8	s^{-1}	Gajendiran and Buist (2011)
$k_{3/4/7}$	15/5/10	s^{-1}	Gajendiran and Buist (2011)
k_{cat_K}/k_{cat_P}	27/16	s^{-1}	Gajendiran and Buist (2011)
K_{m_K}/K_{m_P}	10/15	μM	Gajendiran and Buist (2011)
MLCP	7.5	μM	Gajendiran and Buist (2011)

4 Simulation of a Healthy Stomach

This section presents the simulation of the electro-chemomechanical excitation-contraction process arising in porcine SW tissue under physiological conditions. The modeling approach proposed in Sect. 3 is applied to a geometry of a porcine stomach, see Sect. 4.1, to investigate gastric motor activity.

4.1 Geometrical Model of the Stomach

The geometry of the stomach, see Fig. 3, was created by recording a real porcine stomach via an optical system consisting of 2 CCD cameras. The optical information of different angles was used to generate the three-dimensional geometry in a post-processing step. For more details, the reader is referred to Böl et al. (2013), where a similar procedure was realized to reconstruct a skeletal muscle geometry. After smoothing the outer three-dimensional surface, an offset was created towards the inside of the stomach based on constant representative thicknesses. As indicated in Sect. 3.3, two layers, namely the mucosal layer ($T_{MCL} = 2.9$ mm, dark grey) and the muscular layer ($T_{MSL} = 4.6$ mm, light grey), are considered. Both layers are dominated by an anisotropic arrangement of longitudinal and circumferential orientated fibers, which represent the collagen fibers in the mucosal layer and smooth muscle fibers in the muscular layer. To mimic physiological boundary conditions, finite spring elements (stiffness: $k_{spring} = 10^{-2}$ mN/mm) were attached to the edges of the greater and lesser curvature, representing the fixation via ligaments (compare with Sect. 2.1 and Fig. 1b). In addition, symmetric boundary conditions were applied accordingly, and the end surfaces of the esophageal entrance A_e and pyloric ending A_p were constrained in the normal direction. Finally, a pressure, which increases

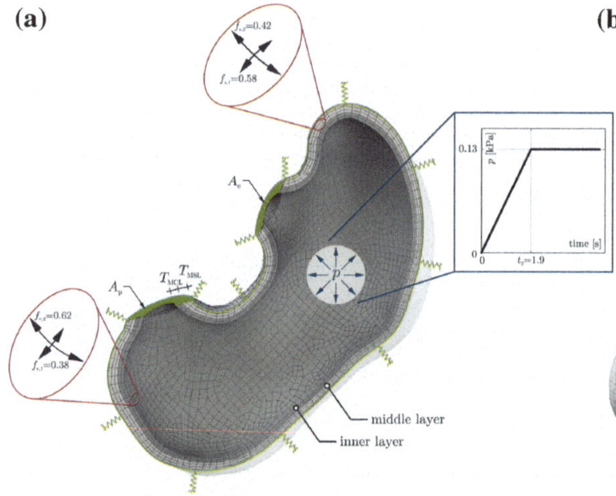

Fig. 3 Illustration of a porcine, geometric stomach model: (a) two layers (MCL, MSL) were considered for the simulations. The MCL features passive, anisotropic mechanical characteristics with collagen fiber fractions in longitudinal ($f_{c,1}$) and circumferential ($f_{c,2}$) directions. For the active anisotropic MSL, SMC fiber fractions in the longitudinal ($f_{s,1}$) and circumferential ($f_{s,1}$) directions were considered and are illustrated by double-sided arrows. Here, according to Eq. (5), $\partial\mathcal{B}_\varphi$ represents the greater and lesser curvature and the surfaces at the esophageal and pyloric ending, and $\partial\mathcal{B}_\sigma$ denotes the internal surface. For the discretization of the half system, 8,140 linear mixed brick elements (Q1P0) with 10,236 nodes were used. (b) Four different locations (P_1: fundus, P_2: upper corpus, P_3: middle corpus, P_4: antrum) were chosen for tracking the electro-chemical behavior of the filled stomach (Sect. 4.2)

linearly to $p = 0.13$ kPa and is then kept constant over the time of simulation, acts perpendicular at the internal surface of the stomach and simulates the filling of the stomach during ingestion, see Fig. 3a.

4.2 *Temporal Behavior of* Φ_{ICC}, Φ_{SMC}, *and* $[Ca^{2+}]$

Figure 4 illustrates the temporal evolutions of Φ_{ICC}, Φ_{SMC}, and $\left[Ca^{2+}\right]$ for four different positions $P_{1/2/3/4}$, see Fig. 3b, across the filled stomach. After reaching constant internal pressure at 1.9 s, simulation continued for an additional 125.1 s. The plot in Fig. 4a shows the temporal progress of Φ_{ICC}. In all four locations Φ_{ICC} features a RMP of $\Phi_{ICC} = -70$ mV and increases to a peak potential of -20 to -15 mV followed by a plateau-like decrease. Eight slow waves can be seen in P_2, which represents the dominant pacing site. The eighth slow wave in P_1 and P_3 can not be fully observed, because of a phase shift, which indicates that Φ_{ICC} first increases in P_2. In P_1, P_2, and P_3 the slow wave show a dominant frequency of above 3.5 cpm, thus, in these regions entrainment leads to one dominant frequency. This frequency is slightly higher than the value set for the highest frequency (3 cpm), because of the

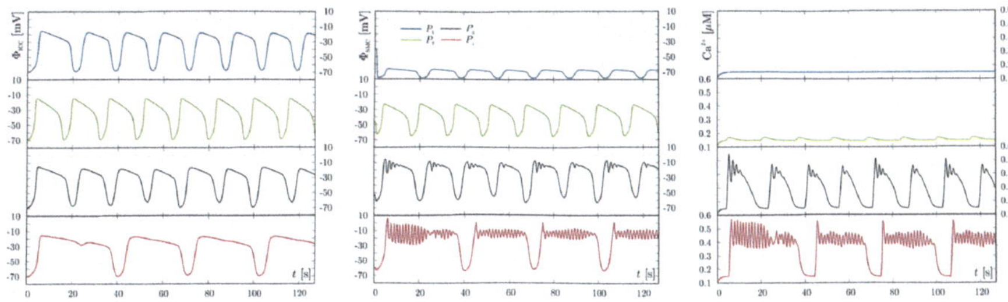

Fig. 4 Temporal behavior of Φ_{ICC}, Φ_{SMC}, and $[Ca^{2+}]$ at four different locations, see Fig. 3b, of the filled stomach

mechanical feedback-mechanism for the ICC slow wave frequency. However, in P_4 the frequency is significantly lower, indicating a failed entrainment in more distal regions.

Figure 4b shows the temporal behavior of Φ_{SMC}. At each point the timing of the SMC slow waves coincides with the ones of the ICC. The electrical tone of the fundus (P_1), which was initially set to -40 mV, has decreased to approximately -70 mV, with small oscillations. Generally, the amplitude of the SMC slow waves increases in aboral direction. While Φ_{SMC} at P_2 does not reach -20 mV, at P_3 the PP of the SMC membrane potential reaches values around -10 mV. Every time a SMC slow wave at P_3 reaches the peak potential, the progress is followed by a damped oscillation, and only then does the plateau-like decrease appear. In this regard, the number of peaks superimposed on the PP slightly changes for every slow wave, but the pattern repeats for every two slow wave cycles, which is approximately every 30 s. In the distal stomach (P_4), Φ_{SMC} reaches a peak potential of 0 mV at the beginning of the plateau potential. Hereafter, Φ_{SMC} features several peaks around -10 mV. The peaks become smaller in amplitude at first, but then increase again. This is because a new slow wave emerging from the corpus (P_3) reaches the distal stomach, where Φ_{SMC} is still elevated, compare also with the following Sect. 4.3. This is also the reason why the frequency in this region is lower than in the other regions.

In Fig. 4c the temporal behavior of the intracellular calcium concentration of the SMC is illustrated. In the fundus (P_1), $[Ca^{2+}]$ remains at its resting value of $[Ca^{2+}]_{rest} = 0.15$ μM. At P_2 the calcium concentration only slightly increases to 0.16 μM in a phasic manner, virtually remaining at resting levels. At P_3, every time the SMC membrane potential increases, $[Ca^{2+}]$ increases, with a small delay, from $[Ca^{2+}]_{rest}$ to a peak concentration of 0.45 to 0.5 μM, with slightly damped oscillations of peaks superimposed on the plateau. This increase is even higher in the antrum, with peak concentrations of 0.55 μM for the first peak and 0.45 μM for the following peaks. The mechanical feedback-mechanism for the SMC plateau potential Φ^{p}_{SMC} is the reason for the damped oscillation of peaks superimposed on the SMC plateau potential and calcium increase.

4.3 Spatio-Temporal Propagation of the Main Variables Across the Stomach Wall

Figure 5 illustrates the spatio-temporal propagation of the ICC membrane potential Φ_{ICC}, the SMC membrane potential Φ_{SMC}, the calcium concentration $\left[\text{Ca}^{2+}\right]$ in the SMC, and the von Mises stress σ_{vM} at the discrete time point $t_{1/2/3/4/5/6/7/8} = 0/1.9/68.7/73.2/74.9/76.95/80/83.625$ s.

Hereby, t_1 is the reference configuration and t_2 is the time where the maximal pressure is reached, so these two time steps are shown to illustrate the deformation of the stomach during filling. Within the time range of t_3 to t_8, approximately one slow wave cycle is illustrated. This cycle starts at $t_3 = 68.7$ s to ensure stabilization of the electro-chemomechanical processes. To better illustrate the contractions of the SW, the deformation is scaled by a factor of 1.3. The ICC slow wave starts at t_3 with an increase of Φ_{ICC} at the upper corpus at the greater curvature, being the dominant pacing site. The remnants of the previous slow wave can be seen in the proximal fundus and distal antrum. In the following time steps, these remnants are vanishing, while the new slow wave fully propagates in circumferential direction and, slightly slower, in longitudinal direction. When the increase in Φ_{ICC} reaches the antrum, the ICC membrane potential starts to decrease at the dominant pacing site, which can be seen in t_5, t_6 and t_7. At t_8, a new slow wave originates again at the dominant pacing site, while the old one starts to vanish in the fundus and antrum.

For the SMC membrane potential Φ_{SMC}, the RMP gradient can be seen at the beginning of the simulation (t_1), whereas the RMP decreases in the fundus after inflation (t_2), due to the mechanical feedback. Then, at t_3, Φ_{SMC} starts to increase

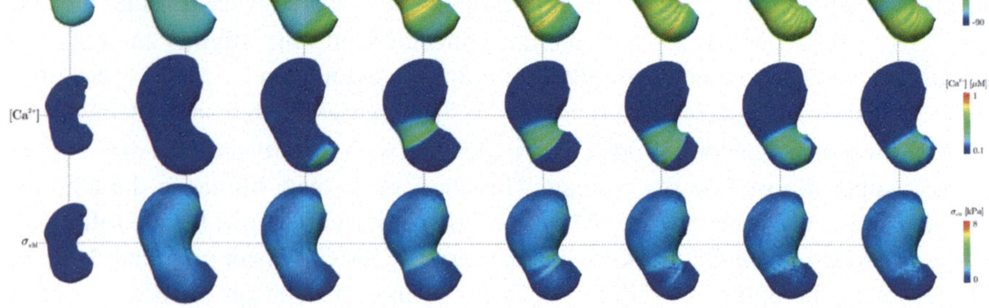

Fig. 5 Spatio-temporal progress of the main variables Φ_{ICC}, Φ_{SMC}, $\left[\text{Ca}^{2+}\right]$, and σ_{vM} during the active stomach contraction illustrated at the discrete time point $t_{1/2/3/4/5/6/7/8} = 0/1.9/68.7/73.2/74.9/76.95/80/83.625$ s

in the corpus, being in line with the increase in Φ_{ICC}. A small region in the distal antrum shows an increase as well from the previous slow wave. In the fundus, the SMC membrane potential remains constant and low at nearly -80 mV for the next time steps. However, at t_4 a ring of depolarization emerges in the corpus, which moves distally in the following time steps. In fact, several peaks, where Φ_{SMC} increases up to 0 mV or higher, emerge in the area of SMC membrane depolarization, propagating distally with this area. The highest peak is at the front of the slow waves, while the following peaks become lower. These peaks have already been mentioned in Sect. 4.2. At t_7 the SMC membrane potential decreases at the dominant pacing site. At t_8, the decrease propagates in every direction, while Φ_{SMC} increases again in the upper corpus.

Every time Φ_{SMC} increases in the corpus and antrum, $\left[\text{Ca}^{2+}\right]$ increases as well, leading to a distally propagating ring of elevated calcium concentration. This ring gets relatively large in thickness as it propagates distally. This is because of the increasing number of peaks of high Φ_{SMC} in this direction. These peaks can also be distinguished for $\left[\text{Ca}^{2+}\right]$ when viewing the third row in Fig. 5 in more detail.

The increase of the von Mises stress σ_{vM} is highest at t_4 in the region of the ring of SMC depolarization. The active stress increases here, and a contraction can be seen due to the elevated calcium concentration, which restricts the stomach circumferentially. This results in an hourglass-shaped deformation of the filled stomach. The increase of σ_{vM} propagates in aboral direction at time steps t_5 and t_6, while it becomes weaker. However, a slight increase can be observed in the antrum, as well as a deformation which reduces the capacity of the antrum. Because of the thick ring of elevated calcium concentration, the deformation is not expressed as a small circumferential restriction, as it is in time step t_4.

4.4 Isochronal Activation Map of the SMC Membrane Potential

To investigate the propagation of the activation of the muscular layer of the SW in more detail, an isochronal map of the SMC electrical activation across the SW is illustrated in Fig. 6. The plot starts at a time of 73.2 s and ends at 78.7 s, with time increments of 0.5 s in between. Thus, the time of observation lies in the range of the slow wave cycle investigated in the previous results. The SMC is considered electrically activated as soon as Φ_{SMC} reaches values of -30 mV or higher, which is illustrated as a colored area for each time step. The grey region represents the area of gastric tissue where Φ_{SMC} is below -30 mV, and thus, not electrically activated in every considered time step. The activation is illustrated with respect to the reference configuration, i.e., the deformation is not shown. As can be seen in Fig. 6, most of the muscular layer in the corpus is activated after 73.2 s, which represents the beginning of a new slow wave, which originated at the dominant pacing site. At the same time, a small region in the distal antrum below the pylorus is also activated,

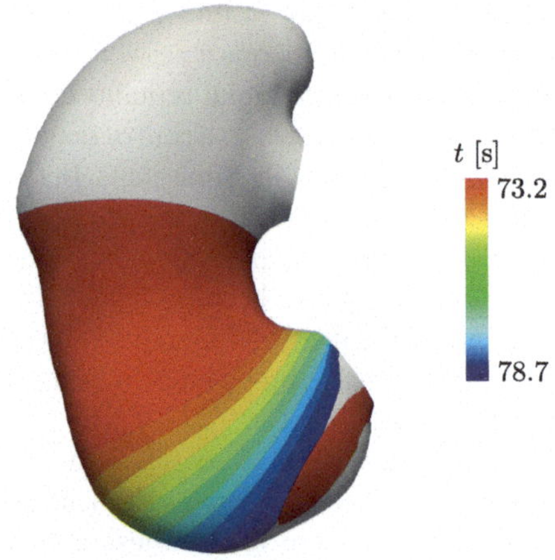

Fig. 6 Isochronal time plot for the SMC membrane potential of the healthy stomach. The plot shows which region of the SW is electrically activated (threshold: $\Phi_{SMC} \geq -30$ mV) at time steps of 0.5 s

which corresponds to the slowly vanishing previous slow wave. As can be seen in the following time steps, the slow wave propagates distally in an organized fashion.

5 Discussion

In the present work a three-dimensional multi-field model of the excitation-contraction characteristics in gastric smooth muscle tissue is presented, to get a better understanding of the electro-chemomechanical process during the activation of gastric contractions. Within the finite element method, whole-organ simulations on a three-dimensional numerical model of a porcine SW were performed to predict the spatio-temporal propagation of the ICC and SMC membrane potential (electrical fields), the intracellular calcium concentration (chemical field), and the generation of active contraction (mechanical field). To investigate the electro-chemomechanical behavior of the SW during filling by a bolus of food, an internal pressure was applied, acting perpendicular on the inner surface of the SW. Thus, the phenomenologically-based mechanical feedback of the model could be investigated.

Generally, the stomach can expand to a maximum volume of 2 to 4 liters. At rest, normal intragastric pressure is below 1 kPa, and can increase up to 4 kPa when distended (Wali and Munnur 2013). The capacity of the filled stomach in the simulation amounts to nearly 2 liters at an intragastric pressure of 1.3 kPa. After inflation, the simulation demonstrates a mechanically induced decrease in the electrical tone in the SMC in the fundus, especially around the oesophagus, without contractions. The relaxed SW in this region enables the stomach to accumulate large amounts of incoming food (Azpiroz 1994). At the dominant pacing site in the corpus at the greater curvature, the dominant frequency of the ICC slow waves, initially set to 3 cpm, increased due to the inflation of the stomach. A dominant ICC slow wave

6.2 Gastric Modeling as Computational Support in Diagnostics

This section aims to clarify what can be expected of future gastric models in diagnostic analyses. In a first step, the predictive power of our model in terms of a numerical support in diagnostics is tested. For this purpose, a stomach after surgery will be simulated. The postsurgical stomach features an incision in the SW, which is expected to evoke disturbances in the ICC electrical slow wave propagation and entrainment. Again, an internal pressure simulates the filling of the postsurgical stomach, and the electro-chemomechanical outcome will be compared with the healthy stomach of Sect. 4. At the end of this section, the results will be discussed, as well as the use of biophysically-based models as alternative approaches for computational tools in diagnostics.

6.2.1 Simulation of a Postsurgical Stomach

The postsurgical stomach was obtained by creating a hole in the SW, which represents the excision of gastric tissue due to, e.g., the removal of small gastrointestinal stromal tumors (GISTs) (Novitsky et al. 2006). This was done by deleting some elements, whereby it was made sure to obtain the same number of nodes on both sides of the hole. The configuration was named reference 1, see Fig. 7a. The closure of the excision was generated by a displacement-controlled simulation prescribing the opposite nodes at the hole to move towards each other. Hereafter, we obtained the modified reference configuration (reference 2), see Fig. 7b, where the internal pressure was applied in the actual simulation. During the actual simulation, the nodes with the same coordinates at the incision were tied with respect to the displacement degrees of freedom. The initial fiber orientations and the location-dependent parameters were set with respect to reference 1. The modification of the reference configuration for the simulation of the postsurgical stomach resulted in a non-flux condition at the nodes of the incision, representing a disturbance in the ICC electrical network after surgery.

Figure 8 shows the spatio-temporal evolution of the four main variables (Φ_{ICC}, Φ_{SMC}, $\left[\text{Ca}^{2+}\right]$, and σ_{vM}) for the postsurgical stomach at the same discrete time points as in Sect. 4.3.

The first row tracks the propagation of a slow wave in the ICC-network, emerging at the dominant pacing site at t_3. It is striking that the front of the distally propagating slow wave slightly shifts towards the incision at t_4 and t_5. At t_5, the incision hinders the propagation of the depolarization. As a consequence, at t_6, the tissue under the incision depolarizes in fully circumferential direction because of the depolarization at the sides. Hereafter, the slow wave propagates further in distal direction, whereby the direction has somehow changed to that in the healthy stomach, cp. Fig. 5. The same shifted behavior can be observed for the peaks of the increase of Φ_{SMC} at t_4 and t_5. At t_6, the Φ_{SMC} peak potentials propagate only at the sides of the incision.

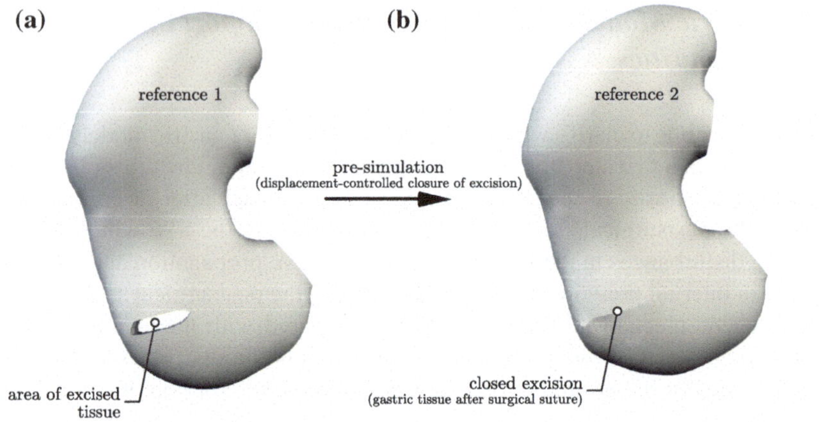

Fig. 7 Modification of the reference configuration for the simulation of the postsurgical stomach: (a) the excision was obtained by the deletion of elements (reference 1); (b) with a displacement-controlled simulation, the excision in the stomach tissue was closed (reference 2)

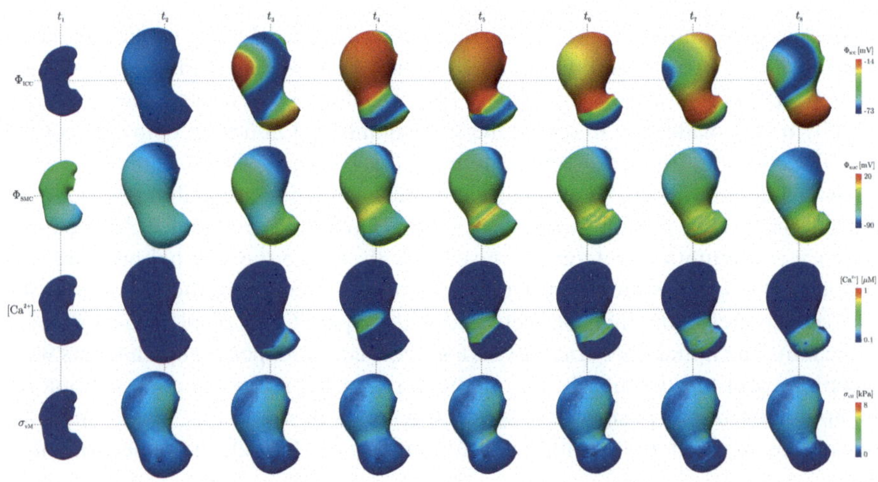

Fig. 8 Spatio-temporal progress of the main variables Φ_{ICC}, Φ_{SMC}, $[\mathrm{Ca}^{2+}]$, and σ_{vM} during the active stomach contraction of the postsurgical stomach

As a result, the propagation pattern of the peaks in the antrum at t_7 and t_8 is not organized anymore, in comparison to the healthy stomach. This is an indicator, that the deformation in the antrum must be unorganized as well, as the peaks are due to the mechanical feedback of the model. However, the deformations are too small to distinguish them from the healthy contractions. Instead, the area of elevated calcium concentration is shifted at t_5 leading to a slight deviation from the original hourglass-deformation observed for the healthy stomach between t_4 and t_5. This also appears for the active stress at t_5 and t_6 in the last panel.

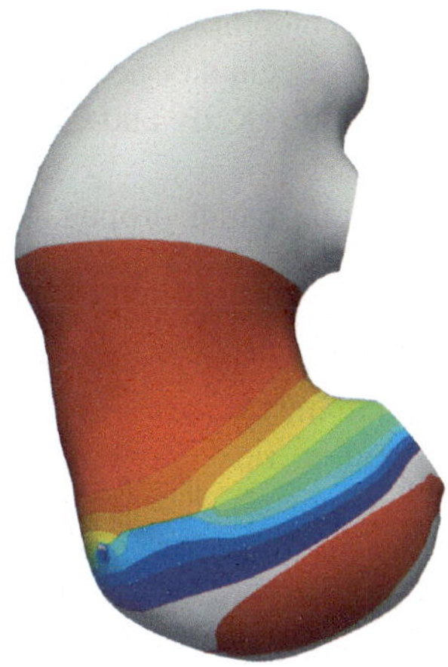

Fig. 9 Isochronal time plot for the SMC membrane potential of the postsurgical stomach. The plot shows which region of the SW is electrical activated (threshold: $\Phi_{SMC} \geq -30$ mV) at time steps of 0.5 s

Figure 9 shows an isochronal time plot of the SMC membrane potential across the postsurgical stomach. The time discretization is the same as in Fig. 6 of the isochronal activation map of the healthy stomach. Again, the muscular layer was considered activated when Φ_{SMC} reached a value of -30 mV or higher. The propagation of the SMC membrane potential above the incision already differs from that of the healthy stomach of Sect. 4.4 in Fig. 6. The front of the slow wave is slightly shifted, as observed earlier, and the region directly above the incision is activated earlier than the tissue in circumferential direction of this region. Due to the zero-flux condition at the incision, the SMC slow wave propagates only at the sites of the incision at time steps of 75.2 and 75.7 s. In the next second, the increase of the electrical SMC membrane potential propagates in circumferential direction under the incision. Thus, a ring wave creates again, which propagates further in distal direction. However, when comparing the propagation direction with that of the healthy stomach, the direction has slightly changed, which was also observed earlier. The change of direction can also bee seen when comparing the previous slow wave of the healthy, cp. Fig. 6, and the postsurgical stomach, cp. Fig. 9, at 73.2 s.

6.2.2 Discussion of the Postsurgical Stomach and Future Perspective

Failure of gastric tissue electrophysiology can lead to several gastric dysrhythmias and motility disorders, and causes for this failure can be manifold. Abnormal pacemaker activity of ICCs, loss of ICCs, or other damages in the ICC-network, due to gastric surgery, or neuronal or myogenic dysfunctions, are said to be probable causes for a failed entrainment and disruptive slow wave propagation (Kelly and Code 1971; Sarna et al. 1972; Hinder and Kelly 1977; O'Grady et al. 2012; Oh and Pasricha 2013; Du et al. 2015). Herein, different types of gastric arrhythmias have been observed based on high-resolution spatial mapping (O'Grady et al. 2011). One example is that of an incomplete or complete conduction block, which can be evoked by a failure of ICC pacemaker activity or, as it is the case in our simulations, by a cut in the gastric tissue (Kelly and Code 1971; Sarna et al. 1972; Hinder and Kelly 1977; Du et al. 2015).

The rapid circumferential propagation distal to the incision, see Figs. 8 and 9, recreates a ring of depolarization, which can further propagate in longitudinal direction. This anisotropic conduction regenerates the propagation pattern of electrical slow waves and has already been observed experimentally and numerically in Du et al. (2015). Herein the excision was modeled by defining an ellipsoidal region with no slow waves and no flux. However, they did not consider any effects related to the closure of the excision. Further, no deformations were considered in this study. In our simulation, the excised tissue was closed, representing the 'suture' of gastric excisions, see Fig. 7. Because of the closing, the ICC frequency gradient was shifted together with the tissue, resulting in a shift of the ring of depolarization in both ICC and SMC membrane potential. Consequently, the hourglass-shaped deformation of the postsurgical stomach slightly differed from the healthy stomach. Another observation is the unorganized pattern of SMC membrane peak potentials distal to the incision, which indicates an impaired motility pattern in this region.

The simulation of the postsurgical stomach showed that impaired gastric motility can be caused by excisions of the SW, evoking a disturbance in the electrical ICC network. As one example, impaired motility may result in delayed gastric emptying, which is a common postoperative diagnosis (Bar-Natan et al. 1996).

Although our model is based on a phenomenological approach to simulate gastric tissue electrophysiology, it can replicate several effects of gastric surgery on electrical slow wave propagation across the stomach, observed for animals and humans (Kelly and Code 1971; Sarna et al. 1972; Hinder and Kelly 1977; Du et al. 2015). However, biophysically-based gastric models, as described in Sect. 6.1, are the preferred numerical tools in diagnostic simulations, as they correlate gastric motility with the underlying events of ICC network electrophysiology. Further, due to the detailed micro-scale description of ICC pacemaker activity, such a model can be used for the simulation of possible pharmacological treatments, particularly targeting the ICCs, which may be used to treat a disturbed slow wave propagation (Huizinga et al. 1997). With the calculation of the deformation at hand, the mechanical outcome can be investigated as well.

Appendix: Experimental Investigations

To determine the mechanical parameters for the constitutive models of the mucosal and muscular layer in Sect. 3.3, see Table 1, active and passive experiments on fresh porcine tissue strips (16×8 mm) from the proximal stomach were conducted. The experiments were grouped into strips featuring the entire wall structure and strips whose mucosal layer had been removed so that only the muscular layer remained for testing. Both types of structures were tested in longitudinal ($n = 5$) orientation (i.e., parallel to the greater curvature) and in circumferential ($n = 5$) orientation (perpendicular to the longitudinal orientation), so that a total sample size of $n = 20$ was tested. The protocol can be taken from Klemm et al. (2020). Briefly, for the active stress-stretch response of the muscular layer, the tissue strips were electrically stimulated with alternating pulses with a current, frequency, and pulse width of 1 A, 100 Hz, and 5 ms (van Mastrigt and Glerum 1985; Borsdorf et al. 2019) via two platinum field electrodes at a constant solution temperature of 32 ± 0.1 °C. After an equilibration period of 30 min at L_s (slack length, see Klemm et al. 2020), the strips were stimulated isometrically for about 23 s every 7 min (to avoid muscle fatigue effects) until a steady state force (deviation $< 5\%$ of maximum isometric force F_{max}) was reached. Within each interval, a preconditioning protocol of 150 cycles with a frequency of 1 Hz (amplitude: 0.03 strip length) was applied. A series of 18-24 isometric contractions (length increments: 0.1 L_s in ascending order), starting from an initial sample length of 0.8 L_s, were conducted. The active force was calculated as the difference between the maximum total strip force (during simulation) and the passive force (immediately before simulation). The muscle length at which F_{max} was produced was defined as the optimal muscle length λ_{opt}.

For the passive stress-stretch response of the entire wall, the same procedure was applied to measure the passive forces. Based on the force values, the passive stress-stretch relations were determined using the mean cross-sectional areas. The results are presented in terms of the mean and the standard deviations. The mean active stress-stretch responses are provided in Fig. 10 as black curves and the standard deviations are presented as grey shaded areas. By comparing the active stress-stretch response in the (a) longitudinal and (b) circumferential directions, it is noticeable that the optimum stretch in the circumferential direction is larger, whereby a smaller stress is achieved.

The passive force-stretch and stress-stretch relations of complete strips and those only featuring the muscular layer are presented in Fig. 11, where the typical exponential response can be seen in both tissue strips. Regardless of the stresses or the forces, the entire strip shows isotropic behavior (Fig. 11a, b), which is surprising considering the inhomogeneous microstructure in the fundus (Bauer et al. 2020). In contrast, a certain anisotropic behavior can be seen in the mechanical responses of the muscle layer (Fig. 11c, d), which in turn reflects the inhomogeneous microstructure (Bauer et al. 2020). Thus, when comparing both tissue structures, it seems that the mucosal layer balances the mechanical anisotropy of the muscularis.

The layer-specific data of the mucosal layer was obtained by subtracting the measured passive forces of the muscular layer from the entire wall. Based on the collected experimental data, the mechanical parameters in Table 1 for both layers were identified by minimizing the objective function

$$\mathcal{O}(p^K) = \frac{1}{n}\sum_{i=1}^{n}\sqrt{\frac{\left(F_i^{\text{exp},K} - F_i^{\text{sim},K}\right)^2}{\left(F^{\text{mean},K}\right)^2}} \quad \text{with} \quad F^{\text{mean},K} = \frac{1}{n}\sum_{i=1}^{n} F_i^{\text{exp},K}. \tag{34}$$

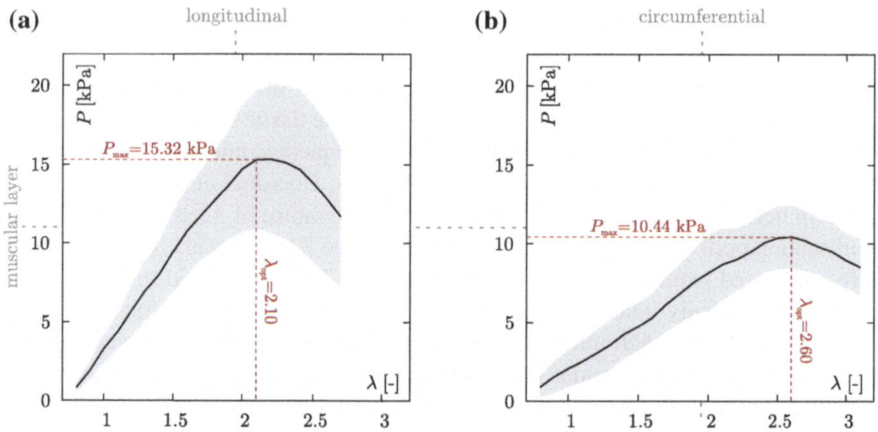

Fig. 10 Active stress-stretch responses of 10 gastric muscular tissue samples in the (a) longitudinal and (b) circumferential directions, see also Klemm et al. (2020) (® Copyright 2020 Elsevier B.V.). The solid curves indicate mean values, whereas the shaded areas depict the standard deviations. During all the experiments, the forces F and the strip length changes Δl were measured and converted to engineering stress $P = F/A_0$ (A_0 denotes the average cross-sectional area of the strip measured in the initial and unloaded situations) and stretch $\lambda = 1 + \Delta l / l_0$ (l_0 denotes the length of the strip measured in the initial, unloaded situation), respectively

Herein, $F_i^{\mathrm{sim},K}$, $F_i^{\mathrm{exp},K}$, and $F^{\mathrm{mean},K}$ are the simulated and experimentally measured layer-specific force values evaluated at n deformation increments and the corresponding experimental mean value, respectively. Further, K is associated with three different sets of parameters, which is one set for the mucosal layer's passive behavior and two sets for the muscular layer's active and passive behavior, respectively.

References

Akhmadeev, N.R., Miftahof, R.: Stress-strain distribution in the human stomach. Int. J. Dec. Nat. Ecodyn. **5**, 90–107 (2010)

Aliev, R.R., Richards, W., Wikswo, J.P.: A simple nonlinear model of electrical activity in the intestine. J. Theor. Biol. **204**, 21–28 (2000)

Azpiroz, F.: Control of gastric emptying by gastric tone. Dig. Dis. Sci. **39**, 18S-19S (1994)

Bar-Natan, M., Larson, G.M., Stephens, G., Massey, T.: Delayed gastric emptying after gastric surgery. Am. J. Surg. **172**, 24–28 (1996)

Bauer, M., Morales-Orcajo, E., Klemm, L., Seydewitz, R., Fiebach, V., Siebert, T., Böl, M.: Biomechanical and microstructural characterisation of the porcine stomach wall: location- and layer-dependent investigations. Acta Biomater. **102**, 83–99 (2020)

Böl, M., Schmitz, A.: A coupled chemomechanical model for smooth muscle contraction. In: Computer Models in Biomechanics, pp. 63–75. Springer (2013)

Böl, M., Schmitz, A., Nowak, G., Siebert, T.: A three-dimensional chemo-mechanical continuum model for smooth muscle contraction. J. Mech. Behav. Biomed. Mater. **13**, 215–229 (2012)

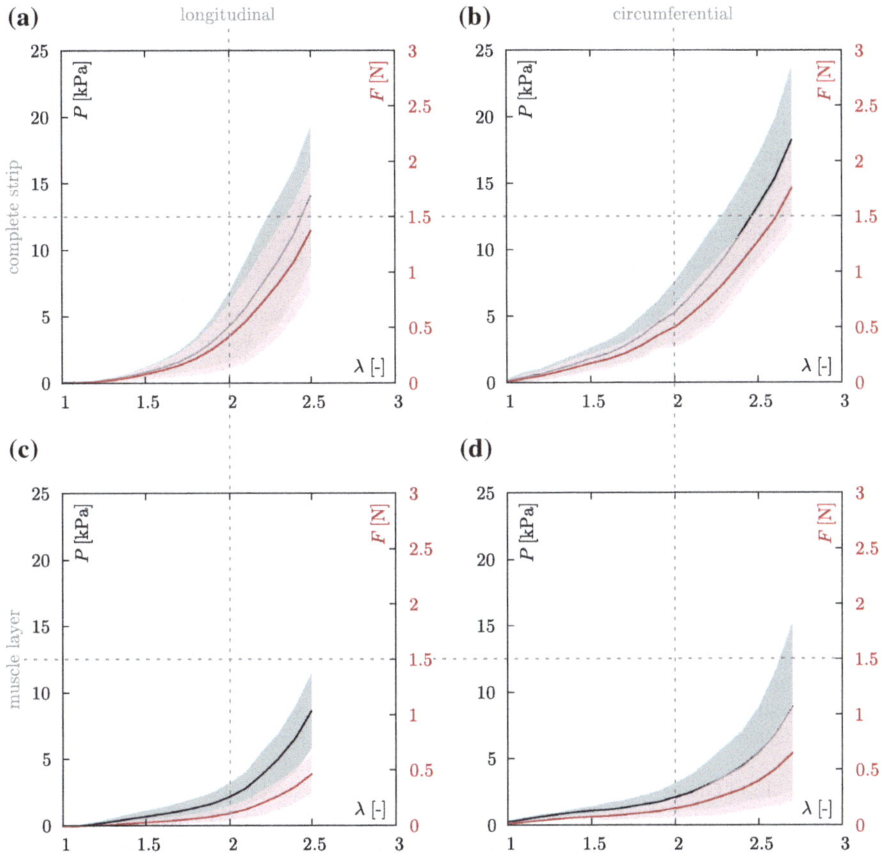

Fig. 11 Passive force-stretch (red) and stress-stretch (black) responses of 20 gastric tissue samples (complete strip and muscle layer) in longitudinal and circumferential directions, see also Klemm et al. (2020) (® Copyright 2020 Elsevier B.V.). The solid curves indicate the mean values, whereas the shaded areas depict the standard deviations

Böl, M., Leichsenring, K., Weichert, C., Sturmat, M., Schenk, P., Blickhan, R., Siebert, T.: Three-dimensional surface geometries of the rabbit soleus muscle during contraction - input for biomechanical modelling and its validation. Biomech. Model. Mechanobiol. **12**, 1205–1220 (2013)

Borsdorf, M., Tomalka, A., Stutzig, N., Morales Orcajo, E., Böl, M., Siebert, T.: Locational and directional dependencies of smooth muscle properties in pig urinary bladder. Front. Physiol. **10**, 63 (2019)

Brandstaeter, S., Gizzi, A., Fuchs, S.L., Gebauer, A.M., Aydin, R.C., Cyron, C.J.: Computational model of gastric motility with active-strain electromechanics. ZAMM – J. Appl. Math. Mech. / Z. Angew. Math. Mech. **98**, 2177–2197 (2018)

Buist, M.L., Cheng, L.K., Sanders, K.M., Pullan, A.J.: Multiscale modelling of human gastric electric activity: can the electrogastrogram detect functional electrical uncoupling? Exp. Physiol. **91**, 383–390 (2006)

Buist, M.L., Corrias, A., Poh, Y.C.: A model of slow wave propagation and entrainment along the stomach. Ann. Biomed. Eng. **38**, 3022–3030 (2010)

Cannon, W.B.: The movement of the stomach studied by means of the Röntgen rays. Am. J. Physiol.-Leg. Cont. **1**, 359–382 (1898)
Chen, J.D., Schirmer, B.D., McCallum, R.W.: Serosal and cutaneous recordings of gastric myoelectrical activity in patients with gastroparesis. Am. J. Physiol.-Leg. Cont. **266**, G90-8 (1994)
Chen, J.D., Lin, Z., Pan, J., McCallum, R.W.: Abnormal gastric myoelectrical activity and delayed gastric emptying in patients with symptoms suggestive of gastroparesis. Dig. Dis. Sci. **41**, 1538–1545 (1996)
Cheng, L.K., Buist, M.L., Yassi, R.S., Richards, W.O., Bradshaw, L.A., Pullan, A.J. (eds.): A model of the electrical activity of the stomach: from cell to body surface. In: Proceedings of the 25th Annual International Conference of the IEEE Eng. Med. Biol. Soc. (IEEE Cat. No.03CH37439) (2003)
Cheng, L.K., Komuro, R., Austin, T.M., Buist, M.L., Pullan, A.J.: Anatomically realistic multiscale models of normal and abnormal gastrointestinal electrical activity. World J. Gastroenterol. **13**, 1378–1383 (2007)
Cherubini, C., Filippi, S., Gizzi, A.: Electroelastic unpinning of rotating vortices in biological excitable media. Phys. Rev. E **85**, 031915 (2012)
Corrias, A., Buist, M.L.: A quantitative model of gastric smooth muscle cellular activation. Ann. Biomed. Eng. **35**, 1595–1607 (2007)
Corrias, A., Buist, M.L.: Quantitative cellular description of gastric slow wave activity. Am. J. Physiol.-Gastroint. Liver Physiol. **294**, G989–95 (2008)
Corrias, A., Pathmanathan, P., Gavaghan, D.J., Buist, M.L.: Modelling tissue electrophysiology with multiple cell types: applications of the extended bidomain framework. Integr. Biol. (Camb) **4**, 192–201 (2012)
Du, P., Li, S., O'Grady, G., Cheng, L.K., Pullan, A.J., Chen, J.D.Z.: Effects of electrical stimulation on isolated rodent gastric smooth muscle cells evaluated via a joint computational simulation and experimental approach. Am. J. Physiol.-Gastroint. Liver Physiol. **297**, G672–G680 (2009)
Du, P., O'Grady, G., Davidson, J.B., Cheng, L.K., Pullan, A.J.: Multiscale modeling of gastrointestinal electrophysiology and experimental validation. Crit. Rev. Biomed. Eng. **38**, 225–254 (2010)
Du, P., Gao, J., O'Grady, G., Cheng, L.K.: A simplified biophysical cell model for gastric slow wave entrainment simulation. Conf. Proc. IEEE Eng. Med. Biol. Soc. **2013**, 6547–6550 (2013a)
Du, P., O'Grady, G., Gao, J., Sathar, S., Cheng, L.K.: Toward the virtual stomach: progress in multiscale modeling of gastric electrophysiology and motility. Wiley Interdiscip. Rev. Syst. Biol. Med. **5**, 481–493 (2013b)
Du, P., Hameed, A., Angeli, T.R., Lahr, C., Abell, T.L., Cheng, L.K., O'Grady, G.: The impact of surgical excisions on human gastric slow wave conduction, defined by high-resolution electrical mapping and in silico modeling. Neurogastroenterol. Motil. **27**, 1409–1422 (2015)
Edwards, F.R., Hirst, G.D.S.: An electrical analysis of slow wave propagation in the guinea-pig gastric antrum. J. Physiol. **571**, 179–189 (2006)
Farrugia, G.: Interstitial cells of Cajal in health and disease. Neurogastroenterol. Motil. **20**(Suppl 1), 54–63 (2008)
Ferrua, M.J., Singh, R.P.: Modeling the fluid dynamics in a human stomach to gain insight of food digestion. J. Food Sci. **75**, R151–R162 (2010)
Fukuta, H., Kito, Y., Suzuki, H.: Spontaneous electrical activity and associated changes in calcium concentration in guinea-pig gastric smooth muscle. J. Physiol. **540**, 249–260 (2002)
Gajendiran, V., Buist, M.L.: A quantitative description of active force generation in gastrointestinal smooth muscle. Int. J. Numer. Method. Biomed. Eng. **27**, 450–460 (2011)
Gregersen, H., Gilja, O.H., Hausken, T., Heimdal, A., Gao, C., Matre, K., Ødegaard, S., Berstad, A.: Mechanical properties in the human gastric antrum using b-mode ultrasonography and antral distension. Am. J. Physiol.-Gastroint. Liver Physiol. **283**, G368–G375 (2002)
Guyton, A.C., Hall, J.E.: Textbook of medical physiology: Online access + interactive extras: studentconsult.com, 11. ed. Elsevier Saunders, Philadelphia, Pa. (2006)

Hinder, R.A., Kelly, K.A.: Human gastric pacesetter potential. Site of origin, spread, and response to gastric transection and proximal gastric vagotomy. Am. J. Surg. **133**, 29–33 (1977)

Hirst, G.D.S., Edwards, F.R.: Electrical events underlying organized myogenic contractions of the guinea pig stomach. J. Physiol. **576**, 659–665 (2006)

Hirst, G.D.S., Garcia-Londoño, A.P., Edwards, F.R.: Propagation of slow waves in the guinea-pig gastric antrum. J. Physiol. **571**, 165–177 (2006)

Hoermann, J.M., Bertoglio, C., Kronbichler, M., Pfaller, M.R., Chabiniok, R., Wall, W.A.: An adaptive hybridizable discontinuous galerkin approach for cardiac electrophysiology. Int. J. Numer. Method. Biomed. Eng. **34**, e2959 (2018)

Holzapfel, G.A.: Biomechanics of soft tissue. In: Lemaitre, J. (ed.), The Handbook of Materials Behavior Models. Volume III, Multiphysics Behaviors, Chapter 10, Composite Media, Biomaterials, pp. 1057–1071. Academic Press, Boston (2001)

Huizinga, J.D., Lammers, W.J.E.P.: Gut peristalsis is governed by a multitude of cooperating mechanisms. Am. J. Physiol.-Gastroint. Liver Physiol. **296**, G1–8 (2009)

Huizinga, J.D., Thuneberg, L., Vanderwinden, J.M., Rumessen, J.J.: Interstitial cells of Cajal as targets for pharmacological intervention in gastrointestinal motor disorders. Trends Pharmacol. Sci. **18**, 393–403 (1997)

Hurtado, D.E., Rojas, G.: Non-conforming finite-element formulation for cardiac electrophysiology: an effective approach to reduce the computation time of heart simulations without compromising accuracy. Comput. Mech. **61**, 485–497 (2018)

Johnson, L.R.: Physiology of the Gastrointestinal Tract, 5th edn. Academic Press, London (2012)

Kasai, Y., Tsutsumi, O., Taketani, Y., Endo, M., Iino, M.: Stretch-induced enhancement of contractions in uterine smooth muscle of rats. J. Physiol. **486**(Pt 2), 373–384 (1995)

Kelly, K.A., Code, C.F.: Canine gastric pacemaker. Am. J. Physiol.-Leg. Cont. **220**, 112–118 (1971)

Kim, S.J., Ahn, S.C., Kim, J.K., Kim, Y.C., So, I., Kim, K.W.: Changes in intracellular Ca2+ concentration induced by L-type Ca2+ channel current in guinea pig gastric myocytes. Am. J. Physiol. **273**, C1947-1956 (1997)

Kirber, M.T., Guerrero-Hernández, A., Bowman, D.S., Fogarty, K.E., Tuft, R.A., Singer, J.J., Fay, F.S.: Multiple pathways responsible for the stretch-induced increase in ca2+ concentration in toad stomach smooth muscle cells. J. Physiol. **524**(Pt 1), 3–17 (2000)

Klemm, L., Seydewitz, R., Borsdorf, M., Siebert, T., Böl, M.: On a coupled electro-chemomechanical model of gastric smooth muscle contraction. Acta Biomater. **109**, 163–181 (2020)

Kraichely, R.E., Farrugia, G.: Mechanosensitive ion channels in interstitial cells of Cajal and smooth muscle of the gastrointestinal tract. Neurogastroenterol. Motil. **19**, 245–252 (2007)

Kwiatek, M.A., Steingoetter, A., Pal, A., Menne, D., Brasseur, J.G., Hebbard, G.S., Boesiger, P., Thumshirn, M., Fried, M., Schwizer, W.: Quantification of distal antral contractile motility in healthy human stomach with magnetic resonance imaging. J. Magn. Reson. Imaging **24**, 1101–1109 (2006)

Lin, A.S.H., Buist, M.L., Smith, N.P., Pullan, A.J.: Modelling slow wave activity in the small intestine. J. Theor. Biol. **242**, 356–362 (2006)

Masaoka, T., Tack, J.: Gastroparesis: current concepts and management. Gut Liver **3**, 166–173 (2009)

Miftahof, R.N.: Biomechanics of the Human Stomach. Springer, Cham (2017)

Morgan, K.G., Muir, T.C., Szurszewski, J.H.: The electrical basis for contraction and relaxation in canine fundal smooth muscle. J. Physiol. **311**, 475–488 (1981)

Moriya, M., Miyazaki, E.: Force-velocity characteristics of stomach muscle: a comparison between longitudinal and circular muscle strips. Comp. Biochem. Physiol., Part A Mol. Integr. Physiol. **81**, 531–537 (1985)

Natali, A.N., Audenino, A.L., Artibani, W., Fontanella, C.G., Carniel, E.L., Zanetti, E.M.: Bladder tissue biomechanical behavior: experimental tests and constitutive formulation. J. Biomech. **48**, 3088–3096 (2015)

Notivol, R., Coffin, B., Azpiroz, F., Mearin, F., Serra, J., Malagelada, J.R.: Gastric tone determines the sensitivity of the stomach to distention. Gastroenterology **108**, 330–336 (1995)

Novitsky, Y.W., Kercher, K.W., Sing, R.F., Heniford, B.T.: Long-term outcomes of laparoscopic resection of gastric gastrointestinal stromal tumors. Ann. Surg. **243**, 738–45 (2006)

O'Grady, G., Du, P., Cheng, L.K., Egbuji, J.U., Lammers, W.J.E.P., Windsor, J.A., Pullan, A.J.: Origin and propagation of human gastric slow-wave activity defined by high-resolution mapping. Am. J. Physiol.-Gastroint. Liver Physiol. **299**, G585–92 (2010)

O'Grady, G., Egbuji, J.U., Du, P., Lammers, W.J.E.P., Cheng, L.K., Windsor, J.A., Pullan, A.J.: High-resolution spatial analysis of slow wave initiation and conduction in porcine gastric dysrhythmia. Neurogastroenterol. Motil. **23**, e345-55 (2011)

O'Grady, G., Angeli, T.R., Du, P., Lahr, C., Lammers, W.J.E.P., Windsor, J.A., Abell, T.L., Farrugia, G., Pullan, A.J., Cheng, L.K.: Abnormal initiation and conduction of slow-wave activity in gastroparesis, defined by high-resolution electrical mapping. Gastroenterology **143**, 589-598.e3 (2012)

Oh, J.H., Pasricha, P.J.: Recent advances in the pathophysiology and treatment of gastroparesis. Neurogastroenterol. Motil. **19**, 18–24 (2013)

Ozaki, H., Stevens, R.J., Blondfield, D.P., Publicover, N.G., Sanders, K.M.: Simultaneous measurement of membrane potential, cytosolic Ca2+, and tension in intact smooth muscles. Am. J. Physiol. **260**, C917-925 (1991)

Poh, Y.C., Beyder, A., Strege, P.R., Farrugia, G., Buist, M.L.: Quantification of gastrointestinal sodium channelopathy. J. Theor. Biol. **293**, 41–48 (2012)

Pullan, A., Cheng, L., Yassi, R., Buist, M.: Modelling gastrointestinal bioelectric activity. Prog. Biophys. Mol. Biol. **85**, 523–550 (2004)

Rocha, B.M., Dos Santos, R.W., Igreja, I., Loula, A.F.D.: Stabilized hybrid discontinuous galerkin finite element method for the cardiac monodomain equation. Int. J. Numer. Meth. Biomed. Eng. **36**, e3341 (2020)

Sanders, K.M.: Regulation of smooth muscle excitation and contraction. Neurogastroenterol. Motil. **20**(Suppl 1), 39–53 (2008)

Sanders, K.M., Koh, S.D., Ward, S.M.: Interstitial cells of Cajal as pacemakers in the gastrointestinal tract. Annu. Rev. Physiol. **68**, 307–343 (2006)

Sanders, K.M., Ward, S.M., Koh, S.D.: Interstitial cells - regulators of smooth muscle function. Physiol. Rev. **94**, 859–907 (2014a)

Sanders, K.M., Ward, S.M., Koh, S.D.: Interstitial cells: regulators of smooth muscle function. Physiol. Rev. **94**, 859–907 (2014b)

Sarna, S.K., Daniel, E.E., Kingma, Y.J.: Effects of partial cuts on gastric electrical control activity and its computer model. Am. J. Physiol.-Leg. Cont. **223**, 332–340 (1972)

Sathar, S., Trew, M.L., Du, P., O'Grady, G., Cheng, L.K.: A biophysically based finite-state machine model for analyzing gastric experimental entrainment and pacing recordings. Ann. Biomed. Eng. **42**, 858–870 (2014)

Sathar, S., Trew, M.L., OGrady, G., Cheng, L.K.: A multiscale tridomain model for simulating bioelectric gastric pacing. IEEE Trans. Biomed. Eng. **62**, 2685–2692 (2015)

Schmitz, A., Böl, M.: On a phenomenological model for active smooth muscle contraction. J. Biomech. **44**, 2090–2095 (2011)

Schulze, K.: Imaging and modelling of digestion in the stomach and the duodenum. Neurogastroenterol. Motil. **18**, 172–183 (2006)

Schulze-Delrieu, K., Shirazi, S.S.: Pressure and length adaptations in isolated cat stomach. Am. J. Physiol.-Gastroint. Liver Physiol. **252**, G92–G99 (1987)

Seydewitz, R., Menzel, R., Siebert, T., Böl, M.: Three-dimensional mechano-electrochemical model for smooth muscle contraction of the urinary bladder. J. Mech. Behav. Biomed. Mater. **75**, 128–146 (2017)

Streutker, C.J., Huizinga, J.D., Driman, D.K., Riddell, R.H.: Interstitial cells of Cajal in health and disease. part i: Normal ICC structure and function with associated motility disorders. Histopathology **50**, 176–189 (2007)

Tack, J., Caenepeel, P., Piessevaux, H., Cuomo, R., Janssens, J.: Assessment of meal induced gastric accommodation by a satiety drinking test in health and in severe functional dyspepsia. Gut **52**, 1271–1277 (2003)

van Mastrigt, R., Glerum, J.J.: Electrical stimulation of smooth muscle strips from the urinary bladder of the pig. J. Biomed. Eng. **7**, 2–8 (1985)

Vogalis, F., Publicover, N.G., Hume, J.R., Sanders, K.M.: Relationship between calcium current and cytosolic calcium in canine gastric smooth muscle cells. Am. J. Physiol.-Leg. Cont. **260**, C1012-8 (1991)

Wali, A., Munnur, U.: The patient with a full stomach. In: Benumof and Hagberg's Airway Management, pp. 705–722. Elsevier (2013)

Wilson, R.L., Stevenson, C.E.: Anatomy and physiology of the stomach. In: Yeo, C.J. (ed.) Shackelford's Surgery of the Alimentary Tract, pp. 634–646. Philadelphia (2019)

Won, K.J., Sanders, K.M., Ward, S.M.: Interstitial cells of Cajal mediate mechanosensitive responses in the stomach. Proc. Natl. Acad. Sci. U. S. A. **102**, 14913–14918 (2005)

Yeoh, J.W., Corrias, A., Buist, M.L.: A mechanistic model of a PDFGFRα(+) cell. J. Theor. Biol. **408**, 127–136 (2016)

Additional Challenges in Solid (Bio)mechanics

Classification of Biomechanical Models: The Wrong Battle Between Phenomenological and Structural Approaches, the Partly Underestimated Strength of Phenomenology and Challenges for Future (Clinical) Applications

Christian Bleiler and Oliver Röhrle

Starting a new job in a completely new field of research and in a country as far away from Europe as one can imagine, in my case at the Auckland Bioengineering Institute at the University of Auckland in New Zealand, it was a matter of weeks, maybe even only days until your name kept popping up over and over again! It was obvious, there is a world-renowned expert in Europe that I must to get to know and meet. That was in 2004. Since then, we met in numerous occasions – a journey of scientific excitement, pleasure and joy. From my point of view, however, I must admit that I might not have realized your biggest accomplishment until returning to Europe and realizing the enormous kind of groundbreaking work you did in establishing the field of computational/continuum biomechanics here in Europe! Your work made many things for me easier. Many thanks for your hard work, your unremitting efforts, for all your impact you left on me and the community and all the great memories. All the best for the next 60 years!

Oliver

Abstract In continuum biomechanics, models are typically classified into models appealing to phenomenological or so-called structural modeling approaches. However, this categorization often starts from the model itself and does not necessarily focus on the desired application and purpose of the utilized model. Particularly with regard to the future extension and the transfer of biomechanical models to clinical problems, this sometimes seems to be insufficient. In order to enhance interdisciplinary applicability, more universal and accurate designations of models should be

C. Bleiler · O. Röhrle (✉)
Institute for Modelling and Simulation of Biomechanical Systems,
Pfaffenwaldring 5a, 70569 Stuttgart, Germany
e-mail: roehrle@simtech.uni-stuttgart.de

G. Sommer et al. (eds.), *Solid (Bio)mechanics: Challenges of the Next Decade*,
Studies in Mechanobiology, Tissue Engineering and Biomaterials 24,
https://doi.org/10.1007/978-3-030-92339-6_14

sought. Above all, one should clearly state the aim of a specific model, e.g., whether it is intended to reproduce experimental data *in silico* or has explanatory intentions beyond that. Besides that, one should also clearly state the definition of the concept of phenomenology, which has in the context of mechanical models often a too negative connotation. Being clear about the aim and purpose and the hereby concluded choice of model(ling approach) helps to formulate models that have an optimal balance between complexity—and thus computational cost—and desired benefit. Herein, we aim to provide a concise overview, necessary definitions and implications on two short examples taken from the broad field of (continuum) biomechanics, namely, the modeling of collagen fibers and skeletal muscles.

1 Introduction

For several decades now, the broad field of continuum mechanics has found its way into applications in biology and biomechanics. In addition to modeling and understanding real processes, the clinical application of biomechanical simulations is now increasingly coming into focus. In this context, it is crucial that the scope, the level of detail and the desired benefit and goal of a model are clearly stated. This is particularly important in an interdisciplinary context when scientists from different fields of research work together. Understandably, however, models developed by engineers are often categorized using relatively non-standardized, engineering terms. Many such terms can be found in the literature, such as *ad hoc*, *biophysical*, *empirical*, *micromechanical*, *micromechanically-based*, *microstructural*, *microstructurally-based*, *multiscale*, *phenomenological*, *physiological*, *single-scale*, or *structure-based*, as well as more general namings such as *computational*, *mathematical*, *mechanical*, or *multiphysics* models.[1]

These terms are not always clearly defined and often overlap in meaning. This means that certain models can be attributed to several of these terms. However, two fundamentally different model approaches emerge. We refer to them as type A and type B models, see Fig. 1. Such a classification is, however, anything but clear and the respective terms of a model type should under no circumstances be understood as synonyms. Roughly speaking, models of type A aim at a pure description of observed phenomena, whereas models of type B typically include more levels of detail and often have explanatory intentions. However, the individual terms within the model types can still have very specific meanings and should be used with caution to avoid ambiguity and imprecise descriptions of models. For more clarity, however, we should first explain a few important terms in more detail, starting with the model terms 'phenomenological' and 'structural' in the next section.

[1] In the continuum-mechanical context, the term 'model' usually refers to a constitutive material model that complements the physical balance equations and enables the calculation of initial-boundary-value problems. However, the following considerations on models are not limited to the field of continuum (bio)mechanics and can be conceived in a more general way. The underlying ideas apply equally, for instance, to zero-dimensional models in the field of (mechano)biology.

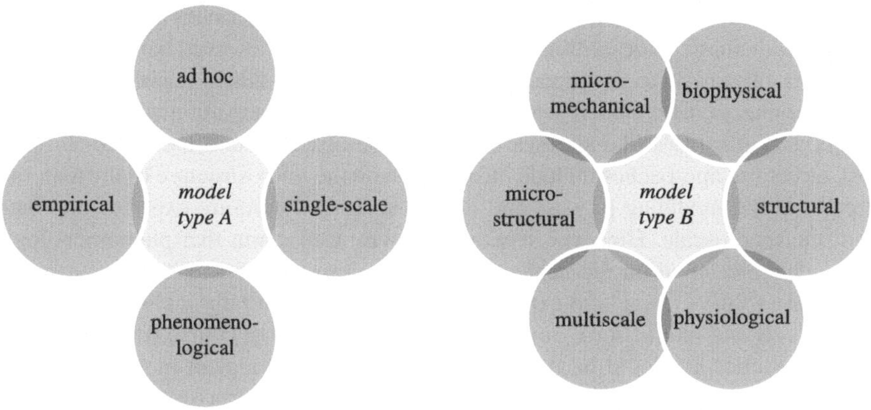

Fig. 1 A coarse identification of two different model types in continuum (bio)mechanics

2 A Model-Based ('Engineering') Point of View: Phenomenological and Structural

A very common classification, especially in biomechanics, distinguishes between phenomenological and structural model approaches. A clear definition of the former can be found in Woo et al. (1993):

> We refer to phenomenological models as those that do not have explicit parameters related to the microstructure of the tissue. This category includes a number of models, from those that are derived simply by curve fitting experimental data, to rigorously formulated continuum models. (Woo et al. 1993)

Another definition has been formulated by McMullin (1968):

> [...][A] phenomenological model appears to be an arbitrarily-chosen mathematically-expressed correlation of physical parameters from which the empirical laws of some domain can be derived. (McMullin 1968)

On the other hand, Woo et al. (1993) defines structural model approaches as follows:

> Structural models are based on known (or assumed) behavior of the constituents of the tissue. The mechanical responses of the individual components are then combined or generalized to produce a description of gross mechanical behavior. [...]they include parameters which are directly related to the structure of the tissue. (Woo et al. 1993)

These definitions are essentially based on the consideration of different length scales and the (self-evident) presence of a so-called microstructure of the material on smaller scales. The behavior of the material—that appears homogeneous on the macroscale—results from the interaction of different constituents/components—the heterogeneous microstructure—on the microscale.[2] Phenomenological approaches are purely based on the (mathematical) description of the macroscopically observed behavior and are indispensably linked to the process of calibrating the models to experimental data. In this connection, the macroscale is identified as the observation scale, at which effects and quantities like stresses or strains are observed and measured. In contrast, structural approaches include information on the microstructure of the material directly in the modeling process and do not necessarily require experimental data from the macroscale. However, it would be wrong to claim that phenomenological models are based on the neglection of the microstructure. On the contrary, by calibrating ('curve fitting') to experimental data—that naturally includes all small-scale effects and interactions—microstructural effects are of course fully taken into account. Rather, it should be made clear that phenomenological models take into account the microstructure in an *implicit* manner, whereas structural models aim at an *explicit* resolution of individual components of the material. This justifies some of the terms presented in Fig. 1, such as the term single-scale for phenomenological approaches, and multiscale, microstructural and micromechanical for structural approaches.

The above summarized definitions tie a very tight corset around the concept of phenomenological models, as they *must not* include parameters that are directly related to the microstructure and they *must* be based on the calibration to macroscopic experimental results. Further, this classification is strongly based on a model-oriented and not application-oriented point of view. Not necessarily the purpose or the benefit, but the model itself is the central structure of the classification.

Possible reasons that determine the choice between a phenomenological and a structural modeling approach are illustrated in Fig. 2. Therein, reasons (i)–(iii) are mainly linked to insufficient experimental data on the macroscale. If any (or even several) of those reasons holds true, the use of a phenomenological, single-scale modeling approach and the meaningfulness of the thereon-based results become limited and questionable. Structural, multiscale modeling approaches serve as a remedy to this problem. Instead of making use of experimental data only on the observation scale, a multiscale approach also takes into account data from smaller scales. Ideally, this shifts the constitutive modeling process and the necessary calibration step from the data-poor macroscale to data-rich(er) smaller scales. Of course, *a multiscale approach only makes sense if the experimental data on the smaller scales is of better quality*. While a multiscale approach due to reasons (i)–(iii) is rather a remedy to overcome the experimental limitations, reason (iv) is a more active motivation to utilize multiscale models. There, the material response on smaller scales and the influence of the microstructure on the macroscopic material response are of interest. Under these circumstances, the observation scale is no longer clearly defined, because phenomena on at least two scales are then of interest and the interconnections

[2] Although the characteristic length scale of inhomogeneities and constituents in a lot of applications is in fact in the order of micrometers (μm), the prefixes micro and macro do not imply a direct connection to absolute sizes, but rather stand for small and large by referring to their Greek roots *mikrós* and *makrós*, respectively. The microscale can therefore be as well in the range of, e.g., nanometers without the necessity to refer to it as nanoscale.

Reasons to switch from phenomenological to structural models

(i) if no reliable experimental data on the macroscopic observation scale exists to properly calibrate the phenomenological model,
- comprehensive multi-axial data required for anisotropic materials
- rate-dependent experimental data required to assess viscous effects

(ii) if the experimental data only describes one specimen or subject and is not representative for the whole material type,
- large intra- and inter-subject variability particularly in soft biological tissue
- differences between healthy and diseased biological tissue

(iii) if the conditions of the experimental setup do not represent the conditions under which the model shall be investigated,
- *in vivo* versus *ex vivo* conditions for biological tissues
- possible dehydration of tissue samples and associated changes of mechanical properties

(iv) if the small-scale components are of explicit interest.
- strongly localized material behavior (e.g., plasticity) that requires the resolution of the microstructure
- constructing an optimal macroscopic behavior by specifically manipulating small-scale components
- predicting the macroscopic response due to structural variations on smaller scales

Fig. 2 Motivation to switch to structural models

between the scales move into focus. Hence, to be able to explain a certain macroscopically observed material behavior, multiscale modeling approaches can be applied as sophisticated tool for a deeper investigation of phenomena across the scales. Finally, a multiscale approach in this context always has to be accompanied with appropriate upscaling and homogenization steps, such that the results become available on the macroscopic observation scale. Examples of such models can be found in many areas of biomechanics, such as the multiphase modeling of cement injection into vertebral bodies, see, e.g., Bleiler et al. (2015), the structurally-based modeling of myocardium and arteries, see, e.g., Holzapfel and Ogden (2009, 2010), or the multiscale modeling of skeletal muscles and general homogenization approaches by the authors, see, e.g., Röhrle et al. (2008), Röhrle et al. (2012), Wang et al. (2013) Heidlauf et al. (2016), Bleiler et al. (2019), Schmid et al. (2019), Röhrle et al. (2019) and Bleiler et al. (2021).

In the next section, a slightly different, rather philosophical and more application-based point of view for the classification of models is outlined.

3 An Application-Based Point of View: Phenomenological and Explanatory

In addition to the model terms commonly used in engineering and biomechanical applications, see Fig. 1, a variety of other designations can be found if the concept of a model is taken more broadly and the topic is explored more from a philosophical point of view. For instance, an extensive list of terms can be found in the article of Frigg and Hartmann (2020):

> Probing models, phenomenological models, computational models, developmental models, explanatory models, impoverished models, testing models, idealized models, theoretical models, scale models, heuristic models, caricature models, exploratory models, didactic models, fantasy models, minimal models, toy models, imaginary models, mathematical models, mechanistic models, substitute models, iconic models, formal models, analogue models, and instrumental models (Frigg and Hartmann 2020)

A detailed analysis of all these terms is beyond the scope of this section, but the list highlights the wide range of names for models. Above all these terms there is a very fundamental level of categorization, in which models are classified into so-called *representative*, *idealized* and *fictional*. Very clear definitions for representative (or representational) models have been formulated by Craver (2006):

> [...]representational models, that is, those that scientists construct as more or less abstract descriptions of a real system. (Craver 2006)

as well as by Frigg and Hartmann (2020):

> they[representative models] represent a selected part or aspect of the world, which is the model's target system. (Frigg and Hartmann 2020)

In turn, idealized models are described by Frigg and Hartmann (2020) as follows:

> Idealized models are models that involve a deliberate simplification or distortion of something complicated with the objective of making it more tractable or understandable. (Frigg and Hartmann 2020)

A common example for an idealized model is the neglection of friction in a mechanical system. However, here a certain difficulty in distinguishing between representational and idealized models is already apparent. In particular, there is a certain lack of clarity in the descriptions as to whether any model assumption is already an idealization. Yet, since model assumptions are indispensable in the formulation of mechanical models (that are usually not based on first principles), such a statement would make it impossible to formulate representative mechanical models and would therefore go too far.

Finally, Bokulich (2011) describes the third model form as follows:

> [...] by fictional model I mean simply a model that represents the world by means of fictional entities, states, or processes (and to distinguish fictional models from idealized models-specifically fictional entities or processes that are not related to the true ones in the world by what might be thought of as a distortion or series of successive cases). (Bokulich 2011)

Here too, the distinction between idealized and fictional models is at times difficult. We assume, however, that in biomechanics one rather finds idealized models and not purely fictional models without direct reference to a real application.

Once this first level of model classification has been introduced, a second level of distinction can be identified. To motivate this, we consult two statements of Craver (2006): 'Not all models are explanatory. Some models are data summaries.' as well as 'There is a widely accepted distinction between merely *modeling* a mechanism's behavior and *explaining* it.' Here we apparently distinguish between models that have explanatory capabilities and those which model and replicate certain behaviors ('data summaries'). Consequently, the latter approach can be identified as a rather phenomenological one, which leads us to the distinction between *phenomenological* and *explanatory* models, see also Bokulich (2011). This pairing seems to be very similar to that mentioned in the previous section, where we distinguished between phenomenological and structural models. However, the distinction considered here is still quite different, which can be illustrated by the following definitions. According to Bokulich (2011),

> A *phenomenological model* is only of instrumental value to the scientist. Often–though not exclusively–they are constructed via an ad hoc fitting of the model to the empirical data. Phenomenological models are useful for making predictions, but they do not purport to give us any genuine insight into the way the world is. (Bokulich 2011)

whereas

> An *explanatory model*, by contrast, does aim to give genuine insight into the way the world is. (Bokulich 2011)

The difference between the two model types is thus the ability to go beyond the mere reproduction of results and to provide explanatory insights into the underlying mechanisms. Of course, it can be stated that such explanatory models in the context of biomechanical problems usually lead to the formulation of multiscale approaches. One might therefore end up again with the distinction between phenomenological and structural approaches (or between other terms belonging to type A or type B in Fig. 1). However, the crucial difference is that the multiscale approach is the essential feature of structural models (an end in itself), whereas in the case of explanatory models it is a possible consequence (a means to an end). This means that the here presented classification proceeds from an application-based point of view, because it only shows itself through the application of the model and whether it has explanatory intentions or not. On the other hand, a purely phenomenological model has the characteristic of a blackbox that takes some input (calibrated parameters) and provides some output, without providing explanations on the underlying mechanism.

As far as explanatory models are concerned, a further classification is possible, which is based on the way underlying mechanisms are described. According to Machamer et al. (2000) and Craver (2006), one can distinguish between *how-possibly* and *how-actually* models. They are defined as follows:

> *How-possibly models* (unlike merely phenomenal models) are purported to explain, but they are only loosely constrained conjectures about the mechanism that produces the explanandum

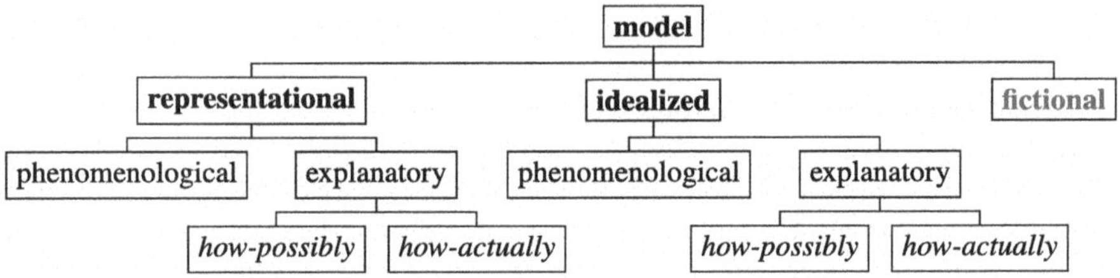

Fig. 3 Scheme of model classifications from a 'philosophical' point of view

> phenomenon. They describe how a set of parts and activities might be organized such that they exhibit the explanandum phenomenon. (Craver 2006)

and

> *How-actually models* describe real components, activities, and organizational features of the mechanism that in fact produces the phenomenon. They show how a mechanism works, not merely how it might work. (Craver 2006)

The first form thus describes models in which possible mechanisms are postulated and assumed, while the second form demands knowledge of them. Applied to mechanical problems, this means that how-actually models provide precise knowledge of the underlying causes of a macroscopic material response, e.g., by examining microscopic image data and exactly identifying the relevant microstructural components and their interplay. This last classification is very crucial as a preliminary consideration during the formulation of a model. In particular, it is important to consider whether the underlying mechanisms of the macroscopic behavior that is sought to be described are known at all. In this connection, however, it is directly evident that 'complete' descriptions of a material behavior across all scales are hardly possible and that certain simplifications must always be made. Craver (2006) commented: '[...]*ideally complete descriptions of a mechanism*. Such models include all of the entities, properties, activities, and organizational features that are relevant to every aspect of the phenomenon to be explained.' He further states: 'In fact, such descriptions would include so many potential factors that they would be unwieldy for the purposes of prediction and control and utterly unilluminating to human beings.' We may therefore assume that models usually lie in a range between the how-possibly and how-actually models. Machamer et al. (2000) refer to this intermediate range as *how-plausibly* models. In this context, it also has to be emphasized that an explanatory model is not improved by adding more details, but only by adding more *relevant* details, see Craver and Kaplan (2020).

The three levels of classification introduced in this section are finally summarized in Fig. 3. In this connection, it may seem somewhat confusing that explanatory approaches can be found in both representative and idealized models. In this regard, also Frigg and Hartmann (2020) posed the question:

> Some models explain. But how can they fulfill this function given that they typically involve idealizations? Do these models explain *despite* or *because* of the idealizations they

> involve? Does an explanatory use of models presuppose that they represent, or can non-representational models also explain? (Frigg and Hartmann 2020)

Here, however, it is necessary to point out once again a certain ambiguity in the distinction between representative models and idealized models, especially with regard to how model assumptions are assessed. It makes sense that also idealized models (that, e.g., neglect viscous behavior due to quasi-static loading conditions) in mechanical applications can be explanatory.

Now that we have looked at the classification of models from a slightly different angle in this section, we should take another closer look at the concept of phenomenology. This follows in the next section.

4 The Strength (and Omnipresence) of Phenomenology

In the previous two sections, phenomenological models were introduced as counterparts to structural and explanatory modeling approaches. In both cases, the phenomenological models had the character of 'simpler approaches'. This is an accurate description if 'simple' means less complex and easier to handle in numerical environments. However, it should by no means be seen as a synonym for 'worse' or 'inferior'. The mechanical description of materials is not possible at all without the concept of phenomenology. This section therefore gives a brief overview of the strength (and omnipresence) of phenomenology.

As a start we look at the general definition of the term phenomenology given by Smith (2018):

> The *Oxford English Dictionary* presents the following definition: 'Phenomenology. a. The science of phenomena as distinct from being (ontology). b. That division of any science which describes and classifies its phenomena. From the Greek *phainomenon*, appearance.' In philosophy, the term is used in the first sense, amid debates of theory and methodology. In physics and philosophy of science, the term is used in the second sense, albeit only occasionally. (Smith 2018)

In particular, the first ('a.') meaning shows that a detailed study of the concept of phenomenology would take us far into the realm of philosophy. Beyond the use of the term in physics and mechanics, phenomenology actually represents a complete disciplinary field in philosophy. It is especially linked to the name of Edmund Husserl.[3] For our purposes, however, the term *phenomena* (hence, *appearance*) mentioned in the definition is particularly interesting and—unsurprisingly—represents the central notion and stands in the middle of the concept of phenomenology, see also Bogen and

[3] Edmund Husserl (1859–1938) was an Austrian-German philosopher and mathematician. Interestingly, both the well-known philosopher Franz Brentano and the mathematician Karl Weierstraß were among his academic teachers. Husserl's work and contributions are essential in establishing phenomenology as an independent philosophical school. Main parts of his research work were published as a complete edition named *Husserliana*. See, e.g., Husserl (1999) for an English translation of one of his works.

Woodward (1988) and Bogen (2011). Further, the notion of phenomena is equally important in modern continuum mechanics, as stated by Truesdell and Noll (2004) in their famous magnum opus:

> The task of the theorist is to bring order into the chaos of the phenomena of nature, to invent a language by which a class of these phenomena can be described efficiently and simply. (Truesdell and Noll 2004)

Summarizing, the observation of a phenomena has to be the starting point for a phenomenological approach. In a mechanical problem, e.g., such a phenomenon can be an experimentally *measured* stress-strain curve (hence, an observed material behavior). Then, for the basic problem of material theory, which is finding appropriate constitutive formulations, this means that the description of the phenomenon 'stress-strain curve' by means of suitable mathematical functions is a proper phenomenological approach. As a consequence, such an approach *cannot be incorrect or insufficient*. A phenomenological approach guarantees an unbiased and objective view on the things as they are, it does not require any further subjective opinion of the observer and thus avoids incorrect modeling. It can be said that a phenomenological model does not directly describe a material, but an observed behavior of that material. In this context, it is also irrelevant *why* the material behavior shows itself in this way. That is also emphasized by Smith and Smith (1995) who stated:

> Phenomenology [. . .] is above all a *descriptive* enterprise, a theory of appearances, of symptoms, as contrasted with those disciplines which deal in *causal explanation*, and with what lies behind the appearances. (Smith and Smith 1995)

Hence, a phenomenological description does not aim to explain why phenomena appear the way they appear. However, this can be seen as an advantage, if it is made clear that such explanations are not only not the aim but also *not necessary* at all. Moreover, the term 'symptom' in the last quote is interesting, too. To illustrate this with an example from medicine: One starts by examining a patient and describing an observed symptom, and not by describing a disease. Applied to mechanical problems, one starts with an observed material behavior and formulates a model based on it, and not *vice versa*. This way of thinking is crucial in finding the best possible models.

If one wants to find the underlying causes of a particular phenomenon, Machamer et al. (2000) states that 'Mechanisms are sought to explain how a phenomenon comes about or how some significant process works'. Without going into the philosophical controversies surrounding the exact definition of what a mechanism is, we here adopt the pragmatic definition of Illari and Williamson (2012):

> A mechanism for a phenomenon consists of entities and activities organized in such a way that they are responsible for the phenomenon. (Illari and Williamson 2012)

Looking at this from a mechanical point of view, the term 'entities and activities' immediately brings to mind the classical multiscale problem with its different interacting components on smaller scales. This connection becomes even clearer by looking at Table 1, which is adopted from Darden (2006, 2008). Although not directly related to mechanical problems, this list could be used for exactly those. An observed

Table 1 Origin of phenomena: features of mechanism (modified from Darden (2006, 2008))

Phenomenon		
Components		Contextual locations
Spatial arrangement	*Temporal arrangement*	• Location within a hierarchy
• Localization	• Order	• Location within a series
• Structure	• Rate	
• Orientation	• Duration	
• Connectivity	• Frequency	
• Compartmentalization		

phenomenon appears as the result of the spatial and temporal interaction of certain structures (components) on smaller scales. These considerations make it clear, of course, that the investigation of structures on smaller scales is in turn based on the observation of certain phenomena and their necessary description on these scales. At a certain point one will always fall back on phenomenological descriptions in mechanical considerations. Even if a multiscale, explanatory model describes a macroscopic phenomenon in a very detailed manner and incorporates effects from smaller scales, e.g., on the order of micrometers or even nanometers, it is in a physical sense still 'macroscopic' and relies on phenomenological concepts like pressure, volume, temperature, heat, work and mass. The designation 'non-phenomenological' for a model would therefore be somewhat doubtful. Consequently, the designation of a model as phenomenological or explanatory only makes sense in relation to a certain phenomenon (an observed material behavior) at a certain scale and not to a material with all its appearances across scales.

5 An Exemplary 'Symptom-Based' Modeling Workflow

If we now summarize some of the investigations and explanations presented so far, the following questions can be formulated for the creation and setup of models:

Questions for model setup (e.g., in clinical applications)

- What is the purpose of the model?
  - What phenomenon shall be reproduced/described?
  - What is the observation scale?
  - What is the desired data output?
  - Is the aim to merely reproduce/perform real experiments in-silico or to gain further knowledge and insights from the model?
- What data can be used as input?

- On which scales is input data available?
- How reliable and representative is the input data?
- Must/can further experiments be performed for data acquisition?

Sequentially answering those questions and processing the requirements list will lead to the model that fits best. This can be any kind of model, such as a phenomenological model for the pure mathematical description of a stress-stretch curve, or an explanatory (multiscale) model that contains detailed descriptions of smaller scales and that can not only reproduce but also explain macroscopic effects. Yet, it must be stressed that the choice of a particular model is a direct consequence of the problem and the available data. One tries to present a clearly formulated phenomenon or to solve a specific problem by first looking at a 'symptom' and then formulates the appropriate model based on this. The quality of the model is not assessed by its complexity or its ability to explain underlying mechanisms, but by the ability to provide the desired output data. Thus, none of the different model approaches can a priori be regarded as better or worse. In particular, we emphasize that explanatory multiscale models are not better than phenomenological (single-scale) approaches by definition. The former demand for considerable effort in the modeling process, such as the need for reliable upscaling and homogenization methods, and usually increase the complexity of the resulting models. In turn, the higher complexity eventually entails higher computational costs when solving actual problems. Further, it always has to be reminded that modeling on more than one scale plus capturing interscale effects usually results in the need of more modeling assumptions. While a single-scale approach smears all small-scale effects and we might thus say that it accounts for them in an intrinsic manner, a multiscale approach actually has to resolve small-scale effects and requires more information and more knowledge about a system. In any case, one should always keep in mind the paradigm *as simple as possible, but not simpler*[4] during the formulation of a specific model in order to avoid unnecessary complexity and computational costs.

6 Example I: Collagen Fiber Modeling

In order to clarify the previous considerations, we will now look in this first example at the modeling of collagen fibers. In this connection, collagenous structures can be

[4] This saying—in this or similar forms—is often attributed to Albert Einstein, but that cannot be clearly proven. However, it is assumed that it is a highly simplified version of his verified statement 'It can scarcely be denied that the supreme goal of all theory is to make the irreducible basic elements as simple and as few as possible without having to surrender the adequate representation of a single datum of experience.' from his work Einstein (1934). It is, for a continuum-mechanist, interesting to note that in the same work Einstein also formulated a brief justification for continuum field theories: '[...]in a continuum theory, the atomistic character could be satisfactorily expressed by integral propositions without localizing the particles which constitute the atomistic system.'.

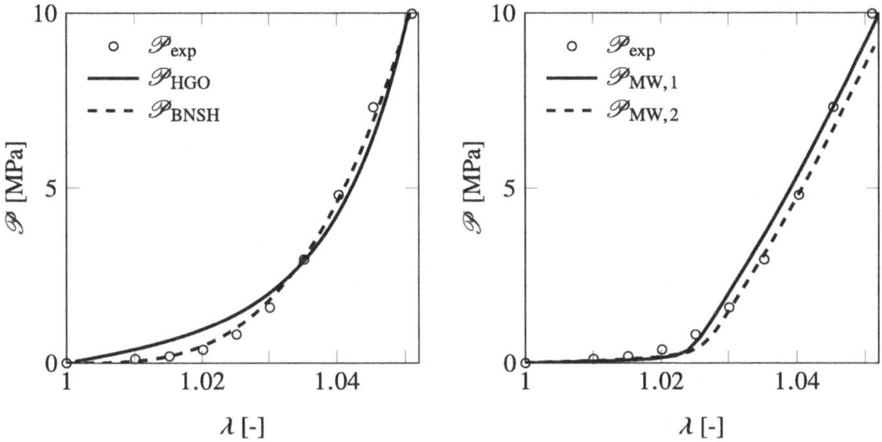

Fig. 4 Modeling of collagen fiber bundles. Experimental stress-stretch data, $\mathscr{P}_{\text{exp}}$, from Hansen et al. (2002). Left: best-fit curves for $\mathscr{P}_{\text{HGO}}$ ($k_1 = 9.058\,\text{MPa}$, $k_2 = 151.2$, $R^2 = 0.9787$) and $\mathscr{P}_{\text{BNSH}}$ ($\alpha_1 = 1.489\,\text{GPa}$, $\alpha_2 = 4.171$, $R^2 = 0.9971$). Right: results $\mathscr{P}_{\text{MW},1}$ and $\mathscr{P}_{\text{MW},2}$ from the multiscale model by Marino and Wriggers (2017)

found in nearly every biological tissue, such as arteries or muscles, and represent a key contributor to their mechanical properties. If a collagen fiber (or a bundle of collagen fibers) is subjected to a tensile stretch, say λ, in its longitudinal direction, a typical J-like stress-stretch curve can be observed. For instance, the nominal stress that was measured in experiments by Hansen et al. (2002) is shown in Fig. 4, indicated as $\mathscr{P}_{\text{exp}}$. This means that the material behavior of a collagen fiber bundle shows itself as an observed phenomenon by means of experimental measurements. We now obtain a phenomenological description by calibrating suitable mathematical functions to the experimental results. It does not matter why the phenomenon appears in this way and no knowledge of the underlying mechanisms is required. For example, in the context of hyperelasticity, the exponential strain-energy function by Holzapfel et al. (2000), given by $\mathscr{W}_{\text{HGO}}(\lambda) = k_1 \exp\{k_2 \langle \lambda^2 - 1 \rangle^2 - 1\}/(2k_2)$, as well as the formulation by Balzani et al. (2006), given by $\mathscr{W}_{\text{BNSH}}(\lambda) = \alpha_1 \langle \lambda^2 - 1 \rangle^{\alpha_2}$, are particularly useful. Therein, $\langle x \rangle = (x + |x|)/2$ and each of these two energy functions contains two fitting parameters: k_1 and k_2 for $\mathscr{W}_{\text{HGO}}$ and α_1 and α_2 for $\mathscr{W}_{\text{BNSH}}$. The calibration of the corresponding nominal stress formulations $\mathscr{P}_{\text{HGO}} = \partial_\lambda \mathscr{W}_{\text{HGO}}$ and $\mathscr{P}_{\text{BNSH}} = \partial_\lambda \mathscr{W}_{\text{BNSH}}$ to the experimental results $\mathscr{P}_{\text{exp}}$ provides the curves as shown in Fig. 4 (left). From this it can be seen that both energy functions are very well suited for the phenomenological description of the experimental results. They can be easily integrated into numerical simulation tools.

A phenomenological modeling approach has no intention to explain why the experimental stress-stretch curve looks as it does and what mechanisms are responsible for it. In fact: it cannot, does not want to and does not have to explain it. In contrast, an explanatory model aims to identify the mechanisms and components that are responsible for the observed material behavior. A look into the microstructure of collagen fibers helps here. Starting on the smallest (considered) length scale, collagen fibers consist of triple helical tropocollagen molecules with a diameter of about 1.5 nm. Multiple molecules are arranged in staggered arrays and form collagen fibrils with diameters of about 50–250 nm. The molecules interact on the fibril-level through intermolecular covalent cross-links (between two molecules) and so-called weak bonds. Finally, bundles of densely packed fibrils form the collagen fiber, which usually appears in a crimped form in the unloaded reference configuration. This crimped shape has a major influence on the observed J-shaped stress-stretch curve with its flat toe region at lower stretches (when the fiber and underlying structures are crimped) and a stiffening at higher stretches (when the fiber is straightened). More details can be found, e.g., in the textbook of Fratzl (2008). Explanatory models that include these mechanisms across several length scales and provide the resulting macroscopic material response were formulated by, e.g., Buehler (2008), Maceri et al. (2010) and Marino and Wriggers (2017). Results of the latter are shown in Fig. 4 (right). A multiscale model such as the one by Marino and Wriggers (2017) involves microstructural components directly and is usually based on model parameters with direct physical meaning (i.e., no fitting parameters). Of course, this can be very advantageous in biomechanical applications, since it allows, e.g., to describe the material behavior due to microstructural changes caused by diseases. In this context, Craver (2006) commented that explanatory models '[. . .]show how the system would behave under a wider range of interventions than do phenomenal models, and so they can be used to answer more w-questions', whereby the term 'w-questions' refers to the work of Woodward (2003) and means 'what-if-things-had-been-different'. In contrast, Craver (2006) states that a phenomenological model is 'a model that is useful only within a narrow range of conditions (such as health, proper functioning, or the absence of disturbing outside forces) but that fails outside of those narrow conditions'. Hence, a more complex multiscale model can also be applied to scenarios for which no macroscopic experiments exist. However, this complexity is usually associated with additional numerical effort. In the example dealt with here, it could be seen that both phenomenological and explanatory models describe the observed collagen behavior very well and that the choice between these two approaches must depend on the application and the availability of experimental data.

7 Example II: Skeletal Muscle Modeling

The second example deals with skeletal muscle modeling and will be discussed here only briefly. While only uniaxial deformations are usually observed and described for the collagen fibers considered in the previous section, a multi-axial deformation state (that can be expressed by the deformation gradient $\boldsymbol{F}$) and the resulting observation of anisotropic material behavior is of interest when examining muscle tissue. This means that the anisotropic behavior results directly from the observation of experiments. In the context of hyperelastic continuum mechanics, this leads to the identification of the material symmetry group $\mathcal{MG} \subseteq \mathcal{O}(3)$ under which the strain

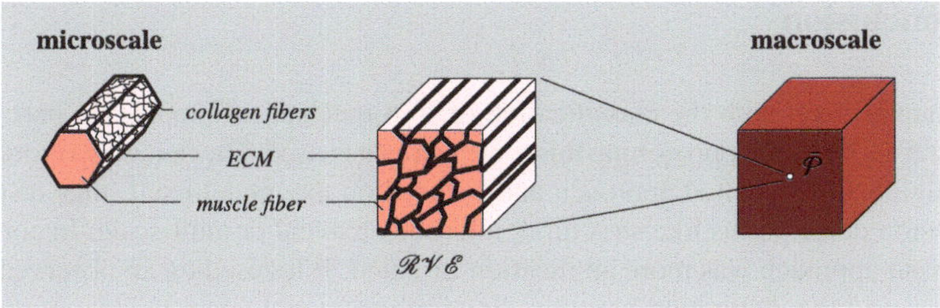

Fig. 5 A multiscale framework for the modeling of skeletal muscle tissue. Relevant microstructural constituents, like the extracellular matrix (ECM), are collected in a representative volume element ($\mathscr{RVE}$) and determine the effective behavior at a macroscopic material point $\bar{\mathcal{P}}$

energy $\mathscr{W}(\boldsymbol{F}\,\boldsymbol{Q}^T) = \mathscr{W}(\boldsymbol{F})$ becomes invariant with respect to referential transformations $\boldsymbol{Q} \in \mathcal{MG}$, see also Holzapfel (2000). Thus, a sufficient number of multi-axial experiments is essential, not only for the identification of $\mathcal{MG}$, but also for the subsequent calibration of strain-energy functions $\mathscr{W}(\boldsymbol{F})$. In skeletal muscle modeling, however, the presence of muscle fibers in the tissue often resulted in the assumption that $\mathcal{MG}$ represents the transverse isotropy group and that the material is stiffest in the direction of the muscle fiber. This led to the use of strain-energy functions for fiber-reinforced materials, but beyond that, these were often only fitted to uniaxial experimental data. Experiments such as the one by Takaza et al. (2013) show, however, that muscle can be stiffer in the transverse-to-the-fiber direction and the fiber-reinforced model therefore provides a false statement. Yet, this is not a weakness of a supposedly phenomenological model, because the procedure is simply not observation-based. Rather, it represents an incorrect explanatory approach with a wrong how-possibly assumption. Either one rigorously formulates a phenomenological model based on observations (experimental data) or one postulates mechanisms in a how-possibly model and clearly names it as such.

Finally, a last point is illustrated by means of the multiscale muscle model by Bleiler et al. (2019). This model is based on the description of microstructural components, such as the extracellular matrix and the helical arrangement of collagen fibers, and provides the macroscopic behavior of muscle tissue by suitable homogenization and upscaling steps, see Fig. 5. There one can see that the collagen fibers appear on the microscale in this observation and that their description and inclusion are decisive for the prediction of the macroscopic muscle behavior. In turn, the description of the collagen fibers brings us back to the considerations of the previous section. Thus, even in an explanatory multiscale model, the question arises whether individual structures should (or have to) be described on smaller scales by phenomenological or structural/explanatory models. The designation of an explanatory model therefore only makes sense with a clear reference to a certain scale, whereas the description of individual mechanisms in such a model naturally relies again on phenomenological approaches. Hence, the idea of phenomenology is always there.

8 Conclusion

This chapter dealt with the classification of (bio)mechanical models. In particular, two different ways of approaching this topic were discussed. The first was a more classical engineering-oriented approach and focused on the model itself, thus resulting in common descriptions like structural, micromechanical or multiscale. In contrast, the second approach was more application-oriented. It focused on an observed phenomenon and the distinction between direct, phenomenological model approaches and such that explain (or postulate) the underlying mechanisms of a phenomenon. Further, an attempt was made to clarify the meaning and importance of phenomenology in mechanics and to avoid the designation of phenomenological models as inaccurate or insufficient. In mechanics, phenomenological and explanatory approaches represent two different ways of describing observed (material) behavior and neither is better or worse *per se*. The focus in basic research is usually on the detailed investigation of complex processes across length and time scales, thus leading to detailed multiscale models, while in real (e.g., clinical) applications one might be more interested in the fast and reliable generation of desired output data for given input data, for which phenomenological models are well suited. In this context, analytical microstructurally-motivated constitutive models such as the ones proposed by Professor Holzapfel are particularly valuable, as they combine the advantages of simple usability and the presence of microstructural parameters with physical meaning. Examples are for instance the widely used model by Gasser et al. (2006) for dispersed collagen fiber orientations as well as the constitutive models for arteries proposed by Holzapfel and Ogden (2010) and for cardiac muscle tissue by Holzapfel and Ogden (2009). The latter model, e.g., has made decisive steps towards clinical applications in the context of the Living Heart Project, see, e.g., Baillargeon et al. (2014) and Peirlinck et al. (2021).

In the context of classifying models, it makes sense to refer to (bio)mechanical models in basic research by means of terms like structural or micromechanical, particularly if the model itself is the focus of a work. However, if one is more concerned with answering specific questions (like describing some measured material behavior), such terms are often not helpful, especially in an interdisciplinary environment. In this case, it is sensible to name models according to their abilities in relation to that specific question and to use terms like explanatory or the mentioned distinction between how-possibly- and how-actually-models.

Acknowledgements This research was partially funded by the Deutsche Forschungsgemeinschaft (DFG, German Research Foundation) under Germany's Excellence Strategy—EXC 2075-390740016 and by the Federal Ministry of Education and Research—BMBF as part of the project '3DFoot' (funding number: 01EC1907B)

References

Baillargeon, B., Rebelo, N., Fox, D.D., Taylor, R.L., Kuhl, E.: The living heart project: a robust and integrative simulator for human heart function. Eur. J. Mech. A-Solid **48**, 38–47 (2014)

Balzani, D., Neff, P., Schröder, J., Holzapfel, G.A.: A polyconvex framework for soft biological tissues. Adjustment to experimental data. Int. J. Solids Struct. **43**, 6052–6070 (2006)

Bleiler, C., Wagner, A., Stadelmann, V.A., Windolf, M., Köstler, H., Boger, A., Gueorguiev-Rüegg, B., Ehlers, W., Röhrle, O.: Multiphasic modelling of bone-cement injection into vertebral cancellous bone. Int. J. Numer. Meth. Bio. **31**(1), 37–57 (2015)

Bleiler, C., Ponte Castañeda, P., Röhrle, O.: A microstructurally-based, multi-scale, continuum-mechanical model for the passive behaviour of skeletal muscle tissue. J. Mech. Behav. Biomed. **97**, 171–186 (2019)

Bleiler, C., Ponte Castañeda, P., Röhrle, O.: Tangent second-order homogenisation estimates for incompressible hyperelastic composites with fibrous microstructures and anisotropic phases. J. Mech. Phys. Solids **147**, 104251 (2021)

Bogen, J.: 'Saving the phenomena' and saving the phenomena. Synthese **182**, 7–22 (2011)

Bogen, J., Woodward, J.: Saving the phenomena. Philos. Rev. **97**, 303–352 (1988)

Bokulich, A.: How scientific models can explain. Synthese **180**, 33–45 (2011)

Buehler, M.J.: Nanomechanics of collagen fibrils under varying cross-link densities: atomistic and continuum studies. J. Mech. Behav. Biomed. **1**, 59–67 (2008)

Craver, C.F.: When mechanistic models explain. Synthese **153**, 355–376 (2006)

Craver, C.F., Kaplan, D.M.: Are more details better? On the norms of completeness for mechanistic explanations. Brit. J. Philos. Sci. **71**, 287–319 (2020)

Darden, L.: Reasoning in Biological Discoveries Essays on Mechanisms, Interfield Relations, and Anomaly Resolution, 4th edn. Cambridge University Press, Cambridge (2006)

Darden, L.: Thinking again about biological mechanisms. Philos. Sci. **75**, 958–969 (2008)

Einstein, A.: On the method of theoretical physics. Philos. Sci. **1**, 163–169 (1934)

Fratzl, P.: Collagen: Structure and Mechanics. Springer, New York (2008)

Frigg, R., Hartmann, S.: Models in Science. In: Zalta, E.N. (ed.) The Stanford Encyclopedia of Philosophy, spring, 2020th edn. Metaphysics Research Lab, Stanford University (2020)

Gasser, T.C., Ogden, R.W., Holzapfel, G.A.: Hyperelastic modelling of arterial layers with distributed collagen fibre orientations. J. Roy. Soc. Interface **3**, 15–35 (2006)

Hansen, K.A., Weiss, J.A., Barton, J.K.: Recruitment of tendon crimp with applied tensile strain. J. Biomech. Eng. **124**, 72–77 (2002)

Heidlauf, T., Klotz, T., Rode, C., Altan, E., Bleiler, C., Siebert, T., Röhrle, O.: A multi-scale continuum model of skeletal muscle mechanics predicting force enhancement based on actin-titin interaction. Biomech. Model. Mechanobiol. **15**, 1423–1437 (2016)

Holzapfel, G.A.: Nonlinear Solid Mechanics. A Continuum Approach for Engineering. Wiley, Chichester (2000)

Holzapfel, G.A., Ogden, R.W.: Constitutive modelling of passive myocardium: a structurally based framework for material characterization. Philos. T. R. Soc. A **367**, 3445–3475 (2009)

Holzapfel, G.A., Ogden, R.W.: Constitutive modelling of arteries. P. Roy. Soc. A **466**, 1551–1597 (2010)

Holzapfel, G.A., Gasser, T.C., Ogden, R.W.: A new constitutive framework for arterial wall mechanics and a comparative study of material models. J. Elasticity **61**, 1–48 (2000)

Husserl, E.: The Idea of Phenomenology. A Translation of Die Idee der Phänomenologie Husserliana II. Springer, Dordrecht (1999)

Illari, P.M., Williamson, J.: What is a mechanism? Thinking about mechanisms across the sciences. Eur. J. Philos. Sci. **2**, 119–135 (2012)

Maceri, F., Marino, M., Vairo, G.: A unified multiscale mechanical model for soft collagenous tissues with regular fiber arrangement. J. Biomech. **43**, 355–363 (2010)

Machamer, P., Darden, L., Craver, C.F.: Thinking about mechanisms. Philos. Sci. **67**, 1–25 (2000)

Marino, M., Wriggers, P.: Finite strain response of crimped fibers under uniaxial traction: an analytical approach applied to collagen. J. Mech. Phys. Solids **98**, 429–453 (2017)

McMullin, E.: What do physical models tell us? In: Van Rootselaar, B., Staal, J.F. (eds.) Logic, Methodology and Philosophy of Science III, Studies in Logic and the Foundations of Mathematics, vol. 52, pp. 385–396. North-Holland Publishing Company, Amsterdam (1968)

Peirlinck, M., Sahli Costabal, F., Yao, J., Guccione, J.M., Tripathy, S., Wang, Y., Ozturk, D., Segars, P., Morrison, T.M., Levine, S., Kuhl, E.: Precision medicine in human heart modeling. Biomech. Model, Mechanobiol (2021)

Röhrle, O., Davidson, J.B., Pullan, A.J.: Bridging scales: a three-dimensional electromechanical finite element model of skeletal muscle. SIAM J. Sci. Comput. **30**(6), 2882–2904 (2008)

Röhrle, O., Davidson, J.B., Pullan, A.J.: A physiologically based, multi-scale model of skeletal muscle structure and function. Front. Physiol. **3**(358) (2012)

Röhrle, O., Yavuz, U.Ş., Klotz, T., Negro, F., Heidlauf, T.: Multiscale modeling of the neuromuscular system: coupling neurophysiology and skeletal muscle mechanics. Wires. Syst. Biol. Med. **11**(6) (2019)

Schmid, L., Klotz, T., Siebert, T., Röhrle, O.: Characterization of electromechanical delay based on a biophysical multi-scale skeletal muscle model. Front. Physiol. **10**, 1270 (2019)

Smith, B., Smith, D.W.: Introduction. In: Barry, S., Smith, D.W. (eds.) The Cambridge Companion to Husserl, pp. 1–44. Cambridge University Press (1995)

Smith, D.W.: Phenomenology. In: Zalta, E.N. (ed.) The Stanford Encyclopedia of Philosophy, summer, 2018th edn. Metaphysics Research Lab, Stanford University (2018)

Takaza, M., Moerman, K.M., Gindre, J., Lyons, G., Simms, C.K.: The anisotropic mechanical behaviour of passive skeletal muscle tissue subjected to large tensile strain. J. Mech. Behav. Biomed. **17**, 209–220 (2013)

Truesdell, C., Noll, W.: The Non-Linear Field Theories of Mechanics, 3rd edn. Springer, Berlin, Heidelberg (2004)

Wang, Y.K., Nash, M.P., Pullan, A.J., Kieser, J.A., Röhrle, O.: Model-based identification of motion sensor placement for tracking retraction and elongation of the tongue. Biomech. Model. Mechanobiol. **12**(2), 383–399 (2013)

Woo, S.L.Y., Johnson, G.A., Smith, B.A.: Mathematical modeling of ligaments and tendons. J. Biomech. Eng. **115**, 486–473 (1993)

Woodward, J.: Making Things Happen: A Theory of Causal Explanation. Oxford University Press, New York (2003)

A Contribution to the Medication-Induced Treatment of Brain–Tumor Metastases

Wolfgang Ehlers, Patrick Schröder, and Arndt Wagner

It must have been around 1998, when I met Professor Gerhard A. Holzapfel for the first time. In those days, we, as well as many others, have been very much interested in the evolution of the scientific field of continuum biomechanics. Much time has elapsed since then, and Gerhard has become one of the leading persons in biomechanical research. Based on his strong commitment in the field, he succeeded to convince the publishing house Springer to found 'Biomechanics and Modeling in Mechanobiology', one of the world-leading journals on biomechanical research that he is leading as the editor-in-chief since 2001. This and many outstanding scientific recognitions paved his scientific career.
Now, at the age of 60, it is time to contemplate what was and what might be in the years to come.
All the best for you.

Yours, Wolfgang

Abstract Cancer is one of the most serious diseases for human beings, especially, when metastases come into play. In the present chapter, the example of lung-cancer metastases in the brain is used to discuss the basic problem of cancer growth and atrophy as a result of both nutrients and medications. As the brain itself is a soft tissue that is saturated by blood and interstitial fluid, the biomechanical description of the problem is based on the Theory of Porous Media enhanced by the results of medication tests carried out on in-vitro experiments on cancer-cell cultures. Based on the theoretical and experimental results, the existence of cancer cells beyond a certain threshold is included in the description by so-called mass-production terms added to the mass of the solid skeleton. Furthermore, the mass interaction of nutrients and medical drugs between the solid and the interstitial fluid and its influence on proliferation, necrosis and apoptosis of cancer cells is taken into consideration. As a

W. Ehlers (✉) · P. Schröder · A. Wagner
Institute of Applied Mechanics, University of Stuttgart, Stuttgart, Germany
e-mail: Wolfgang.Ehlers@mechbau.uni-stuttgart.de

G. Sommer et al. (eds.), *Solid (Bio)mechanics: Challenges of the Next Decade*, Studies in Mechanobiology, Tissue Engineering and Biomaterials 24,
https://doi.org/10.1007/978-3-030-92339-6_15

result, the overall model is appropriate for the description of brain-tumor treatment combined with stress and deformation induced by cancer growth in the skull.

1 Introduction

For human beings, cancer is one of the most leading causes of death, cf. the World Cancer Report of 2014 (Wild and Stewart 2014) by Wild and Stewart and the related webpage[1] of the World Health Organization (WHO). Basically, cancer arises from normal cells that are transformed into tumor cells. Although cancer cells are produced in a very small number every day, these cells are usually eliminated by the immune system. Generally, we are talking about cancer if a malignant tumor is formed during a multistage process from a pre-cancerous lesion. The individual processes are routinely subdivided into the famous hallmarks of cancer, cf. Hanahan and Weinberg (2011). The genesis is caused by the interaction between a person's genetic factors and external agents, such as physical (e.g., ionising radiation), chemical (e.g., tobacco smoke) or biological (e.g., virus infections) carcinogens.

The problem with cancer is the rapid creation of abnormal cells, resulting in a space-demanding growth process. This leads to an invasion of adjoining parts of the body and, even more serious, to a spreading towards other organs and a creation of metastases. In this regard, metastases represent a major cause of death from cancer.

A numerical description of cancer therapies must catch the basic features of the human body, especially, of the organ under consideration. As mammals consist of hard and soft tissues governed by a porous organic solid that is percolated by interstitial fluid and blood both in separated compartments but interacting in capillary vessels between arteries and veins, a biomechanical description is naturally based on the Theory of Porous Media (TPM). When cancer comes into play, the cancer cells are sticking to the porous solid, while nutrients, medical drugs and further components come along with the blood and the interstitial fluid.

In the present chapter, the above porous-media model is applied to lung-cancer metastases that have been proliferating within a brain tissue.

2 Continuum Mechanics of Brain Tumors

2.1 The TPM in Brief

Based on the brain model by Ehlers and Wagner (2015), the continuum-mechanical description of growth and atrophy of brain-tumor metastases needs the mass and momentum balances of a set of components. These are the healthy or the cancerous

[1] https://www.who.int/news-room/fact-sheets/detail/cancer, retrieved 31 January 2020.

brain material, the blood and the interstitial fluid in different compartments and further ingredients that might be part of the fluid components. Thus, the model φ is composed of

$$\varphi = \bigcup_\alpha \varphi^\alpha = \varphi^S \cup \varphi^B \cup \varphi^I, \qquad \text{where} \quad \alpha = \{S,\ B,\ I\}. \tag{1}$$

Therein, φ^S stands for the porous organic solid, φ^B for the blood and φ^I for the interstitial fluid.

From the above composition, volume fractions n^α can formally be defined by the fraction of a volume element $\mathrm{d}v^\alpha$ of φ^α over the total volume element $\mathrm{d}v$ of φ. Thus,

$$n^\alpha := \frac{\mathrm{d}v^\alpha}{\mathrm{d}v} \qquad \text{with} \qquad \mathrm{d}v = \sum_{\alpha=S,\,B,\,I} \mathrm{d}v^\alpha. \tag{2}$$

This definition naturally includes the sum of all volume fractions to yield the so-called saturation condition

$$\sum\nolimits_\alpha n^\alpha = 1. \tag{3}$$

The governing set of balance equations is given by the mass and momentum balances of all basic constituents, solid, blood and interstitial fluid via

$$\begin{aligned} (\rho^\alpha)'_\alpha + \rho^\alpha \operatorname{div} \overset{\prime}{\mathbf{x}}_\alpha &= \hat{\rho}^\alpha, \\ \rho^\alpha \overset{\prime\prime}{\mathbf{x}}_\alpha &= \operatorname{div} \mathbf{T}^\alpha + \rho^\alpha \mathbf{g} + \hat{\mathbf{p}}^\alpha \end{aligned} \tag{4}$$

with the constraining side conditions

$$\sum_\alpha \hat{\rho}^\alpha = 0 \qquad \text{and} \qquad \sum_\alpha (\hat{\mathbf{p}}^\alpha + \hat{\rho}^\alpha \overset{\prime}{\mathbf{x}}_\alpha) = \mathbf{0}, \tag{5}$$

compare Ehlers (2002). In the above equations, $\rho^\alpha = n^\alpha \rho^{\alpha R}$ is the partial density of φ^α given as the product of the intrinsic (respectively effective or real) density $\rho^{\alpha R}$ and the volume fraction n^α, $\mathbf{T}^\alpha$ is the partial Cauchy stress tensor, and $\mathbf{g}$ the gravitation vector, while $\hat{\rho}^\alpha$ and $\hat{\mathbf{p}}^\alpha$ are the so-called density production and direct momentum production terms coupling the mass and momentum balances of solid and pore fluids. The constraints (5) result from the fact that the continuum-mechanical system as a whole is a so-called closed system, while the individual components are open systems, such that they can mutually interact, compare, e.g., Ehlers (2002).

Since the TPM provides a continuum-mechanical view onto porous-media problems, all terms have to be understood as local means representing the local averages of their microscopic counterparts. In particular, the density production represents an increase of partial density driven by chemo-physical processes like phase transformations or chemical reactions, while the direct momentum production exhibits the volumetric average of the local contact forces acting at the local interfaces (pore walls) between the solid and the pore fluids. In addition to the above, $\operatorname{div}(\cdot)$ is the divergence operator corresponding to the gradient operator $\operatorname{grad}(\cdot) = \partial(\cdot)/\partial\mathbf{x}$ with $\mathbf{x}$ as the local position vector. Furthermore, $\overset{\prime}{\mathbf{x}}_\alpha$ and $\overset{\prime\prime}{\mathbf{x}}_\alpha$ are the local velocity and acceleration terms of φ^α, while $(\cdot)'_\alpha$ is the material time derivative following the motion of φ^α.

In the present modeling process, all components are assumed to be materially incompressible meaning that their intrinsic densities $\rho^{\alpha R}$ remain constant under isothermal conditions assuming a constant temperature of 37 °C, such that energy balances can be ignored. For the solid skeleton, this basically means that, although the intrinsic solid density remains constant, the partial solid density $\rho^S = n^S \rho^{SR}$ can vary through variations of the solid volume fraction n^S and the solid density production $\hat{\rho}^S$.

2.2 Model Constitution

Describing tumor-related processes, we make the assumption that the solid skeleton is composed of brain cells φ^{SB} and tumor or metastatic cancer cells φ^{ST}. It is furthermore assumed that the tumor is in an early avascular stage such that it is not supplied by blood vessels and receives nutrients only from the surrounding tissue through the interstitial fluid. As a result of these assumptions, the composition (1) of the model has to be specified by

$$\varphi = \bigcup_\alpha \varphi^\alpha = \varphi^S \cup \varphi^B \cup \varphi^I, \qquad \text{where} \begin{cases} \varphi^S = \varphi^{SB} \cup \varphi^{ST}, \\ \varphi^B = \varphi^{BL}, \\ \varphi^I = \varphi^{IL} \cup \varphi^{IN} \cup \varphi^{IC} \cup \varphi^{IV} \cup \varphi^{ID}, \end{cases} \tag{6}$$

compare also with Fig. 1. It is seen from (6) and Fig. 1 that the situation here basically describes two pore fluids φ^β with $\beta = \{B, I\}$. Based on the assumption that the blood has conveyed its ingredients to the interstitial fluid φ^I, such that φ^B maintains a single blood liquid φ^{BL}, φ^I is a mixture of multiple components $\varphi^{I\gamma}$ with $\gamma = \{L, N, C, V, D\}$. Thus, φ^I consists of φ^{IL} as the interstitial-fluid solvent with nutrients φ^{IN}, mobile cancer cells φ^{IC}, vascular endothelial growth factors φ^{IV} and therapeutic drugs φ^{ID} as solutes.

For materially incompressible constituents with constant $\rho^{\alpha R}$, the mass balance $(4)_1$ reduces to the volume balance

$$(n^\alpha)'_\alpha + n^\alpha \operatorname{div}(\overset{\prime}{x})_\alpha = \hat{n}^\alpha, \qquad \text{where} \qquad \hat{\rho}^\alpha = \hat{n}^\alpha \rho^{\alpha R} \tag{7}$$

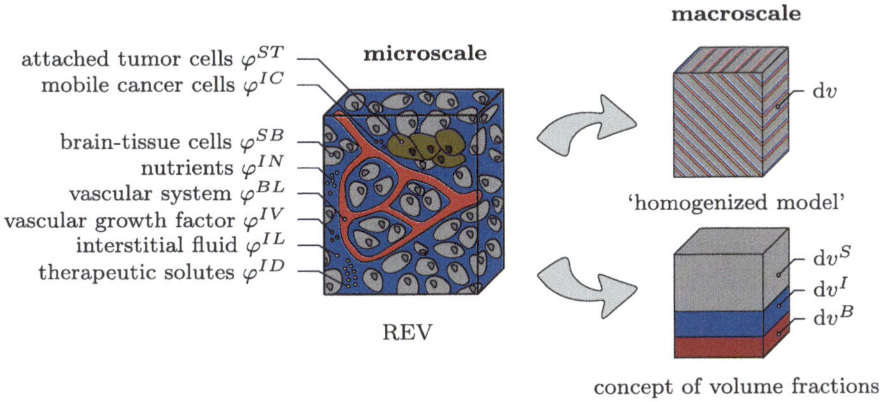

Fig. 1 Representative elementary volume with exemplarily displayed micro-structure of tumor-affected brain tissue and macroscopic multiphasic and multicomponental modeling approach

has been used on the basis that an addition of equal mass does not change the material density $\rho^{\alpha R}$ but changes the volume fraction of φ^{α} through $\hat{n}^{\alpha}$. As a result of (7), the volume balances of solid, blood and interstitial fluid read

$$\begin{aligned} (n^S)'_S + n^S \operatorname{div} \overset{\prime}{\mathbf{x}}_S &= \hat{n}^S, \\ (n^B)'_S + \operatorname{div}(n^B \mathbf{w}_B) + n^B \operatorname{div} \overset{\prime}{\mathbf{x}}_B &= 0, \\ (n^I)'_S + \operatorname{div}(n^I \mathbf{w}_I) + n^I \operatorname{div} \overset{\prime}{\mathbf{x}}_I &= \hat{n}^I, \end{aligned} \tag{8}$$

where $\mathbf{w}_\beta = \overset{\prime}{\mathbf{x}}_\beta - \overset{\prime}{\mathbf{x}}_S$ defines the seepage velocity of φ^β. Note that this description of the fluid balances is necessary for the numerical treatment of porous-media problems, since the solid material is treated by a Langrangean description on the basis of the solid displacement vector $\mathbf{u}_S = \mathbf{x} - \mathbf{x}_0$ with $\mathbf{x}_0$ as the solid location vector in the reference configuration at time $t = t_0$, while the pore fluids are described by a modified Eulerian description with the fluid seepage motion given relatively to the deforming skeleton.

From the mass-balance side condition $(5)_1$, the volume productions $\hat{n}^S$ and $\hat{n}^I$ are coupled via

$$\hat{n}^S \rho^{SR} = \hat{n}^I \rho^{IR}, \quad \text{such that} \quad \hat{n}^S = -\frac{\rho^{IR}}{\rho^{SR}} \hat{n}^I. \tag{9}$$

In addition to this, the solid volume balance can be integrated analytically to yield

$$n^S = n_0^S \exp\left(\int_{t_0}^{t} \frac{\hat{n}^S}{n^S} \, \mathrm{d}t \right) (\det \mathbf{F}_S)^{-1} =: n_{0g}^S \, (\det \mathbf{F}_S)^{-1}, \tag{10}$$

where n^S_{0g} reduces to the standard reference solid volume fraction n^S_0 at t_0, when $\hat{n}^S$ is excluded from the modeling approach. Following this, n^S_{0g} can be interpreted as the growth-dependent reference solid volume fraction that is related to the current volume fraction n^S through the inverse determinant of the solid's deformation gradient $\mathbf{F}_S$.

Concerning the above volume balances, it seems necessary to add the following remark. As ρ^{BR}, the effective densities ρ^{SR} and ρ^{IR} have been assumed to be constant as a result of mechanical incompressibility at constant temperature, although there can be an intake of mass through $\hat{n}^S$ and $\hat{n}^I$. During cancer proliferation, cancer cells are added to the brain skeleton exhibiting approximately the same density as healthy brain material, such that ρ^{SR} = const. is justified. On the other hand, ρ^{IR} is a mixture of various components that can be expressed as

$$\rho^{IR} = \sum_{\gamma} \rho_I^{I\gamma} \quad \text{with} \quad \rho_I^{I\gamma} = c_m^{I\gamma} M_m^{\gamma}. \tag{11}$$

This equation expresses the effective density ρ^{IR} of the interstitial-fluid mixture as the sum of its partial densities $\rho_I^{I\gamma}$ defined with respect to the interstitial-fluid volume, cf. Ehlers (2009), while the different $\rho_I^{I\gamma}$ are themselves defined through their molar concentrations $c_m^{I\gamma}$ expressed in mol/m^3 and molar masses M_m^{γ} expressed in kg/mol. As the partial densities $\rho_I^{I\delta}$ of the solutes $\varphi^{I\delta}$ with $\delta = \{N,\ C,\ V,\ D\}$ are generally negligible with respect to the partial density ρ_I^{IL} of the interstitial-fluid solvent φ^{IL}, it is justified to use

$$\rho^{IR} \approx \rho_I^{IL}, \tag{12}$$

thus also supporting the assumption of constant ρ^{IR} at constant temperature for a materially incompressible interstitial fluid.

In addition to the effective density ρ^{IR} and the partial densities $\rho_I^{I\gamma}$ defined with respect to the interstitial-fluid volume, further terms can be introduced with respect to the overall volume. These terms are

$$\rho^I = n^I \rho^{IR} \quad \text{and} \quad \rho^{I\gamma} = n^I \rho_I^{I\gamma}. \tag{13}$$

Furthermore, it is seen from (10) that the volume production $\hat{n}^S$ or the density production $\hat{\rho}^S = \hat{n}^S \rho^{SR}$, respectively, plays a crucial role for the description of tumor growth and regression. Initially, the existence of some single cancer cells is described by their concentration. During the cancer-cell proliferation and the further infiltration into the brain tissue, a volumetric growth occurs. Thus, the cancer-cell component is subsequently described by a volume fraction after exceeding a certain critical cancer-cell concentration. To obtain this volume fraction, the effective cancer-cell density ρ^{ICR} is computed by use of (11) and (13):

$$\rho^{ICR} = \frac{n^I \rho_I^{IC}}{n^{IC}}. \tag{14}$$

Following the assumption that $\rho^{ICR} = \rho^{SR} = \text{const.}$, (14) can be solved with respect to n^{IC} yielding

$$
\begin{aligned}
n^{IC} &= \frac{n^I \rho_I^{IC}}{\rho^{ICR}} \quad \text{and} \quad n^{ST} = \frac{n^I c_m^{IC} M_m^C}{\rho^{STR}}, \\
&\text{where } n^{IC} = n^{ST} \text{ and } \rho^{ICR} = \rho^{STR} \equiv \rho^{SR}
\end{aligned}
\tag{15}
$$

have been used. In (15), the exchange of the superscript from $(\cdot)^{ICR}$ to $(\cdot)^{STR}$ has been used to point out that the cancer cells φ^{IC} of the interstitial-fluid mixture aggregate during cancer-cell proliferation and infiltrate into the brain tissue, thus initiating a local volumetric metastasis growth. As the volume fractions n^{IC} and n^{ST} are identical, the local metastasis volume element $\mathrm{d}v^{ST}$ equals the cancer-cell volume element $\mathrm{d}v^{IC}$. Note in passing that a certain critical cancer-cell concentration

$$
c_m^{IC} \geq \tilde{c}_m^{IC} \tag{16}
$$

is necessary, such that cancer cells transform to metastases.

In (16), the concentration threshold $\tilde{c}_m^{IC}$ indicates the micrometastatic switch related to a critical cancer-cell number. Typically, micrometastases are characterized by a specific diameter size of approximately 500 μm, which can be related to a number of 28 900 cells, cf. Freyer and Sutherland (1986). This yields a cancer-cell-concentration threshold of $\tilde{c}_m^{IC} = 3.5709 \times 10^{-18}$ mol/m³.

In addition to the volume balances (8) of the solid skeleton φ^S, the bulk fluid φ^B and the bulk-fluid mixture φ^I, concentration balances

$$
\begin{aligned}
& n^I (c_m^{I\gamma})'_S + \operatorname{div}(n^I c_m^{I\gamma} \mathbf{w}_{\gamma I}) + c_m^{I\gamma} \operatorname{div}(n^B \mathbf{w}_B) + c_m^{I\gamma} \operatorname{div}(\mathbf{u}_S)'_S - \\
& -c_m^{I\gamma} \left(\frac{\hat{\rho}^S}{\rho^{SR}} + \frac{\hat{\rho}^{I\gamma}}{\rho_I^{I\gamma}} \right) = 0
\end{aligned}
\tag{17}
$$

of the interstitial-fluid components $\varphi^{I\gamma}$ are provided. Furthermore, the overall momentum balance is needed yielding

$$
\mathbf{0} = \operatorname{div} \sum_\alpha \mathbf{T}^\alpha + \rho\, \mathbf{g} + \sum_\alpha \hat{\mathbf{p}}^\alpha, \tag{18}
$$

where inertia terms have been neglected, such that (18) represents the quasi-static form of the momentum balance of the overall system. Furthermore, the overall density ρ or the so-called mixture density is defined as

$$
\rho = \sum_\alpha n^\alpha \rho^{\alpha R} \tag{19}
$$

with ρ^{IR} from (11).

The equation system described above is only appropriate to solve tumor-relevant problems, once it has been closed by convenient constitutive equations for the Cauchy stress tensors $\mathbf{T}^\alpha$, the direct momentum productions $\hat{\mathbf{p}}^\alpha$ and the volume productions $\hat{n}^\alpha$ or the density productions $\hat{\rho}^\alpha$, respectively. In contrast to the Cauchy stresses that have to be found for all constituents, direct momentum productions $\hat{\mathbf{p}}^\alpha$ and density productions $\hat{\rho}^\alpha$ are constraint through (5), such that

$$\hat{\rho}^S = -\sum_\beta \hat{\rho}^\beta \quad \text{and} \quad \hat{\mathbf{p}}^S = -\sum_\beta (\hat{\mathbf{p}}^\beta + \hat{\rho}^\beta \mathbf{w}_\beta). \tag{20}$$

2.3 *Constitutive Setting*

Based on standard techniques for the evaluation of thermodynamical restrictions, compare Ehlers (2002, 2009, 2018), Ehlers and Wagner (2015) and related work, the following constitutive equations are found for stresses and direct momentum productions:

$$\begin{aligned}
&\bullet \text{ solid and fluid stresses: } \begin{cases} \mathbf{T}^S = -n^S p\,\mathbf{I} + \mathbf{T}^S_{E\,\text{biomech}}, \\ \mathbf{T}^B = -n^B p^{BR}\,\mathbf{I}, \\ \mathbf{T}^I = -n^I p^{IR}\,\mathbf{I}, \end{cases} \\
&\bullet \text{ momentum productions: } \begin{cases} \hat{\mathbf{p}}^S = -(\hat{\mathbf{p}}^B + \hat{\mathbf{p}}^I) - \hat{\rho}^I \mathbf{w}_I, \\ \hat{\mathbf{p}}^B = p^{IR} \operatorname{grad} n^B - p^{\text{dif}} s^B \operatorname{grad} n^S + \hat{\mathbf{p}}^B_{E\,\text{dis}}, \\ \hat{\mathbf{p}}^I = p^{IR} \operatorname{grad} n^I + \sum_\gamma \hat{\mathbf{p}}^{I\gamma}_{E\,\text{dis}}. \end{cases}
\end{aligned} \tag{21}$$

In these equations, p is the pore pressure that is obtained from Dalton's law, cf. Dalton (1802), via $p = s^B p^{BR} + s^I p^{IR}$ with $s^\beta = n^\beta / (\sum_\beta n^\beta)$. Therein, the saturations s^β act as volume fractions of the pore space. Furthermore $p^{\text{dif}} = p^{BR} - p^{IR}$ is the difference of the effective pressures between the blood and the interstitial fluid. Finally, $\mathbf{T}^S_{E\,\text{biomech}}$ is the biomechanically induced solid extra stress reflecting the growth- and deformation-dependent material behavior, while the extra momentum productions $\hat{\mathbf{p}}^B_{E\,\text{dis}}$ and $\hat{\mathbf{p}}^{I\gamma}_{E\,\text{dis}}$ depend on the seepage velocities $\mathbf{w}_\beta$ of φ^β. In case of $\hat{\mathbf{p}}^{I\gamma}_{E\,\text{dis}}$, further terms come into play depending on the diffusion velocities $\mathbf{d}_{I\gamma} = \mathbf{w}_{I\gamma} - \mathbf{w}_I$ of the species $\varphi^{I\gamma}$ in the interstitial fluid mixture φ^I. For details, compare Ehlers (2009).

In addition to the constitutive equations (21), further results are needed for the description of mass-production terms resulting in

$$\hat{\rho}^S + \hat{\rho}^I = 0 \quad \text{with} \quad \hat{\rho}^I = \hat{\rho}^{IL} + \hat{\rho}^{IC} + \hat{\rho}^{IN} + \hat{\rho}^{IV} + \hat{\rho}^{ID}, \tag{22}$$

compare $(5)_1$, (8) and (9). As φ^I is a mixture of the liquid solvent φ^{IL} and solutes $\varphi^{I\gamma}$, $\hat{\rho}^I$ has to be constituted formally via $(22)_2$. While $\hat{\rho}^{IL}$ of the liquid solvent does usually not exist, the reminder $\hat{\rho}^{I\gamma}$ of mass-production terms splits into gains $\hat{\rho}^{I\gamma}_{\oplus\varepsilon}$ and losses $\hat{\rho}^{I\gamma}_{\ominus\varepsilon}$, where the second subscript ε is a placeholder for the source of a specific process. Furthermore, as nutrients supply the cancer cells with energy for basal reactions as well as for their cancer proliferation, metastatic proliferation $\hat{\rho}^{ST}_{\oplus IN}$ corresponds to its nutrient consumption $\hat{\rho}^{IN}_{\ominus SC}$. Following Guppy et al. (2002), it is moreover assumed that nutrients are only consumed by cancer cells during the proliferation process and that all further substances needed for cell proliferation are available in a sufficient amount.

In malnutrition states, meaning undersupply of nutrients, the solid skeleton (cancer and regular cells) will undergo a necrotic process via $\hat{\rho}^{ST}_{\ominus IN}$ and $\hat{\rho}^{SB}_{\ominus IN}$. In addition, the apoptosis reaction $\hat{\rho}^{ST}_{\ominus ID}$ based on the effect of the therapeutic agent is also considered resulting in a mass-loss term (see Fig. 2). In conclusion, the mass-production term of the solid skeleton is found as

$$\hat{\rho}^S = \hat{\rho}^{ST}_{\oplus IN} - \hat{\rho}^{ST}_{\ominus IN} - \hat{\rho}^{ST}_{\ominus ID} - \hat{\rho}^{SB}_{\ominus IN}, \tag{23}$$

where the remainder of substances compared to (22) has been omitted as elements of mass productions in the present study. For the description of the individual mass-production terms, Monod-like kinetics are used in accordance to Ambrosi and Presziosi (2002), Shelton (2011) and Gray et al. (2013). Following this, growth and atrophy terms due to nutrient supply and drug medication can be modeled via

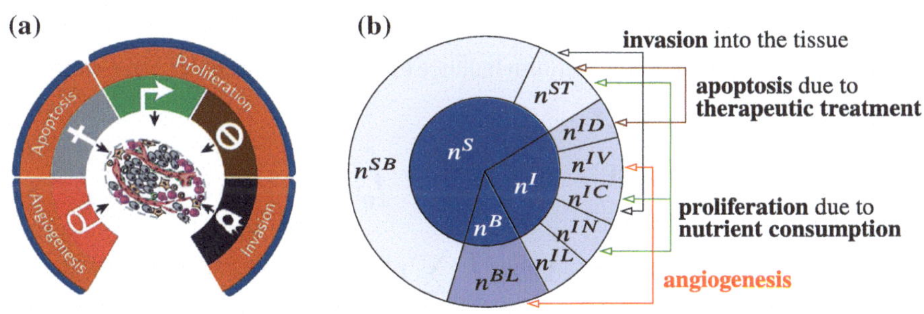

Fig. 2 (a) selected cancer processes adopted from Hanahan and Weinberg (2011) and (b) related mass (volume) interactions in the metastasis model. Proliferation appears through nutrient consumption, whereas atrophy is either related to insufficient nutrient supply (necrosis) or apoptosis which can be caused by the presence of a drug. These processes interchange volume (mass) between the solid and the interstitial fluid

$$
\begin{aligned}
\hat{\rho}^{ST}_{\oplus IN} &= \rho^{ST}\,\mu_{IN\,\max}\,\frac{c^{IN}_m - \bar{c}^{IN}_m}{K_{gr} + (c^{IN}_m - \bar{c}^{IN}_m)},\\
\hat{\rho}^{ST}_{\ominus IN} &= \rho^{ST}\,\tilde{\mu}_{ST\,\max}\,\frac{\tilde{c}^{IN}_m - c^{IN}_m}{K_{gr} + (\tilde{c}^{IN}_m - c^{IN}_m)},\\
\hat{\rho}^{ST}_{\ominus ID} &= \rho^{ST}\,\mu_{ST\,\max}\,\frac{c^{ID}_m - \bar{c}^{ID}_m}{K_{gr} + (c^{ID}_m - \bar{c}^{ID}_m)},\\
\hat{\rho}^{SB}_{\ominus IN} &= \rho^{SB}\,\mu_{SB\,\max}\,\frac{\tilde{c}^{IN}_m - c^{IN}_m}{K_{gr} + (\tilde{c}^{IN}_m - c^{IN}_m)}.
\end{aligned}
\tag{24}
$$

Therein, $\mu_{IN\,\max}$, $\tilde{\mu}_{ST\,\max}$, $\mu_{ST\,\max}$ and $\mu_{SB\,\max}$ characterize the maximal reaction rates, while K_{gr} indicates the concentration of c^{IN}_m or c^{ID}_m, respectively, at which the maximal reaction rate reaches half of its value. Finally, $\bar{c}^{IN}_m$, $\tilde{c}^{IN}_m$ and $\bar{c}^{ID}_m$ are threshold concentrations governing the mass-production process.

2.4 *Weak Form of the Governing Equations Describing the Finite Element Analysis of Metastasis Atrophy*

For use of the finite element analysis (FEA) for the numerical computation of proliferation, necrosis and apoptosis processes, the basic governing equations have to be transformed into their weak forms. As the problem is generally highly coupled through the simultaneous action of the overall momentum balance, the fluid mass balances and the concentration balances of the interstitial-fluid species, a fully coupled algorithm is strove for, where the system of weak equations is solved simultaneously within the Bubnov-Galerkin method.

Starting with the overall momentum balance (18) tested by $\delta\mathbf{u}_S$, the corresponding weak form reads

$$
\begin{aligned}
\mathcal{G}_u = &\int_\Omega \Big(\sum_\alpha \mathbf{T}^\alpha\Big)\cdot \operatorname{grad}\delta\mathbf{u}_S\,\mathrm{d}v - \int_\Omega \rho\,\mathbf{g}\cdot\delta\mathbf{u}_S\,\mathrm{d}v - \\
&- \int_\Gamma \bar{\mathbf{t}}\cdot\delta\mathbf{u}_S\,\mathrm{d}a - \int_\Omega \sum_\alpha (\hat{\rho}^\alpha\,\overset{\prime}{\mathbf{x}}_\alpha)\cdot\delta\mathbf{u}_S\,\mathrm{d}v = 0,
\end{aligned}
\tag{25}
$$

where $\bar{\mathbf{t}} = \sum_\alpha \mathbf{T}^\alpha\mathbf{n}$ is the total external load vector acting at the Neumann boundary with outward-oriented unit normal vector $\mathbf{n}$.

On the basis of (8), the volume balances of the overall interstitial fluid and the blood are tested by $\delta p^{\beta R}$ and yield in their weak forms

$$
\begin{aligned}
\mathcal{G}_{p^{\beta R}} = & \int_\Omega \Big\{ (n^\beta)'_S + n^\beta \mathrm{div}(\mathbf{u}_S)'_S - \hat{n}^\beta \Big\} \, \delta p^{\beta R} \, \mathrm{d}v - \\
& - \int_\Omega n^\beta \mathbf{w}_\beta \cdot \mathrm{grad} \, \delta p^{\beta R} \, \mathrm{d}v + \int_\Gamma \overline{v}^\beta \, \delta p^{\beta R} \, \mathrm{d}a = 0,
\end{aligned} \tag{26}
$$

where $\overline{v}^\beta = n^\beta \mathbf{w}_\beta \cdot \mathbf{n}$ is the volumetric efflux of the particular pore fluid φ^β.

Finally, the relevant dissolved species of the interstitial-fluid mixture, cancer cells, nutrients and drugs, are described by the weak formulations of their concentration balances (17) tested by $\delta c_m^{I\gamma}$ via

$$
\begin{aligned}
\mathcal{G}_{c_m^{I\gamma}} = & \int_\Omega \Big\{ n^I (c_m^{I\gamma})'_S + c_m^{I\gamma} \, \mathrm{div}(\mathbf{u}_S)'_S + c_m^{I\gamma} \, \mathrm{div}(n^B \mathbf{w}_B) - \\
& - c_m^{I\gamma} \Big(\frac{\hat{\rho}^S}{\rho^{SR}} + \frac{\hat{\rho}^{I\gamma}}{\rho_I^{I\gamma}} \Big) \Big\} \, \delta c_m^{I\gamma} \, \mathrm{d}v - \int_\Omega n^I c_m^{I\gamma} \mathbf{w}_{\gamma I} \cdot \mathrm{grad} \delta c_m^{I\gamma} \, \mathrm{d}v + \\
& + \int_\Gamma \bar{j}^{I\gamma} \, \delta c_m^{I\gamma} \, \mathrm{d}a = 0.
\end{aligned} \tag{27}
$$

Therein, $\bar{j}^{I\gamma} = n^I c_m^{I\gamma} \mathbf{w}_{\gamma I} \cdot \mathbf{n}$ is the molar efflux of $\varphi^{I\gamma}$ across the Neumann boundary.

The above weak equations are sufficient to compute the primary variables of the model. These are the solid displacement $\mathbf{u}_S$, the effective pressures p^{BR} and p^{IR} of the blood and the interstitial fluid mixture and the relevant concentrations $c_m^{I\gamma}$ of the considered interstitial-fluid species. The secondary variables, such as stresses and momentum productions, can be obtained as functions of these terms, compare Ehlers (2009) and Ehlers and Wagner (2015).

2.5 Parameter Identification Based on Experimental Data

Due to the large number of model parameters, there is a high uncertainty for quantitative results of numerical computations. To some extend, this problem can be solved by taking data obtained from laboratory experiments into consideration. For a successful model calibration, it is furthermore useful to start the procedure with a sensitivity analysis and to go on in a second step with an optimization of the crucial parameters.

The following parameter-identification process is based on the parameter-sampling method by Morris (1991) and includes the sensitivity measure of Campolongo et al. (2007). In this context, the deviation between the model and the data is denoted as model output y. The dependency of y on each parameter p_i is evaluated using the elementary effect

$$EE(p_i) = \frac{y_1(p_1, ..., p_i + \Delta, p_k) - y_2(p_1, ..., p_i, p_k)}{\Delta}, \tag{28}$$

introduced by Morris (1991) that has further been improved in Campolongo et al. (2007). Therein, the change of the model output y is computed by modifying one of the k parameters p_i with a specific step size Δ chosen as $\Delta = 1/(l-1)$ with l as the number of parameter changes. Then, the two model outputs y_1 and y_2 are compared to each other in (28).

This process is repeated until all parameters have been changed, cf. Morris (1991) or King and Perera (2013). The final quantity of interest in determining the importance of a parameter is achieved by the absolute mean

$$\mu_i^* = \sum_i \mid EE(p_i) \mid /r, \tag{29}$$

where r is the number of so-called trajectories within the parameter space. The calculated mean μ_i^* related to a parameter allows for the comparison between the other parameters. Consequently, the parameters with the highest values are considered as most significant, cf. King and Perera (2013). However, the introduced method allows only for a sensitivity-based ordering of the parameters. The actual choice of relevant parameters needs still to be decided manually by the user.

In a second step, the parameter-optimization strategy is based on the maximum likelihood estimation, cf. Myung (2003). To relate the simulation results y_{simu} to the measured data y_{data}, a logarithmic error $\log \epsilon = \log(y_{\text{data}}) - \log(y_{\text{simu}})$ is introduced. This error is assumed to be Gaussian distributed resulting in the likelihood function

$$\mathcal{L} = \sum_i \log\left(\frac{1}{\sigma_i \sqrt{2\pi}}\right) - \frac{(y_{\text{data},i} - y_{\text{simu},i})^2}{2\sigma_i^2}. \tag{30}$$

Therein, $\mathcal{L}$ is the summed likelihood function including the data at i and measurement including the variance σ_i of the data. In order to optimize the crucial model parameters, the likelihood function can be minimized, using, e.g., the commercial software `Matlab` with its optimizer `fmincon`.

Based on in-vitro experimental investigations carried out on lung-cancer-cell cultures by the group of Markus Morrison,[2] Fig. 3 displays the result of the above parameter identification and optimization process that has been applied to cancer-cell apoptosis induced by the medical drug TRAIL (tumor necrosis factor-related apoptosis-inducing ligand). In particular, five experiments have been realized to investigate the response of the cancer-cell line NCI-H4600 to $\text{Db}_{\alpha\text{EGFR}}$-scTRAIL, where TRAIL concentrations of $c_m^{ID} \in [10^{-10}, 10^{-9}, 10^{-8}, 10^{-7}, 10^{-6}]$ mol/m^3 have been applied, also compare Schröder et al. (2019). Concerning Fig. 3, the dark red dots display the experimental results for the five concentrations c_m^{ID}, while the light red dots are the mean values of the experimental data at the particular c_m^{ID}.

[2] Markus Morrison is currently the head of the Institute of Cell Biology and Immunology at the University of Stuttgart.

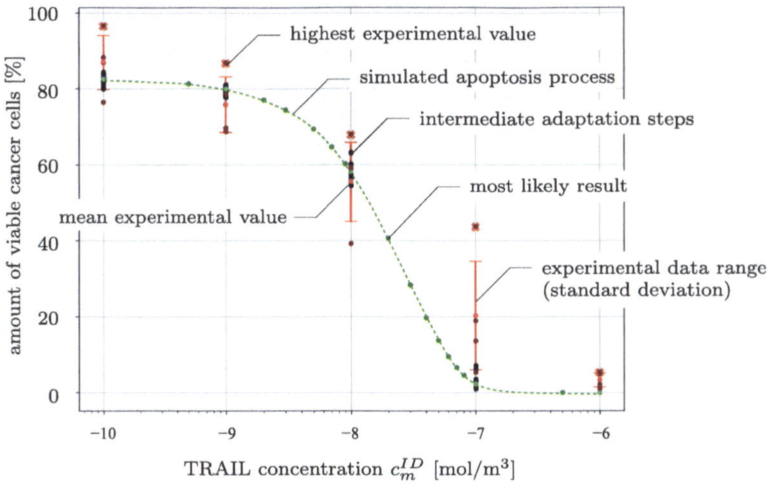

Fig. 3 Experimental data (single viable cell ratios) at five TRAIL concentrations c_m^{ID} depicted in dark red with crosses as maximum values. Mean viable cancer-cell data given in light red with standard deviation as vertical bars. The simulation output resulting from the optimization process is given in black including the best result in light green. Additional simulation outputs corresponding to the best parameter sets are given in dark green

Furthermore, the red crosses exhibit the highest experimental values, and the vertical red bars are the standard deviations. During the optimization process, the black dots are produced as intermediate steps towards the optimal parameter of each of the five c_m^{ID} displayed as green dots. These results are sufficient to get further non-experimentally obtained data represented as dark green dots at concentrations that did not belong to the five concentrations above. The dashed green line links the most likely results. Finally, it is seen that the green line as the most likely result of the experiments fits the experimental data well as it quantitatively correlates with the data obtained from the realized in vitro experiments.

3 Numerical Example of Apoptosis Due to Therapeutic Treatment

As an illustration of the above procedure, the following numerical example has been set up for the simulation of an experimental-data-based apoptosis process, where the medication represents an immediate interaction with the cancer cells of the metastases with TRAIL.

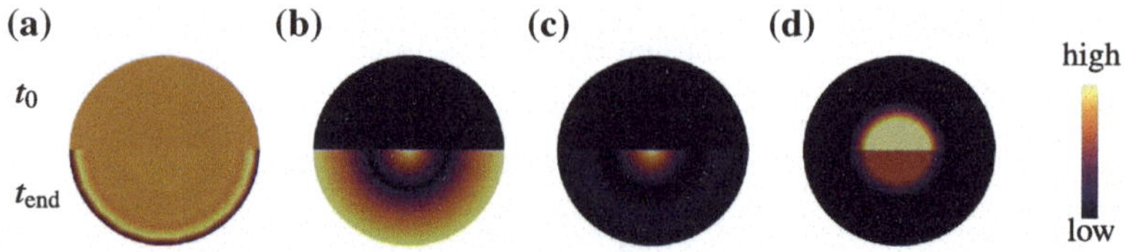

Fig. 4 Simulation of apoptosis. The upper half of the circular simulation domain shows the conditions at the initial time t_0, while the lower half shows qualitative results at the final state at $t_{end} - t_0 = 24$ h with (a) drug consumption indicated by the concentration $c_m^{ID} \in [6.3 \cdot 10^{-8},\ 1.1 \cdot 10^{-7}]$ in mol/m^3, (b) outward-oriented radial solid deformation $\mathbf{u}_S \in [-3.3 \cdot 10^{-4},\ 0\,]$ in m, (c) outward-oriented radial seepage velocity $\mathbf{w}_I \in [-4.8 \cdot 10^{-9},\ -3.0 \cdot 10^{-12}]$ in m/s, and (d) metastases volume fraction $n^{ST} \in [0,\ 0.1]$ in [-]

The numerical example is based on a circular specimen with a diameter of $d = 0.2$ m. Within the FEA, the numerical tool Pandas[3] is used, where an implicit Euler time integration has been implied together with a spatial discretization by 954 Taylor-Hood elements resulting in 2 899 nodes. Quadratic approximation functions are used for the solid displacement $\mathbf{u}_S$ and linear approximation functions for the blood and interstitial-fluid pressures, p^{BR} and p^{IR}, as well as for the relevant molar concentrations $c_m^{I\gamma}$. In order to simplify the complex coupled computations, the blood volume fraction n^B has been kept constant and only the following interstitial-fluid species have been considered: cancer cells φ^{IC}, nutrients φ^{IN} and medical drugs φ^{ID}, particularly TRAIL.

Starting from a primarily computed tumor proliferation, compare Schröder et al. (2019, 2017), the initial and boundary conditions are as follows: the outer boundary is fixed in space and is furthermore applied with a nutrient concentration of $c_m^{IN} = 1\,\mathrm{mol/m^3}$. The same initial concentration of c_m^{IN} is applied to the whole domain ensuring a sufficient nutrient supply as was guaranteed throughout the experiment. Moreover, the drug concentration is similarly applied proceeding from different initial concentrations $c_m^{ID} \in [10^{-10},\ 10^{-9},\ 10^{-8},\ 10^{-7},\ 10^{-6}]$ in mol/m^3. When the simulation was started, the metastasis volume fraction was set to $n^{ST} = 0.1$ within the inner core of the domain with $d = 0.04$ m. Over a period of 24 h, the drug triggers apoptosis within the metastasis leading to a reduction of cancer cells, compare Fig. 4.

During the computation over the whole time period of 24 h, the metastasis decline is leading to a shrinkage of the solid skeleton that extrudes interstitial fluid and the containing death-receptor agonist into the metastasis.

4 Conclusion

The present contribution is an attempt to numerically describe growth and atrophy processes of cancer cells based on the example of lung-cancer metastases in the

[3] **P**orous Media **a**daptive **n**onlinear finite element solver based on **d**ifferential **al**gebraic **s**ystems, cf. http://www.get-pandas.com.

brain. In the framework of a continuum-mechanical investigation of the problem, the Theory of Porous Media has been used for the description of the coupled behavior of brain tissue with interstitial fluid and blood. Mass-production terms have been included to especially capture the nutrient and drug transport from the interstitial fluid to the brain solid. This delivers the reason for metastasis growth as well as for atrophy, while a first numerical example has been added to carry out the possibilities of the model.

Acknowledgements The authors would like to thank the German Research Foundation (DFG) for the financial support of the project within the Cluster of Excellence in Simulation Technology (ExC 310 and ExC 2075-390740016) at the University of Stuttgart.

References

Ambrosi, D., Presziosi, L.: On the closure of mass balance models for tumor growth. Math. Models Methods Appl. Sci. **12**, 737–754 (2002)

Campolongo, F., Cariboni, J., Saltelli, A.: An effective screening design for sensitivity analysis of large models. Environ. Model Softw. **22**, 1509–1518 (2007)

Dalton, J.: On the expansion of elastic fluids by heat. In: Essay IV of Memoirs of the Literary and Philosophical Society of Manchester, vol. 5, pp. 595–602 (1802)

Ehlers, W.: Foundations of multiphasic and porous materials. In: Ehlers, W., Bluhm, J. (eds.) Porous Media: Theory, Experiments and Numerical Applications, pp. 3–86. Springer, Berlin (2002)

Ehlers, W.: Challenges of porous media models in geo- and biomechanical engineering including electro-chemically active polymers and gels. Int. J. Adv. Eng. Sci. Appl. Math. **1**, 1–24 (2009)

Ehlers, W., Wagner, A.: Multi-component modelling of human brain tissue: a contribution to the constitutive and computational description of deformation, flow and diffusion processes with application to the invasive drug-delivery problem. Comput. Methods. Biomech. Biomed. Engin. **18**, 861–879 (2015)

Ehlers, W.: Effective stresses in multiphasic porous media: a thermodynamic investigation of a fully non-linear model with compressible and incompressible constituents. Geomech. Energy Environ. **15**, 35–46 (2018)

Freyer, J.P., Sutherland, R.M.: Regulation of growth saturation and development of necrosis in emt6/ro multicellular spheroids by the glucose and oxygen supply. Cancer Res. **46**, 3504–3512 (1986)

Gray, G.S.W., Ferrari, M., Decuzzi, P., Schrefler, B.: On computational modeling in tumor growth. Arch. Comput. Methods Eng. **20**, 327–352 (2013)

Guppy, M., Leedman, P., Zu, X., Russell, V.: Contribution by different fuels and metabolic pathways to the total ATP turnover of proliferating MCF-7 breast cancer cells. Biochem. J. **364**, 309–315 (2002)

Hanahan, D., Weinberg, R.A.: Hallmarks of cancer: the next generation. Cell **144**, 646–674 (2011)

King, D.M., Perera, B.: Morris method of sensitivity analysis applied to assess the importance of input variables on urban water supply yield - a case study. J. Hydrol. **477**, 17–32 (2013)

Morris, M.D.: Factorial sampling plans for preliminary computational experiments. Technometrics **33**, 161–174 (1991)

Myung, I.J.: Tutorial on maximum likelihood estimation. J. Math. Psychol. **47**, 90–100 (2003)

Schröder, P., Wagner, A., Stöhr, D., Rehm, M., Ehlers, W.: Variation of different growth descriptions in a metastatic proliferation model. In: von Scheven, M., Keip, M.-A., Karajan, N. (eds.): Proceedings of the 7th GACM Colloquium on Computational Mechanics. Stuttgart, pp. 259–262 (2017)

Schröder, P., Wagner, A., Stöhr, D., Rehm, M., Ehlers, W.: Continuum-mechanical modelling of apoptosis. PAMM. Proc. Appl. Math. Mech. **19**, e201900310 (2019)
Shelton, S.E.: Mechanistic Modeling of Cancer Tumor Growth Using a Porous Media Approach. Master's thesis, University of North Carolina at Chapel Hill (2011)
Wild, C.P., Stewart, B.W.: World Cancer Report 2014. World Health Organization, Geneva, Switzerland (2014)

Global Parameter Identification in Soft Tissues

Jonas Stålhand and Jan-Lucas Gade

The first author met Professor Gerhard A. Holzapfel at the International Centre for Mechanical Sciences in Udine just after the turn of the millennium. Professor Holzapfel organized a series of lectures on the biomechanics of cardiovascular soft tissues, which were a perfect introduction for a young Ph.D. student. Some years later, Professor Holzapfel moved to the Royal Institute of Technology in Stockholm to take up the Chair of Biomechanics. During his time in Sweden, he was the faculty opponent at my Ph.D. defense. Part of my thesis was about modeling of smooth muscles, and it became the embryo to our co-operation. Professor Holzapfel's articles, books, and presentations on biomechanics have been a great source of inspiration to me, and I always look forward to the next visit at his institute. I feel honored to be able to call him both a colleague and a personal friend.

Jonas

Abstract Identification of material parameters for state of the art soft tissue models is a complex process. The objective functions are generally nonlinear and nonconvex, and gradient-based solvers are not guaranteed to find the global solution. Therefore, deterministic global methods are needed to compute the parameters, which give the best fit between the model response and experiments. In this chapter, we discuss parameter identification for soft tissues and present a branch-and-bound algorithm that converges to the global solution.

J. Stålhand (✉) · J.-L. Gade
Solid Mechanics, Department of Management and Engineering,
Linköping University, SE-581 83 Linköping, Sweden
e-mail: jonas.stalhand@liu.se

G. Sommer et al. (eds.), *Solid (Bio)mechanics: Challenges of the Next Decade*,
Studies in Mechanobiology, Tissue Engineering and Biomaterials 24,
https://doi.org/10.1007/978-3-030-92339-6_16

1 Introduction

Arterial mechanics plays a fundamental role in the cardiovascular system's physiology under both normal and diseased conditions. For example, it is well known that large arteries close to the heart distend and store energy during the ejection phase and return this energy to the blood flow in the elastic recoil when the aortic valve closes. This energy transformation prevents a sudden pressure drop in the systemic arteries and allows for a near-constant flow in the capillaries (Humphrey 2002). The elasticity is closely connected to the constituents and their structure in the arterial wall. The main contributors to the mechanical properties are collagen and elastin in the extracellular network and smooth muscle cells (Burton 1954; Roach and Burton 1957). The structure and composition of the constituents can change both as part of a natural remodeling process or through a pathological process. Whatever causes the microscopic alterations will be expressed as changes in the elastic properties of the artery at the macroscopic level. The mechanical properties are therefore important factors for assessing the health of the cardiovascular system as well as for understanding the etiology and development of diseases.

The mechanical response of soft tissues is commonly tested using, for example, extension-inflation tests (Sommer et al. 2010) or uniaxial and biaxial extension tests (Sommer et al. 2015). These experiments can be used to identify material parameters of individual constituents (or a group of constituents). The parameters usually influence the macroscopic tissue response through nonlinear relations and are computed by fitting the model response to the measured tissue response. This minimization problem can be nonconvex, and the parameters identified using standard techniques with gradient-based solvers are not guaranteed to give the best fit. This problem is amplified by a large number of parameters and by using mechanically incomplete measurements, which is typically for in vivo measurements (Stålhand et al. 2004; Stålhand 2009; Gade et al. 2019).

In this chapter, we discuss some important features that impact parameter identification and, in particular, how to use deterministic global methods to identify the true best-fit parameters.

2 Continuum Model for Soft Tissue

Soft tissue is a complex material from an engineering perspective. The composition and structure of its constituents make the tissue anisotropic, heterogeneous, and incompressible. Furthermore, the response to external loads is characterized by large finite deformations, and it is influenced by residual stresses in its unloaded state (Burton 1954; Roach and Burton 1957; Humphrey 2002). As a consequence, standard techniques for linear construction materials under small deformations no longer apply, and more general methods are needed. Continuum mechanics, with its well-defined mathematical structure and flexibility, has become the natural choice for modeling soft tissues.

2.1 Deformation

Consider a piece of soft tissue in a state where it is free from external loads and residual stresses, and let this be the reference configuration $\mathscr{B}_{\mathrm{R}}$. The deformation out of this configuration to a deformed configuration $\mathscr{B}$ subjected to external loads is given by the deformation gradient F. The configuration obtained by removing all external loads is the unloaded configuration and is denoted by $\mathscr{B}_0$. In general, this configuration does not necessarily need to be equivalent to $\mathscr{B}_{\mathrm{R}}$ (Vaishnav and Vossoughi 1983; Fung 1984) since the tissue can still be subjected to residual stresses as long as the stress field satisfies global equilibrium (Ogden 2003). The residual stresses can be accounted for by using a multiplicative decomposition of the deformation gradient (Humphrey 2002), i.e.

$$\mathsf{F} = \mathsf{F}_{\mathrm{e}}\mathsf{F}_{\mathrm{r}}, \tag{1}$$

where F_{e} and F_{r} describe the elastic deformation from $\mathscr{B}_0$ to $\mathscr{B}$ and the residual deformation from $\mathscr{B}_{\mathrm{R}}$ to $\mathscr{B}_0$, respectively. Because the tissue is assumed to be incompressible, the deformation gradient must satisfy the constraint

$$\det \mathsf{F} = 1. \tag{2}$$

In this text, we will only consider the extension-inflation experiment of an arterial segment, but the results are valid for other tests as well. Let the unloaded arterial segment be mounted in a test rig and identify it with the unloaded configuration $\mathscr{B}_0$. The elastic deformation in an extension-inflation test is given by

$$\mathsf{F}_{\mathrm{e}} = \frac{\partial r}{\partial \rho}\mathbf{e}_r \otimes \mathbf{E}_\rho + \frac{r}{\rho}\mathbf{e}_\theta \otimes \mathbf{E}_\vartheta + \frac{\partial z}{\partial \zeta}\mathbf{e}_z \otimes \mathbf{E}_\zeta, \tag{3}$$

where $\{r, \theta, z\}$ and $\{\rho, \vartheta, \zeta\}$ denote the radial, circumferential, and axial coordinates in the deformed and unloaded configurations, respectively. The symbol $\otimes$ is the dyadic product, and the cylindrical base vectors in the deformed and unloaded configurations are $\{\mathbf{e}_r, \mathbf{e}_\theta, \mathbf{e}_z\}$ and $\{\mathbf{E}_\rho, \mathbf{E}_\vartheta, \mathbf{E}_\zeta\}$, respectively.

While the elastic deformation in Eq. (3) is straightforward to define, the form of the residual deformation F_{r} is less obvious. There are essentially two methods in the literature: (i) an experimentally motivated assumption regarding the map between $\mathscr{B}_{\mathrm{R}}$ and $\mathscr{B}_0$ (Choung and Fung 1986; Holzapfel et al. 2000); (ii) a more general map-independent approach, which assumes $\mathscr{B}_{\mathrm{R}}$ to be a virtual configuration that is either a non-Euclidean geometry or a set of infinitely small and incompatible pieces (Humphrey and Rajagopal 2003; Klarbring et al. 2007). The former method has the advantage that residual stresses can be quantified in experiments. For arteries, the classical technique is to place a radial cut through the arterial wall and measure the opening angle (central angle) once the segment has equilibrated (Choung and Fung 1986). By identifying the equilibrated cut-open segment with $\mathscr{B}_{\mathrm{R}}$, the opening angle Φ, together with the axial deformation Λ, describe the residual stress field. Using these quantities, the residual deformation F_r can be written as (Holzapfel et al. 2000)

$$\mathsf{F}_\mathrm{r} = \frac{\partial \rho}{\partial R}\mathbf{E}_\rho \otimes \mathbf{E}_R + \frac{k\rho}{R}\mathbf{E}_\vartheta \otimes \mathbf{E}_\Theta + \Lambda \mathbf{E}_\zeta \otimes \mathbf{E}_Z, \tag{4}$$

where $\{R, \Theta, Z\}$ denote the radial, circumferential, and axial coordinates in the reference configuration, and the corresponding cylindrical base vectors are $\{\mathbf{E}_R, \mathbf{E}_\Theta, \mathbf{E}_Z\}$. The parameter $k = 2\pi/(2\pi - \Phi)$ is introduced because the cut-open segment is closed to a cylinder by the map between $\mathscr{B}_\mathrm{R}$ and $\mathscr{B}_0$. Note that it is possible to generalize the idea behind the method and make cuts in several directions to reveal more complex residual stress distributions (Holzapfel et al. 2007).

Finally, back-substituting Eqs. (3) and (4) into Eq. (1) leads to

$$\mathsf{F} = \frac{\partial r}{\partial R}\mathbf{e}_r \otimes \mathbf{E}_R + \frac{kr}{R}\mathbf{e}_\theta \otimes \mathbf{E}_\Theta + \lambda \mathbf{E}_z \otimes \mathbf{E}_Z, \tag{5}$$

where $\lambda = (\partial z/\partial \zeta)\Lambda$ is the segment's axial stretch. Using Eqs. (2) and (5) allows the deformed radius r to be computed explicitly (Holzapfel et al. 2000), i.e.

$$r(R) = \sqrt{\frac{R^2 - R_\mathrm{i}^2}{k\lambda} + r_\mathrm{i}^2}, \tag{6}$$

where R_i and r_i are the inner radius in the reference and deformed configurations, respectively.

2.2 Constitutive Law

The Cauchy stress tensor for an incompressible and nonlinear material is given by (Holzapfel 2000)

$$\sigma = -p\mathsf{I} + 2\mathsf{F}\frac{\partial \Psi}{\partial \mathsf{C}}\mathsf{F}^\mathrm{T}, \tag{7}$$

where I is the identity tensor, $\mathsf{C} = \mathsf{F}^\mathrm{T}\mathsf{F}$ is the right Cauchy-Green tensor, the superscribed T indicates a transpose, and $\Psi = \Psi(\mathsf{C})$ is the strain energy. The undetermined reaction stress p is associated with the incompressibility in Eq. (2) and must be computed from the boundary conditions. Many strain-energy functions for various soft tissues have been suggested in the literature, see Blatz et al. (1969), Fung et al. (1979), Humphrey and Yin (1986), Takamizawa and Hayashi (1987), Holzapfel et al. (2000), Gasser et al. (2006), and Baek et al. (2007), among others. Herein, we use the Holzapfel-Gasser-Ogden material model proposed in Holzapfel et al. (2000) and take

$$\Psi = c_0 (I_1 - 3) + \frac{c_1}{2c_2} \left[\exp\left(c_2(I_4 - 1)^2\right) - 1 \right] + \frac{c_1}{2c_2} \left[\exp\left(c_2(I_6 - 1)^2\right) - 1 \right], \tag{8}$$

where c_0, c_1, and c_2 are non-negative material constants, the first invariant $I_1 = \text{tr}\,\mathbf{C}$, and the pseudo-invariants $I_4 = \mathbf{M} \cdot \mathbf{CM}$ and $I_6 = \mathbf{N} \cdot \mathbf{CN}$ for the unit vectors $\mathbf{M}$ and $\mathbf{N}$ describing the orientation of the two collagen fiber families in the reference configuration. Based on histological evidence (Holzapfel et al. 2000), these fiber families are usually assumed to be symmetrically arranged around the circumferential direction. Following the original paper, it is assumed that the second and third terms on the right side of Eq. (8) become zero whenever $I_4 < 1$ and $I_6 < 1$, respectively.

2.3 *Equilibrium*

In a quasi-static state, equilibrium implies

$$\text{div}\,\sigma = \mathbf{0}, \tag{9}$$

where div is the divergence operator defined on the deformed configuration (Holzapfel 2000). For a shear free deformation, only the radial direction is non-zero and reads

$$\frac{\partial \sigma_{rr}}{\partial r} = \frac{1}{r} \left(\sigma_{\theta\theta} - \sigma_{rr} \right). \tag{10}$$

By assuming a pressure P on the inner boundary and a traction free outer boundary, Eq. (10) can be integrated to give

$$P = \int_{r_i}^{r_o} \frac{1}{r} \left(\sigma_{\theta\theta} - \sigma_{rr} \right) \mathrm{d}r, \tag{11}$$

where r_o is the outer radius in the deformed configuration.

In addition to the pressure, equilibrium must also hold in the axial direction. The axial force required in equilibrium is given by

$$F_{\text{eq}} = 2\pi \int_{r_i}^{r_o} \sigma_{zz} r \,\mathrm{d}r. \tag{12}$$

This is not the axial force measured in the inflation test, where the lumen pressure also acts on the closed ends. This gives an axial force contribution, which reduces the measured force. It can be shown that the measured reduced axial force is given by (Holzapfel et al. 2000; Humphrey 2002)

$$F = \pi \int_{r_i}^{r_o} \left(2\sigma_{zz} - \sigma_{\theta\theta} - \sigma_{rr} \right) r \,\mathrm{d}r. \tag{13}$$

3 Parameter Identification

The soft tissue model described in Sect. 2 includes a number of unknown parameters related to the material or the geometry. Some of these parameters will not be possible to derive directly from measurements but must be computed through a parameter identification problem. The key components in the parameter identification are control and state variables together with the model. The vector of control variables $\bar{\mathbf{u}}$ excites the tissue and elicits a response in the form of a vector of state variables $\bar{\mathbf{y}}$. The choice of control and state variables depends on the particular experiment. For example, in a stretch-controlled uniaxial extension test of a tissue specimen, the control variable will be the applied stretch, while the state variable is the measured force. The model ϕ takes the control variable $\bar{\mathbf{u}}$ together with a vector of parameters $\boldsymbol{\kappa}$ and generates an estimate $\mathbf{y}$ of the system's state variable $\bar{\mathbf{y}}$. The parameters in $\boldsymbol{\kappa}$ are then computed by tuning the model response to measurements. The tuning process can be expressed as the inverse identification problem: for a given set of N control and state variables pairs $\{\bar{u}_k, \bar{y}_k\}$, find the parameter vector $\boldsymbol{\kappa}^*$ such that the model response $\phi(\bar{u}_k, \boldsymbol{\kappa}) = \bar{y}_k$, for all $k = 1, 2, \ldots, N$.

The number of control and state variable pairs N is usually much larger than the number of parameters making the problem over-determined. In addition, the inverse problem described above is ill-posed since the measurement errors and scattering make it impossible to find a unique set of parameters $\boldsymbol{\kappa}^*$ which satisfy all N pairs simultaneously. To obtain a well-posed problem, the inverse problem can be replaced by a relaxed problem where the deviation between the state variables $\mathbf{y}$ and $\bar{\mathbf{y}}$ is penalized. The solution is then the vector $\boldsymbol{\kappa}^*$, which minimizes the deviation, i.e. gives the best match between the model response and measurements.

3.1 Formulating a Parameter Identification Problem

A common relaxation of the inverse problem is to use the least-squares method. A scalar-valued objective function e is defined using the sum of squared residuals, i.e.

$$e = \sum_{k=1}^{N} \frac{1}{2} \left[\bar{y}_k - \phi_k(\boldsymbol{\kappa}, \bar{u}_k)\right]^2 . \tag{14}$$

The parameters are then obtained by solving the following minimization problem.

Parameter Identification Problem

Given the control and state variable vectors $\bar{\mathbf{u}}$ and $\bar{\mathbf{y}}$, find $\boldsymbol{\kappa}^*$ such that

$$
\begin{aligned}
&\min_{\boldsymbol{\kappa}} e(\boldsymbol{\kappa}, \bar{\mathbf{u}}, \bar{\mathbf{y}}) \\
&\text{s.t. } C_1(\boldsymbol{\kappa}, \bar{\mathbf{u}}, \bar{\mathbf{y}}) = 0, \\
&\qquad \vdots \\
&\quad C_m(\boldsymbol{\kappa}, \bar{\mathbf{u}}, \bar{\mathbf{y}}) = 0, \\
&\quad \boldsymbol{\kappa}^{\mathrm{LB}} \leq \boldsymbol{\kappa} \leq \boldsymbol{\kappa}^{\mathrm{UB}}.
\end{aligned}
\tag{15}
$$

The functions C_i, $i = 1, \ldots, m$, are equality constraints that must be satisfied. The parameter vector $\boldsymbol{\kappa}$ is also bounded by the lower and upper bounds $\boldsymbol{\kappa}^{\mathrm{LB}}$ and $\boldsymbol{\kappa}^{\mathrm{UB}}$, respectively. These bounds can be introduced on physical, mathematical, or modeling grounds. Firstly, certain parameters are naturally restricted in physics. For example, many constitutive functions require non-negative material parameters to ensure a dissipative deformation process (Holzapfel 2000). Secondly, properly chosen lower and upper bounds limit the parameter space and can reduce the minimization problem. In addition, some parameters may require bounds to have a unique solution. This applies to angles in trigonometric functions, for example, where the periodicity causes a non-uniqueness for unbounded angles. Thirdly, bounds can be used to exclude deformation modes, which are non-physical. For example, requiring the axial stretch $\lambda \geq 1$ in Eq. (5) avoids solutions, which can have buckling instabilities.

There are several objective functions e suggested in the literature. A natural choice is to fit the pressure-radius response in Eq. (11) to inflation measurements using

$$
e = \sum_{k=1}^{N} \frac{1}{2} \left[\bar{P}_k - P(\boldsymbol{\kappa}, \bar{r}_k) \right]^2. \tag{16}
$$

This approach was used for in vivo data in Stålhand et al. (2004), Stålhand and Klarbring (2005), and Masson et al. (2011). A major draw-back of the objective function in Eq. (16) is that no consideration is given to the axial response, which can cause non-physiological axial deformations (Stålhand and Klarbring 2005). For in vitro inflation tests, it is possible to tie down the axial deformation by appending the squared residual of the reduced axial force $(\bar{F}_k - F(\boldsymbol{\kappa}, \bar{r}_k))^2$ to the objective function (Sommer and Holzapfel 2012; Ferruzzi et al. 2013). Since the magnitude of the pressure is on the order of 10 kPa while the reduced axial force is about 1 N, a weighting of the two terms is required to ensure that the axial response has a notable influence on the parameters. Instead of using pressure and reduced axial force, Schulze-Bauer and Holzapfel (2003) suggested an objective function for in vivo data based on the stress residuals, i.e.

$$
e = \sum_{k=1}^{N} \left(w \left[\bar{\sigma}_{\theta\theta,k} - \sigma_{\theta\theta}(\boldsymbol{\kappa}, \bar{r}_k) \right]^2 + (1 - w) \left[\bar{\sigma}_{zz,k} - \sigma_{zz}(\boldsymbol{\kappa}, \bar{r}_k) \right]^2 \right), \tag{17}
$$

where the circumferential and axial stresses, respectively, $\bar{\sigma}_{\theta\theta}$ and $\bar{\sigma}_{zz}$ are computed by assuming a thin-walled structure using the Laplace law and the weight factor $w = 1/2$. The objective function was later used in Stålhand (2009) with the circumferential and axial stresses defined at the mid-wall as

$$\bar{\sigma}_{\theta\theta} = \frac{P}{2A}\left(4\pi r_{\mathrm{i}}^2 + A\right), \qquad \bar{\sigma}_{zz} = \frac{\pi r_{\mathrm{i}}^2 P + F}{A}, \tag{18}$$

where A is the cross-sectional area of the arterial wall. The axial force F is not easily measured in vivo and must be obtained by other means. Towards this end, two assumptions are usually introduced. First, the axial force is taken to be independent of the pressure, and, second, the ratio γ between the axial and circumferential stresses is known at a certain pressure level. The stress ratio $\gamma = 0.59$ at 13.3 kPa (mean arterial pressure) was taken from the literature (Schulze-Bauer and Holzapfel 2003).

3.2 Constraints

The parameter identification problem (15) fits the model to the measured state variables, either directly using the pressure in Eq. (16) or implicitly through the stresses in Eq. (17). In addition to the state variables, it is likely that additional knowledge about the tissue's response has been reported but is not considered in the experiments. An example of such behavior is the invariance of the axial stretch λ with respect to the pressure at the in situ axial force F_{is}, see, e.g., Schulze-Bauer and Holzapfel (2003) and Sommer et al. (2010). In principle, it is possible to account for this behavior in the parameter identification by taking one equality constraint in problem (15) equal to

$$C = \left.\frac{\partial \lambda}{\partial P}\right|_{F_{\mathrm{is}}} = 0. \tag{19}$$

On the one hand, an equality constraint restricts the solution to a hypersurface in the solution space, which can notably speed-up the identification. On the other hand, the constitutive model must allow for the behavior in Eq. (19) for a solution to exist (Ogden and Schulze-Bauer 2000). To avoid the problem with a non-existing solution while still letting the invariance influence the parameter identification, it was suggested in Stålhand and Klarbring (2005) to use a relaxed constraint. By inverting the functional dependence in the nominator such that the reduced axial force was invariant of the pressure at the solution, the constraint was relaxed by penalizing the squared difference between two consecutive values on F in the objective function (16), i.e.

$$e = \sum_{k=1}^{N} \frac{1}{2}\left[\bar{P}_k - P(\boldsymbol{\kappa}, \bar{r}_k)\right]^2 + \sum_{k=2}^{N} \frac{\mu}{2}\left[F(\boldsymbol{\kappa}, \bar{r}_k) - F(\boldsymbol{\kappa}, \bar{r}_{k-1})\right]^2, \tag{20}$$

where $\mu > 0$. The objective function was interpreted as a weighted sum in Stålhand and Klarbring (2005) and solved for a suitable μ. Alternatively, Eq. (20) can be viewed as a quadratic penalty function and solved using penalty methods where the constraint is enforced in the limit when μ tends to infinity (Nocedal and Wright 2006).

3.3 Gradient-Based Solvers

The standard approach to solve the minimization problem (15) is to use a gradient-based solver. Starting from a feasible point, these solvers iteratively step towards the minimum by using the gradient of the objective function to find search directions. The solution is obtained when there are no admissible search directions, which gives a reduction in the objective function value (Nocedal and Wright 2006).

Most gradient-based solvers for non-constrained minimization generate a search direction vector $\mathbf{p}$ in each step by the modified Newton method

$$(\mathsf{H} + \tau \mathsf{I})\,\mathbf{p} = -\nabla e, \tag{21}$$

where $\mathsf{H} = \nabla^2 e$ is the Hessian of the objective function, I is the identity matrix, and ∇ denotes the gradient with respect to the parameters in $\boldsymbol{\kappa}$. The multiplier $\tau \geq 0$ is chosen large enough to make the matrix $\mathsf{H} + \tau \mathsf{I}$ positive definite.

Newton methods have a quadratic rate of convergence but require the Hessian. For all but the simplest objective functions, an explicit calculation of the Hessian matrix components $H_{ij} = \partial^2 e / \partial \kappa_i \partial \kappa_j$ is a highly elaborate, time-consuming, and error-prone process. Commonly used alternatives are to compute H numerically using finite differences or to approximate it based on previously computed information, e.g., the Broyden-Fletcher-Goldfarb-Shanno (BFGS) method and the Marquardt approximation (Nocedal and Wright 2006).

For the constrained minimization problem, the situation becomes more complex. First, the method needs to respect the constraints when computing the search direction. Second, the gradient of the objective function ∇e does not need to be zero at the solution if a constraint is active. The solution is to formulate the Lagrangian function

$$\mathscr{L}(\boldsymbol{\kappa}, \mu_1, \ldots, \mu_m) = e(\boldsymbol{\kappa}) - \sum_{i=1}^{m} \mu_i C_i(\boldsymbol{\kappa}), \tag{22}$$

where μ_i are the Lagrangian multipliers for the m constraints in (15). Note that the dependence on other variables in e and C_i has been suppressed for clarity. The modified Newton method in Eq. (21) is then replaced by the Karush-Kuhn-Tucker (KKT) system (Nocedal and Wright 2006)

$$\begin{bmatrix} \mathsf{W} & -\mathsf{A}^{\mathrm{T}} \\ \mathsf{A} & 0 \end{bmatrix} \begin{pmatrix} \mathbf{p} \\ \mathbf{q} \end{pmatrix} = \begin{pmatrix} -\nabla e + \mathsf{A}^{\mathrm{T}} \boldsymbol{\mu} \\ -\mathbf{c} \end{pmatrix}, \tag{23}$$

where $\mathsf{W} = \nabla^2 e - \sum_i \mu_i \nabla^2 C_i$ is the Hessian of Eq. (22), $\mathsf{A}^{\mathrm{T}} = [\nabla C_1, \ldots, \nabla C_m]$ is a matrix with constraint gradients as columns, $\boldsymbol{\mu} = (\mu_1, \ldots, \mu_m)^{\mathrm{T}}$ is the vector of Lagrangian multipliers, and $\mathbf{c} = (C_1, \ldots, C_m)^{\mathrm{T}}$ is the vector of constraints. The computed 'search direction' in Eq. (23) now includes two vectors: $\mathbf{p}$ associated with reductions in the objective function and the increment $\mathbf{q}$ in the Lagrangian parameters needed to remain on the hypersurface defined by the constraints. The coupled equation system in (23) can be solved directly using symmetric indefinite factorization, but it is usually more efficient to use a null-space method and partition the search direction according to $\mathbf{p} = \mathsf{R}\mathbf{p}_R + \mathsf{N}\mathbf{p}_N$, where R and N are the range and null space of A^{T}, respectively. Substituting the partitioned vector $\mathbf{p}$ into Eq. (23) gives after some straightforward manipulations

$$\mathsf{N}^{\mathrm{T}}\mathsf{W}\mathsf{N}\mathbf{p}_N = -\mathsf{N}^{\mathrm{T}}\mathsf{W}\mathsf{R}\mathbf{p}_R - \mathsf{N}^{\mathrm{T}}\nabla e, \tag{24}$$

$$\mathsf{A}\mathsf{R}\mathbf{p}_R = -\mathbf{c}. \tag{25}$$

The system is decoupled, and Eqs. (24) and (25) are used to compute $\mathbf{p}_R$ and $\mathbf{p}_N$. This allows the full vector $\mathbf{p}$ to be constructed and substituted into the first row in Eq. (23). After multiplication by R, the vector $\mathbf{q}$ is solved from

$$(\mathsf{A}\mathsf{R})^{\mathrm{T}}\mathbf{q} = \mathsf{R}^{\mathrm{T}}(\nabla e + \mathsf{W}\mathbf{p}). \tag{26}$$

Equation (26) is usually simplified by taking $\mathsf{R} = \mathsf{A}^{\mathrm{T}}$. This requires the constraint gradients to be linearly independent, but it can be ensured when A^{T} is constructed. Another simplification is to drop the term $\mathsf{W}\mathbf{p}$ since $\mathbf{p}$ tends to zero near the solution, making ∇e the dominant term. This approximation completely decouples $\mathbf{q}$ and $\mathbf{p}$. Finally, the reduced Hessian $\mathsf{N}^{\mathrm{T}}\mathsf{W}\mathsf{N}$ is strictly positive definite in the neighborhood of a local minimum and can, therefore, be approximated using the BFGS method giving a quasi-Newton-like method.

The null-space method described only accounts for equality constraints. It can be modified to handle inequality constraints, however. A straightforward adaptation is by defining a working set, which includes all equality constraints together with all active inequality constraints (satisfied with equality) at the current point. The null-space method can then be used on the working set. As we move along the sequence of search directions, the working set is continuously updated by including or removing inequality constraints based on their Lagrangian multiplier. This method is known as the active-set method (Nocedal and Wright 2006).

3.4 Global Minimum and Heuristic Methods

A gradient-based solver is attracted to a minimum along the sequence of search directions. In this sense, they are guaranteed to find a solution to the minimization problem (15). For nonconvex problems, there may exist multiple points in the parameter space, which fulfill the requirements of a solution. These points are collectively referred to as local minima. An implication of several minima is that a second initial guess, sufficiently far from the first, is likely to give a different solution in another region of the parameter space (Stålhand 2009; Badel et al. 2012; Gade et al. 2019). This obviously raises the question, which solution to use.

Mathematically it makes sense to define the solution as the vector $\boldsymbol{\kappa}^*$ that gives the lowest objective function value. This point is called the global minimum and gives the best fit between the model and experimental data. In principle, it is, therefore, possible to identify all local minima and compare their objective function values to obtain the global solution. This is not feasible in practice, however, since it requires knowledge of the approximative position for the local minima such that the initial guesses can be taken close enough to be within the basin of attraction for each minimum.

Several heuristics can be used to generate candidates for the global minimum, e.g., a pre-minimization with a derivative-free method such as pattern search or a genetic algorithm to find points in close proximity to local minima. These points are then provided to a gradient-based solver as hot-starts. Another popular method, which is naturally suited for parallel computing, is to seed a large number of initial guesses throughout the parameter space following some rules and run a gradient-based algorithm from each point (Smoljkić et al. 2015; Wittek et al. 2016; Gade et al. 2019). Despite credible arguments for convergence to a minimum, there are no guarantees that heuristic methods can identify the global minimum in a deterministic way (Horst and Tuy 1996). Nonetheless, it has been shown that seeding a large enough number of initial guesses works well in practice (Gade et al. 2019).

3.5 Global Solvers

There are deterministic methods, which can guarantee convergence to a global solution in a finite number of iterations. A common feature of all these methods is that they require knowledge about the problem structure, which makes the algorithms highly problem-dependent. In addition, the number of iterations and the memory required to identify the solution are usually very large, and global methods have previously been avoided. The fast growth in computer performance combined with algorithm development and parallel computing are starting to make global methods interesting for engineering applications.

A relatively general and popular approach is the branch-and-bound method (Adjiman et al. 1998a, b; Esposito and Floudas 1998). In this method, the global minimum is computed in a finite number of iterations by using successive partitioning of the parameter space Ω and exclusion of regions, which cannot contain the solution. The problem (15) is replaced by two sub-problems: (i) an upper problem providing an upper bound on the global minimum; (ii) a lower problem providing a lower bound on the same minimum. The upper problem can be taken to be the original nonconvex minimization problem, and the lower problem is a convex relaxation of the original problem. The relaxation must be formulated such that it is a convex underestimator of the objective function on part of the parameter space. In principle, any convex function that underestimates e can be used, provided it generates increasingly tighter underestimators for successive partitions. The efficiency of the branch-and-bound method is highly dependent on the tightness, however, and the choice of the underestimator is crucial for the performance.

The branch-and-bound method keeps track of subsets $\Omega_n \subseteq \Omega$, which are candidates for containing the global solution, and stores these regions in a list of active regions $\mathscr{A}$. In each step, the method removes one region Ω_n from $\mathscr{A}$ and partitions it into the new regions ω_1 and ω_2 ($\Omega_n = \omega_1 \cup \omega_2$) according to some rule. This is referred to as the branching step. For each region ω_m ($m = 1, 2$), two steps are performed: (i) The upper problem is solved on ω_m giving the minimum $\boldsymbol{\kappa}^*_{\text{UP}}$. This can be done using a gradient-based solver as described in Sect. 3.3, for example. The objective function value $e(\boldsymbol{\kappa}^*_{\text{UP}})$ is compared to the global variable UB containing the lowest upper bound on all previously examined regions, and the variable UB is updated if $e(\boldsymbol{\kappa}^*_{\text{UP}}) < UB$. (ii) The lower problem is solved on ω_m using a gradient-based solver, which gives the solution $\boldsymbol{\kappa}^*_{\text{LP}}$ together with the lower bound $LB_m = e(\boldsymbol{\kappa}^*_{\text{LP}})$ on the minimum, or minima, in the region ω_m. Because the lower problem is convex, the gradient-based solver will always identify the best-fit solution to the lower problem on ω_m. If $LB_m > UB$, the current region ω_m cannot contain the global solution and it is excluded from further analysis. Otherwise, ω_m and its lower bound LB_m are added to the list of active regions $\mathscr{A}$. This is referred to as the bounding step.

Finally, the method checks if the global solution has been reached by computing the relative difference $|UB - LB|/UB$, where LB is the smallest of all lower bounds in the list of active regions $\mathscr{A}$. If the relative difference is less than a user-defined tolerance ε, the method has converged to the global solution $\boldsymbol{\kappa}^*$. Otherwise, the branch-and-bound steps above are repeated.

4 Branch-and-Bound Identification for an Inflation Experiment

In this section, we describe how the branch-and-bound method can be used to identify a set of material parameters in a simple arterial inflation experiment. Pressure-strain data is taken from Badel et al. (2012) and converted to stress-stretch data using values in the paper together with Eqs. $(18)_{1,2}$, where the reduced axial force $F = 0$. The N axial and circumferential stretch samples, $\bar{\lambda}_{\theta,k}$ and $\bar{\lambda}_{z,k}$, are taken as the control variables, and the axial and circumferential stress samples, $\bar{\sigma}_{\theta\theta,k}$ and $\bar{\sigma}_{zz,k}$, as the state variables. The material parameters are collected in the vector $\boldsymbol{\kappa} = (c_0,\ c_1,\ c_2)^{\mathrm{T}}$, and the identification is performed using the objective function, as defined in Eq. (17).

The model stresses are computed using Eqs. (5), (7), and (8) by assuming transmurally homogeneous conditions and neglecting residual stresses. The latter allows us to identify $\mathscr{B}_0$ with $\mathscr{B}_{\mathrm{R}}$. The fiber directions are given by $\mathbf{M} = \cos\beta\mathbf{E}_\Theta + \sin\beta\mathbf{E}_Z$ and $\mathbf{N} = \cos\beta\mathbf{E}_\Theta - \sin\beta\mathbf{E}_Z$, where β is the fiber angle with respect to the circumferential direction. The two non-zero stress components are

$$\sigma_{\theta\theta} = 2\left(\lambda_\theta^2 - \frac{1}{\lambda_\theta^2\lambda_z^2}\right)c_0 + 4\lambda_\theta^2(I-1)(\cos\beta)^2 c_1 \exp\left[c_2(I-1)^2\right], \tag{27}$$

$$\sigma_{zz} = 2\left(\lambda_z^2 - \frac{1}{\lambda_\theta^2\lambda_z^2}\right)c_0 + 4\lambda_z^2(I-1)(\sin\beta)^2 c_1 \exp\left[c_2(I-1)^2\right], \tag{28}$$

where λ_θ and λ_z are the circumferential and axial stretch, respectively, and $I = I_4 = I_6 = \lambda_\theta^2(\cos\beta)^2 + \lambda_z^2(\sin\beta)^2$.

The major difficulty in the branch-and-bound method is the construction of the convex underestimator. These are generally constructed by a symbolic reformulation of the original problem where nonconvex terms are transformed to a simpler type using bilinear, fractional, and componentwise convex terms (Adjiman et al. 1998b). Suitable underestimators for these simpler terms can then be found in the literature, see, e.g., Adjiman et al. (1998b), Esposito and Floudas (1998), Keller (2015), and Najman et al. (2019). Towards this end, we first remove the nonlinearity from the objective function and reformulate the minimization to the upper problem.

Upper Problem for the Branch-and-Bound Solver

$$\begin{aligned}
\min_{\boldsymbol{\kappa},\boldsymbol{\Sigma}_{\theta\theta},\boldsymbol{\Sigma}_{zz}} \quad & \sum_{k=1}^{N}\left(w\left[\bar{\sigma}_{\theta\theta,k} - \Sigma_{\theta\theta,k}\right]^2 + (1-w)\left[\bar{\sigma}_{zz,k} - \Sigma_{zz,k}\right]^2\right) \\
\text{s.t.} \quad & \sigma_{\theta\theta}(\boldsymbol{\kappa},\bar{\lambda}_{\theta,1},\bar{\lambda}_{z,1}) - \Sigma_{\theta\theta,1} = 0, \\
& \vdots \\
& \sigma_{\theta\theta}(\boldsymbol{\kappa},\bar{\lambda}_{\theta,N},\bar{\lambda}_{z,N}) - \Sigma_{\theta\theta,N} = 0, \\
& \sigma_{zz}(\boldsymbol{\kappa},\bar{\lambda}_{\theta,1},\bar{\lambda}_{z,1}) - \Sigma_{zz,1} = 0, \\
& \vdots \\
& \sigma_{zz}(\boldsymbol{\kappa},\bar{\lambda}_{\theta,N},\bar{\lambda}_{z,N}) - \Sigma_{zz,N} = 0, \\
& \boldsymbol{\kappa}_n^{\mathrm{LB}} \le \boldsymbol{\kappa} \le \boldsymbol{\kappa}_n^{\mathrm{UB}}, \\
& -\infty < \Sigma_{\theta\theta,k},\ \Sigma_{zz,k} < \infty.
\end{aligned} \tag{29}$$

In problem (29), $\boldsymbol{\Sigma}_{\theta\theta} = (\Sigma_{\theta\theta,1}, \ldots, \Sigma_{\theta\theta,N})^{\mathrm{T}}$, $\boldsymbol{\Sigma}_{zz} = (\Sigma_{zz,1}, \ldots, \Sigma_{zz,N})^{\mathrm{T}}$, and $\boldsymbol{\kappa}_n^{\mathrm{LB}}$ and $\boldsymbol{\kappa}_n^{\mathrm{UB}}$ are the lower and upper bound, respectively, on the parameters in the region $\Omega_n \subseteq \Omega$. The objective function is now convex in the variables $\Sigma_{\theta\theta,k}$ and $\Sigma_{zz,k}$, but it comes at the cost of introducing $2N$ nonlinear equality constraints. To simplify the notation when the convex underestimators are constructed, we introduce the factors

$$f_{0,k} = (I_k - 1)^2 \tag{30}$$

and

$$\begin{cases} f_{1,k} = 2\left(\lambda_{\theta,k}^2 - \dfrac{1}{\lambda_{\theta,k}^2 \lambda_{z,k}^2}\right), & f_{2,k} = 4\lambda_{\theta,k}^2 (\cos\beta)^2 (I_k - 1), \\ f_{3,k} = 2\left(\lambda_{z,k}^2 - \dfrac{1}{\lambda_{\theta,k}^2 \lambda_{z,k}^2}\right), & f_{4,k} = 4\lambda_{z,k}^2 (\sin\beta)^2 (I_k - 1). \end{cases} \tag{31}$$

This allows us to rewrite the model stresses in Eqs. (27) and (28) according to

$$\sigma_{\theta\theta} = c_0 f_{1,k} + c_1 f_{2,k} \exp(c_2 f_{0,k}), \tag{32}$$

$$\sigma_{zz} = c_0 f_{3,k} + c_1 f_{4,k} \exp(c_2 f_{0,k}). \tag{33}$$

Let us consider the stress in Eq. (32). The first term is linear in c_0 and does not require any relaxation since it is already convex. Note that the factors $f_{0,k}, \ldots, f_{4,k}$ do not impact the convexity of the material parameters. The second term is of the type '$x \exp(y)$' and is componentwise convex (the term is convex in x when y is kept fixed, and vice versa). An efficient relaxation of these componentwise convex terms is obtained by introducing the auxiliary variable $W_k = c_1 \exp(c_2 f_{0,k})$ together with the linear inequality constraints (Najman et al. 2019)

$$(c_1 - c_1^{\mathrm{LB}}) \exp(c_2^{\mathrm{LB}} f_{0,k}) + c_1^{\mathrm{LB}} \exp(c_2 f_{0,k}) - W_k \leq 0, \tag{34}$$

$$(c_1 - c_1^{\mathrm{UB}}) \exp(c_2^{\mathrm{UB}} f_{0,k}) + c_1^{\mathrm{UB}} \exp(c_2 f_{0,k}) - W_k \leq 0, \tag{35}$$

where the superscripts LB and UB indicate lower and upper bounds on the variable, respectively. The same auxiliary variables W_k appear in both model stresses, and we do not need to relax any other terms. Collecting the auxiliary variables in the vector $\mathbf{W} = (W_1, \ldots, W_N)^{\mathrm{T}}$, it is now straightforward to formulate the lower problem using Eqs. (32) to (35).

Lower Problem for the Branch-and-Bound Solver

$$\begin{aligned}
\min_{\boldsymbol{\kappa},\mathbf{W},\boldsymbol{\Sigma}_{\theta\theta},\boldsymbol{\Sigma}_{zz}} \quad & \sum_{k=1}^{N}\left(w\left[\bar{\sigma}_{\theta\theta,k}-\Sigma_{\theta\theta,k}\right]^2+(1-w)\left[\bar{\sigma}_{zz,k}-\Sigma_{zz,k}\right]^2\right)\\
\text{s.t. } & c_0 f_{1,1}+f_{2,1}W_1-\Sigma_{\theta\theta,1}=0,\\
& \vdots\\
& c_0 f_{1,N}+f_{2,N}W_N-\Sigma_{\theta\theta,N}=0,\\
& c_0 f_{3,1}+f_{4,1}W_1-\Sigma_{zz,1}=0,\\
& \vdots\\
& c_0 f_{3,N}+f_{4,N}W_N-\Sigma_{zz,N}=0,\\
& (c_1-c_1^{\mathrm{LB}})\exp(c_2^{\mathrm{LB}}f_{0,1})+c_1^{\mathrm{LB}}\exp(c_2 f_{0,1})-W_1\le 0,\\
& \vdots\\
& (c_1-c_1^{\mathrm{LB}})\exp(c_2^{\mathrm{LB}}f_{0,N})+c_1^{\mathrm{LB}}\exp(c_2 f_{0,N})-W_N\le 0,\\
& (c_1-c_1^{\mathrm{UB}})\exp(c_2^{\mathrm{UB}}f_{0,1})+c_1^{\mathrm{UB}}\exp(c_2 f_{0,1})-W_1\le 0,\\
& \vdots\\
& (c_1-c_1^{\mathrm{UB}})\exp(c_2^{\mathrm{UB}}f_{0,N})+c_1^{\mathrm{UB}}\exp(c_2 f_{0,N})-W_N\le 0,\\
& \boldsymbol{\kappa}_n^{\mathrm{LB}}\le\boldsymbol{\kappa}\le\boldsymbol{\kappa}_n^{\mathrm{UB}},\\
& \mathbf{W}_n^{\mathrm{LB}}\le\mathbf{W}\le\mathbf{W}_n^{\mathrm{UB}},\\
& -\infty<\Sigma_{\theta\theta,k},\,\Sigma_{zz,k}<\infty.
\end{aligned} \tag{36}$$

Therein, $\mathbf{W}_n^{\mathrm{LB}}$ and $\mathbf{W}_n^{\mathrm{UB}}$ are vectors of the lower and upper bounds on the auxiliary variables in the region $\Omega_n \subseteq \Omega$. The lower and upper bounds are obtained by substituting, respectively, $\{c_1^{\mathrm{LB}}, c_2^{\mathrm{LB}}\}$ and $\{c_1^{\mathrm{UB}}, c_2^{\mathrm{UB}}\}$ in W_k.

The choice of variables or parameters to branch is very important for the performance of the algorithm, and several strategies are available (Adjiman et al. 1998a; Esposito and Floudas 1998). Studying the mathematical formulation of the lower problem often gives valuable insights to which branching strategy to use. In general, parameters, which give an improvement of the underestimators should be branched. The parameter c_0 participates only linearly in the lower problem, and branching on it does not affect the underestimator. In contrast, the parameters c_1 and c_2 participate through a componentwise convex term and will impact the underestimator. Consequently, only c_1 and c_2 are considered for branching in this problem. To select which of c_1 and c_2 to branch on, we use a straightforward approach and use a least-reduced-axis rule (Adjiman et al. 1998a). First, compute the region-reduction measures

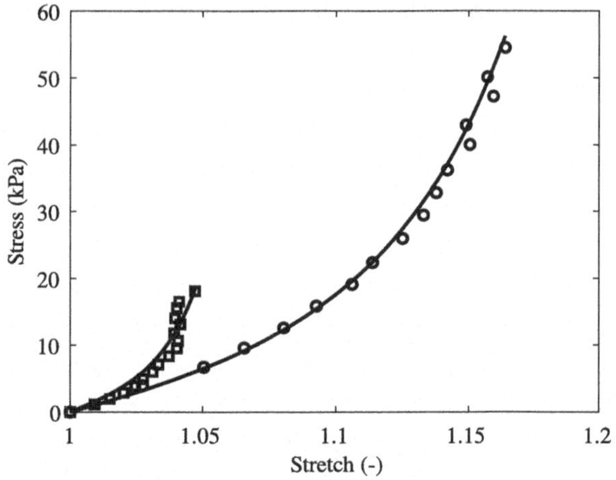

Fig. 1 Stress-stretch response in the axial (squares) and circumferential (circles) directions using data from Badel et al. (2012). The solid curves represent the model response for the parameters $\boldsymbol{\kappa}^*$ in Table 1

$$\Delta_j = \frac{(\boldsymbol{\kappa}_n^{\mathrm{UB}})_j - (\boldsymbol{\kappa}_n^{\mathrm{LB}})_j}{(\boldsymbol{\kappa}^{\mathrm{UB}})_j - (\boldsymbol{\kappa}^{\mathrm{LB}})_j}, \tag{37}$$

where the index $j = 2, 3$ indicates the vector component, and, second, choose the largest region-reduction value and bisect the region on this parameter.

The parameter identification was solved in Matlab (The Mathworks, Natick, Massachusetts, US) by implementing the branch-and-bound method with the upper problem given by (29) and the lower problem given by (36). The weight in the objective function (17) was taken to be $w = 1/2$, and the collagen fiber angle was set to $\beta = 32°$. In each iteration, both problems were solved using an active-set method, see Sect. 3.3. To speed up the minimization, analytical expressions for the gradient and Hessian were computed using Maple (Maplesoft, Waterloo, Ontario, Canada) and provided to the active-set method. Branching was performed on the parameters c_1 and c_2 using the least-reduced-axis rule and the region-reduction measures in Eq. (37). The upper and lower bounds on the parameter vector were taken from Badel et al. (2012), i.e. $0.3\,\mathrm{kPa} \leq c_0 \leq 50\,\mathrm{kPa}$, $0.5\,\mathrm{kPa} \leq c_1 \leq 100\,\mathrm{kPa}$, and $0.1 \leq c_2 \leq 100$. The tolerance on the relative difference was set to $\varepsilon = 0.01$. The algorithm is given in the Appendix.

The Matlab code was run on a standard PC, and the results are presented in Table 1 and Fig. 1. The parameter identification was also performed using a heuristic method for comparison. The heuristic method seeded ten initial guesses for $\boldsymbol{\kappa}$ in the parameter space and used these as starting points for the same gradient-based solver.

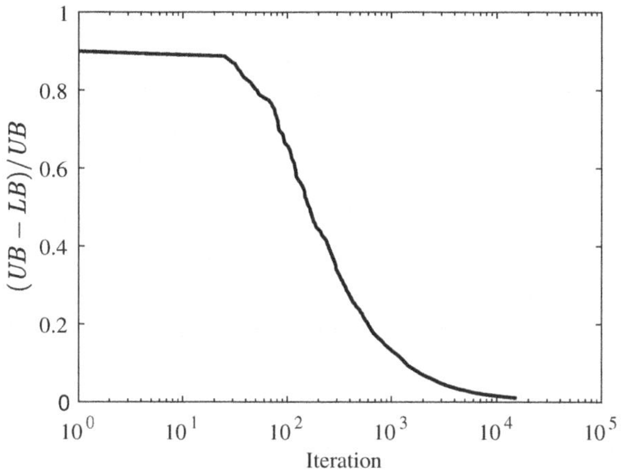

Fig. 2 Relative difference $(UB - LB)/UB$ as a function of the branch-and-bound iteration

Table 1 Results from the parameter identification using the branch-and-bound method and a heuristic method

Method	c_0 (kPa)	c_1 (kPa)	c_2 (-)	Time (s)	Initial guesses
Branch-and-bound	0.39	21.26	10.51	991	–
Heuristic method	0.39	21.26	10.51	0.4	10

5 Discussion

Global methods are usually very expensive compared to gradient-based solvers. This is evident from the results in Fig. 2 and Table 1. The heuristic method used a gradient-based solver starting from ten initial guesses. It took 0.41 s to find the global minimum and, on average, 31 iterations per initial guess. In contrast, the branch-and-bound algorithm took more than 2400 times longer to identify the same solution and required more than 15000 iterations. In addition, global solvers are also demanding from a memory point-of-view. The list of active regions for the identification problem in Sect. 4 included almost 13000 active regions when the relative difference reached the threshold $\varepsilon = 0.01$. For larger parameter identification problems, the number of active regions may grow beyond 10^6 and becomes a serious memory issue. Fortunately, the branch-and-bound method is relatively easy to parallelize, allowing for the work to be distributed on several cores.

The branch-and-bound method introduced in Sect. 4 makes use of a gradient-based solver to compute the lower and upper problems in each step. It, therefore, makes sense to adapt the identification problem to maximize its efficiency. One important property is the scaling of the objective function e. If the changes in one of the parameters produce a much larger variation in e than the others, the problem is poorly scaled, and the convergence can be slow. This is often the case for soft tissue-like materials, which includes both a linear and an exponential term, e.g., the HGO-model in Eq. (8), see Aggarwal (2017). One remedy can be to scale the parameters such that their order of magnitude becomes similar (Gade et al. 2019). Another important factor is the computation of the Hessian for the KKT system in Eq. (23). If possible, it is recommended to provide analytical expressions for both the gradient and the Hessian rather than to approximate them using, e.g., finite differences or the BFGS update. Computer algebra systems like Maple or SageMath (www.sagemath.org) can be useful towards this end.

The identification of the three material parameters was a relatively limited problem. For example, the only nonlinear term which requires a convex underestimator is $c_1 \exp(c_2(I-1)^2)$. Furthermore, the parameters c_1 and c_2 enter the componentwise convex term explicitly, and only the auxiliary variables W_k are needed. In more advanced problems, parameters may enter nonlinear terms in a more complex way. This is the case if also the fiber angle β is sought in the branch-and-bound problem. By taking $c_3 = (\cos\beta)^2$, the circumferential stress in Eq. (32) can be written as

$$\sigma_{\theta\theta} = c_0 f_{1,k} + 4\lambda_{\theta,k}^2 c_1 c_3 \bar{I}_k \exp(c_2 \bar{I}_k^2), \tag{38}$$

where $\bar{I}_k = \lambda_{\theta,k}^2 c_3 + \lambda_{z,k}^2 (1-c_3) - 1$ become new auxiliary variables enforced as linear equality constraints. The terms $c_1 c_3 \bar{I}_k$ and $c_2 \bar{I}_k^2$ are trilinear and componentwise convex terms, respectively, for which additional auxiliary variables must be introduced and relaxed. This results in more linear inequality constraints and a much larger minimization problem. Also, the axial stress in Eq. (33) must be reformulated analogously.

6 Conclusion

Deterministic global methods that are guaranteed to converge to the global solution are computationally costly. However, the development of computer hardware and algorithms have now reached a point where these methods can be applied to parameter identification problems in biomechanics.

Acknowledgements The authors wish to thank Dr. Carl-Johan Thore for his support on global minimization. This work was financially supported by the Swedish Research Council, grant no. 621-2014-4165.

Appendix

The pseudo code for the branch-and-bound solver used in Sect. 4 is presented below.

Input : sets of control variables $\{\bar{\lambda}_{\theta,k}, \bar{\lambda}_{z,k}\}$ and state variables $\{\bar{\sigma}_{\theta\theta,k}, \bar{\sigma}_{zz,k}\}$; parameter space bounds $\Omega = [\boldsymbol{\kappa}^{\mathrm{LB}}, \boldsymbol{\kappa}^{\mathrm{UB}}]$; ε; *iterMax*
Output: global solution for $\boldsymbol{\kappa}^*$
Initialize optimization
  `solve` upper problem (29) on region Ω using a random starting point
    $\boldsymbol{\kappa}^* = \operatorname{argmin} e_{\mathrm{UP}}(\boldsymbol{\kappa})$; $UB = e_{\mathrm{UP}}(\boldsymbol{\kappa}^*)$
  `determine` $\mathbf{W}^{\mathrm{LB}}$ and $\mathbf{W}^{\mathrm{UB}}$ in region Ω
  `solve` lower problem (36) in region Ω using a random starting point
    $\boldsymbol{\kappa}^*_{\mathrm{LP}} = \operatorname{argmin} e_{\mathrm{LP}}(\boldsymbol{\kappa})$; $LB = e_{\mathrm{LP}}(\boldsymbol{\kappa}^*_{\mathrm{LP}})$
  `add` LB, $\boldsymbol{\kappa}^{\mathrm{LB}}$, $\boldsymbol{\kappa}^{\mathrm{UB}}$ to $\mathscr{A}$
  `set` counter $iter^{\mathrm{BB}} = 0$
while $|UB - LB|/UB > \varepsilon$ and $iter^{\mathrm{BB}} < iterMax$ **do**
  $iter^{\mathrm{BB}} = iter^{\mathrm{BB}} + 1$
  `get and remove` region Ω_n with lowest LB_n from $\mathscr{A}$
  `partition` region Ω_n into ω_1 and ω_2 based on Eq. (37)
  for $m = 1$ **to** 2 **do**
    `solve` upper problem (29) on $\omega_m = [\boldsymbol{\kappa}_m^{\mathrm{LB}}, \boldsymbol{\kappa}_m^{\mathrm{UB}}]$ using a random starting point
      $\boldsymbol{\kappa}^*_{\mathrm{UP}} = \operatorname{argmin} e_{\mathrm{UP}}(\boldsymbol{\kappa})$
    if $e_{\mathrm{UP}}(\boldsymbol{\kappa}^*_{\mathrm{UP}}) < UB$ **then**
      $UB = e_{\mathrm{UP}}(\boldsymbol{\kappa}^*_{\mathrm{UP}})$; $\boldsymbol{\kappa}^* = \boldsymbol{\kappa}^*_{\mathrm{UP}}$
    end
    `determine` $\mathbf{W}^{\mathrm{LB}}$ and $\mathbf{W}^{\mathrm{UB}}$ in region $\omega_m = [\boldsymbol{\kappa}_m^{\mathrm{LB}}, \boldsymbol{\kappa}_m^{\mathrm{UB}}]$
    `solve` lower problem (36) on $\omega_m = [\boldsymbol{\kappa}_m^{\mathrm{LB}}, \boldsymbol{\kappa}_m^{\mathrm{UB}}]$ using a random starting point
      $\boldsymbol{\kappa}^*_{\mathrm{LP}} = \operatorname{argmin} e_{\mathrm{LP}}(\boldsymbol{\kappa})$; $LB_n = e_{\mathrm{LP}}(\boldsymbol{\kappa}^*_{\mathrm{LP}})$
    if $e_{\mathrm{LP}}(\boldsymbol{\kappa}^*_{\mathrm{LP}}) < UB$ **then**
      `add` LB_n, ω_m to $\mathscr{A}$
    end
  end
end

References

Adjiman, C.S., Androulakis, I.P., Floudas, C.A.: A global optimization method, α BB, for general twice-differentiable constrained NLPs – II. Implementation and computational results. Comput. Chem. Eng. **22**, 1159–1179 (1998a)

Adjiman, C.S., Dallwig, S., Floudas, C.A., Neumaier, A.: A global optimization method, αBB, for general twice-differentiable constrained NLPs - I. Theoretical advances. Comput. Chem. Eng. **22**, 1137–1158 (1998b)

Aggarwal, A.: An improved parameter estimation and comparison for soft tissue constitutive models containing an exponential function. Biomech. Model. Mechanobiol. **16**, 1309–1327 (2017)

Badel, P., Avril, S., Lessner, S., Sutton, M.: Mechanical identification of layer-specific properties of mouse carotid arteries using 3D-DIC and a hyperelastic anisotropic constitutive model. Comput. Methods Biomech. Biomed. Engin. **15**, 37–48 (2012)

Baek, S., Gleason, R.L., Rajagopal, K.R., Humphrey, J.D.: Theory of small on large: potential utility in computations of fluid-solid interactions in arteries. Comput. Methods Appl. Mech. Eng. **196**, 3070–3078 (2007)

Blatz, P.J., Mae Chu, B., Wayland, H.: On the mechanical behavior of elastic animal tissue. Trans. Soc. Rheo. **13**, 83–102 (1969)

Burton, A.C.: Relation of structure to function of the tissues of the wall of blood vessels. Physiol. Rev. **34**, 619–642 (1954)

Choung, C.J., Fung, Y.C.: On residual stresses in arteries. J. Biomech. **108**, 189–192 (1986)

Esposito, W.R., Floudas, C.A.: Global optimization in parameter estimation of nonlinear algebraic models via the error-in-variables approach. Ind. Eng. Chem. Res. **37**, 1841–1858 (1998)

Ferruzzi, J., Bersi, M.R., Humphrey, J.D.: Biomechanical phenotyping of central arteries in health and disease: advantages of and methods for murine models. Ann. Biomed. Eng. **41**, 1311–1330 (2013)

Fung, Y.C.: Biodynamics: Circulation. Springer, New York (1984)

Fung, Y.C., Fronek, K., Patitucci, P.: Pseudoelasticity of arteries and the choice of its mathematical expression. Am. J. Physiol. **237**, H620–H631 (1979)

Gade, J.L., Stålhand, J., Thore, C.J.: An in vivo parameter identification method for arteries: numerical validation for the human abdominal aorta. Comput. Methods Biomech. Biomed. Engin. **22**, 426–441 (2019)

Gasser, T.C., Ogden, R.W., Holzapfel, G.A.: Hyperelastic modelling of arterial layers with distributed collagen fibre orientations. J. R. Soc. Interface **3**, 15–35 (2006)

Holzapfel, G.A.: Nonlinear Solid Mechanics. A Continuum Approach for Engineering. Wiley, Chichester (2000)

Holzapfel, G.A., Gasser, T.C., Ogden, R.W.: A new constitutive framework for arterial wall mechanics and a comparative study of material models. J. Elast. **61**, 1–48 (2000)

Holzapfel, G.A., Sommer, G., Auer, M., Regitnig, P., Ogden, R.W.: Layer-specific 3D residual deformations of human aortas with non-atherosclerotic intimal thickening. Ann. Biomed. Eng. **35**, 530–545 (2007)

Horst, R., Tuy, H.: Global Optimization: Deterministic Approaches, 3rd edn. Springer, Berlin, Heidelberg (1996)

Humphrey, J.D.: Cardiovascular Solid Mechanics. Cells, Tissues, and Organs. Springer, New York (2002)

Humphrey, J.D., Rajagopal, K.R.: A constrained mixture model for arterial adaptations to a sustained step change in blood flow. Biomech. Model. Mechanobiol. **2**, 109–126 (2003)

Humphrey, J.D., Yin, F.C.P.: Fiber-induced material behavior in composites. Mech. Res. Commun. **13**, 277–283 (1986)

Keller, A.A.: Convex underestimating relaxation techniques for nonconvex polynomial programming problems: computational overview. J. Mech. Behav. Biomed. Mater. **24**, 129–143 (2015)

Klarbring, A., Olsson, T., Stålhand, J.: Theory of residual stresses with application to an arterial geometry. Arch. Mech. **59**, 341–364 (2007)

Masson, I., Beaussier, H., Boutouyrie, P., Laurent, S., Humphrey, J.D., Zidi, M.: Carotid artery mechanical properties and stresses quantified using in vivo data from normotensive and hypertensive humans. Biomech. Model. Mechanobiol. **10**, 867–882 (2011)

Najman, J., Bongartz, D., Mitsos, A.: Convex relaxations of componentwise convex functions. Comput. Chem. Eng. **130**, 106527 (2019)

Nocedal, J., Wright, S.: Numerical Optimization, 2nd edn. Springer, New York (2006)

Ogden, R.W.: Nonlinear elasticity, anisotropy, material stability and residual stresses in soft tissue. In: Holzapfel, G.A., Ogden, R.W. (eds.) Biomechanics of Soft Tissue in Cardiovascular Systems, pp. 65–108. Springer, Wien, New York (2003). CISM Courses and Lectures no. 441

Ogden, R.W., Schulze-Bauer, C.A.J.: Phenomenological and structural aspects of the mechanical response of arteries. In: Casey, J., Bao, G. (eds.) Mechanics in Biology. The American Society of Mechanical Engineers (ASME). New York (2000). AMD-Vol. 242/BED-Vol. 46, pp. 125–140

Roach, M.R., Burton, A.C.: The reason for the shape of the distensibility curves of arteries. Canad. J. Biochem. Physiol. **35**, 681–690 (1957)

Schulze-Bauer, C.A.J., Holzapfel, G.A.: Determination of constitutive equations for human arteries from clinical data. J. Biomech. **36**, 165–169 (2003)

Schulze-Bauer, C.A.J., Mörth, C., Holzapfel, G.A.: Passive biaxial mechanical response of aged human iliac arteries. J. Biomech. Eng. **125**, 395–406 (2003)

Smoljkić, M., Vander Sloten, J., Segers, P., Famaey, N.: Non-invasive, energy-based assessment of patient-specific material properties of arterial tissue. Biomech. Model. Mechanobiol. **14**, 1045–1056 (2015)

Sommer, G., Holzapfel, G.A.: 3D constitutive modeling of the biaxial mechanical response of intact and layer-dissected human carotid arteries. J. Mech. Behav. Biomed. Mater. **5**, 116–128 (2012)

Sommer, G., Regitnig, P., Költringer, L., Holzapfel, G.A.: Biaxial mechanical properties of intact and layer-dissected human carotid arteries at physiological and supra-physiological loadings. Am. J. Physiol. Heart Circ. Physiol. **298**, H898–H912 (2010)

Sommer, G., Haspinger, D.C., Andrä, M., Sacherer, M., Viertler, C., Regitnig, P., Holzapfel, G.A.: Quantification of shear deformations and corresponding stresses in the biaxially tested human myocardium. Ann. Biomed. Eng. **43**, 2234–2348 (2015)

Stålhand, J.: Determination of human arterial wall parameters from clinical data. Biomech. Model. Mechanobiol. **8**, 141–148 (2009)

Stålhand, J., Klarbring, A.: Aorta in vivo parameter identification using an axial force constraint. Biomech. Model. Mechanobiol. **3**, 191–199 (2005)

Stålhand, J., Klarbring, A., Karlsson, M.: Towards in vivo material identification and stress estimation. Biomech. Model. Mechanobiol. **2**, 169–186 (2004)

Takamizawa, K., Hayashi, K.: Strain energy density function and uniform strain hypothesis for arterial mechanics. J. Biomech. **20**, 7–17 (1987)

Vaishnav, R.N., Vossoughi, J.: Estimation of residual strains in aortic segments. In: Hall, C.W. (ed.) Biomedical Engineering II: Recent Developments, pp. 330–333. Pergamon Press, New York (1983)

Wittek, A., Derwich, W., Karatolios, K., Fritzen, C.P., Vogt, S., Schmitz-Rixen, T., Blase, C.: A finite element updating approach for identification of the anisotropic hyperelastic properties of normal and diseased aortic walls from 4D ultrasound strain imaging. J. Mech. Behav. Biomed. Mater. **58**, 122–138 (2016)

Modeling Failure and Fracture in Soft Biological Tissues

Konstantin Y. Volokh

I was neither a student nor a collaborator of Gerhard A. Holzapfel. I learned about him through his very popular constitutive models of soft tissue. I use and appreciate these models. Science still moves forward by small steps that we call new ideas. Not every researcher – even among the able ones – succeeds in proposing something new. Gerhard A. Holzapfel succeeded. Congratulations Gerhard and go ahead!

Kosta

Abstract Soft biological tissues are exposed to moderately large stretches and they are prone to failure and fracture. Failure means the onset of damage and fracture means the damage localization into cracks with their subsequent propagation. There are various approaches to modeling failure and fracture and none of them is superior yet. The description of failure and fracture remains the main challenge in the general field of mechanics of materials for a century. Despite the enormous research effort the progress is mild. In this chapter, our recent work on the topic is briefly reviewed. Our approach is based on two physical assumptions, which avoid the introduction of internal variables. First, we assume that the number and energy of molecular bonds are bounded in a representative volume and, consequently, the macroscopic strain–energy function should also be bounded in the constitutive law. This notion leads to the introduction of energy limiters, which are calibrated in standard tests. Second, we assume that broken bonds are diffused during the fracture process. Such an assumption directly leads to a consideration of the coupled deformation–mass–sink problem. Mathematically, the coupling provides a regularized formulation for modeling crack propagation.

K. Y. Volokh (✉)
Faculty of Civil and Environmental Engineering, Technion – Israel Institute of Technology, 3200003 Haifa, Israel
e-mail: cvolokh@technion.ac.il

G. Sommer et al. (eds.), *Solid (Bio)mechanics: Challenges of the Next Decade*, Studies in Mechanobiology, Tissue Engineering and Biomaterials 24, https://doi.org/10.1007/978-3-030-92339-6_17

1 Introduction

Human body comprises soft structures attached to the skeleton. These structures constantly undergo deformation and, sometimes, they can fail and fracture. Tear of meniscus, dissection of arteries, rupture of aneurysms are just a few examples of soft biological tissues that fail and fracture under mechanical deformation. Structural analysis can be helpful for understanding and predicting the load tolerance of parts of human body. It should not be missed that biological tissues are soft and active and their mechanical behavior is strongly nonlinear from both geometric and physical standpoints. Below we review some recent advances in the modeling failure and fracture of soft materials with emphasis on biological tissues. Background information can be found in Volokh (2019a).

2 Failure

Traditional structural analysis is based on the assumption that materials do not fail. Such assumption requires special restrictions on the material models, e.g., polyconvexity, strong ellipticity, Baker-Ericksen inequalities, etc. (Truesdell and Noll 2004). Of course, analytical and numerical methods for the analysis of non-failing materials are relatively simple. However, real materials fail. To describe the failure continuum damage mechanics (CDM) was developed[1] (Simo 1987; Govindjee and Simo 1991; Johnson and Beatty 1993; Miehe 1995; De Souza Neto et al. 1998; Ogden and Roxburgh 1999; Holzapfel 2000; Menzel and Steinmann 2001; Guo and Sluys 2006; De Tommasi et al. 2008; Dal and Kaliske 2009; Li and Holzapfel 2019). The idea of CDM is to decrease material stiffness via an additional damage variable. Such a variable does not have an appealing physical interpretation and, because of that, it is called internal variable. Despite the vague meaning, the use of internal variables pays off in cases where a description of the gradual accumulation of damage is necessary. If the damage is abrupt rather than gradual then there is no need in internal variables and it is only necessary to bound the strain-energy function (Volokh 2013). Such approach of bounding the strain energy with a limiter and its implications are considered in the present section.

2.1 *Energy Limiter*

Structural analysis means solution of the initial boundary-value problem (IBVP), which includes the linear momentum balance: $\rho\ddot{\mathbf{y}} = \text{Div}\mathbf{P}$; the angular momentum balance: $\mathbf{P}\mathbf{F}^{\mathrm{T}} = \mathbf{F}\mathbf{P}^{\mathrm{T}}$; the constitutive law: $\mathbf{P} = \partial\psi/\partial\mathbf{F}$; and kinematics: $\mathbf{F} = \partial\mathbf{y}/\partial\mathbf{x}$. Here $\mathbf{x} \in \Omega_0$ and $\mathbf{y}(\mathbf{x}) \in \Omega$ denote initial and current positions of a material point

[1] We refer to the works on CDM in which finite strains were considered.

accordingly; ρ is the referential mass density; $\ddot{\mathbf{y}}$ is the acceleration; $\mathbf{P}$ is the first Piola-Kirchhoff stress tensor; and ψ is the strain-energy function. Traction boundary conditions on $\partial\Omega_0$ are given: $\mathbf{Pn} = \bar{\mathbf{t}}$, where $\bar{\mathbf{t}}$ is a prescribed traction and $\mathbf{n}$ is a unit outward normal to $\partial\Omega_0$; or, alternatively, placements are given on $\partial\Omega_0$: $\mathbf{y} = \bar{\mathbf{y}}$. In addition, initial conditions in Ω_0 complete formulation of IBVP: $\mathbf{y}(t = 0) = \mathbf{y}_0$ and $\dot{\mathbf{y}}(t = 0) = \mathbf{v}_0$.

The choice of the strain-energy function ψ is an art. However, a restriction should be imposed on ψ if we wish to include failure in the material description. Indeed, the number of molecules in the body or any part of it is bounded and, consequently, the energy that molecular bonds can accumulate is bounded. Thus, the strain-energy function must be bounded or limited. Various ways to limit the strain energy can be proposed. We use the upper incomplete gamma function (Volokh 2007, 2010), $\Gamma[s, x] = \int_x^\infty t^{s-1}e^{-t}\mathrm{d}t$, in order to bound the strain-energy function, i.e.

$$\psi(\mathbf{F}) = \phi m^{-1}\Gamma[m^{-1}, W(\mathbf{1})^m\phi^{-m}] - \phi m^{-1}\Gamma[m^{-1}, W(\mathbf{F})^m\phi^{-m}]. \quad (1)$$

Here the first term on the right-hand side corresponds to the failure energy and the second term corresponds to the elastic energy. Also, $W(\mathbf{F})$ is the strain-energy function without failure; ϕ is the energy limiter—average bond energy (Volokh 2007); $\mathbf{1}$ is the identity tensor; and m is a material parameter.

Deformation increase beyond a critical threshold leads to the decrease of the elastic energy, which numerically vanishes and the strain energy approaches the failure energy. To make the process irreversible a modification of (1) is necessary (Volokh 2014); however, the irreversibility is important when damage localization is considered and we postpone its consideration to the section on fracture.

The strain-energy function (1) yields the constitutive law

$$\mathbf{P} = \partial\psi/\partial\mathbf{F} = \exp[-W^m\phi^{-m}]\partial W/\partial\mathbf{F}, \quad (2)$$

which does not include any gamma function and only the exponential factor makes difference between the present formulation and traditional hyperelasticity for the intact material behavior.

Examples of the stress-stretch curves for two aneurysm tissues are shown in Fig. 1, where the Cauchy stress tensor $\boldsymbol{\sigma} = J^{-1}\mathbf{PF}^{\mathrm{T}}$ is used with $J = \det\mathbf{F}$. The intact strain energy is chosen as follows: $W = c_1(\mathbf{F} : \mathbf{F} - 3) + c_2(\mathbf{F} : \mathbf{F} - 3)^2$. Material is incompressible: $J = 1$. Constants for aneurysm tissues 1 and 2 are fitted to the experimental data from Volokh and Vorp (2008) and Raghavan and Vorp (2000) accordingly, see Table 1.

The reader should note that the limit points and downhill branches on the stress-stretch curves are a direct consequence of the limited strain energy. Without limiting the strain energy we would get stresses going to infinity with the increasing stretch, which is evidently nonphysical. We should also note that damage localizes after the limit point, while it is implied that the state of the deformation state is uniform in uniaxial tests. To overcome this interpretation difficulty the reader might imagine

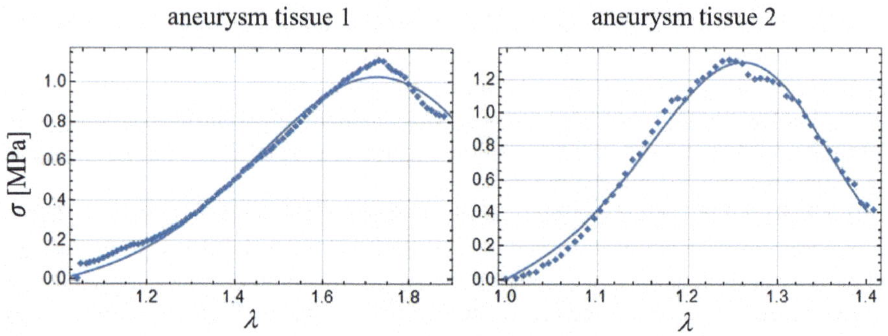

Fig. 1 Cauchy stress versus stretch in uniaxial tension for two aneurysm tissues: solid curves denote modeling and diamonds denote experimental data. Reprinted from Volokh and Vorp (2008) and Volokh (2015), with permission from Elsevier and Springer accordingly

Table 1 Material constants

	c_1 [MPa]	c_2 [MPa]	ϕ [MPa]	m [-]
Aneurysm tissue 1	0.103	0.18	0.402	1
Aneurysm tissue 2	0.52	3.82	0.255	1

that the experiment is done inside a black box. The only measurable quantities are stresses and stretches of a specimen at the entry and exit of the box. Thus, stretches can increase with the decreasing stresses after damage and rupture. Specific localization of damage is of minor importance in the case of repeatability of the experimental results.

In the following subsections we consider various applications of the proposed formulation concerned with the prediction of failure.

2.2 Cavitation

Remodeling of soft tissues can be accompanied by morphological changes that can lead to appearance of micro-voids. Cavitation is defined as a sudden irreversible expansion of micro-voids into visible macroscopic voids. Such a failure mode is often a predecessor of cracks in soft materials.

Mathematically, the void expansion in isotropic material can be described by the following integral formula

$$p(\lambda_a) = \int_1^{\lambda_a} \frac{1}{\lambda^3 - 1} \frac{d\hat{\psi}}{d\lambda} d\lambda, \tag{3}$$

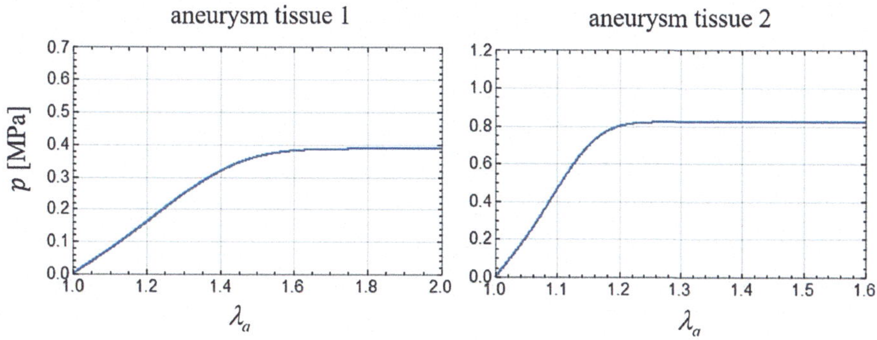

Fig. 2 Hydrostatic tension versus hoop stretch for void growth. Reprinted from Volokh (2015), with permission from Springer

where p is the remote hydrostatic tension; λ is the hoop stretch; $\lambda_a = a/A$ with A and a denoting the initial and current radius of the void accordingly; and the strain energy is expressed in terms of the principal stretches $\lambda_1, \lambda_2, \lambda_3$ for incompressible material: $\hat{\psi}(\lambda) = \psi(\lambda_1, \lambda_2, \lambda_3) = \psi(\lambda^{-2}, \lambda, \lambda)$.

Substituting two tissue models described above in (3), it is possible to draw curves shown in Fig. 2 (Volokh 2015).

The curves reach the tension limit with the increasing hoop stretch. At this limit the void expansion becomes unstable because it does not require a pressure increase. This is the critical cavitation tension. It cannot be overstated that the critical tension can only be reached for material models incorporating a failure description (Volokh 2011; Lev and Volokh 2016). Remarkably, unstable growth of cavities can start under critical stresses that are considerably less than the aneurysm strength. Also, we should note that the state of the hydrostatic tension triggering the unstable cavity expansion can occur in the vicinity of rigid inclusions. Thus, tissue calcification can be a qualitative indicator of a possible onset of rupture.

2.3 *Calcification*

Calcification is an abnormal accumulation of calcium salts in soft tissue causing it to harden. We modeled tension of aneurysm material including stiff calcified particles (Volokh and Aboudi 2016). Particularly, we analyzed the effect of the varying amount of calcification (10, 40, and 70%), i.e. the relative volume of the hard inclusion within the periodic elementary cell, on the tissue stiffness and strength, see Fig. 3.

We found that the increase of the relative volume of calcium particles unconditionally led to the stiffening of the tissue. However, the strength did not increase in the most considered cases—it could significantly decrease. The drop of the strength varied from 10 to 40% and more. This finding emphasizes the difference between

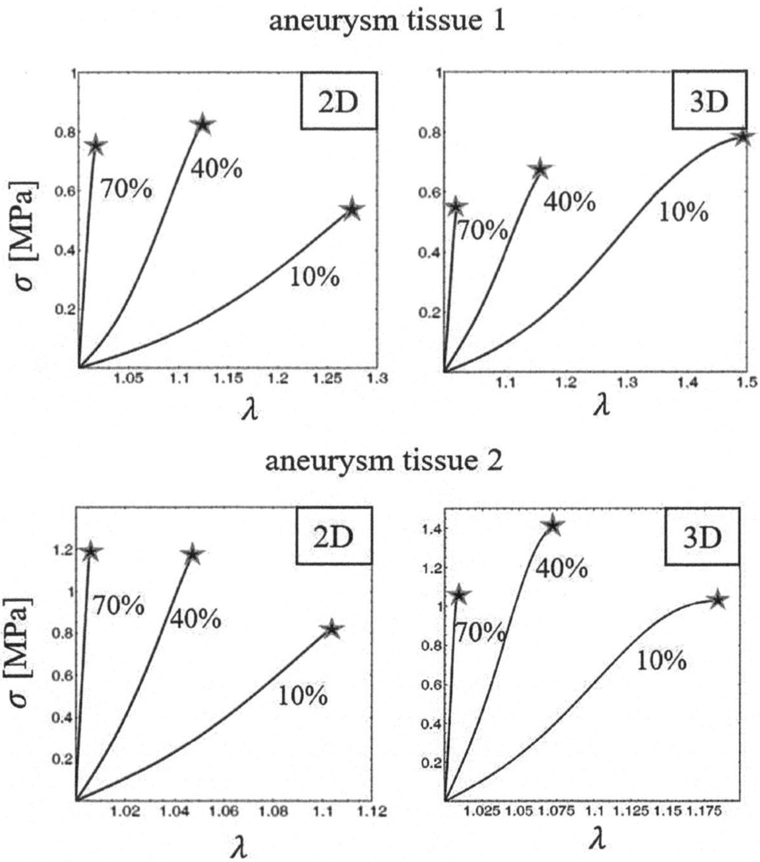

Fig. 3 Cauchy stress versus stretch in uniaxial tension for aneurysm tissues with varying calcification. Stars denote points beyond which static solution does not exist—strength. Reprinted from Volokh and Aboudi (2016), with permission from Springer

the concepts of stiffness and strength. The strength of a composite is significantly influenced by the locally inhomogeneous deformation. Hard particles can be stress concentrators amplifying the likelihood of the local material failure.

Besides, the hard particles restrain deformation in their vicinity creating a state of hydrostatic tension which, in its turn, can trigger cavitation with the subsequent fracturing. Figure 4 shows the maps of the triaxiality ratio defined by formula $\mathrm{tr}\boldsymbol{\sigma}(27\mathrm{dev}\boldsymbol{\sigma} : \mathrm{dev}\boldsymbol{\sigma}/2)^{-1/2}$, where 'tr' and 'dev' denote the trace and the deviatoric part of a second-order tensor, respectively. High triaxiality ratios appear at the poles of hard inclusions. These are locations where damage starts.

It is worth emphasizing that smaller calcified particles might be more dangerous from the standpoint of strength. Consequently, the tissues at initial stages of calcification might be more prone to rupture!

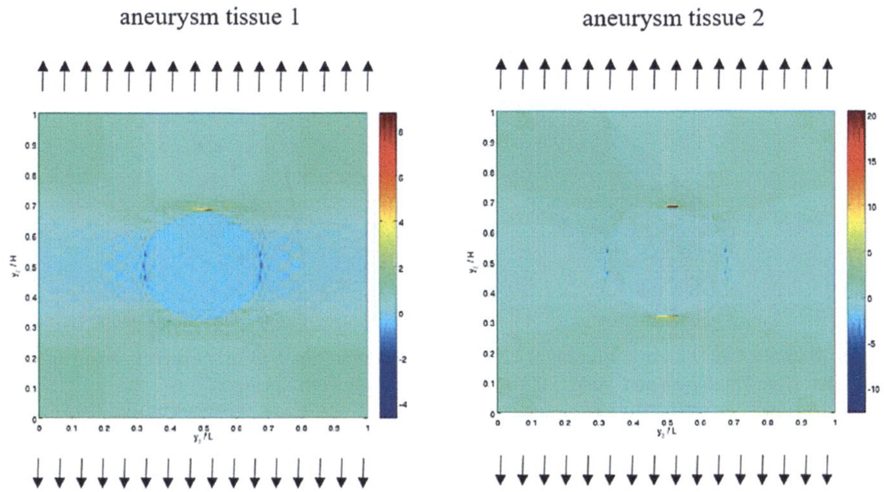

Fig. 4 Triaxiality ratio $\mathrm{tr}\sigma(27\mathrm{dev}\sigma : \mathrm{dev}\sigma/2)^{-1/2}$ distribution for 10% calcification for aneurysm tissues at the failure load. Reprinted from Volokh and Aboudi (2016), with permission from Springer

The obtained results have limitations because an ideally periodic distribution of calcified particles was assumed in computations while in reality the distribution is random. Thus, additional research in stochastic mechanics of failure analysis is required.

2.4 *Crack Direction*

The idea of predicting the onset of the damage and the the direction of its localization—crack—can be qualitatively described as follows. Intact material can propagate superposed plane waves defined by vector $\mathbf{r}g(\mathbf{s} \cdot \mathbf{y} - vt)$, where $\mathbf{r}$ and $\mathbf{s}$ are the unit vectors in the directions of the wave polarization and wave propagation, respectively, and v is the wave speed. However, cracks become barriers on the way of the wave propagation. Within cracks a material is damaged and, consequently, a superposed wave cannot run in the direction perpendicular to the damage localization. The latter notion prompts the idea to find the crack direction: it is necessary to find a direction in which the superposed wave has zero speed (Volokh 2017a; Mythravaruni and Volokh 2018, 2019, 2020; Volokh 2019b).

Mathematically, the condition of the zero wave speed can be written as

$$\rho v^2 = s_j s_l r_i r_k F_{js} F_{lr} \frac{\partial^2 \psi}{\partial F_{is} \partial F_{kr}} = 0, \tag{4}$$

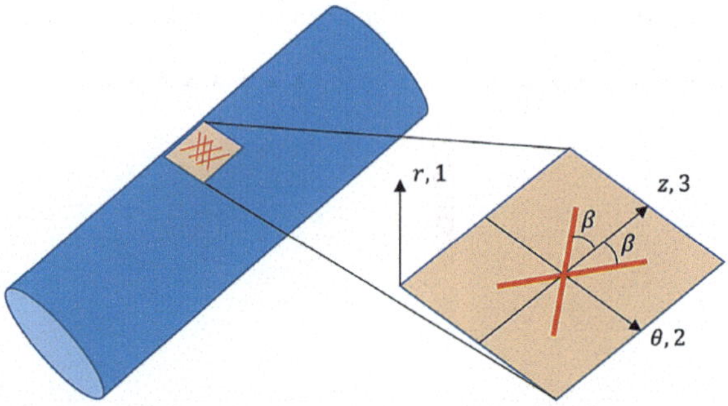

Fig. 5 Radial (1 or r), circumferential (2 or θ), and longitudinal (3 or z) directions in artery. Reprinted from Raghavan and Vorp (2000), with permission from Springer

Table 2 Material constants

c [kPa]	k_1 [kPa]	k_2 [–]	ϕ [kPa]	m [–]	β [°]
75	1500	0.03	95	1.2	45.8

where, accounting for (1),

$$\frac{\partial^2 \psi}{\partial F_{is} \partial F_{kr}} = \left(\frac{\partial^2 W}{\partial F_{is} \partial F_{kr}} - m W^{m-1} \phi^{-m} \frac{\partial W}{\partial F_{kr}} \frac{\partial W}{\partial F_{is}} \right) \exp[-W^m \phi^{-m}]. \quad (5)$$

We illustrate the use of this condition via example of arterial tissue, which is anisotropic due to two families of oriented bundles of collagen fibers, see Fig. 5.

Following Volokh (2019b), we consider the strain energy in the form

$$\begin{aligned} \psi = c(\mathbf{F} : \mathbf{F} - 3)/2 + 2\phi m^{-1}\Gamma[m^{-1}, 0] - \phi m^{-1}\Gamma[m^{-1}, W_4^m \phi^{-m}] \\ -\phi m^{-1}\Gamma[m^{-1}, W_6^m \phi^{-m}], \end{aligned} \quad (6)$$

where

$$W_{4,6} = k_1\{\exp[k_2 \langle |\mathbf{F}\mathbf{m}_{4,6}|^2 - 1 \rangle] - 1\}/(2k_2), \quad [\mathbf{m}_{4,6}] = [0, \pm \sin\beta, \cos\beta]^{\mathrm{T}}. \quad (7)$$

Here, triangle Macaulay brackets are used to account for the fiber response in tension only: $\langle x \rangle = x$ for positive x and $\langle x \rangle = 0$ otherwise. Unit vectors $\mathbf{m}_4$ and $\mathbf{m}_6$ denote two families of collagen fibers. Based on the experimental data kindly provided by the Institute of Biomechanics at Graz University of Technology, it was possible to calibrate the model (see Fig. 6), and the material constants are given in Table 2 (Volokh 2019b).

Since it is assumed that the material is incompressible, we consider the propagation of the superimposed transverse or shear waves in the plane that is tangent to the artery, see Fig. 5. These waves are defined by the following mutually orthogonal

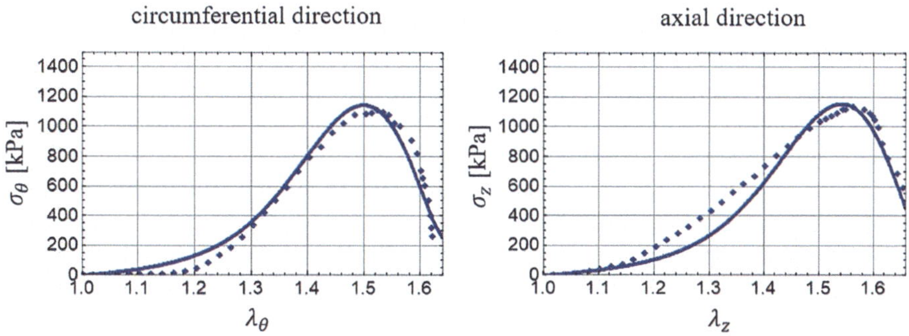

Fig. 6 Stress-stretch curves for arterial tissue: solid curves for the theory and diamonds for the experimental data. Reprinted from Volokh (2019b), with permission from Springer

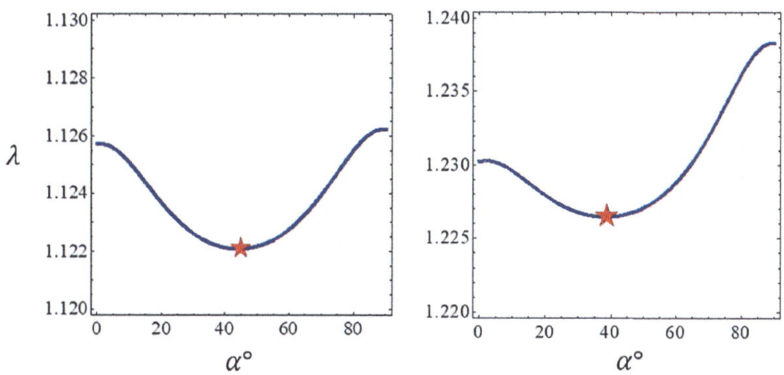

Fig. 7 Equibiaxial tension (left) and pure shear (right): $\rho v^2(\lambda, \alpha) = 0$ (stars show minima). Reprinted from Volokh (2019b), with permission from Springer

unit vectors: $[\mathbf{s}] = [0, \sin\alpha, \cos\alpha]^{\mathrm{T}}$ and $[\mathbf{r}] = [0, \cos\alpha, -\sin\alpha]^{\mathrm{T}}$. We also limit our considerations by equibiaxial and pure shear deformations. In these cases only one stretch λ describes the deformation. Thus, we have only two independent variables λ and α and (4) takes on the form: $\rho v^2(\lambda, \alpha) = 0$. Curves which are given by this equation are presented in Fig. 7. Starred points in Fig. 7 give the minimum stretches λ at which materials fail and damage starts localizing. Angles α correspond to the minima and provide directions perpendicular to the damage localization in the current configuration. The latter notion makes it possible to predict the direction of the future crack. In the considered examples,the directions of the cracks are aligned with the direction of the fibers. This finding agrees with the reported experiments by Luo et al. (2016) where the authors concluded that 'the direction of the rupture … is aligned with the direction of maximum stiffness', which is the direction of fibers.

In the considered example, we used the assumption of material incompressibility. Such assumption can be effective for the creation of analytical solutions, as Ronald Rivlin has demonstrated at length in his work. However, from our perspective there is an unfavorable side to the incompressibility assumption: it suppresses the consideration of longitudinal or pressure waves. Such waves might give important predictions concerning the onset of cracks and they cannot be ignored (Mythravaruni and Volokh 2019). The incompressibility constraint can turn into a Trojan Horse in the analysis!

In the light of remarks concerning the incompressibility enforcement we refer to (Mythravaruni and Volokh 2020) where the constraint was relaxed and both superimposed transverse and longitudinal waves were considered as well as fiber the dispersion was introduced in the constitutive setting. The vanishing pressure wave speed predicted cracks in the direction perpendicular to the tension in the uniaxial tension and pure shear. The vanishing shear wave speed predicted cracks in the direction inclined to tension in the uniaxial tension and pure shear. Equibiaxial stretching can lead to the appearance of cracks in any direction despite the anisotropy of material. The inclined cracks oriented along the bundles of collagen fibers were experimentally found and reported in Sugita and Matsumoto (2017).

3 Fracture

Hyperelastic models with energy limiters described above cannot be directly used for the modeling of damage localization and propagation. The reason is that the numerical solution of the IBVP becomes mesh-dependent. The thickness of the damaged area or crack is equal to the size of a cell in the spatial discretization mesh. Reducing the cell size one reduces the thickness of the damage localization area, which, in its turn, reduces the energy dissipated during the fracture process. Ultimately, the dissipated energy vanishes with mesh refinement. Thus, fracture can theoretically occur without any dissipated energy. The latter possibility is physically meaningless, of course. The described pathology is inherent in any continuum damage theory and not only the theory described above.

A way to suppress the pathological mesh sensitivity is to introduce strong discontinuity defined by a cohesive surface with a traction-separation constitutive law. In this way, we immediately have two new surfaces representing crack. Unfortunately, the surface nucleation, orientation, branching and arrest require extra criteria, which are not a part of constitutive description. Even worse, cohesive surface formulations presume a simultaneous use of different constitutive models: one for the surface traction-separation and another for the intact bulk material. Correspondence between these two constitutive theories is not readily available.

A more attractive way to suppress the pathological mesh sensitivity would be a regularization of continuum damage theories preventing from zero energy fractures. For the latter purpose, nonlocal continuum formulations were invented where a characteristic length was incorporated to limit the size of the spatial damage localization (Pijaudier-Cabot and Bazant 1987; Lasry and Belytschko 1988; Peerlings et al. 1996; Francfort and Marigo 1998; Silling 2000). In some of these formulations, e.g., the internal damage variable is described by an additional differential equation of the reaction-diffusion type. In this equation, the highest spatial derivatives of the damage variable are scaled by a characteristic length providing solutions of the boundary layer type. Such layer is interpreted as a diffused crack of finite thickness. A special choice of the regularizing equation, called phase-field approach, gained popularity in recent years (Hofacker and Miehe 2012; Borden et al. 2012; Denli et al. 2020). It is claimed that the phase-field formulation yields convergence of the diffused crack to the surface of discontinuity under the decrease of the characteristic length, which is interpreted as a varying numerical parameter. However, an analysis of the simple uniaxial tension shows that, in the phase-filed formulation, the characteristic length is a fixed physical parameter linked to material strength.[2]

The regularization strategy rooted in nonlocal continua formulations is attractive because it is coherent mathematically. Regrettably, the generalized nonlocal continua theories are based on the physical assumption of long-range particle interactions, while the actual particle interactions are short-range involving only the closest neighbors. Therefore, the physical basis for the nonlocal models appears disputable. A more physically based treatment of the pathological mesh sensitivity should probably include multi-physics coupling as we show below.

3.1 Material Sink

We can see crack surfaces and we rightfully conclude that these surfaces are a result of material separation. However, we usually take another logical step and assume that the separation surfaces is a result of the debonding of two adjacent atomic or molecular layers, see Fig. 8, left. The latter assumption is the simplest and, therefore, speaks to intuition. However, this assumption contradicts routine observations that cracks are visible to the naked eye, see Fig. 8, center. Indeed, if the separation was between two adjacent atomic layers, then we would not see closed cracks because our eye can only distinguish objects on the micrometer scale and not angstroms. Thus, the crack surfaces are not created by two adjacent atomic layers—they are created by a massive bond breakage spread over a region with a characteristic length l, Fig. 8, right.

The process of the bond breakage is diffusive rather than confined to two atomic planes! Some atoms fly out of the bulk material. Generally, we cannot see them because of their very small amount, as compared to the bulk. Sometimes, we can see them—remember the dust of fracturing brittle concrete. The characteristic length of the damaged region is large enough for observing small disintegrated pieces of concrete.

The notion of material sink within the characteristic small region gives rise to a mathematical formulation in which momenta and mass balance are coupled (Volokh 2017b). We add the mass balance in Ω_0 to the governing equations

[2] For example, the authors rightfully note in Borden et al. (2012) that 'although the length-scale parameter associated with the phase-field approximation is introduced as a numerical parameter it is, in fact, a material parameter that influences the critical stress at which crack nucleation occurs.'

Fig. 8 Left: idealized crack with zero thickness; center: visible closed crack in unloaded tire; right: realistic bulk crack with finite thickness l

$$\mathrm{Div}\mathbf{s} + \xi = 0, \tag{8}$$

where $\mathbf{s}$ and ξ are the referential mass flux and source (sink) accordingly.

The mass balance on the boundary $\partial\Omega_0$ provides the natural boundary condition

$$\mathbf{s} \cdot \mathbf{n} = 0. \tag{9}$$

We note that the mass balance is used in the reduced form $\mathrm{Div}\mathbf{s} + \xi = 0$ instead of $\dot{\rho} = \mathrm{Div}\mathbf{s} + \xi$ because we are only interested in pre- and post- cracked states, while the transition process of the bond rupture is so fast that it can be neglected. Such simplification is analogous to consideration of the buckling process in thin-walled structures, in which pre- and post–buckled states are analyzed by using a time-independent approach and the fast dynamic transition to the buckled state is ignored.

Skipping details given in Volokh (2017b), we define constitutive laws for the stress

$$\mathbf{P} = (\rho/\rho_0)\partial W/\partial \mathbf{F}, \tag{10}$$

the mass sink

$$\xi = \beta\rho_0 H(\gamma)\exp[-W^m \phi^{-m}] - \beta\rho, \tag{11}$$

and the mass flux

$$\mathbf{s} = \kappa H(\gamma)\exp[-W^m \phi^{-m}] J(\mathbf{F}^{\mathrm{T}}\mathbf{F})^{-1}\partial\rho/\partial\mathbf{x}, \tag{12}$$

where $\rho_0 = \rho(t = 0)$ is the initial density; $\beta > 0$ and $\kappa > 0$ are material constants; $H(\gamma)$ is a unit step function, i.e. $H(\gamma) = 0$ if γ is negative and $H(\gamma) = 1$ otherwise; the switch parameter $\gamma \in (-\infty, 0]$ is necessary to prevent from the material healing and it is defined by the evolution equation $\dot{\gamma} = -H(\epsilon - \rho/\rho_0)$, $\gamma(t = 0) = 0$ where $0 < \epsilon \ll 1$ is a dimensionless precision constant.

It is worth noting that the elasticity with energy limiters emerges as a particular case of the present formulation when there is no damage localization via diffusion of broken bonds. Indeed, if mass flux and sink are not active, $\mathbf{s} = \mathbf{0}$ and

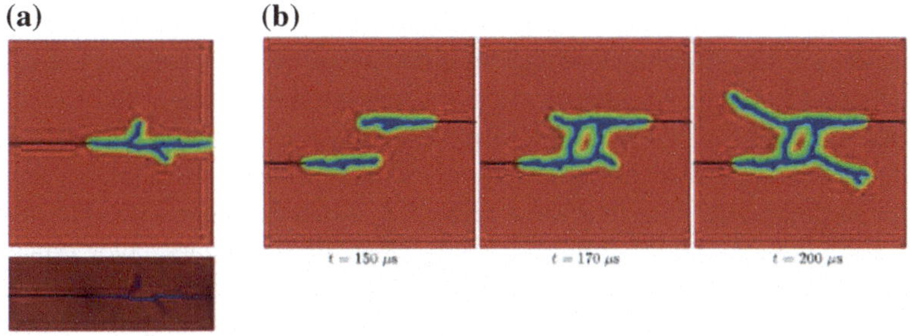

Fig. 9 (a) Propagation of Mode 1 crack in aneurysm tissue 2 in current (top) and referential (bottom) configurations; (b) crack bridging and kinking. Reprinted from Faye et al. (2019), with permission from Springer

$\xi = 0$, then we calculate from (11): $\rho/\rho_0 = H(\gamma)\exp[-W^m\phi^{-m}]$. Since the irreversibility is not important in this case, we set $H(\gamma) \equiv 1$ and further simplify: $\rho/\rho_0 = \exp[-W^m\phi^{-m}]$. Substitution of the latter formula in (10) yields (2).

3.2 Dynamic Fracture

The onset and localization of damage are usually related to the loss of the static stability. Beyond the static instability point the process becomes dynamic. The latter is the reason why most cracks propagate dynamically unless they are highly restrained. We implemented the material sink formulation presented above in analysis of dynamic crack propagation in aneurysm tissue 2 (Faye et al. 2019).

We note that the substitution of (11) and (12) in (8) yields the following second-order partial differential equation with respect to the referential mass density ρ, i.e.

$$l^2\text{Div}\{H(\gamma)\exp[-W^m\phi^{-m}]J(\mathbf{F}^{\mathrm{T}}\mathbf{F})^{-1}\partial\rho/\partial\mathbf{x}\} + \rho_0 H(\gamma)\exp[-W^m\phi^{-m}] - \rho = 0. \tag{13}$$

Remarkably, we do not need to know material constants κ and β separately. We only need to know their ratio, which defines the characteristic length: $l^2 = \kappa/\beta$. The length is a small multiplier for the highest (second) spatial derivative of the mass density and, consequently, it causes solution of the boundary layer type. This layer regularizes the crack thickness and suppresses the pathological mesh sensitivity.

Results of modeling propagation of a single crack and bridging of two cracks are shown in Fig. 9. These simulations led to the following interesting conclusions.

First, the inertia forces play a crucial role at the tip of the propagating crack. If inertia is not canceled together with the material stiffness, then cracks tend to nonphysically widen with the increasing speed of their propagation. Most existing models of cracks completely ignore this fact and they do not cancel inertia when they cancel stiffness. Only very recently, phase-field modelers started recognizing the importance of canceling inertia (Chen et al. 2017; Agrawal and Dayal 2017). Needless to say, the simultaneous cancelation of stiffness and inertia are a direct consequence of the material sink formulation presented in this chapter.

Second, the proposed material sink formulation enables the suppression the strong or classical pathological mesh sensitivity associated with the zero energy fracture. The latter is due to the fact that the augmented initial boundary-volume problem enforces the characteristic length and solutions of the boundary layer type. Such layer, associated with the crack thickness, does not vanish under mesh refinement.

Third, we observed a weak mesh sensitivity, which we defined as the effect of the mesh shape and size on the specific crack pattern. We found that various meshes caused slightly different crack patterns for the same amount of dissipated energy. The weak mesh sensitivity remained even after a significant mesh refinement, which showed that the regularized formulations were not a universal recipe as many would expect. The weak mesh sensitivity is similar to the effect of structural inhomogeneities in real materials, which affect the crack path depending on the specific sample under consideration. Though all samples are made of the same material they have various microstructural patterns and, consequently, slightly different propagating cracks.

4 Conclusions

We showed that physically reasonable assumptions of the bounded bond energy and diffused bond breakage were enough for macroscopic analysis of the onset of material damage and its localization into cracks with their subsequent propagation.

The energy of molecular bonds is always bounded because the number of molecules is finite. This automatically implies that the macroscopic strain-energy function must be limited. If the strain energy is limited then the stress cannot approach infinity with increasing strain, the stress must be limited and drop to zero. The limiting stress indicates the onset of material instability or failure, which can be interpreted as inability of material to bear further load. The phenomena of cavitation, strength under calcification, and even direction of possible cracks in soft tissues can be understood based on the approach of elasticity with energy limiters.

Failure is the onset of the damage process. Fracture is the development of damage, its localization, and propagation. Microscopically, the process of fracture means bond breakage. The breakage can hardly be confined to two adjacent molecular layers. Broken bonds are diffused and, because of that, material can be lost locally. Material sink during fracture prompts a macroscopic continuum mechanics formulation based on coupled momenta and mass balance. Such formulation naturally provides regularization of crack simulations suppressing the pathological mesh sensitivity.

Assumptions and methods considered above allow avoiding the use of internal variables, and hence make the theory appealing from the physics standpoint.

Acknowledgements This research was supported by the ISRAEL SCIENCE FOUNDATION (grant No. 394/20).

References

Agrawal, V., Dayal, K.: Dependence of equilibrium Griffith surface energy on crack speed in phase-field models for fracture coupled to elastodynamics. Int. J. Fract. **207**, 243–249 (2017)

Borden, M.J., Verhoosel, C.V., Scott, M.A., Hughes, T.J.R., Landis, C.M.: A phase-field description of dynamic brittle fracture. Comp. Methods Appl. Mech. Eng. **217–220**, 77–95 (2012)

Chen, C.H., Bouchbinder, E., Karma, A.: Instability in dynamic fracture and the failure of the classical theory of cracks. Nat. Phys. **13**, 1186 (2017)

Dal, H., Kaliske, M.: A micro-continuum-mechanical material model for failure of rubberlike materials: application to ageing-induced fracturing. J. Mech. Phys. Solids **57**, 1340–1356 (2009)

De Souza Neto, E.A., Peric, D., Owen, D.R.J.: Continuum modeling and numerical simulation of material damage at finite strains. Arch. Comp. Methods Eng. **5**, 311–384 (1998)

De Tommasi, D., Puglisi, G., Saccomandi, G.: Localized vs diffuse damage in amorphous materials. Phys. Rev. Lett. **100**, 085502 (2008)

Denli, F.A., Gültekin, O., Holzapfel, G.A., Dal, H.: A phase-field model for fracture of unidirectional fiber-reinforced polymer matrix composites. Comput. Mech. **65**, 1149–1166 (2020)

Faye, A., Lev, Y., Volokh, K.Y.: The effect of local inertia around the crack tip in dynamic fracture of soft materials. Mech. Soft Mater. **1**, 4 (2019)

Francfort, G.A., Marigo, J.J.: Revisiting brittle fracture as an energy minimization problem. J. Mech. Phys. Solids **46**, 1319–1342 (1998)

Govindjee, S., Simo, J.C.: A micro-mechanically based continuum damage model of carbon black-filled rubbers incorporating the Mullins effect. J. Mech. Phys. Solids **39**, 87–112 (1991)

Guo, Z., Sluys, L.: Computational modeling of the stress-softening phenomenon of rubber like materilas under cyclic loading. Eur. J. Mech. A/Solids **25**, 877–896 (2006)

Hofacker, M., Miehe, C.: Continuum phase field modeling of dynamic fracture: variational principles and staggered FE implementation. Int. J. Fract. **178**, 113–129 (2012)

Holzapfel, G.A.: Nonlinear Solid Mechanics. A Continuum Approach for Engineering, Wiley, Chichester (2000)

Johnson, M.A., Beatty, M.F.: A constitutive equation for the Mullins effect in stress controlled in uniaxial extension experiments. Cont. Mech. Therm. **5**, 301–318 (1993)

Lasry, D., Belytschko, T.: Localization limiters in transient problems. Int. J. Solids Struct. **24**, 581–597 (1988)

Lev, Y., Volokh, K.Y.: On cavitation in rubberlike materials. J. Appl. Mech. **83**, 044501 (2016)

Li, K., Holzapfel, G.A.: Multiscale modeling of fiber recruitment and damage with a discrete fiber dispersion method. J. Mech. Phys. Solids **126**, 226–244 (2019)

Luo, Y., Duprey, A., Avril, S., Lu, J.: Characteristics of thoracic aortic aneurysm rupture in vitro. Acta. Biomater. **42**, 286–295 (2016)

Menzel, A., Steinmann, P.: A theoretical and computational framework for anisotropic continuum damage mechanics at large strains. Int. J. Solids Struct. **38**, 9505–9523 (2001)

Miehe, C.: Discontinuous and continuous damage evolution in Ogden-type large-strain elastic materials. Eur. J. Mech. A/Solids **14**, 697–720 (1995)

Mythravaruni, P., Volokh, K.Y.: Failure of rubber bearings under combined shear and compression. J. Appl. Mech. **85**, 074503 (2018)

Mythravaruni, P., Volokh, K.Y.: On incompressibility constraint and crack direction in soft solids. J. Appl. Mech. **86**, 101004 (2019)

Mythravaruni, P., Volokh, K.Y.: On the onset of cracks in arteries. Mol. Cell Biomech. **17**, 1–17 (2020)

Ogden, R.W., Roxburgh, D.G.: A pseudo-elastic model for the Mullins effect in filled rubber. Proc. R . Soc. Lond. Ser. A **455**, 2861–2877 (1999)

Peerlings, R.H.J., de Borst, R., Brekelmans, W.A.M., de Vree, J.H.P.: Gradient enhanced damage for quasi-brittle materials. Int. J. Num. Methods Eng. **39**, 3391–3403 (1996)

Pijaudier-Cabot, G., Bažant, Z.P.: Nonlocal damage theory. J. Eng. Mech. **113**, 1512–1533 (1987)

Raghavan, M.L., Vorp, D.A.: Toward a biomechanical tool to evaluate rupture potential of abdominal aortic aneurysm: identification of a finite strain constitutive model and evaluation of its applicability. J. Biomech. **33**, 475–482 (2000)

Silling, S.A.: Reformulation of elasticity theory for discontinuities and long-range forces. J. Mech. Phys. Solids **48**, 175–209 (2000)

Simo, J.C.: On a fully three-dimensional finite strain viscoelastic damage model: formulation and computational aspects. Comput. Methods Appl. Mech. Eng. **60**, 153–173 (1987)

Sugita, S., Matsumoto, T.: Local distribution of collagen fibers determines crack initiation site and its propagation direction during aortic rupture. Biomech. Model. Mechnobiol. **17**, 577–587 (2017)

Truesdell, C., Noll, W.: The Non-Linear Field Theories of Mechanics. Springer, Berlin (2004)

Volokh, K.Y.: Hyperelasticity with softening for modeling materials failure. J. Mech. Phys. Solids **55**, 2237–2264 (2007)

Volokh, K.Y., Vorp, D.A.: A model of growth and rupture of abdominal aortic aneurysm. J. Biomech. **41**, 1015–1021 (2008)

Volokh, K.Y.: On modeling failure of rubberlike materials. Mech. Res. Commun. **37**, 684–689 (2010)

Volokh, K.Y.: Cavitation instability in rubber. Int. J. Appl. Mech. **3**, 29311 (2011)

Volokh, K.Y.: Review of the energy limiters approach to modeling failure of rubber. Rubber Chem. Technol. **86**, 470–487 (2013)

Volokh, K.Y.: On irreversibility and dissipation in hyperelasticity with softening. J. Appl. Mech. **81**, 074501 (2014)

Volokh, K.Y.: Cavitation instability as a trigger of aneurysm rupture. Biomech. Model. Mechanobiol. **14**, 1071–1079 (2015)

Volokh, K.Y., Aboudi, J.: Aneurysm strength can decrease under calcification. J. Mech. Behav. Biomed. Mater. **57**, 164–174 (2016)

Volokh, K.Y.: Loss of ellipticity in elasticity with energy limiters. Eur. J. Mech. A/Solids **63**, 36–42 (2017a)

Volokh, K.Y.: Fracture as a material sink. Mater. Theory **1**, 3 (2017b)

Volokh, K.Y.: Mechanics of Soft Materials. Springer, Singapore (2019a)

Volokh, K.Y.: Constitutive model of human artery adventitia enhanced with a failure description. Mech. Soft Mater. **1**, 8 (2019b)

Stretchable Fibrous Materials with Different Micro-Geometries of Wavy Fibers

Mor Ben-Or Frank and Gal deBotton

Twenty years ago, I supervised a young graduate student, Ilia Hariton, who was about to complete his M.Sc. thesis. Ilia wished to continue his studies towards a Ph.D. in Biomechanics, and while we were discussing his options, we studied some of the recent works in this field. Among the most intriguing and advanced works we encountered were the ones by Gerhard and his colleagues. Since back then, I was rather new in the field of Biomechanics, I contacted Gerhard and offered him to join me and supervise Ilia's work together. Although we have never met before, I received a warm and positive response. After a few emails, we decided to join forces and, in Gerhard's words, to explore 'a micro-mechanical constitutive approach to describe the mechanical behavior of biological soft tissues'. During our work, I found that Gerhard loves to work in the field of biomechanics, to collaborate, to explore new ideas, to share his knowledge with others, and to listen to them too. In fact, our work was a real fusion of our expertise in the fields of Biomechanics and Micromechanics. I am glad that twenty years afterward, Gerhard is creative than ever, and I am looking forward to his next contributions and our future collaboration.

Gal

Abstract The mechanical behavior of collagenous tissues crucially depends on the fibers' micro-geometry. Herein, we systematically analyze this ubiquitous class of tissues. Finite element simulations of periodic cells enable us to relate the macroscopic tissue response to the mechanical processes of the underlying microstructure. We examine cells with a single thick fiber and cells with a bundle of seven thin fibers while keeping the fraction of the fiber substance identical in both cells. Models with shallow and wrinkled fibers are examined in both cases. Plane stress conditions are imposed, where the stretching direction is aligned with or inclined to the fiber direction. The latter mimics the loading state in the arterial wall. To even better mimic

M. Ben-Or Frank · G. deBotton (✉)
Ben-Gurion University of the Negev, Beer-Sheva, Israel
e-mail: debotton@bgu.ac.il

G. Sommer et al. (eds.), *Solid (Bio)mechanics: Challenges of the Next Decade*,
Studies in Mechanobiology, Tissue Engineering and Biomaterials 24,
https://doi.org/10.1007/978-3-030-92339-6_18

the physiological state, we superimpose a 10% prestretch in the transverse direction. Overall, the stress-stretch curves are bilinear with a low initial slope that converges into a steeper slope as the fibers straighten. This recruitment process becomes even more pronounced under aligned loading. Our expectation that the thick fiber model will be stiffer than the seven fibers bundle model due to its larger bending stiffness turned out to be inaccurate. Analysis of the stresses in the surrounding matrix material revealed that this counterintuitive occurrence is due to a through-matrix interaction between the fibers in the bundle.

1 Introduction

The microstructure of a material dictates its mechanical behavior. Of particular importance is the class of biological tissues, where the microstructure is constructed to fit their intended functioning. For example, the wavy nature and orientation of the collagen fibers in the outermost layer of arteries (tunica adventitia) are responsible for the mechanical response of the tissue under physiological loading conditions (Chen et al. 2011). In a similar manner, modern additive manufacturing techniques, which attract the attention of the engineering community nowadays, enable us to produce materials whose mechanical response is governed by the underlying microstructure. In recent years, stretchable fibrous materials, which are able to undergo large deformations together with extensive strength and durability, attract high interest in the research fields aerospace, biomechanics, and stretchable electronics (Choi et al. 2007; Chortos and Bao 2014; Pratt et al. 2001). Thus, the study of the micro-to-macro relationship is essential for understanding and designing such materials. In particular, the unique behavior of soft tissues that stiffen at large strains is attributed to the recruitment of their wavy collagen fibers. Furthermore, in the development of scaffolds made of synthetic and natural polymers, fibers with a distinct pattern of periodicity are produced (Anumandla and Gibson 2006; Li et al. 2005). These fiber-reinforced materials provide extremely high elastic moduli and strength properties along with extended flexibility. In this work, we systematically analyze this class of stretchable reinforced solids with emphasis on soft fibrous tissues.

At the microscopic level, collagenous tissues are composed of an extrafibrilar matrix (i.e. elastin) and collagen fibers. Broadly speaking, the former controls the flexibility of the tissues, and the latter provides mechanical firmness, strength, and toughness. At the macroscopic level, these tissues behave anisotropically, whereby the stiffness along specific directions is determined by the collagen fiber dispersion (Calvo et al. 2007). Recent studies assert that the fibers are crimped in the resting state of the tissue. Thus, the initial stiffness of the tissues originates from the elastin. At higher loading and strain levels, the collagen fibers straighten, and hence, more load is transferred from the elastin to the collagen. This process of progressive recruitment of the crimped collagen fibers at increasing strains results in the familiar macroscopic exponential-like stiffening. The stiffening of fibrous tissues at large strains is crucial for their functioning (Humphrey et al. 2014).

The first models that described this stiffening of stretchable collagenous tissues were phenomenological. For example, Chuong and Fung (1983) proposed an isotropic stress-strain relationship in terms of an exponential strain-energy function. In later models, the internal structure of the tissues was accounted for by adaptation of anisotropic models (Cacho et al. 2007; Chen et al. 2011; Gasser et al. 2006; Holzapfel et al. 2015; Weisbecker et al. 2015; Zulliger et al. 2004). An attempt to account for the interactions between the fibers and the matrix was made in Sopakayang and Holzapfel (2017). These are all phenomenological models that account for fiber orientation and waviness in terms of fitted macroscopic parameters.

However, the underlying microstructural and mechanical tissue details must be accounted for to better understand and more accurately model the macroscopic tissue behavior. Hence, micromechanically motivated models incorporate explicit microstructural and mechanical parameters. In this context, we recall a few recent experimental efforts in which the microstructure of the studied tissues was recorded (Rezakhaniha et al. 2012; Schriefl et al. 2012). In particular, high-resolution images of the fiber networks were taken by different microscopy techniques. These experimental works provide crucial data for modeling, analyzing, and predicting the response of these highly heterogeneous materials. Thus, important microstructural parameters, such as the amplitude, the diameter, the contour, and period lengths of wavy collagen fibers as well as their overall volume fraction can be accurately extracted and directly employed in micromechanical models.

Micromechanical constitutive models that account for the various structural parameters were developed to describe and predict the macroscopic behavior of fibrous materials. The majority of the models available in the literature assume a linear elastic behavior of the fibers (Górski 2011; Liang et al. 2006). Garnich and Karami (2004) and Karami and Garnich (2005) developed periodic micromechanical models with different degrees of fiber waviness in a single lamina using varying values of the ratio between the amplitude and the wavelength. Karami and Garnich (2005) also considered a laminate with different fiber waviness and incorporated a failure mechanisms in the model. Velmurugan et al. (2014) studied the influence of the waviness of discontinuous fibers on the effective properties. Since these models were developed within the framework of linear elasticity, they cannot capture the mechanical response of soft biological tissues that undergo large deformations.

Recently, different models were developed that are able to capture the macroscopic *nonlinear* material behavior originating from the geometrical changes that the fibers undergo at the microscopic level during the tissue deformation. In general, these micromechanical models are more appropriate when referring to soft matters and tissues in particular. Kao et al. (2011) modeled the pulmonary artery considering an orthotropic distribution of wavy collagen fibers. Fallah et al. (2016) developed a two-dimensional periodic model with wavy layers. In de Botton and Oren (2013) a periodic unit cell with two fiber families was considered to determine the overall response of a coronary artery. Therein, the collagen fibers were modeled to be straight, and the stiffening of the fibers was captured by assigning a hardening model to the fibers instead of a more accurate representation of the waviness pattern (Gent 1996). However, these models were limited to planar loading conditions and thus did not capture the spatial (3D) response, which would be a more realistic in vivo scenario. Hence, in the current work, we focus on the behavior of 3D continuous wavy fibers embedded in a soft matrix and use the finite element (FE) method to analyze their mechanical response under different loading conditions.

2 Theoretical Background

Considering the mapping of a body that is transforming from one configuration to another, the position vector of a material point in the reference configuration β_0 is $\mathbf{X}$, and its position vector in the deformed configuration β is $\mathbf{x}$. The deformation of the body is characterized by the mapping

$$\mathbf{x} = \chi(\mathbf{X}). \tag{1}$$

The deformation gradient

$$\mathbf{F} = \frac{\partial \chi(\mathbf{X})}{\partial \mathbf{X}} \tag{2}$$

is invertible and hence non-singular. Accordingly,

$$J(\mathbf{X}) \equiv \det \mathbf{F}(\mathbf{X}) = \frac{\mathrm{d}v}{\mathrm{d}V} > 0, \tag{3}$$

where J, the determinant of $\mathbf{F}$, is equal to the volume ratio of an infinitesimal material element in the deformed and the reference configurations, respectively. If the material is incompressible, as commonly assumed for soft biological tissues under normal physiological conditions, then $J \equiv 1$. We also recall that the stretch of a material element is

$$\lambda = \sqrt{\widehat{\mathbf{N}} \cdot \mathbf{F}^{\mathrm{T}} \mathbf{F} \widehat{\mathbf{N}}}, \tag{4}$$

where $\widehat{\mathbf{N}}$ is unit a vector along the material element in the reference configuration.

In the sequel, we restrict our attention to the class of hyperelastic materials whose constitutive relations are given in terms of a strain-energy function $\psi(\mathbf{F})$ such that the first Piola-Kirchhoff stress tensor $\mathbf{P}$ can be computed as

$$\mathbf{P} = \frac{\partial \psi(\mathbf{F})}{\partial \mathbf{F}}. \tag{5}$$

For incompressible materials, because of the kinematic constraint $J \equiv 1$, a pressure-like term p has to be subtracted from the stress, i.e.

$$\mathbf{P} = \frac{\partial \psi(\mathbf{F})}{\partial \mathbf{F}} - p\mathbf{F}^{-\mathrm{T}}. \tag{6}$$

This pressure-like term can be determined from the boundary conditions imposed on the body. The Cauchy stress tensor $\boldsymbol{\sigma}$ is related to the first Piola-Kirchhoff stress tensor via

$$\boldsymbol{\sigma} = \frac{1}{J}\mathbf{P}\mathbf{F}^{\mathrm{T}}. \tag{7}$$

In the absence of body forces, the equilibrium equation reads

$$\nabla_{\mathbf{x}} \cdot \boldsymbol{\sigma} = \mathbf{0}. \tag{8}$$

To effectively examine the behavior of heterogeneous materials, one needs to characterize its *macroscopic* behavior in terms of relevant information stemming from the microstructural level. To this end, a homogenization theory for finite deformations was introduced in Hill (1972) and Ogden (1974). Following these pioneering works, we assume that *homogeneous boundary conditions* of the form

$$\mathbf{x} = \mathbf{F}_0\mathbf{X} \tag{9}$$

are applied on the boundary of the heterogeneous body, where $\mathbf{F}_0$ is a constant matrix with $\det \mathbf{F}_0 > 0$. For incompressible materials, $\det \mathbf{F}_0 = 1$. In addition, we assume that at the interfaces between the phases composing the material, both the displacements and tractions are continuous. Consequently, it can be shown that the average deformation gradient is

$$\overline{\mathbf{F}} \equiv \int_{\beta_0} \mathbf{F}(\mathbf{X})\,\mathrm{d}V = \mathbf{F}_0 \tag{10}$$

and the average first Piola-Kirchhoff stress tensor is

$$\overline{\mathbf{P}} \equiv \frac{1}{V}\int_{\beta_0} \mathbf{P}(\mathbf{X})\,\mathrm{d}V. \tag{11}$$

According to the Hill-Mandel identity, when a heterogeneous body is subjected to a homogeneous incremental deformation $\dot{\mathbf{x}} = \dot{\mathbf{F}}_0\mathbf{X}$, then

$$\frac{1}{V}\int_{\beta_0} \dot{\mathbf{F}} : \mathbf{P}\,\mathrm{d}V = \dot{\overline{\mathbf{F}}} : \overline{\mathbf{P}}, \tag{12}$$

where $\dot{\overline{\mathbf{F}}} = \dot{\mathbf{F}}_0$, according to Eq. (10). For hyperelastic materials, this identity enables to determine the effective strain-energy function

$$\widetilde{\psi}(\overline{\mathbf{F}}) = \inf_{\mathbf{F}\in\mathcal{K}(\overline{\mathbf{F}})} \left\{ \frac{1}{V}\int_{\beta_0} \psi(\mathbf{F}, \mathbf{X})\,\mathrm{d}V \right\}, \tag{13}$$

where $\mathcal{K}(\overline{\mathbf{F}}) \equiv \{\mathbf{F} \mid \mathbf{F} = \partial \chi(\mathbf{X})/\partial \mathbf{X}, \mathbf{X} \in \beta_0;\ \chi(\mathbf{X}) = \overline{\mathbf{F}}\mathbf{X}, \mathbf{X} \in \partial\beta_0\}$ is the set of admissible deformations (Hill 1972). Hence, together with Eq. (11), the macroscopic constitutive behavior is

$$\overline{\mathbf{P}} = \frac{\partial \widetilde{\psi}(\overline{\mathbf{F}})}{\partial \overline{\mathbf{F}}}, \tag{14}$$

and it can be further shown that the average Cauchy stress tensor is

$$\overline{\boldsymbol{\sigma}} = \frac{1}{J}\overline{\mathbf{P}}\,\overline{\mathbf{F}}^{\mathrm{T}}. \tag{15}$$

Ultimately, if all the constituting phases are incompressible, the relation

$$\overline{\boldsymbol{\sigma}} = \frac{\partial \widetilde{\psi}(\overline{\mathbf{F}})}{\partial \overline{\mathbf{F}}}\overline{\mathbf{F}}^{\mathrm{T}} - \bar{p}\mathbf{I} \tag{16}$$

is satisfied at the macroscopic level, where $\bar{p}$ is a macroscopic pressure-like term.

We assume that the microstructure can be characterized in terms of a periodic unit cell that repeats itself. Accordingly, the macroscopic response of the material is determined by applying periodic boundary conditions (PBCs) on the faces of the unit cell. As can be seen in Fig. 1, we consider a unit cell whose dimensions are $T \times S \times S$ along the principal *material* directions (X, Y, Z) of the periodic unit cell. Thus, the displacements of all nodes at the right face $(X = T)$ and at the left face $(X = 0)$ of the unit cell are related via

$$\begin{aligned} u_x^{\mathrm{Right}} - u_x^{\mathrm{Left}} &= (\bar{F}_{11} - 1)T, \\ u_y^{\mathrm{Right}} - u_y^{\mathrm{Left}} &= \bar{F}_{21}T, \\ u_z^{\mathrm{Right}} - u_z^{\mathrm{Left}} &= \bar{F}_{31}T. \end{aligned} \tag{17}$$

The displacements of all nodes at the front face $(Y = S)$ and at the back face $(Y = 0)$ of the unit cell are related via

$$\begin{aligned} u_x^{\mathrm{Front}} - u_x^{\mathrm{Back}} &= \bar{F}_{12}S, \\ u_y^{\mathrm{Front}} - u_y^{\mathrm{Back}} &= (\bar{F}_{22} - 1)S, \\ u_z^{\mathrm{Front}} - u_z^{\mathrm{Back}} &= \bar{F}_{32}S. \end{aligned} \tag{18}$$

Finally, the displacements of all nodes at the top face $(Z = S)$ and at the bottom face $(Z = 0)$ of the unit cell are related via

$$\begin{aligned} u_x^{\mathrm{Top}} - u_x^{\mathrm{Bottom}} &= \bar{F}_{13}S, \\ u_y^{\mathrm{Top}} - u_y^{\mathrm{Bottom}} &= \bar{F}_{23}S, \\ u_z^{\mathrm{Top}} - u_z^{\mathrm{Bottom}} &= (\bar{F}_{33} - 1)S. \end{aligned} \tag{19}$$

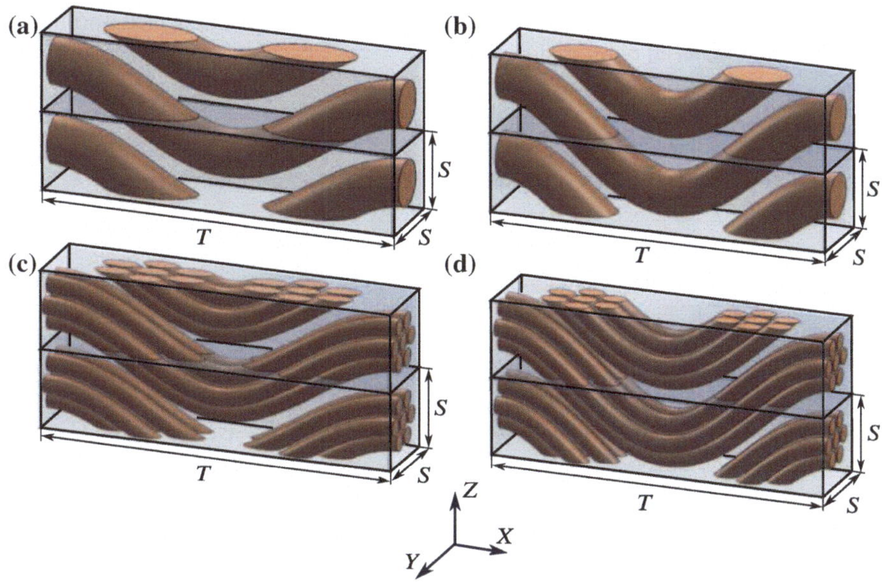

Fig. 1 Two repetitions of the four different periodic unit cell models that are analyzed in this work with dimensions $T \times S \times S$ along the *principal material directions* (X, Y, Z) in the reference configuration. Thereby, the collagen inside the unit cell (orange) was modeled either as a single fiber or as a fiber bundle consisting of seven individual collagen fibers having the same fiber volume fraction as the single fiber model. For both models, recruitment stretches of $\lambda_c = 1.1$ ((a) and (c)) and $\lambda_c = 1.2$ ((b) and (d)) were considered by modifying the fiber waviness

For loading cases where the principal stretch directions of the macroscopic Cauchy-Green tensor $\overline{\mathbf{C}} = \overline{\mathbf{F}}^{\mathrm{T}}\overline{\mathbf{F}}$ are not aligned with the principal axes of the unit cell, we find it convenient to distinguish between a *material coordinate system* and a *lab coordinate system* (Ben-Or Frank et al. 2019). For the specific class of fibrous materials on which we focus in this work, the principal material directions are denoted by the longitudinal direction along the fibers, the transverse direction that is normal to the fibers in their plane, and the out-of-plane direction, which is perpendicular to the fiber plane. The left-handed *material coordinate system* used in this work is sketched together with the unit cell in Fig. 1, where the longitudinal, out-of-plane, and transverse directions are aligned with the X-, Y-, and Z-directions, respectively. The *lab coordinate system* (θ, ρ, ζ), which is associated with the circumferential, radial, and axial directions of a cylinder, is the one in which we define the applied boundary conditions in terms of the macroscopic deformation gradient

$$[\overline{\mathbf{F}}_{\mathrm{lab}}] = \begin{bmatrix} \lambda_\theta & 0 & 0 \\ 0 & \lambda_\rho & 0 \\ 0 & 0 & \lambda_\zeta \end{bmatrix}, \tag{20}$$

where $\lambda_\theta, \lambda_\zeta$, and $\lambda_\rho = 1/(\lambda_\theta\lambda_\zeta)$ are the principal stretches. The induced deformation gradient in the material coordinate system

$$\overline{\mathbf{F}}_{\mathrm{m}} = \mathbf{Q}\overline{\mathbf{F}}_{\mathrm{lab}}\mathbf{Q}^{\mathrm{T}} \tag{21}$$

is applied on the unit cell in the material coordinate system, where

$$[\mathbf{Q}] = \begin{bmatrix} \cos\alpha & 0 & \sin\alpha \\ 0 & 1 & 0 \\ -\sin\alpha & 0 & \cos\alpha \end{bmatrix} \tag{22}$$

is a proper orthogonal rotation matrix, and α is the angle between the circumferential direction and the longitudinal fiber direction. Finally, once the average Cauchy stress tensor $\overline{\boldsymbol{\sigma}}_{\mathrm{m}}$ is determined according to Eqs. (14) and (15) by averaging the Cauchy stresses within the unit cell, the Cauchy stress tensor in the lab system is computed by

$$\overline{\boldsymbol{\sigma}}_{\mathrm{lab}} = \mathbf{Q}^{\mathrm{T}}\overline{\boldsymbol{\sigma}}_{\mathrm{m}}\mathbf{Q}. \tag{23}$$

3 Analysis and Simulations

Periodic unit cells representing harmonic wavy collagen fibers were designed and analyzed using the FE method. Thereby, the fiber direction coincides with the X-axis, and in the reference state, their transversal waviness is defined via the function

$$Z(X) = A \sin\left(\frac{2\pi X}{T}\right), \tag{24}$$

where A and T are the amplitude and the wavelength, respectively. In the reference state, the periodic wavy fibers and the surrounding matrix can be represented within a unit cell occupying the domain $0 \le X \le T$, $0 \le Y \le S$, and $0 \le Z \le S$, where T is the size of the unit cell in the longitudinal direction, and S is its size in the transverse and the out-of-plane directions (see Fig. 1). The recruitment stretch λ_{c} is the ratio between the contour length l of the fiber and the wavelength T, i.e.

$$\lambda_{\mathrm{c}} = \frac{l}{T} = \frac{1}{T}\int_0^T \sqrt{1 + \left(\frac{\mathrm{d}Z}{\mathrm{d}X}\right)^2}\,\mathrm{d}X. \tag{25}$$

In total, four different unit cells were constructed involving single fibers (Fig. 1a, c) and bundles of seven individual smaller fibers (Fig. 1b, d). For clarity, two repetitions of the modeled unit cells are shown in Fig. 1. The recruitment stretches of the unit cells in Fig. 1a, b are $\lambda_{\mathrm{c}} = 1.1$, and $\lambda_{\mathrm{c}} = 1.2$ for the unit cells in Fig. 1c, d. For convenience, we refer to the shallower model with $\lambda_{\mathrm{c}} = 1.1$ as the *rippled* model

Table 1 Geometrical and structural parameters of the four periodic unit cell models

	λ_c (–)	Single fiber model		Seven fibers model	
		C_f (–)	d (μm)	C_f (–)	d (μm)
Rippled fibers	1.1	0.3	13.23	0.3	5.0
Crimped fibers	1.2				

and denote the model with $\lambda_c = 1.2$ as the *crimped* model. The fiber volume fraction C_f is related to the recruitment stretch λ_c via

$$C_f = \frac{\pi d^2 \lambda_c}{4 S^2} N, \tag{26}$$

where d is the fiber diameter, and N is the number of fibers in the unit cell. A hyperelastic incompressible neo-Hookean free-energy function

$$\psi(\mathbf{F}) = \frac{\mu}{2}(I_1 - 3) \tag{27}$$

is assigned for both the matrix and the fibers, where μ is the shear modulus, and $I_1 = \mathrm{tr}\,\mathbf{C}$ is the first invariant of the right Cauchy-Green tensor. This simple non-hardening model is specifically chosen to highlight the role of the recruitment process due to the straightening of the fibers. The shear moduli of the fibers and the matrix material are $\mu^{(f)} = 18.22\,\mathrm{MPa}$ and $\mu^{(m)} = 0.271\,\mathrm{MPa}$, respectively (Miyazaki and Hayashi 1999; Muiznieks and Keeley 2013). The structural and the geometrical parameters used in all four models are summarized in Table 1.

We recall that in common applications involving fibrous materials, the out-of-plane loads (in the material coordinate system) are negligible. Accordingly, in the sequel, we assume *plane stress* conditions in both, the material and the lab coordinate system. Recalling Eq. (16), this implies that $\bar{p}$ is determined via the expression for the normal stress in the out-of-plane direction, i.e. $\bar{\sigma}_{yy} = \partial\widetilde{\psi}(\overline{\mathbf{F}})/\partial\bar{F}_{yk}\bar{F}_{yk} - \bar{p} = 0$. In this work, we distinguish and analyze two loading conditions, the aligned and the inclined loading, as described in the following subsections.

3.1 Aligned Loading

First, we examine the response of the fibrous material to a loading state in which the macroscopic principal stretch directions are aligned with the material directions. Thus, $\alpha = 0°$ in Eq. (22) and the *lab coordinate system* coincides with the *material coordinate system*. As we are primarily interested in the analysis of the recruitment process, i.e. the stage in which the fibers straighten and become the main load-bearing elements, we examine the material response in the range $0.9 \leq \lambda_\theta \leq 1.3$. That is from

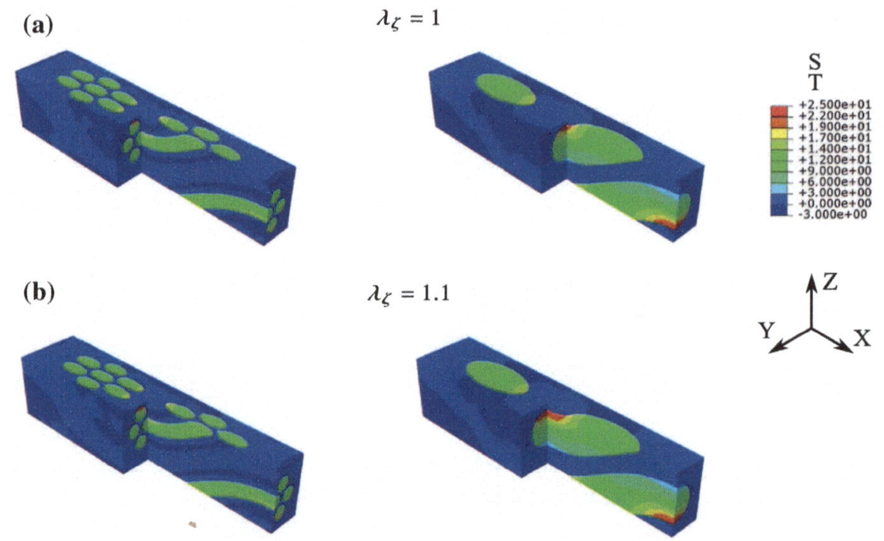

Fig. 2 Contour plots of the longitudinal Cauchy stress distribution σ_{xx} in the *rippled unit cells* ($\lambda_c = 1.1$) at a circumferential stretch of $\lambda_\theta = 1.3$: (a) without a prestretch in the transverse direction; (b) with a transversal prestretch of $\lambda_\zeta = 1.1$

approximately 10% compression in the fiber direction up to 30% extension at which the fibers are almost straight. To examine the effect of the transverse loading, we analyze two cases: (i) without a prestretch in the transverse direction, i.e. $\lambda_\zeta = 1$; (ii) with a transversal prestretch of $\lambda_\zeta = 1.1$.

We begin with the analysis of the stresses at the microscopic level and subsequently examine the corresponding overall stresses, which enables to trace the occurrences at the macroscopic level in terms of those at the microscopic level. The longitudinal Cauchy stress distributions σ_{xx} in the four unit cells are presented for a circumferential stretch of $\lambda_\theta = 1.3$ in Figs. 2 and 3. In all cases, the stresses in the fibers are one order of magnitude larger than those in the surrounding matrix. This is due to the fact that at stretches beyond the recruitment stretch, the fibers, which are typically stiffer than the surrounding matrix material, straighten and carry most of the load. The loading scenario captured in this numerical example, in which the fibers are stretched to the point where they are almost straight and, as they are recruited, start to carry the longitudinal load, is a common physiological in vivo situation. Moreover, also with common expectations, are the bending stresses that develop in the fibers as they are being stretched. The variations in the longitudinal stresses across the fiber cross-sections highlight this bending process, whereas the mean values of the stresses can be associated with the pure fiber extension (stretching).

To highlight these stress variations, the longitudinal Cauchy stresses across the fiber(s) are carefully analyzed along two individual paths at the peak of their harmonic wave, as denoted in Fig. 4. The corresponding longitudinal Cauchy stress distributions for the rippled and the crimped fiber models are presented in Figs. 5

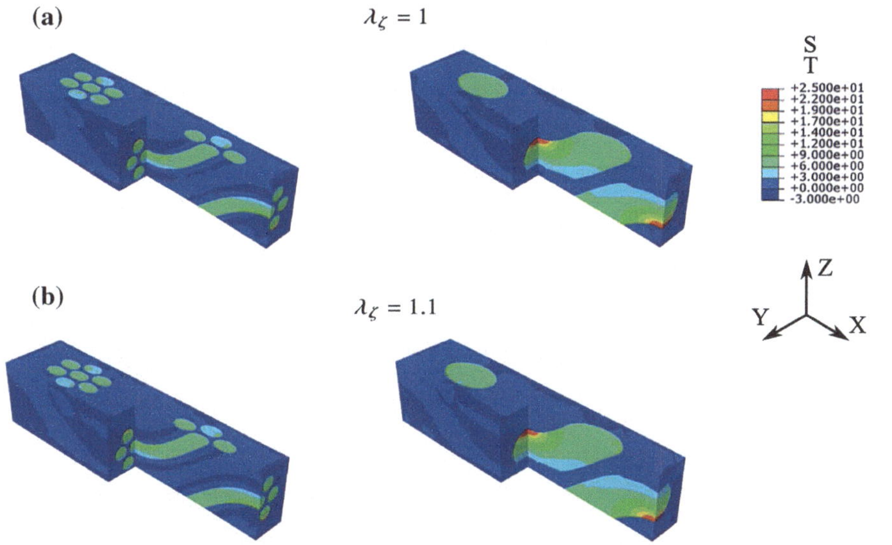

Fig. 3 Contour plots of the longitudinal Cauchy stress distribution σ_{xx} in the *crimped unit cells* ($\lambda_c = 1.2$) at a circumferential stretch of $\lambda_\theta = 1.3$: (a) without a prestretch in the transverse direction; (b) with a transversal prestretch of $\lambda_\zeta = 1.1$

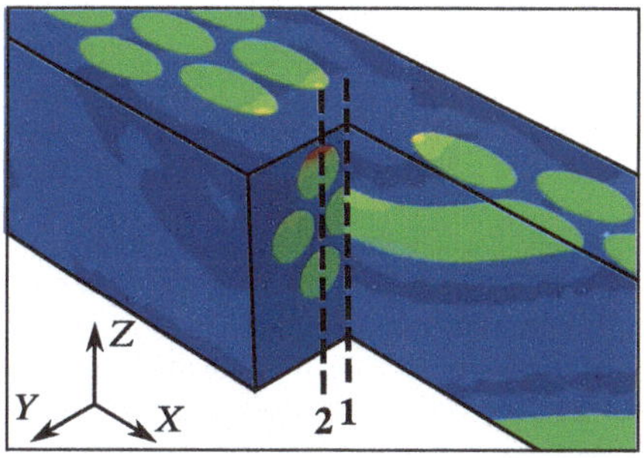

Fig. 4 Definition of two pathways along which the longitudinal Cauchy stress distributions σ_{xx} are analyzed in case of the single fiber model and the seven fibers model: (1) through the middle of the central fiber; (2) through the middle of the top and bottom fibers. Both pathways are parallel to the Z-direction and are located at the peak of their harmonic wave

and 6. For the seven fibers model, we show the stresses along the pathways 1 and 2, which pass through the central fiber and the top as well as bottom fibers, respectively. For the single fiber model, we depict the stresses exclusively along path 1, which passes through the center of the fiber.

Under both loading conditions, in both the rippled and the crimped fiber model, the Cauchy stress distributions σ_{xx} along the cross-sections are similar regardless of the prestretch along the ζ-direction, i.e. for $\lambda_\zeta = 1$ and 1.1. This reveals that the stretching and bending mechanisms are similar for both loading cases under aligned

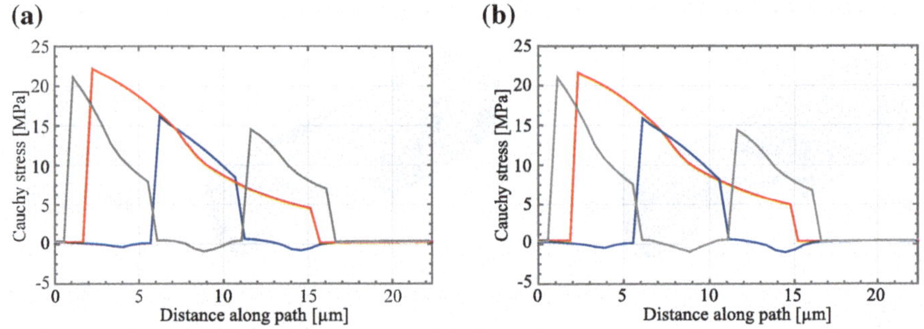

Fig. 5 Longitudinal Cauchy stress distribution σ_{xx} along the defined pathways through the cross-sections of the *rippled fiber models* ($\lambda_c = 1.1$) while stretched at $\lambda_\theta = 1.3$: (a) without a prestretch in the transverse direction; (b) with a transversal prestretch of $\lambda_\zeta = 1.1$. The red curve represents the response of the single fiber model along path 1, while the blue and gray curves correspond to the stress response of the seven fibers model along paths 1 and 2, respectively

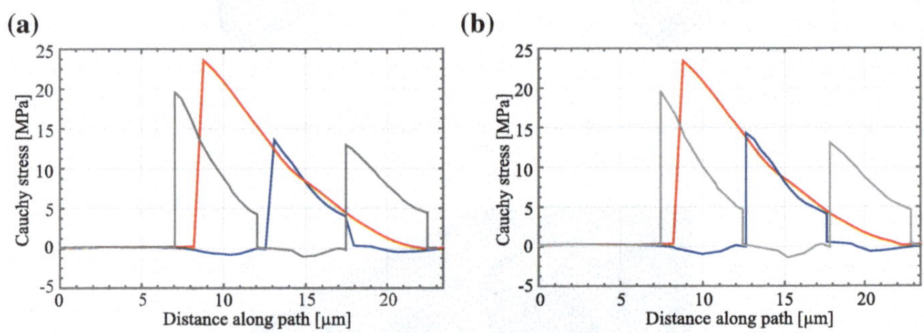

Fig. 6 Longitudinal Cauchy stress distribution σ_{xx} along the defined pathways through the cross-sections of the *crimped fiber models* ($\lambda_c = 1.2$) while stretched at $\lambda_\theta = 1.3$: (a) without a prestretch in the transverse direction; (b) with a transversal prestretch of $\lambda_\zeta = 1.1$. The red curve represents the response of the single fiber model along path 1, while the blue and gray curves correspond to the stress response of the seven fibers model along paths 1 and 2, respectively

loading conditions. A comparison of the stresses in the rippled model and the crimped model reveals that in the rippled fiber model, the mean value of the stresses in the fibers is larger than their variation. In the single fiber model, the mean stress is about 12 MPa, whereas half the difference between the maximal and minimal stresses in the fiber is less than 10 MPa. In the crimped case, the intensity of the stress variation is similar to their mean value, which is approximately 12 MPa. We recall that the mean value of the stress is associated with the elongation of the fiber, whereas the variation of the stress is associated with the bending of the fiber due to its straightening. Thus, the differences between the stress distributions in the two models highlight the fact that in the rippled model, the fibers are almost straight and have already completed their recruitment process, while the fibers in the crimped model are wavier at this

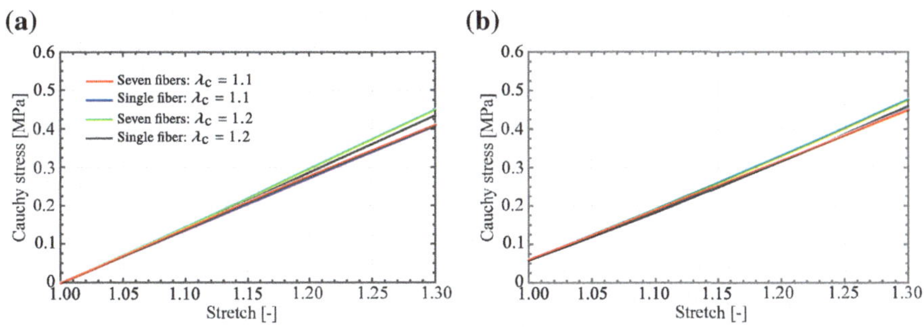

Fig. 7 Average longitudinal Cauchy stresses σ_{xx} in the surrounding matrix as a function of the longitudinal stretch λ_x for the single fiber model and the seven fibers model having different recruitment stretches λ_c: (a) without a prestretch in the transverse direction; (b) with a transversal prestretch of $\lambda_\zeta = 1.1$

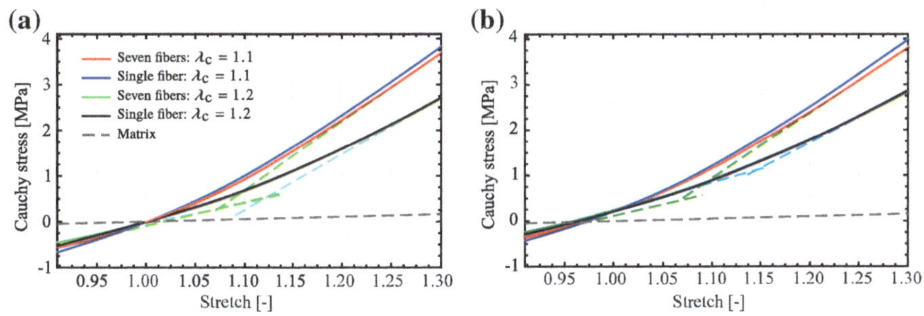

Fig. 8 Average longitudinal Cauchy stresses σ_{xx} in the unit cell as a function of the longitudinal stretch λ_x for the single fiber model and the seven fibers model having different recruitment stretches λ_c: (a) without a prestretch in the transverse direction; (b) with a transversal prestretch of $\lambda_\zeta = 1.1$. The green and cyan dashed line segments represent the initial and the final slopes of the nonlinear fiber stress response

stage of deformation, and thus the stresses associated with the fiber elongation are not the dominant ones yet.

Contrary to materials with straight fibers, we find that due to the fiber waviness, the stresses in the surrounding matrix are not negligible and have a substantial contribution to the macroscopic response. To further investigate this effect, we analyze by means of Fig. 7 the evolution of the average longitudinal Cauchy stresses σ_{xx} in the matrix material for the different fiber models, waviness, and boundary conditions as functions of the longitudinal stretch. The magnitudes of these stresses are almost identical in the rippled and crimped unit cells. Nonetheless, the average stresses in the matrix of the crimped unit cells are somewhat larger, suggesting that the interactions among the fibers are associated with the fiber waviness.

Next, we discuss the macroscopic response of the fibrous tissue. Therefore, the variations of the average longitudinal Cauchy stress versus the longitudinal stretch can be traced in Fig. 8. We note that the stiffening of the fibrous material, which accounts for the non-hardening behavior of both phases, is solely due to the straightening of the fibers. In the unloaded state, the fibers are wavy, they straighten under tension and fold under compression. When being extended from the unloaded state, the initial contribution of the fibers to the overall stress response is rather small. The fibers are recruited with an increase of the applied deformation. Clearly, in the rippled unit cells, the fibers are recruited faster than those in the crimped cells, as reflected by the larger stresses developing in the model with $\lambda_c = 1.1$. The hardening effect can be approximated as a bilinear behavior with an initial lower slope that smoothly increases to a larger one as the stretch increases. To assist the identification of the slopes, we include dashed straight line segments in Fig. 8 that follow the initial and the final slopes of the stress-stretch curves. The bilinear response is exhibited under both loading conditions, i.e. $\lambda_\zeta = 1$ and $\lambda_\zeta = 1.1$. However, the stresses in the $\lambda_\zeta = 1.1$ case are slightly larger as the fibers are prestretched in the transverse direction.

Although the fibers are the primary load bearing components inside the fibrous tissue, the contribution of the surrounding matrix is not negligible. The average longitudinal Cauchy stresses in the matrix of the rippled and the crimped fiber model at $\lambda_\theta = 1.1$ are approximately 14.9% and 21.1% of the overall macroscopic stress, respectively. In a similar manner, when $\lambda_\theta = 1.3$, the average longitudinal Cauchy stresses in the matrix of the rippled and the crimped fiber model are approximately 11.5% and 16.4% of the overall macroscopic stress, respectively. These ratios are substantially larger than the ratio between the shear moduli of the matrix and the collagen fibers, which is 1.5%, indicating a considerable contribution of the matrix to the overall macroscopic stress response due to the interactions among the wavy fibers.

Note that the response of the rippled model with a single fiber is slightly stiffer than that of the rippled model with seven fibers. In accordance with the common notion, this is due to the enhanced bending stiffness of the single fiber, which is proportional to its diameter. However, the responses of the crimped models with one and seven fibers in the range $1 < \lambda_\theta < 1.3$ are almost identical. This is due to two competing mechanisms that contribute to the stiffening of the material. The first is associated with the bending of the fibers, a mechanism more pronounced in the single fiber model. However, in the crimped model, the larger interactions among the fibers result in larger stresses in the matrix phase and contribute to the macroscopic stiffening. In the crimped case, the contribution of the second mechanism in the seven fibers model is equal to that of the first mechanism in the single fiber model, and hence their responses are almost identical.

3.2 Inclined Loading

In a manner similar to the one followed in the previous section, macroscopic deformations were imposed on the boundaries of the four unit cells, this time with the maximum principal stretch direction at a 45° inclination angle relative to the fiber

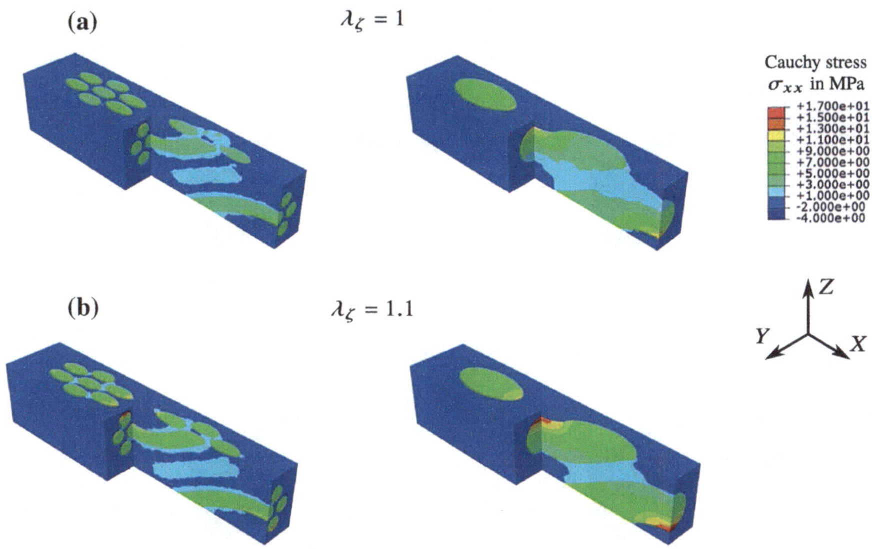

Fig. 9 Contour plots of the longitudinal Cauchy stress distribution σ_{xx} in the *rippled unit cells* ($\lambda_c = 1.1$) at a circumferential stretch of $\lambda_\theta = 1.3$: (a) without a prestretch in the axial direction; (b) with an axial prestretch of $\lambda_\zeta = 1.1$

direction, i.e. $\alpha = 45°$ in Eq. (22). Again, the periodic unit cells were loaded such that the fibrous tissue is circumferentially stretched in the range $0.9 \leq \lambda_\theta \leq 1.3$ with and without considering an axial prestretch of $\lambda_\zeta = 1.1$. We note that when $\lambda_\zeta = 1.1$, Eq. (20) represents a typical physiological in vivo deformation state of the arterial tissue, where the fibers are inclined relative to the circumferential direction, and the tissue is prestretched in the axial direction. According to Eq. (4), in the aligned loading case, the stretch of the fiber λ_x is identical to the macroscopic (lab) stretch λ_θ. Thus, when subjected to $\lambda_\theta = 1.3$, the fibers in both the rippled model and the crimped model are fully recruited. In the inclined loading case, due to the rotation of the fiber, its longitudinal stretch λ_x is smaller than the macroscopic (lab) stretch λ_θ. Hence, in this case, the fibers in the rippled model are fully recruited, while in the crimped model, they are only partially recruited.

The distributions of the longitudinal Cauchy stress σ_{xx} in the four unit cell models are presented in terms of contour plots in Figs. 9 and 10. In accordance with the aligned loading case, the longitudinal Cauchy stresses in the fibers are larger than those in the surrounding matrix material. However, due to the fact that in this case, the fibers straighten while being rotated towards the direction of the principal stretch, their straightening rate is lower. As a result, the stresses in the fibers are smaller than the ones observed in the aligned loading case. On the other hand, the magnitude of the stresses in the matrix is larger.

In analogy with Figs. 5 and 6 for the aligned loading case, Figs. 11 and 12 show the variations of the longitudinal Cauchy stress σ_{xx} along the pathways 1 and 2, as

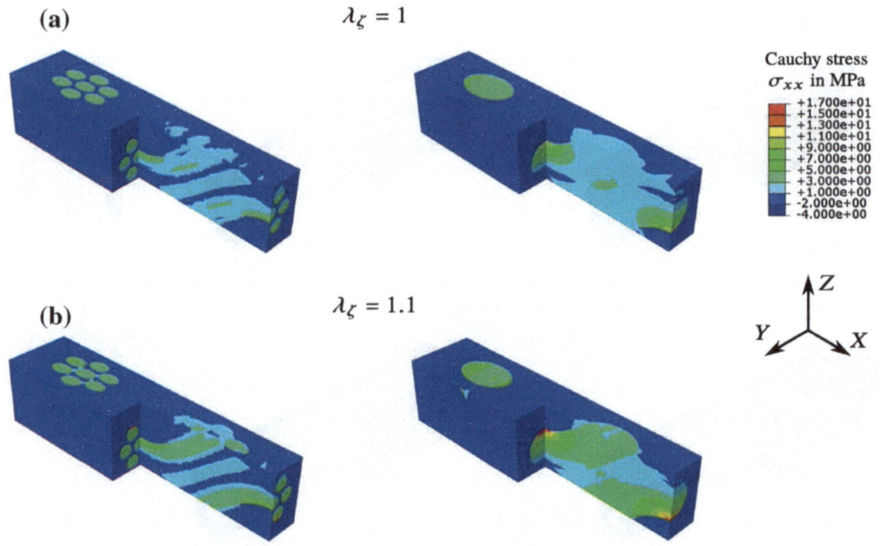

Fig. 10 Contour plots of the longitudinal Cauchy stress distribution σ_{xx} in the *crimped unit cells* ($\lambda_c = 1.2$) at a circumferential stretch of $\lambda_\theta = 1.3$: (a) without a prestretch in the axial direction; (b) with an axial prestretch of $\lambda_\zeta = 1.1$

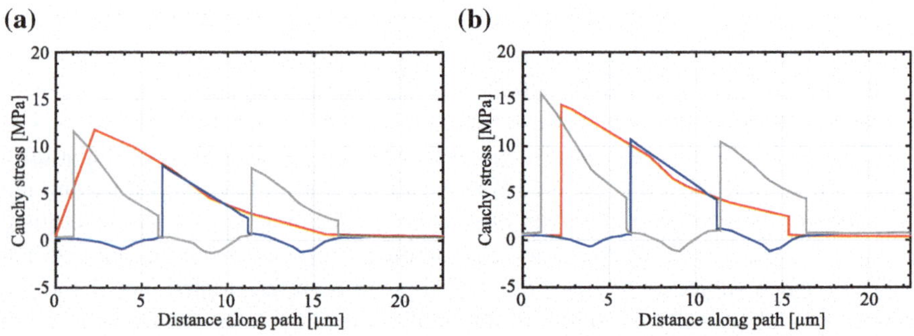

Fig. 11 Longitudinal Cauchy stress distribution σ_{xx} along the defined pathways through the cross-sections of the *rippled fiber models* ($\lambda_c = 1.1$) while stretched at $\lambda_\theta = 1.3$: (a) without a prestretch in the axial direction; (b) with an axial prestretch of $\lambda_\zeta = 1.1$. The red curve represents the response of the single fiber model along path 1, while the blue and gray curves correspond to the stress response of the seven fibers model along paths 1 and 2, respectively

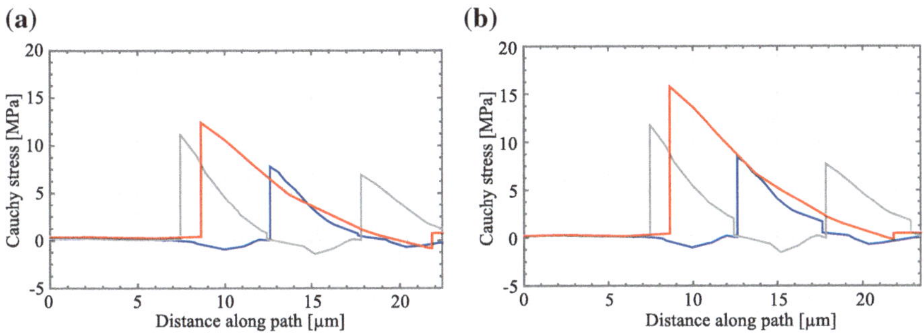

Fig. 12 Longitudinal Cauchy stress distribution σ_{xx} along the defined pathways through the cross-sections of the *crimped fiber models* ($\lambda_c = 1.1$) while stretched at $\lambda_\theta = 1.3$: (a) without a prestretch in the axial direction; (b) with an axial prestretch of $\lambda_\zeta = 1.1$. The red curve represents the response of the single fiber model along path 1, while the blue and gray curves correspond to the stress response of the seven fibers model along paths 1 and 2, respectively

introduced in Fig. 4. Again, we recall that the mean value of the longitudinal stresses in the fiber is associated with its elongation, and the variations in the longitudinal stresses along the fiber cross-section correspond to its bending. We note the difference in the longitudinal Cauchy stress distributions with and without considering an axial prestretch. While prestretched, the mean longitudinal Cauchy stresses in both the rippled and the crimped fibers reach a value of about 7 MPa, and when they are not prestreched this value is reduced to 5 MPa. This highlights the fact that with prestretching, the fibers elongate more, but under inclined loading, this elongation is smaller due to the underlying fiber rotation. The variations of the longitudinal stresses along the fiber cross-sections are in the same order of magnitude as the mean values, indicating that the fibers are wavier than those in the aligned loading case.

Similar to the aligned loading case, in Fig. 13, we present the evolution of the average longitudinal Cauchy stresses σ_{xx} in the surrounding matrix material for the different fiber models, waviness, and the boundary conditions as functions of the longitudinal stretch λ_x. In the rippled fiber model and the crimped fiber model, the magnitudes of these stresses are larger than the ones in the aligned loading case. The average longitudinal Cauchy stresses in the matrix are larger in the unit cells containing seven fibers, except for the prestreched rippled unit cell. This finding highlights the competition between the bending mechanism and the interactions among the individual fibers, indicating that in almost all cases, the interactions among the fibers have a substantial effect on the macroscopic stress response.

The macroscopic responses of the four models in terms of the variations of the average longitudinal Cauchy stress σ_{xx} versus the longitudinal stretch λ_x can be traced in Fig. 14. To emphasize the significant difference in the responses under aligned and inclined loadings, we present these plots on the same scale as the ones in Fig. 8. Both, the bilinear behavior and the difference between the responses of the rippled models and the crimped models are less obvious in comparison with the corresponding aspects under aligned loading. To assist the identification of the mild

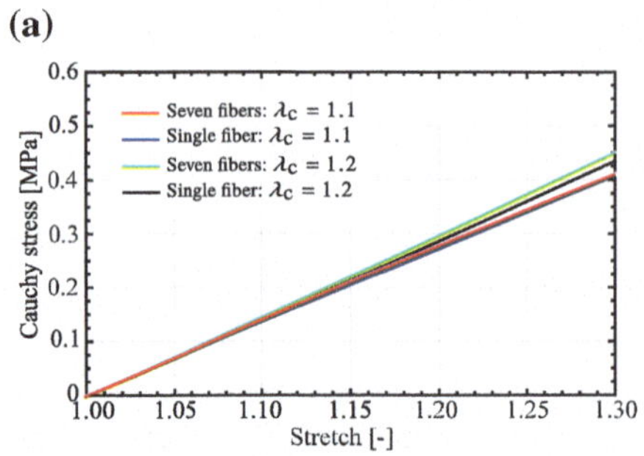

Fig. 13 Average longitudinal Cauchy stress σ_{xx} in the surrounding matrix material as a function of the longitudinal stretch λ_x for the single fiber model and the seven fibers model having different recruitment stretches λ_c: (a) without a prestretch in the axial direction; (b) with an axial prestretch of $\lambda_\zeta = 1.1$

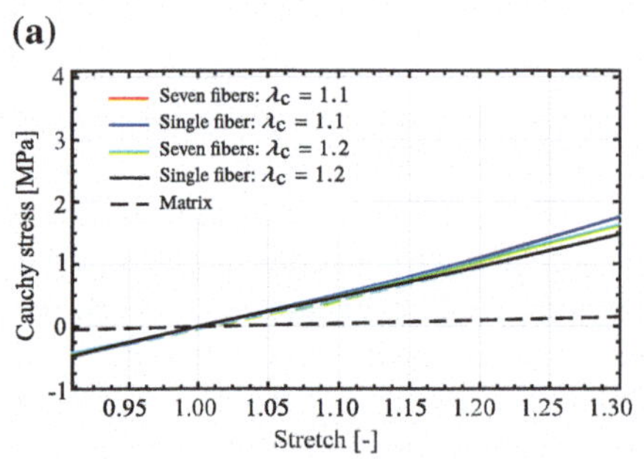

Fig. 14 Average longitudinal Cauchy stresses σ_{xx} in the unit cell and the surrounding matrix as a function of the longitudinal stretch λ_x for the single fiber model and the seven fibers model having different recruitment stretches λ_c: (a) without a prestretch in the axial direction; (b) with an axial prestretch of $\lambda_\zeta = 1.1$. The green and cyan dashed line segments represent the initial and the final slopes of the nonlinear fiber stress response

slopes, we include dashed straight line segments that follow the initial and the final slopes of the stress-stretch curves. This highlights the stiffening of the material due to the straightening and the rotation of the fibers.

Further, the macroscopic stress response is correlated with the stresses developing in the matrix due to the interactions among the fibers. This indicates that these interactions have a substantial effect on the overall stress response, and in this case, the competition between the bending and the interactions among the fibers is further enhanced. In the crimped models, the response of the unit cell with the seven fibers is stiffer than the one with the single fiber for both $\lambda_\zeta = 1$ and 1.1. This highlights the severe interactions among the fibers due to their waviness. In the rippled models, the bending mechanism is more pronounced for $\lambda_\zeta = 1.1$ as we observe that the response of the single fiber model is stiffer than that of the seven fibers model. However, the difference in the stress responses between the single and the seven fibers models is negligible for $\lambda_\zeta = 1$.

4 Conclusions

To accurately characterize the mechanical response of fibrous materials, we developed micromechanically motivated models of wavy fibers. For example, the collagen fibers in the adventitial layer of the artery should be modeled by taking into consideration their level of waviness, configuration, and orientation. To examine the role of these parameters, four FE models were constructed, involving a single fiber or bundles of seven fibers with two levels of waviness for each model by setting the recruitment parameter λ_c to 1.1 or 1.2. The models were analyzed under aligned and inclined loading conditions.

An important outcome of the fiber waviness is the overall bilinear response. Indeed, under aligned loading conditions, a clear bilinear trend was observed. This is due to the fiber recruitment process in which, depending on the degree of waviness, the fibers are recruited and stiffen with increasing tissue extension. This mechanism is captured by our micromechanical models, where the recruitment is associated with the straightening of the fibers. The results demonstrate that the response of the rippled fiber model is stiffer than the one exhibited by the crimped fiber model. This happens due to the straightening of the fibers at smaller stretches, and thus, the fiber stiffening is faster. Under inclined loading, similar but more moderate bilinear behaviors were observed. This is because, in the inclined loading case, the fibers straighten while they rotate towards the major principal direction of the applied loading.

We recall that the straightening of the fibers is correlated with their bending. Accordingly, it is anticipated that due to its almost three-fold larger diameter and enhanced bending stiffness, the response of the single fiber model will be stiffer than the seven fiber model. However, interestingly and counter-intuitively, we find a reverse trend in some cases, where the models with the seven fibers admit a stiffer response. Thus, our micromechanical models shed light on the crucial role of the matrix and the interactions between the individual wavy fibers. Throughout, we observe the competing effects of the through-matrix fiber interactions and the fiber bending stiffness on the overall stress response. This is most pronounced in relation to the stresses developing in the matrix of the crimped models, where the interactions between the fibers are significant. We note that in biological tissues, the interactions between the fibers may be through physical or chemical cross-links between the fibers. In fact, we observe that the interactions between the fibers through the extracellular matrix are reminiscent of the interactions through the cross-links between the fibers (Ben-Or Frank 2020; Ben-Or Frank et al. 2021).

We stress that our proposed micromechanical modeling approach can be used to, but is not limited to, analyze biological tissues. Histological information of fibrous tissues with wavy fibers can be incorporated in the suggested models, and their mechanical behavior can be predicted and compared with experimental measurements (Ben-Or Frank et al. 2019). Furthermore, this approach can be applied in various lightweight industrial structures undergoing large deformations and with inherent high stiffness damping. This is crucial for cutting-edge technologies aimed towards reducing energy consumption and high performance in airplanes with morphing wings, stretchable electronic and wearable devices in which structures based on wavy fibers are desired (Liu et al. 2017; Pratt 1999; Wen et al. 2015).

References

Anumandla, V., Gibson, R.F.: A comprehensive closed form micromechanics model for estimating the elastic modulus of nanotube-reinforced composites. Compos. Part A Appl. Sci. Manuf. **37**, 2178–2185 (2006)

Cacho, F., Elbischger, P.J., Rodríguez, J.F., Doblaré, M., Holzapfel, G.A.: A constitutive model for fibrous tissues considering collagen fiber crimp. Int. J. Non-Linear Mech. **42**, 391–402 (2007)

Calvo, B., Peña, E., Martinez, M.A., Doblaré, M.: An uncoupled directional damage model for fibred biological soft tissues. Formulation and computational aspects. Int. J. Numer. Meth. Engng. **69**, 2036–2057 (2007)

Chen, H., Liu, Y., Zhao, X., Lanir, Y., Kassab, G.S.: A micromechanics finite-strain constitutive model of fibrous tissue. J. Mech. Phys. Solids **59**, 1823–1837 (2011)

Choi, W.M., Song, J., Khang, D.Y., Jiang, H., Huang, Y.Y., Rogers, J.A.: Biaxially stretchable 'wavy' silicon nanomembranes. Nano Lett. **7**, 1655–1663 (2007)

Chortos, A., Bao, Z.: Skin-inspired electronic devices. Mater. Today **17**, 321–331 (2014)

Chuong, C.J., Fung, Y.C.: Three-dimensional stress distribution in arteries. J. Biomech. Eng. **105**, 268–274 (1983)

deBotton, G., Oren, T.: Analytical and numerical analyses of the micromechanics of soft fibrous connective tissues. Biomech. Model. Mechanobiol. **12**, 151–166 (2013)

Fallah, A., Ahmadian, M.T., Firozbakhsh, K., Aghdam, M.M.: Micromechanics and constitutive modeling of connective soft tissues. J. Mech. Behav. Biomed. Mater. **60**, 157–176 (2016)

Garnich, M.R., Karami, G.: Finite element micromechanics for stiffness and strength of wavy fiber composites. J. Compos. Mater. **38**, 273–292 (2004)

Gasser, T.C., Ogden, R.W., Holzapfel, G.A.: Hyperelastic modelling of arterial layers with distributed collagen fibre orientations. J. R. Soc. Interface **3**, 15–35 (2006)

Gent, A.N.: A new constitutive relation for rubber. Rubber Chem. Technol. **69**, 59–61 (1996)

Górski, R.: Elastic properties of composites reinforced by wavy carbon nanotubes. Mech. Control **30**, 203–212 (2011)

Hill, R.: On constitutive macro-variables for heterogeneous solids at finite strain. Proc. R. Soc. Lond. A **326**, 131–147 (1972)

Holzapfel, G.A., Niestrawska, J.A., Ogden, R.W., Reinisch, A.J., Schriefl, A.J.: Modelling non-symmetric collagen fibre dispersion in arterial walls. J. R. Soc. Interface **12**, 20150188 (2015)

Humphrey, J.D., Dufresne, E.R., Schwartz, M.A.: Mechanotransduction and extracellular matrix homeostasis. Nat. Rev. Mol. Cell Bio. **15**, 802–812 (2014)

Kao, P.H., Lammers, S.R., Tian, L., Hunter, K., Stenmark, K.R., Shandas, R., Qi, H.J.: A microstructurally driven model for pulmonary artery tissue. J. Biomech. Eng. **133**, 051002 (2011)

Karami, G., Garnich, M.: Effective moduli and failure considerations for composites with periodic fiber waviness. Compos. Struct. **67**, 461–475 (2005)

Li, M., Mondrinos, M.J., Gandhi, M.R., Ko, F.K., Weiss, A.S., Lelkes, P.I.: Electrospun protein fibers as matrices for tissue engineering. Biomaterials **26**, 5999–6008 (2005)

Liang, Z., Lee, H.K., Suaris, W.: Micromechanics-based constitutive modeling for unidirectional laminated composites. Int. J. Solids Struct. **43**, 5674–5689 (2006)

Liu, Y., Pharr, M., Salvatore, G.A.: Lab-on-skin: a review of flexible and stretchable electronics for wearable health monitoring. ACS Nano **11**, 9614–9635 (2017)

Ben-Or Frank, M.: Predictive micromechanics motivated modeling of fibrous tissues. Ph.D. Thesis, Ben-Gurion University (2020)

Ben-Or Frank, M., Niestrawska, J.A., Holzapfel, G.A., deBotton, G.: Micromechanically-motivated analysis of fibrous tissue. J. Mech. Behav. Biomed. Mater. **96**, 69–78 (2019)
Ben-Or Frank, M., Holzapfel, G.A., deBotton, G.: The influence of a cross-linked structure on the mechanical response of fibrous fissues (2021), in preparation
Miyazaki, H., Hayashi, K.: Tensile tests of collagen fibers obtained from the rabbit patellar tendon. Biomed. Microdevices **2**, 151–157 (1999)
Muiznieks, L.D., Keeley, F.W.: Molecular assembly and mechanical properties of the extracellular matrix: a fibrous protein perspective. Biochim. Biophys. Acta **1832**, 866–875 (2013)
Ogden, R.W.: On the overall moduli of non-linear elastic composite materials. J. Mech. Phys. Solids **22**, 541–553 (1974)
Pratt, W.F.: Patterned fiber composites: process, characterization, and damping performance. Ph. D. Thesis, Brigham Young University (1999)
Pratt, W.F., Allen, M.S., Jensen, C.G.: Designing with wavy composites. In: 46th International SAMPE Symposium and Exhibition, pp. 203–215 (2001)
Rezakhaniha, R., Agianniotis, A., Schrauwen, J.T.C., Griffa, A., Sage, D., Bouten, C.V.C., van de Vosse, F.N., Unser, M., Stergiopulos, N.: Experimental investigation of collagen waviness and orientation in the arterial adventitia using confocal laser scanning microscopy. Biomech. Model. Mechanobiol. **11**, 461–473 (2012)
Schriefl, A.J., Zeindlinger, G., Pierce, D.M., Regitnig, P., Holzapfel, G.A.: Determination of the layer-specific distributed collagen fiber orientations in human thoracic and abdominal aortas and common iliac arteries. J. R. Soc. Interface **9**, 1275–1286 (2012)
Sopakayang, R., Holzapfel, G.A.: A constitutive model of ligaments and tendons accounting for fiber-matrix interaction. Int. J. Biol. Biomed. Eng. **11**, 245–249 (2017)
Velmurugan, R., Srinivasulu, G., Jayasankar, S.: Influence of fiber waviness on the effective properties of discontinuous fiber reinforced composites. Comput. Mater. Sci. **91**, 339–349 (2014)
Weisbecker, H., Unterberger, M.J., Holzapfel, G.A.: Constitutive modelling of arteries considering fibre recruitment and three-dimensional fibre distribution. J. R. Soc. Interface **12**, 20150111 (2015)
Wen, L., Shi, Y., Chen, J., Yan, B., Li, F.: Wavy structures for stretchable energy storage devices: structural design and implementation. Chin. Phys. B **25**, 018207 (2015)
Zulliger, M.A., Fridez, P., Hayashi, K., Stergiopulos, N.: A strain energy function for arteries accounting for wall composition and structure. J. Biomech. **37**, 989–1000 (2004)

Configurational Forces in Penetration Processes

Davide Bigoni, Marco Amato, and Francesco Dal Corso

The authors are proud of celebrating Professor Gerhard A. Holzapfel and his outstanding contributions to biomechanics on the occasion of his 60th birthday. The friendship between Davide and Gerhard began twenty years ago when they were planning a joint research project on the mechanics of human brain tissue. This was a very innovative topic at that time, and many researchers did not believe that mechanics is relevant for brain matter. Gerhard and Davide co-supervised a Ph.D. student, Giulia Franceschini, they worked together and published a paper. In that period, Davide enjoyed a long visit to Gerhard's laboratory in Graz, which he remembers with great pleasure. Since then, Davide has had several fruitful interactions and helpful scientific discussions with Gerhard. He looks forward to many more years of learning from Professor Holzapfel's papers and talks and enjoying his delightful friendship.

Davide

Abstract With a loose reference to problems of penetration in biomechanics (for instance, a penetrating through a cell's membrane or a cell sucked with a pipette), the role of configurational forces is investigated during the process in which a compliant intruder is inserted into an elastic structure. For insertion into a rigid constraint, a configurational force proportional to the square of the strain needed to deform the body, which is penetrating, is found. This force has a more complex structure when the compliance of the constraint is kept into account, but in all cases, it tends to expel the penetrating body.

D. Bigoni (✉) · M. Amato · F. Dal Corso
Department of Civil, Environmental and Mechanical Engineering, University of Trento, Via Mesiano 77, 38123 Trento, Italy
e-mail: bigoni@ing.unitn.it

G. Sommer et al. (eds.), *Solid (Bio)mechanics: Challenges of the Next Decade*, Studies in Mechanobiology, Tissue Engineering and Biomaterials 24,
https://doi.org/10.1007/978-3-030-92339-6_19

1 Introduction

Biomechanics is a fast developing multidisciplinary science aimed at challenging the secrets of nature. The biomechanics of soft tissues, including successfully modeling of muscles (Stålhand et al. 2008), blood vessels (Holzapfel et al. 2000, 2004), brain (Franceschini et al. 2006; Mihai et al. 2017), and vascular tissues (Pandolfi and Holzapfel 2008), is recently oriented to a deeper understanding of the mechanics of the cell (Discher et al. 1998; Boal 2002; Deseri and Zurlo 2013; Terzi et al. 2015; Deseri et al. 2016; Daddi-Moussa-Ider et al. 2019), a field strongly related to several timely problems, including the COVID-19 emergency.

Membrane penetration of nano- or microparticles is a common feature in cell mechanics, important from several points of view, including drug delivery and viral entry. In particular, viral entry into a cell during the early stage of infection may occur in different forms. In the case of penetration, the cell's membrane is punctured after attachment of the virus, which in this way injects its contents inside the cytoplasm.

Cells may be subject to large deformations, a situation occurring, for instance, when a cell is drawn up a pipette, as is the case of a human red blood cell (Discher et al. 1994). This process involves a competition between forces tending to inject and eject the cell. Similarly, but at a smaller scale, the influenza A virus can be trapped into a nanotube with an internal diameter of 400 nm (Yuge et al. 2017). Membrane penetration and pipette suction are just two examples of problems involving the insertion of a compliant body inside a soft (the membrane) or a rigid (the pipette) structure. Many other problems of this kind can be listed in biomechanics: endoscopy, insertion of a catheter into a blood vessel or into the urinary tract, or injection of a needle for biopsy or for puncturing tissues (Roxhed et al. 2007).

In the present article, the penetration of a compliant body into an elastic structure (such as a cell's membrane) or into a rigid constraint (as when a cell is drawn up a pipette) is analyzed from a purely mechanical perspective, under simple assumptions. These include restriction to the two-dimensional formulation (a gross, but common, approximation (Daddi-Moussa-Ider et al. 2019)) and absence of: (i) friction; (ii) dissipative forces; (iii) interparticle non-mechanical interactions (e.g., magnetic attraction). The simple mechanical process of penetration schematized in Fig. 1 is addressed, with the specific purpose to explore and present the role of configura-

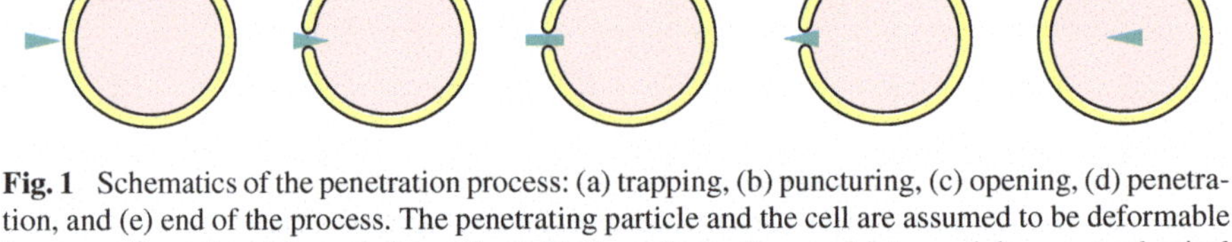

Fig. 1 Schematics of the penetration process: (a) trapping, (b) puncturing, (c) opening, (d) penetration, and (e) end of the process. The penetrating particle and the cell are assumed to be deformable in a two-dimensional formulation, where friction, dissipation, and interparticle non-mechanical interactions are neglected

tional forces, actions that develop when an elastic system can change its configuration through a release of elastic energy. These forces have been introduced by Eshelby (1951) to describe the mechanics of defects and only recently explored in elastic structures subject to bending (Bigoni et al. 2015) and torsion (Bigoni et al. 2014b). Configurational forces in structures lead to unexpected effects, such as possibility of self-encapsulation (Bosi et al. 2015) and expulsion during a penetration process (Bigoni et al. 2014a) due to bending flexibility of inextensible rods.

2 Penetration and Configurational Forces

A simple model is introduced, which may capture essential features of the 'puncturing' and 'penetrating' stages of the intrusion process, shown in Fig. 1. Besides its application to mechanobiology, the model is formulated to introduce the role of the configurational force in the processes of insertion of an elastic body into another.

A deformable elastic layer is considered of initial height h_0, out-of-plane width b_0, and length for the moment left unspecified. This layer has to be inserted between two rigid planar and frictionless constraints, placed at a distance $\bar{h} \leq h_0$ along the x_2-axis when the two linear elastic springs of stiffness k are unloaded, see Fig. 2a. These two rigid plates are prescribed to remain parallel to the x_1-x_3 plane.

The spring stiffness k simply models the elastic stiffness of the structure in which the intruder has to penetrate. For instance, when the structure is a thin elastic ring (as sketched in Fig. 1) of radius R and bending stiffness EJ, the spring stiffness becomes $k = 2EJ/(3\pi R^3)$. The elastic layer is partially inserted (of an amount $l > 0$) between the two frictionless constraints so that a deformed configuration

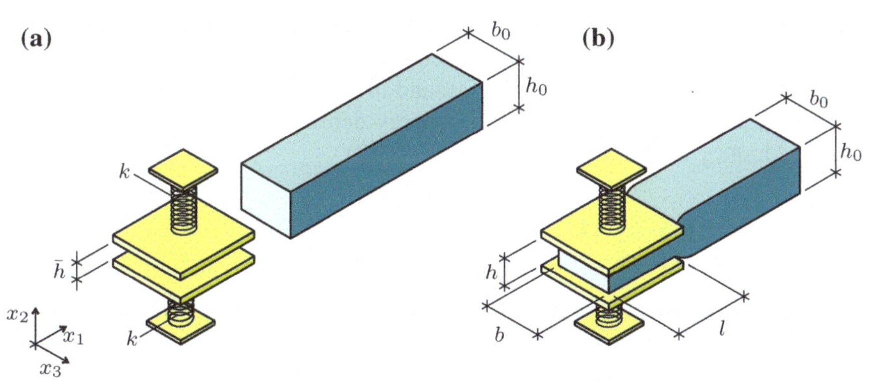

Fig. 2 An elastic layer is inserted between two rigid, flat, and frictionless plates (remaining parallel to the x_1-x_3 plane), each held by a linear spring of elastic stiffness k. The distance between the two rigid plates at rest is $\bar{h}$, smaller than the height h_0 of the elastic layer (a prism of initial cross-section $h_0 \times b_0$) in its unstressed configuration. In the deformed configuration, the prism height is $h \in [\bar{h}, h_0]$, the width is $b \geq b_0$, and the insertion length is l

such as that illustrated in Fig. 2b is realized, where the height of the inserted layer is $h \in [\bar{h}, h_0]$. Although near the edge of the constraint exit, the stress/strain state is highly disuniform, the unloaded configuration is reached at a sufficient distance far from the constraint's exit, while a state of uniaxial compression is approached inside the rigid plates (where the width of the layer becomes $b \geq b_0$). Therefore, the edges of the constraint induce a strong disturbance, but the stress/strain state tends to become uniform far from this point, when the elastic layer is sufficiently long. Since the contact between the layer and the constraint is smooth, one would be tempted to conclude that no horizontal forces are applied to the elastic body along the x_1-axis. This too facile conclusion is missing the forces which develop at the corners of the constraint, representing the 'microscopic' counterpart of the configurational force concept. In particular, the configurational force can be understood as follows. The total potential energy $\mathcal{V}$ of the deformed system represented in Fig. 2 is coincident with the elastic energy $\mathcal{E}$, which is given by the sum of that stored in the springs and in the layer. This energy has to depend on the configurational parameter l, i.e. the amount of layer inside the constraint. If l decreases, the energy $\mathcal{V}(l) = \mathcal{E}(l)$ will decrease too, until becoming null when the layer loses contact with the rigid plates because outside ($l \leq 0$). Eshelby (1951) defined the configurational force P acting on the elastic system as the negative of the partial derivative of the total potential energy $\mathcal{V}(l)$ at equilibrium with respect to the configurational parameter. In the present case, the configurational force reduces to

$$P = -\frac{\mathrm{d}\mathcal{E}(l)}{\mathrm{d}l}, \tag{1}$$

where the '−' sign arises from the fact that on increasing l, the layer moves in the negative direction of the x_1-axis. Therefore, a configurational force emerges, which pushes the layer outside the constraint and can be viewed from a 'microscopic' point of view as the tangential reaction at the frictionless sliding constraint, provided as the resultant of the actions realized at the constraints corners.

The elastic energies stored in the layer and in the springs can be evaluated in an approximate way as follows. The strain-energy density function W for an incompressible Mooney Rivlin material is (Bigoni 2012)

$$W(\lambda_1, \lambda_2, \lambda_3) = \frac{\mu_1}{2}\left(\lambda_1^2 + \lambda_2^2 + \lambda_3^2 - 3\right) - \frac{\mu_2}{2}\left(\frac{1}{\lambda_1^2} + \frac{1}{\lambda_2^2} + \frac{1}{\lambda_3^2} - 3\right), \tag{2}$$

where μ_1 and μ_2 are material constants, while λ_1, λ_2, and λ_3 are the principal stretches, the latter subject to the incompressibility constraint

$$\lambda_1 \lambda_2 \lambda_3 = 1. \tag{3}$$

The moduli μ_1 and μ_2 are subject to the restrictions

$$\mu_1 \geq 0, \qquad \mu_2 \leq 0, \tag{4}$$

and values representative for the behavior of rubber at room temperature are $\mu_1 = 3\,\text{bar}$ and $\mu_2 = -0.3\,\text{bar}$.

Consider a parallelepiped of initial height h_0 and transverse dimensions l_0 and b_0, compressed parallel to the edge h_0 with a uniaxial state of stress, reducing the height to h and enlarging the other two dimensions to l and b. The stretches are

$$\lambda_1 = \frac{l}{l_0}, \qquad \lambda_2 = \frac{h}{h_0}, \qquad \lambda_3 = \frac{b}{b_0}, \tag{5}$$

which, due to loading symmetry ($\lambda_1 = \lambda_3$) and material incompressibility (Eq. (3)), are constrained by

$$\lambda_1 = \frac{l}{l_0} = \lambda_3 = \frac{b}{b_0} = \sqrt{\frac{h_0}{h}}. \tag{6}$$

Therefore, the elastic energy stored in the uniformly deformed parallelepiped corresponds to

$$l_0 b_0 h_0\, W = l\sqrt{\frac{h}{h_0}}\, b_0 h_0 \left[\frac{\mu_1}{2}\left(\frac{h^2}{h_0^2} + 2\frac{h_0}{h} - 3\right) - \frac{\mu_2}{2}\left(\frac{h_0^2}{h^2} + 2\frac{h}{h_0} - 3\right)\right], \tag{7}$$

where the dependency on the parameter l, crucial in the following calculations, has been evidenced.

When the stretch $\lambda_2 = h/h_0$ is close to 1, the energy (Eq. (7)) can be approximated by

$$l_0 b_0 h_0\, W = \frac{3}{2}(\mu_1 - \mu_2)\, l\, b_0\, h_0 \epsilon^2, \tag{8}$$

where ϵ is the infinitesimal strain along the x_2-axis

$$\epsilon = \frac{h}{h_0} - 1, \tag{9}$$

and for small strain, the elastic energy (Eq. (7)) reduces for $\mu_1 = \mu$ and $\mu_2 = 0$ to the strain energy of an isotropic and incompressible linear elastic solid.

The energy, as given in Eq. (7), provides a simple approximation to that stored in the whole deformed elastic layer. The contribution of the highly inhomogeneous zone near the edge of the constraint is completely neglected, but this approximation becomes reasonable when the parts of the layer inside the constraint and outside are sufficiently long. In this way, the variation of the stored energy when a part of the layer is expelled from the constraint corresponds with a good approximation to the final segment of the layer inside the constraint, which is subject to a stress/strain state approximately uniform and corresponding to uniaxial stress.

2.1 Penetration of a Rigid Body

If the penetrated body in which the elastic intruder is inserted is rigid, $k \to \infty$, the height $h = \bar{h}$ is fixed, and no energy is stored within the springs. In these conditions and for the above considerations, the configurational force P can be easily obtained through differentiation of Eq. (7) with respect to the parameter l, i.e.

$$\frac{P}{\mu_1 b_0 h_0} = -\frac{1}{2}\sqrt{\frac{h}{h_0}}\left[\frac{h^2}{h_0^2} + 2\frac{h_0}{h} - 3 - \frac{\mu_2}{\mu_1}\left(\frac{h_0^2}{h^2} + 2\frac{h}{h_0} - 3\right)\right]. \tag{10}$$

Using the conditions stated in Eq. (4), it can be seen from Eq. (10) that the configurational force is negative ($P < 0$), and therefore, it tends to expel the layer from the constraint (in other words, to move it in the positive direction of the x_1-axis).

Equation (10) is a complex function of the elastic energy via the stretch $\lambda_2 = h/h_0$ and shows the independence of the length l for the configurational force P. When the strain ϵ is small, the configurational force P can be approximated as a quadratic expression of the infinitesimal strain ϵ, i.e.

$$\frac{P}{\mu_1 b_0 h_0} = -\frac{3}{2}\left(1 - \frac{\mu_2}{\mu_1}\right)\epsilon^2. \tag{11}$$

2.2 Role of Penetrated Body's Elasticity

In this section, the effect of the elasticity of the structure penetrated by the intruding layer is accounted for by considering each rigid plate suspended by a linear elastic spring of stiffness k, as illustrated in Fig. 2. The total elastic energy $\mathcal{E}$, consisting of the energies stored in the two springs of stiffness k and the elastic layer, is

$$\frac{\mathcal{E}}{\mu_1 b_0 h_0} = \frac{k h_0}{4\mu_1 b_0}\left(\frac{h}{h_0} - \frac{\bar{h}}{h_0}\right)^2 + \frac{l}{2}\sqrt{\frac{h}{h_0}}\left[\frac{h^2}{h_0^2} + 2\frac{h_0}{h} - 3 - \frac{\mu_2}{\mu_1}\left(\frac{h_0^2}{h^2} + 2\frac{h}{h_0} - 3\right)\right]. \tag{12}$$

In this case, the height h is not imposed, as in Eq. (10), but is an unknown, which depends on the relative stiffness between the springs and the elastic layer. To proceed further, the small strain assumption is introduced, together with $\mu_1 = \mu$ and $\mu_2 = 0$. Under these restrictions, the configurational force can be easily calculated as follows. The total elastic energy $\mathcal{E}$ (Eq. (12)) becomes a function of the (small) strain ϵ in the layer and reads

$$\frac{\mathcal{E}(\epsilon, l)}{\mu b_0 h_0} = \frac{k h_0}{4\mu b_0}\left(1 + \epsilon - \frac{\bar{h}}{h_0}\right)^2 + \frac{3}{2} l \epsilon^2. \tag{13}$$

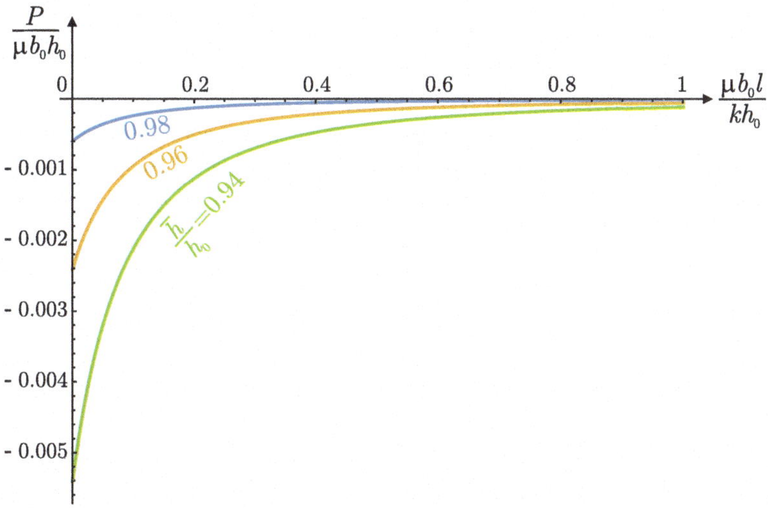

Fig. 3 Configurational force P, made dimensionless through division by $\mu b_0 h_0$, as a function of $\mu b_0 l/(kh_0)$ for three values of the ratio $\bar{h}/h_0 = \{0.94, 0.96, 0.98\}$

The stationarity of the total energy $\mathcal{E}(\epsilon, l)$, as introduced in Eq. (13), with varying the strain ϵ in the layer gives the strain ϵ^* at equilibrium for a given length l, i.e.

$$\epsilon^*(l) = -\frac{1 - \dfrac{\bar{h}}{h_0}}{1 + \dfrac{6\mu b_0 l}{kh_0}} \tag{14}$$

(a negative quantity because of $\bar{h}/h_0 < 1$). Hence, the total elastic energy (Eq. (13)) evaluated at ϵ^* is

$$\frac{\mathcal{E}(\epsilon^*(l), l)}{\mu b_0 h_0} = \frac{kh_0}{4\mu b_0} \frac{\left(1 - \dfrac{\bar{h}}{h_0}\right)^2}{1 + \dfrac{kh_0}{6\mu b_0 l}}. \tag{15}$$

Using Eq. (1), the configurational force P reads

$$\frac{P}{\mu b_0 h_0} = -\frac{3}{2} \left(\frac{1 - \dfrac{\bar{h}}{h_0}}{1 + \dfrac{6\mu b_0 l}{kh_0}} \right)^2 \tag{16}$$

and is reported in Fig. 3 as a function of $\mu b_0 l/(kh_0)$, for three values of the ratio $\bar{h}/h_0$, chosen close to one. At vanishing $\mu b_0 l/(kh_0)$, a finite limit is attained for the force P, i.e.

$$\lim_{\frac{\mu b_0 l}{k h_0} \to 0} \frac{P}{\mu b_0 h_0} = -\frac{3}{2}\left(1 - \frac{\bar{h}}{h_0}\right)^2, \tag{17}$$

which corresponds to the value obtained under the approximation that the penetrated body is rigid (see Eq. (11)), by identifying the distance $\bar{h}$ between the rigid plates at rest with h.

3 Conclusion

Penetration of one body into another is a common process in biomechanics. It has been shown in the present note that configurational forces may play an important role in these processes. The configurational forces can be calculated by considering the stored elastic energy and represent the counterpart of the microscopic actions developing at the moving boundary points, present at the contact between the intruder and the penetrated body.

Acknowledgements Financial support is acknowledged from H2020-MSCA-ITN-2020-LIGHTEN-956547 and ARS01-01384-PROSCAN.

References

Bigoni, D.: Nonlinear Solid Mechanics: Bifurcation Theory and Material Instability. Cambridge University Press (2012)
Bigoni, D., Bosi, F., Dal Corso, F., Misseroni, D.: Instability of a penetrating blade. J. Mech. Phys. Solids **64**, 411–425 (2014a)
Bigoni, D., Dal Corso, F., Misseroni, D., Bosi, F.: Torsional locomotion. Proc. Math. Phys. Eng. Sci. **470**, 20140599 (2014b)
Bigoni, D., Dal Corso, F., Bosi, F., Misseroni, D.: Eshelby-like forces acting on elastic structures: theoretical and experimental proof. Mech. Mater. **80**, 368–374 (2015)
Boal, D.: Mechanics of the Cell. Cambridge University Press, Cambridge (2002)
Bosi, F., Misseroni, D., Dal Corso, F., Bigoni, D.: Self-encapsulation, or the 'dripping' of an elastic rod. Proc. Math. Phys. Eng. Sci. **471**, 20150195 (2015)
Daddi-Moussa-Ider, A., Goh, S., Liebchen, B., Hoell, C., Mathijssen, A.J., Guzmán-Lastra, F., Scholz, C., Menzel, A.M., Löwen, H.: Membrane penetration and trapping of an active particle. J. Chem. Phys. **150**, 064906 (2019)
Deseri, L., Zurlo, G.: The stretching elasticity of biomembranes determines their line tension and bending rigidity. Biomech. Model. Mechanobiol. **12**, 1233–1242 (2013)
Deseri, L., Pollaci, P., Zingales, M., Dayal, K.: Fractional hereditariness of lipid membranes: instabilities and linearized evolution. J. Mech. Behav. Biomed. Mater. **58**, 11–27 (2016)
Discher, D., Mohandas, N., Evans, E.: Molecular maps of red cell deformation: hidden elasticity and in situ connectivity. Science **266**, 1032–1035 (1994)
Discher, D.E., Boal, D.H., Boey, S.K.: Simulations of the erythrocyte cytoskeleton at large deformation. II. Micropipette aspiration. Biophys. J. **75**, 1584–1597 (1998)
Eshelby, J.D.: The force on an elastic singularity. Philos. Trans. Royal Soc. A **244**, 87–112 (1951)

Franceschini, G., Bigoni, D., Regitnig, P., Holzapfel, G.A.: Brain tissue deforms similarly to filled elastomers and follows consolidation theory. J. Mech. Phys. Solids **54**, 2592–2620 (2006)

Holzapfel, G.A., Gasser, T.C., Ogden, R.W.: A new constitutive framework for arterial wall mechanics and a comparative study of material models. J. Elasticity **61**, 1–48 (2000)

Holzapfel, G.A., Sommer, G., Regitnig, P.: Anisotropic mechanical properties of tissue components in human atherosclerotic plaques. J. Biomech. Eng. **126**, 657–665 (2004)

Mihai, L.A., Budday, S., Holzapfel, G.A., Kuhl, E., Goriely, A.: A family of hyperelastic models for human brain tissue. J. Mech. Phys. Solids **106**, 60–79 (2017)

Pandolfi, A., Holzapfel, G.A.: Three-dimensional modeling and computational analysis of the human cornea considering distributed collagen fibril orientations. J. Biomech. Eng. **130**, 061006 (2008)

Roxhed, N., Gasser, T.C., Griss, P., Holzapfel, G.A., Stemme, G.: Penetration-enhanced ultrasharp microneedles and prediction on skin interaction for efficient transdermal drug delivery. IEEE J. Microelectromech. Syst. **16**, 1429–1440 (2007)

Stålhand, J., Klarbring, A., Holzapfel, G.A.: Smooth muscle contraction: mechanochemical formulation for homogeneous finite strains. Prog. Biophys. Mol. Biol. **96**, 465–481 (2008)

Terzi, M.M., Dayal, K., Deseri, L., Deserno, M.: Revisiting the link between lipid membrane elasticity and microscopic continuum models. Biophys. J. **108**, 87a–88a (2015)

Yuge, S., Akiyama, M., Ishii, M., Namkoong, H., Yagi, K., Nakai, Y., Adachi, R., Komatsu, T.: Glycoprotein nanotube traps influenza virus. Chem. Lett. **46**, 95–97 (2017)

Index

G. Sommer et al. (eds.), *Solid (Bio)mechanics: Challenges of the Next Decade*, Studies in Mechanobiology, Tissue Engineering and Biomaterials 24,
https://doi.org/10.1007/978-3-030-92339-6

W

X

The manufacturer's authorised representative in the EU is Springer Nature Customer Service Centre GmbH, Europaplatz 3, 69115 Heidelberg, Germany. If you have any concerns regarding our products, please contact ProductSafety@springernature.com

Printed and bound by CPI Group (UK) Ltd, Croydon, CR0 4YY
30/09/2026
02225496-0015